Infrared and Raman Spectra of Inorganic and Coordination Compounds

Infrared and Raman Spectra of Inorganic and Coordination Compounds

Part A: Theory and Applications in Inorganic Chemistry

Fifth Edition

Kazuo Nakamoto

A WILEY-INTERSCIENCE PUBLICATION
John Wiley & Sons, Inc.
NEW YORK / CHICHESTER / WEINHEIM / BRISBANE / SINGAPORE / TORONTO

This text is printed on acid-free paper.

Library of Congress Cataloging-in-Publication Data:

Nakamoto, Kazuo, 1922–
Infrared and Raman spectra of inorganic and coordination compounds / Kazuo Nakamoto.—5th ed.
p. cm.
"A Wiley-Interscience publication."
Includes bibliographical references.
Contents: pt. A. Theory and applications of inorganic chemistry—pt. B. Applications in coordination, organometallic, and bioinorganic chemistry.
ISBN 0-471-16394-5 (pt. A: cloth: alk. paper).—ISBN 0-471-16392-9 (pt. B: cloth: alk. paper)
1. Infrared spectroscopy. 2. Raman spectroscopy. I. Title.
QD96.I5N33 1997
543′.08583—dc20 96-33456
ISBN 0-471-16394-5 (pt. A)
ISBN 0-471-16392-9 (pt. B)
ISBN 0-471-19406-9 (set)

Printed in the United States of America

10 9 8 7 6 5 4 3 2 1

Contents

Contents of Part B ix

Preface xi

Abbreviations xiii

Section I. Theory of Normal Vibrations 1

I-1. Origin of Molecular Spectra / 1
I-2. Origin of Infrared and Raman Spectra / 6
I-3. Vibration of a Diatomic Molecule / 10
I-4. Normal Coordinates and Normal Vibrations / 16
I-5. Symmetry Elements and Point Groups / 23
I-6. Symmetry of Normal Vibrations and Selection Rules / 27
I-7. Introduction to Group Theory / 37
I-8. The Number of Normal Vibrations for Each Species / 42
I-9. Internal Coordinates / 49
I-10. Selection Rules for Infrared and Raman Spectra / 53
I-11. Structure Determination / 60
I-12. Principle of the **GF** Matrix Method / 63
I-13. Utilization of Symmetry Properties / 71
I-14. Potential Fields and Force Constants / 77
I-15. Solution of the Secular Equation / 80
I-16. Vibrational Frequencies of Isotopic Molecules / 83
I-17. Metal–Isotope Spectroscopy / 85
I-18. Group Frequencies and Band Assignments / 88
I-19. Intensity of Infrared Absorption / 95

I-20. Depolarization of Raman Lines / 97
I-21. Intensity of Raman Scattering / 102
I-22. Principle of Resonance Raman Spectroscopy / 106
I-23. Resonance Raman Spectra / 109
I-24. Vibrational Spectra in Gaseous Phase and Inert Gas Matrices / 113
I-25. Symmetry in Crystals / 117
I-26. Vibrational Analysis of Crystals / 124
I-27. The Correlation Method / 127
I-28. Lattice Vibrations / 132
I-29. Polarized Spectra of Single Crystals / 136
I-30. Vibrational Analysis of Ceramic Superconductors / 141
References / 145

Section II. Applications in Inorganic Chemistry **153**

II-1. Diatomic Molecules / 153
II-2. Triatomic Molecules / 162
II-3. Pyramidal Four-Atom Molecules / 173
II-4. Planar Four-Atom Molecules / 180
II-5. Other Four-Atom Molecules / 184
II-6. Tetrahedral and Square-Planar Five-Atom Molecules / 189
II-7. Trigonal-Bipyramidal and Tetragonal-Pyramidal XY_5 and Related Molecules / 209
II-8. Octahedral Molecules / 214
II-9. XY_7 and XY_8 Molecules / 226
II-10. X_2Y_4 and X_2Y_6 Molecules / 228
II-11. X_2Y_7, X_2Y_8, X_2Y_9, and X_2Y_{10} Molecules / 236
II-12. Metal Cluster Compounds / 240
II-13. Compounds of Boron / 243
II-14. Compounds of Carbon / 248
II-15. Compounds of Silicon and Germanium / 256
II-16. Compounds of Nitrogen / 261
II-17. Compounds of Phosphorous and Other Group VB Elements / 265
II-18. Compounds of Sulfur and Selenium / 271
References / 276

Appendices **321**

I. Point Groups and Their Character Tables / 321
II. Matrix Algebra / 335
III. General Formulas for Calculating the Number of Normal Vibrations in Each Species / 341
IV. Direct Products of Irreducible Representations / 344

V. Number of Infrared- and Raman-Active Stretching Vibrations for MX_nY_m-Type Molecules / 346
VI. Derivation of Equation 12.3 of Section I / 347
VII. The **G** and **F** Matrix Elements of Typical Molecules / 350
VIII. Group Frequency Charts / 356
IX. Correlation Tables / 361
X. Site Symmetry for the 230 Space Groups / 374

Index **383**

Contents of Part B

Contents of Part A vii

Preface xi

Abbreviations xiii

Section III. Applications in Coordination Chemistry 1

III-1. Ammine, Amido, and Related Complexes / 1
III-2. Complexes of Ethylenediamine and Related Ligands / 14
III-3. Complexes of Pyridine and Related Ligands / 23
III-4. Complexes of Bipyridine and Related Ligands / 30
III-5. Metalloporphyrins / 39
III-6. Nitro and Nitrito Complexes / 48
III-7. Lattice Water and Aquo and Hydroxo Complexes / 53
III-8. Complexes of Alkoxides, Alcohols, Ethers, Ketones, Aldehydes, Esters, and Carboxylic Acids / 57
III-9. Complexes of Amino Acids, EDTA, and Related Ligands / 62
III-10. Infrared Spectra of Aqueous Solutions / 69
III-11. Complexes of Oxalato and Related Ligands / 74
III-12. Complexes of Sulfate, Carbonate, and Related Ligands / 79
III-13. Complexes of β-Diketones / 91
III-14. Complexes of Urea, Sulfoxides, and Related Ligands / 100

III-15. Cyano and Nitrile Complexes / 105
III-16. Thiocyanato and Other Pseudohalogeno Complexes / 116
III-17. Complexes of Carbon Monoxide and Carbon Dioxide / 126
III-18. Nitrosyl Complexes / 149
III-19. Complexes of Dioxygen / 154
III-20. Metal Complexes Containing Oxo Groups / 168
III-21. Complexes of Dinitrogen and Related Ligands / 173
III-22. Complexes of Dihydrogen and Related Ligands / 179
III-23. Halogeno Complexes / 183
III-24. Complexes Containing Metal–Metal Bonds / 190
III-25. Complexes of Phosphorus and Arsenic Ligands / 195
III-26. Complexes of Sulfur and Selenium Ligands / 199
References / 211

Section IV. Applications in Organometallic Chemistry **257**

IV-1. Methylene, Methyl, and Ethyl Compounds / 257
IV-2. Vinyl, Allyl, Acetylenic, and Phenyl Compounds / 263
IV-3. Halogeno, Pseudohalogeno, and Acido Compounds / 265
IV-4. Compounds Containing Other Functional Groups / 273
IV-5. π-Complexes of Olefins, Acetylenes, and Related Ligands / 276
IV-6. Cyclopentadienyl Compounds / 285
IV-7. Cyclopentadienyl Compounds Containing Other Groups / 291
IV-8. Complexes of Other Cyclic Unsaturated Ligands / 297
IV-9. Miscellaneous Compounds / 304
References / 305

Section V. Applications in Bioinorganic Chemistry **319**

V-1. Myoglobin and Hemoglobin / 321
V-2. Ligand Binding to Myoglobin and Hemoglobin / 325
V-3. Cytochromes and Other Heme Proteins / 336
V-4. Hemerythrins / 342
V-5. Hemocyanins / 347
V-6. Blue Copper Proteins / 352
V-7. Iron–Sulfur Proteins / 355
V-8. Interactions of Metal Complexes with Nucleic Acids / 362
References / 369

Index **379**

Preface

The first edition of this book, entitled *Infrared Spectra of Inorganic and Coordination Compounds*, was published in 1963. Since then, it has been revised in 1970, 1978, and 1986 to keep up with ever-increasing new literature. The preparation of the fifth edition was begun in the fall of 1990 and completed in the spring of 1996.

As I emphasized in the Preface of the previous editions, this book is intended to describe fundamental theories of vibrational spectroscopy in a condensed form (Section I) and to illustrate their applications in inorganic (Section II), coordination (Section III), organometallic (Section IV) and bioinorganic chemistry (Section V), using typical examples. In the fifth edition, all of these sections have been updated by adding many new references while omitting those shown to be erroneous by later studies. Furthermore, several subsections have been added in all sections to cover new topics. In particular, I have included many infrared and Raman spectral charts of typical compounds because they provide real "feel," which cannot be expressed by tables. The book has been divided into two volumes: Part A covers Sections I and II, and Part B covers Sections III–V. I felt that this division was justified, because Part A covers basic theory and its applications to relatively simple inorganic compounds, whereas Part B is focused on applications of basic theory to large and complex molecules.

As in the past editions, I have tried to give a broad and balanced coverage of the field. It was clearly impossible, however, to include all significant work in the limited space available. I only hope that any unbalanced presentation will be compensated for by the review articles and reference books which are abundantly quoted throughout this book.

I wish to express my sincere thanks to all who helped me in preparing this edition. I am particularly indebted to Professors D. P. Strommen (Idaho State University), R. A. Condrate (Alfred University), R. S. Czernuszewicz (University of Houston), J. R. Kincaid (Marquette University), and T. Kitagawa (Institute for Molecular Science, Okazaki, Japan). Special thanks are given to Prof. R. S. Czernuszewicz, who kindly drew some figures in Sections I, IV, and V. As I mentioned earlier, this edition contains many spectral charts, which could not have been included without the cooperation of many authors and their publishers. For the sake of uniformity, I simply used reference numbers in the figure captions to cite the source of each chart. I would like to acknowledge the cooperation I received by listing here the publisher and publications quoted in this edition: Academic Press (monographs), American Chemical Society (*Journal of the American Chemical Society, Journal of Physical Chemistry, Inorganic Chemistry*, and *Analytical Chemistry*), American Institute of Physics (*Journal of the Chemical Physics*), Cornell University Press (monograph), Elsevier Science (*Chemical Physics, Chemical Physics Letters, Coordination Chemistry Review, Journal of Molecular Structure, Journal of Organometallic Chemistry, Solid State Communications*, and *Spectrochimica Acta*), John Wiley (monographs and *Journal of Raman Spectroscopy*), Material Research Society (*MBS Bulletin*), Macmillan Magazines (*Nature*), National Academy of Science (monograph), Plenum Press (monograph), The Royal Society of Chemistry (*Journal of the Chemical Society, Dalton Transactions*), The Society for Applied Spectroscopy (*Applied Spectroscopy*), and Verlag der Zeitschrift der Naturforschung (*Zeitschrift für Naturforschung*).

Finally, I would like to thank the staffs of the Science Library and the Chemistry Department of Marquette University for their help in preparing this edition.

KAZUO NAKAMOTO

Milwaukee, Wisconsin

Abbreviations

IR, infrared; R, Raman; RR, resonance Raman; *p*, polarized; *dp*, depolarized; *ap*, anomalous polarization; i.a., inactive.

ν, stretching; δ, in-plane bending or deformation; ρ_w, wagging; ρ_r, rocking; ρ_t, twisting; π, out-of-plane bending. Subscripts a, s, and d denote antisymmetric, symmetric, and degenerate modes, respectively. Approximate normal modes of vibration corresponding to these vibrations are given in Figs. I-25 and I-26.

NCA, normal coordinate analysis.

GVF, generalized valence force field; UBF, Urey-Bradley force field.

M, metal; L, ligand; X, halogen; R, alkyl group or cyclopentadienyl (Cp) or other ring compound.

g, gas; l, liquid; s, solid; m or mat, matrix; sol'n or sl, solution.

Me, methyl; Et, ethyl; Pr, propyl; Bu, butyl; Ph, phenyl; Cp, cyclopentadienyl; OAc, acetate ion; py, pyridine; pic, pycoline; en, ethylenediamine. Abbreviations of other ligands are given when they appear in the text.

In the tables of observed frequencies, values in parentheses are calculated or estimated values unless otherwise stated.

I

THEORY OF NORMAL VIBRATIONS

I-1. ORIGIN OF MOLECULAR SPECTRA

As a first approximation, the energy of the molecule can be separated into three additive components associated with (1) the motion of the electrons in the molecule,* (2) the vibrations of the constituent atoms, and (3) the rotation of the molecule as a whole:

$$E_{\text{total}} = E_{\text{el}} + E_{\text{vib}} + E_{\text{rot}} \tag{1.1}$$

The basis for this separation lies in the fact that electronic transitions occur on a much shorter time scale, and rotational transitions occur on a much longer time scale, than vibrational transitions. The translational energy of the molecule may be ignored in this discussion because it is essentially not quantized.

If a molecule is placed in an electromagnetic field (e.g., light), a transfer of energy from the field to the molecule will occur when Bohr's frequency condition is satisfied:

$$\Delta E = h\nu \tag{1.2}$$

where ΔE is the difference in energy between two quantized states, h is Planck's constant (6.625×10^{-27} erg sec), and ν is the frequency of the light. Here, the frequency is the number of electromagnetic waves in the distance light travels in one second:

*Hereafter the word *molecule* may also represent an *ion*.

$$\nu = \frac{c}{\lambda} \tag{1.3}$$

where c is the velocity of light (3×10^{10} cm sec^{-1}) and λ is the wavelength of the electromagnetic wave. If λ has the units of cm, ν has dimensions of (cm sec^{-1})/cm = sec^{-1}, which is also called "Hertz (Hz)."

The wavenumber ($\tilde{\nu}$) defined by

$$\tilde{\nu} = \frac{1}{\lambda} \tag{1.4}$$

is most commonly used in vibrational spectroscopy. It has the dimension of cm^{-1}. By combining Eqs. (1.3) and (1.4), we obtain:

$$\tilde{\nu} = \frac{1}{\lambda} = \frac{\nu}{c} \quad \text{or} \quad \nu = \frac{c}{\lambda} = c\tilde{\nu} \tag{1.5}$$

Although the dimensions of ν and $\tilde{\nu}$ differ from one another, it is convenient to use them interchangeably. Thus, an expression such as "a frequency shift of 5 cm^{-1}" is used throughout this book.

Using Eq. (1.5), Bohr's condition (Eq. 1.2) is written as

$$\Delta E = hc\tilde{\nu} \tag{1.6}$$

Since h and c are known constants, ΔE can be expressed in units such as*

$$\begin{aligned} 1\ (\text{cm}^{-1}) &= 1.99 \times 10^{-16}\ (\text{erg} \cdot \text{molecule}^{-1}) \\ &= 2.86\ (\text{cal} \cdot \text{mole}^{-1}) \\ &= 1.24 \times 10^{-4}\ (\text{eV} \cdot \text{molecule}^{-1}) \end{aligned}$$

Suppose that

$$\Delta E = E_2 - E_1 \tag{1.7}$$

where E_2 and E_1 are the energies of the excited and ground states, respectively.

*Use conversion factors such as

$$\begin{aligned} \text{Avogadro's number, } N_0 &= 6.023 \times 10^{23}\ (\text{mole}^{-1}) \\ 1\ (\text{cal}) &= 4.1846 \times 10^{7} (\text{erg}) \\ 1\ (\text{eV}) &= 1.6021 \times 10^{-12}\ (\text{erg} \cdot \text{molecule}^{-1}) \\ &= 1.6021 \times 10^{-19}\ (\text{joule} \cdot \text{molecule}^{-1}) \end{aligned}$$

Then, the molecule "absorbs" ΔE when it is excited from E_1 to E_2 and "emits" ΔE when it reverts from E_2 to E_1. Figure I-1 shows the regions of the electromagnetic spectrum where ΔE is indicated in $\tilde{\nu}$, λ, and ν. In this book, we are mainly concerned with vibrational transitions which are observed in infrared (IR) or Raman (R) spectra. These transitions appear in the $10^2 \sim 10^4$ cm^{-1} region, and originate from vibrations of nuclei constituting the molecule. Rotational transitions occur in the $1 \sim 10^2$ cm^{-1} region (microwave region) because rotational levels are relatively close to each other, whereas electronic transitions are observed in the $10^4 \sim 10^6$ cm^{-1} region (UV-visible region) because their energy levels are far apart. However, such division is somewhat arbitrary, for pure rotational spectra may appear in the far-infrared region if transitions to higher excited states are involved, and pure electronic transitions may appear in the near-infrared region if electronic levels are closely spaced.

Figure I-2 illustrates transitions of the three types mentioned for a diatomic molecule. As the figure shows, rotational intervals tend to increase as the rotational quantum number J increases, whereas vibrational intervals tend to decrease as the vibrational quantum number v increases. The dashed line below each electronic level indicates the "zero-point energy" that must exist even at a temperature of absolute zero as a result of Heisenberg's Uncertainty Principle:

$$E_0 = \frac{1}{2}h\nu \tag{1.8}$$

It should be emphasized that not all transitions between these levels are possible. To see whether the transition is "allowed or forbidden," the relevant selection rule must be examined. This, in turn, is determined by the symmetry of the molecule.

As expected from Fig. I-2, electronic spectra are very complicated because they are accompanied by vibrational as well as rotational fine structure. The

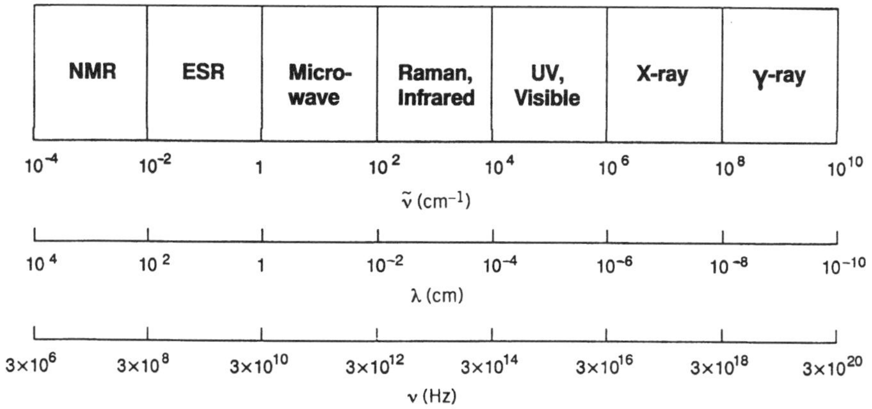

Fig. I-1. Regions of the electromagnetic spectrum and energy units.

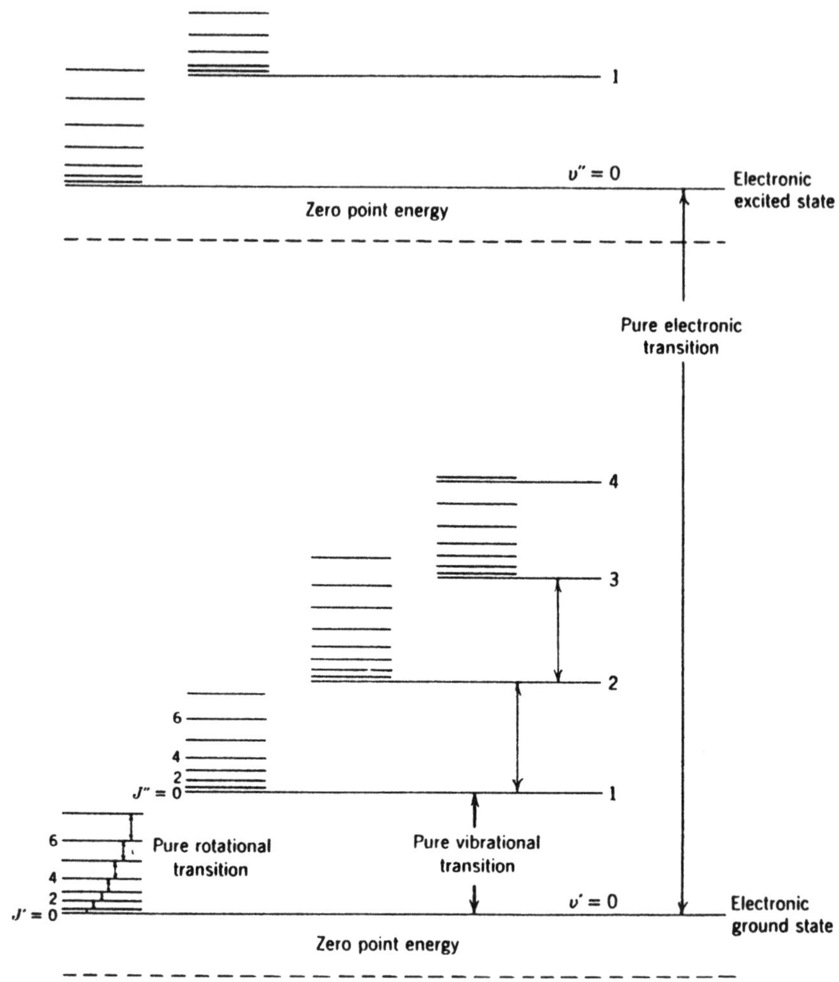

Fig. I-2. Energy level of a diatomic molecule (the actual spacings of electronic levels are much larger, and those of rotational levels are much smaller, than that shown in the figure).

rotational fine structure in the electronic spectrum can be observed if a molecule is simple and the spectrum is measured in the gaseous state under high resolution. The vibrational fine structure of the electronic spectrum is easier to observe than the rotational fine structure, and can provide structural and bonding information about molecules in electronic excited states.

Vibrational spectra are accompanied by rotational transitions. Figure I-3 shows the rotational fine structure observed for the gaseous ammonia molecule. In most polyatomic molecules, however, such a rotational fine structure is not observed because the rotational levels are closely spaced as a result of relatively large moments of inertia. Vibrational spectra obtained in solution do not exhibit

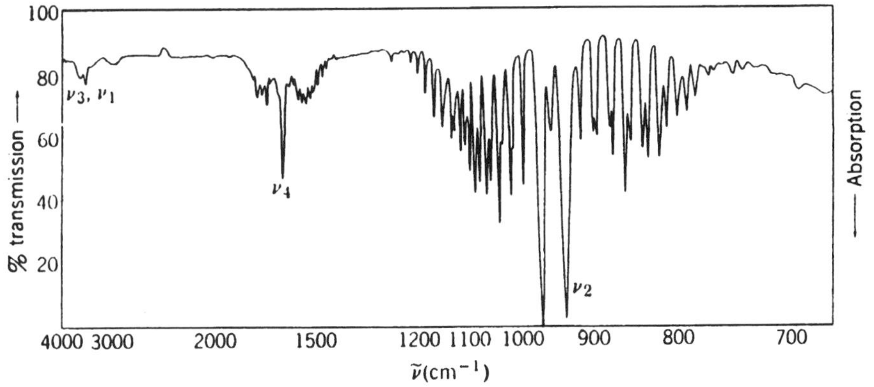

Fig. I-3. Rotational fine structure of gaseous NH_3.

rotational fine structure, since molecular collisions occur before a rotation is completed and the levels of the individual molecules are perturbed differently.

The selection rule allows any transitions corresponding to $\Delta v = \pm 1$ if the molecule is assumed to be a harmonic oscillator (Sec. I-3). Under ordinary conditions, however, only the *fundamentals* that originate in the transition from $v = 0$ to $v = 1$ in the electronic ground state can be observed. This is because the Maxwell–Boltzmann distribution law requires that the ratio of population at $v = 0$ and $v = 1$ states is given by

$$R = \frac{P(v = 1)}{P(v = 0)} = e^{-\Delta E_v/kT} \tag{1.9}$$

where ΔE_v is the vibrational frequency (cm^{-1}) and $kT = 208$ (cm^{-1}) at room temperature. In the case of H_2, $\Delta E_v = 4160$ cm^{-1} and $R = 2.16 \times 10^{-9}$. Thus, almost all molecules are at $v = 0$. However, the population at $v = 1$ increases as ΔE_v becomes small. For example, $R = 0.36$ for I_2 ($\Delta E_v = 213$ cm^{-1}) Then, about 27% of the molecules are at $v = 1$ state, and the transition from $v = 1$ to $v = 2$ can be observed as a "hot band."

In addition to the harmonic oscillator selection rule, another restriction results from the symmetry of the molecule (Sec. I-10). Thus, the number of allowed transitions in polyatomic molecules is greatly reduced. *Overtones and combination bands** of these fundamentals are forbidden by the selection rule. However, they are weakly observed in the spectrum because of the anharmonicity of the vibration (Sec. I-3). Since they are less important than the fundamentals, they will be discussed only when necessary.

*Overtones represent some multiples of the fundamental, whereas combination bands arise from the sum or difference of two or more fundamentals.

I-2. ORIGIN OF INFRARED AND RAMAN SPECTRA

As stated previously, vibrational transitions can be observed as infrared (IR) or Raman spectra.* However, the physical origins of these two spectra are markedly different. Infrared (absorption) spectra originate in photons in the infrared region that are absorbed by transitions between two vibrational levels of the molecule in the electronic ground state. On the other hand, Raman spectra have their origin in the electronic polarization caused by ultraviolet, visible, and near-IR light. If a molecule is irradiated by monochromatic light of frequency ν (laser), then, because of electronic polarization induced in the molecule by this incident beam, the light of frequency ν ("Rayleigh scattering") as well as that of frequency $\nu \pm \nu_i$ ("Raman scattering") is scattered where ν_i represents a vibrational frequency of the molecule. Thus, Raman spectra are presented as

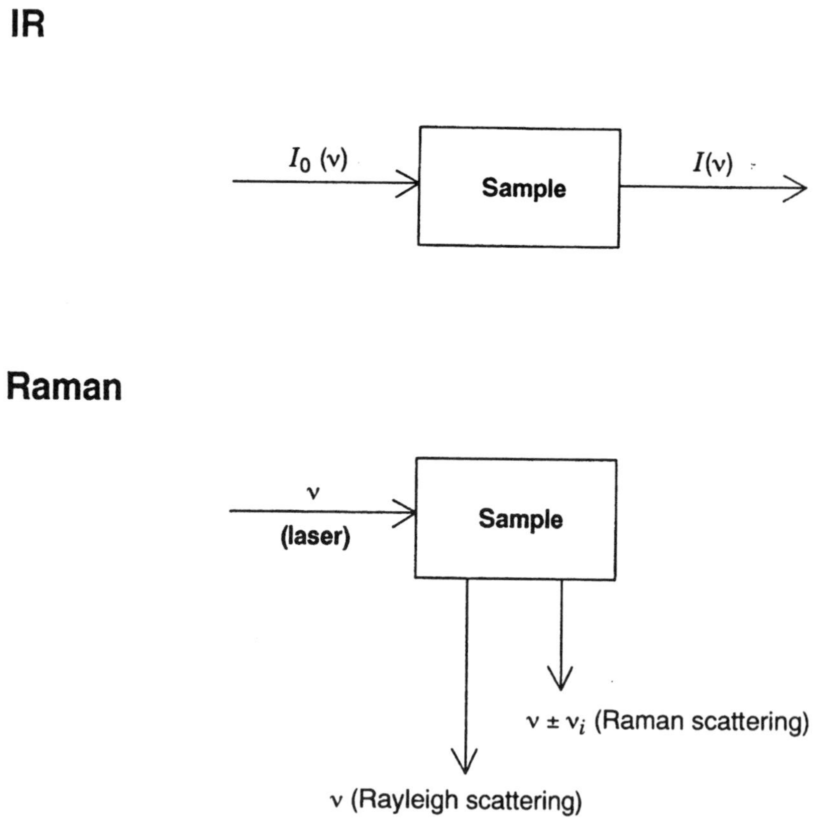

Fig. I-4. Mechanisms of infrared absorption and Raman scattering.

*Raman spectra were first observed by C. V. Raman [*Indian J. Phys.*, **2**, 387 (1928)] and C. V. Raman and K. S. Krishnan [*Nature*, **121**, 501 (1928)].

shifts from the incident frequency in the ultraviolet, visible, and near-IR region. Figure I-4 illustrates the difference between IR and Raman techniques.

Although Raman scattering is much weaker than Rayleigh scattering (by a factor of 10^{-3} to 10^{-5}), it is still possible to observe the former by using a strong exciting source. In the past, the mercury lines at 435.8 nm (22.938 cm^{-1}) and 404.7 nm (24,705 cm^{-1}) from a low-pressure mercury arc were used to observe Raman scattering. However, the advent of lasers revolutionized Raman spectroscopy. Lasers provide strong, coherent monochromatic light in a wide range of wavelengths, as listed in Table I-1. In the case of resonance Raman spectroscopy (Sec. I-22), the exciting frequency is chosen so as to fall inside the electronic absorption band. The degree of resonance enhancement varies as a function of the exciting frequency and reaches a maximum when the exciting frequency coincides with that of the electronic absorption maximum. It is possible to change the exciting frequency continuously by pumping dye lasers with powerful gas or pulsed lasers.

The origin of Raman spectra can be explained by an elementary classical theory. Consider a light wave of frequency ν with an electric field strength E. Since E fluctuates at frequency ν, we can write

$$E = E_0 \cos 2\pi\nu t \tag{2.1}$$

where E_0 is the amplitude and t the time. If a diatomic molecule is irradiated by this light, the dipole moment P given by

TABLE I-1. Some Representative Laser Lines for Raman Spectroscopy

Laser[a]	Mode	Wavelength (nm)	$\tilde{\nu}$ (cm^{-1})
Gas lasers			
Ar-ion	C.W.	488.0 (blue)	20491.8
		514.5 (green)	19436.3
Kr-ion	C.W.	413.1 (violet)	24207.2
		530.9 (green/yellow)	18835.9
		647.1 (red)	15453.6
He-Ne	C.W.	632.8 (red)	15802.8
He-Cd	C.W.	441.6 (blue/violet)	22644.9
Nitrogen	pulsed	337.1 (UV)	29664.7
Excimer (XeCl)		308 (UV)	32467.5
Solid state lasers			
Nd:YAG[b]	C.W. or pulsed	1064 (near-IR)	9398.4
Liquid lasers			
A variety of dye solutions are pumped by strong C.W. or pulsed-laser sources. A wide range from 440 to 800 nm can be covered continuously by choosing proper organic dyes.			

[a] Acronym of "Light Amplification by Stimulated Emission of Radiation."
[b] Acronym of "Neodimium-doped Yttrium Aluminum Garnet."
For more information, see Nakamoto and colleagues[30,32].

$$P = \alpha E = \alpha E_0 \cos 2\pi \nu t \tag{2.2}$$

is induced. Here α is a proportionality constant and is called the *polarizability*. If the molecule is vibrating with frequency ν_i, the nuclear displacement q is written as

$$q = q_0 \cos 2\pi \nu_i t \tag{2.3}$$

where q_0 is the vibrational amplitude. For small amplitudes of vibration, α is a linear function of q. Thus, we can write

$$\alpha = \alpha_0 + \left(\frac{\partial \alpha}{\partial q}\right)_0 q \tag{2.4}$$

Here, α_0 is the polarizability at the equilibrium position, and $(\partial\alpha/\partial q)_0$ is the rate of change of α with respect to the change in q, evaluated at the equilibrium position. If we combine Eqs. 2.2–2.4, we have

$$\begin{aligned} P &= \alpha E_0 \cos 2\pi \nu t \\ &= \alpha_0 E_0 \cos 2\pi \nu t + \left(\frac{\partial \alpha}{\partial q}\right)_0 q_0 E_0 \cos 2\pi \nu t \cos 2\pi \nu_i t \\ &= \alpha_0 E_0 \cos 2\pi \nu t \\ &\quad + \frac{1}{2}\left(\frac{\partial \alpha}{\partial q}\right)_0 q_0 E_0 \{\cos [2\pi(\nu + \nu_i)t] + \cos [2\pi(\nu - \nu_i)t]\} \end{aligned} \tag{2.5}$$

According to classical theory, the first term describes an oscillating dipole which radiates light of frequency ν (Rayleigh scattering). The second term gives the Raman scattering of frequencies $\nu + \nu_i$ (*anti-Stokes*) and $\nu - \nu_i$ (*Stokes*). If $(\partial\alpha/\partial q)_0$ is zero, the second term vanishes. Thus, the vibration is not Raman active unless the polarizability changes during the vibration.

Figure I-5 illustrates the mechanisms of *normal* and *resonance Raman (RR) scattering*. In the former, the energy of the exciting line falls far below that required to excite the first electronic transition. In the latter, the energy of the exciting line coincides with that of an electronic transition.* If the photon is absorbed and then emitted during the process, it is called *resonance fluorescence (RF)*. Although the conceptual difference between resonance Raman scattering and resonance fluorescence is subtle, there are several experimental differences which can be used to distinguish between these two phenomena. For example, in

*If the exciting line is close to but not inside an electronic absorption band, the process is called "preresonance Raman scattering."

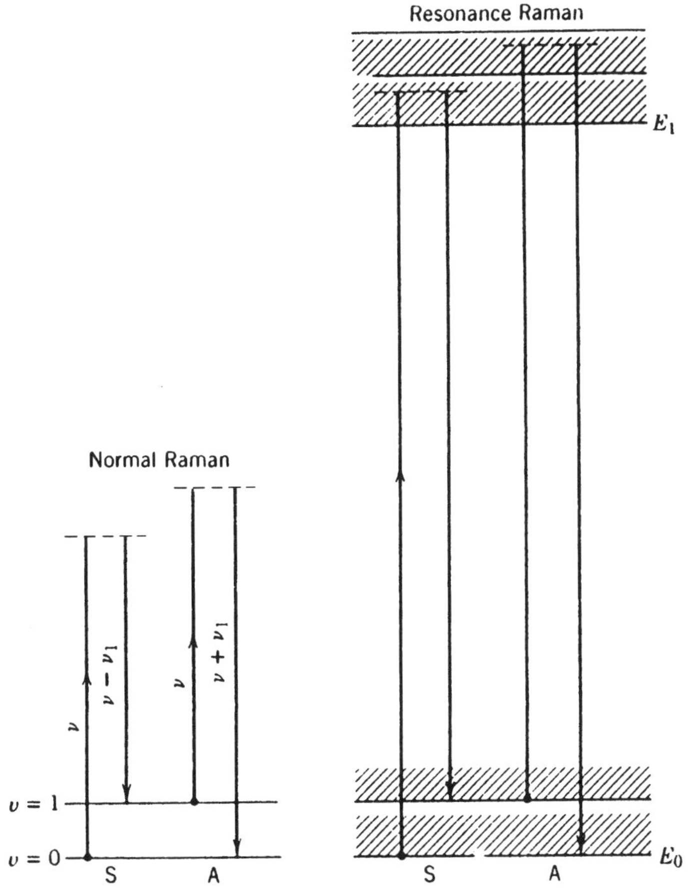

Fig. I-5. Mechanisms of normal and resonance Raman scattering. S and A denote Stokes and anti-Stokes scattering, respectively. The shaded areas indicate the broadening of rotational-vibrational levels in the liquid and solid states (Sec. I-22).

RF spectra all lines are depolarized, whereas in RR spectra some are polarized and others are depolarized. Additionally, RR bands tend to be broad and weak compared with RF bands.[74,75]

In the case of Stokes lines, the molecule at $\upsilon = 0$ is excited to the $\upsilon = 1$ state by scattering light of frequency $\nu - \nu_1$. Anti-Stokes lines arise when the molecule initially in the $\upsilon = 1$ state scatters radiation of frequency $\nu + \nu_1$ and reverts to the $\upsilon = 0$ state. Since the population of molecules is larger at $\upsilon = 0$ than at $\upsilon = 1$ (*Maxwell–Boltzmann distribution law*), the Stokes lines are always stronger than the anti-Stokes lines. Thus, it is customary to measure Stokes lines in Raman spectroscopy. Figure I-6 illustrates the Raman spectrum (below 500 cm^{-1}) of CCl_4 excited by the blue line (488.0 nm) of an argon-ion laser.

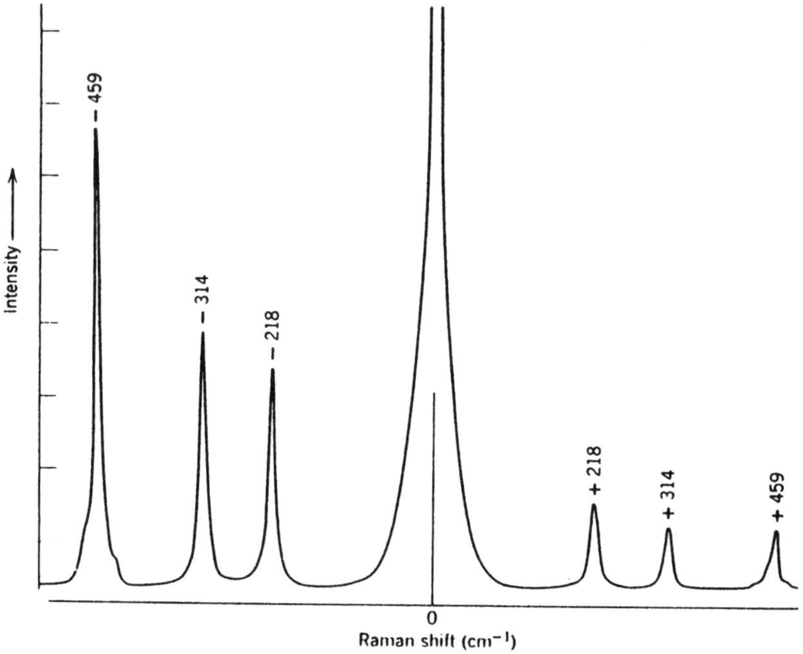

Fig. I-6. Raman spectrum of CCl_4 (488.0 nm excitation).

I-3. VIBRATION OF A DIATOMIC MOLECULE

Through quantum mechanical considerations,[4,5] the vibration of two nuclei in a diatomic molecule can be reduced to the motion of a single particle of mass μ, whose displacement q from its equilibrium position is equal to the change of the internuclear distance. The mass μ is called the *reduced mass* and is represented by

$$\frac{1}{\mu} = \frac{1}{m_1} + \frac{1}{m_2} \tag{3.1}$$

where m_1 and m_2 are the masses of the two nuclei. The kinetic energy is then

$$T = \frac{1}{2}\mu\dot{q}^2 = \frac{1}{2\mu}p^2 \tag{3.2}$$

where p is the conjugate momentum $\mu\dot{q}$. If a simple parabolic potential function such as that shown in Fig. I-7 is assumed, the system represents a *harmonic oscillator*, and the potential energy is simply given by

$$V = \tfrac{1}{2}Kq^2 \tag{3.3}$$

Here K is the force constant for the vibration. Then the Schrödinger wave equa-

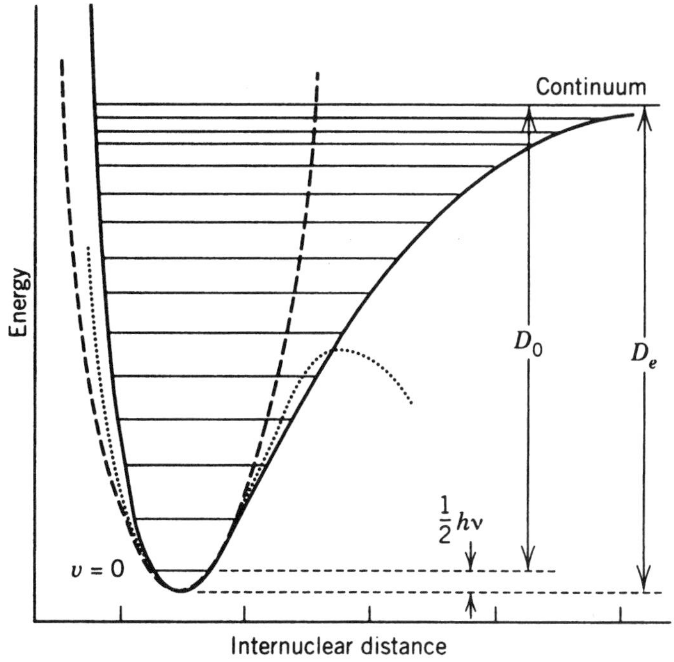

Fig. I-7. Potential energy curves for a diatomic molecule: actual potential (solid line), parabolic potential (dashed line), and cubic potential (dotted line).

tion becomes

$$\frac{d^2\psi}{dq^2} + \frac{8\pi^2\mu}{h^2}\left(E - \frac{1}{2}Kq^2\right)\psi = 0 \tag{3.4}$$

If this equation is solved with the condition that ψ must be single-valued, finite, and continuous, the eigenvalues are

$$E_\upsilon = h\nu(\upsilon + \tfrac{1}{2}) = hc\tilde{\nu}(\upsilon + \tfrac{1}{2}) \tag{3.5}$$

with the frequency of vibration

$$\nu = \frac{1}{2\pi}\sqrt{\frac{K}{\mu}} \quad \text{or} \quad \tilde{\nu} = \frac{1}{2\pi c}\sqrt{\frac{K}{\mu}} \tag{3.6}$$

Here υ is the vibrational quantum number, and it can have the values 0, 1, 2, 3, $\cdots$

The corresponding eigenfunctions are

$$\psi_v = \frac{(\alpha/\pi)^{1/4}}{\sqrt{2^v v!}} e^{-\alpha q^2/2} H_v(\sqrt{\alpha} q) \tag{3.7}$$

where $\alpha = 2\pi\sqrt{\mu K}/h = 4\pi^2\mu\nu/h$, and $H_v(\sqrt{\alpha}q)$ is a Hermite polynomial of the vth degree. Thus the eigenvalues and the corresponding eigenfunctions are

$$\begin{aligned} E_0 &= \tfrac{1}{2}h\nu, & \psi_0 &= (\alpha/\pi)^{1/4}e^{-\alpha q^2/2} \\ E_1 &= \tfrac{3}{2}h\nu, & \psi_1 &= (\alpha/\pi)^{1/4}2^{1/2}qe^{-\alpha q^2/2} \\ &\vdots & &\vdots \end{aligned} \tag{3.8}$$

As Fig. I-7 shows, actual potential curves can be approximated more exactly by adding a cubic term.[2]

$$V = \tfrac{1}{2}Kq^2 - Gq^3 \qquad (K \gg G) \tag{3.9}$$

Then, the eigenvalues become

$$E_v = hc\omega_e(v + \tfrac{1}{2}) - hcx_e\omega_e(v + \tfrac{1}{2})^2 + \cdots \tag{3.10}$$

where ω_e is the wavenumber corrected for *anharmonicity*, and $x_e\omega_e$ indicates the magnitude of anharmonicity. Table II-1*a* of Section II lists ω_e and $x_e\omega_e$ for a number of diatomic molecules. Equation 3.10 shows that the energy levels of the anharmonic oscillator are not equidistant, and the separation decreases slowly as v increases. This anharmonicity is responsible for the appearance of overtones and combination vibrations, which are forbidden in the harmonic oscillator.

The values of x_e and $x_e\omega_e$ can be determined by observing a series of overtone bands in IR and Raman spectra. From Eq. 3.10, we obtain:

$$(E_v - E_0)/hc = v\omega_e - x_e\omega_e(v^2 + v) + \cdots$$

Then,

$$\begin{aligned} \text{Fundamental:} \quad & \tilde{\nu}_1 = \omega_e - 2x_e\omega_e \\ \text{First overtone:} \quad & \tilde{\nu}_2 = 2\omega_e - 6x_e\omega_e \\ \text{Second overtone:} \quad & \tilde{\nu}_3 = 3\omega_e - 12x_e\omega_e \end{aligned}$$

For $H^{35}Cl$, these transitions are observed at 2885.9, 5668.1, and 8347.0 cm^{-1}, respectively, in IR spectrum.[2] Using these values, we find that

$$\omega_e = 2988.9\ \text{cm}^{-1} \quad \text{and} \quad x_e\omega_e = 52.05\ \text{cm}^{-1}$$

As will be shown in Sec. I-23, a long series of overtone bands can be observed when Raman spectra of small molecules such as I_2 and TiI_4 are measured under rigorous resonance conditions. Anharmonicity constants can also be determined from the analysis of rotational fine structures of vibrational transitions.[2]

Since the anharmonicity correction has not been made for most polyatomic molecules, in large part because of the complexity of the calculation, the frequencies given in Section II are not corrected for anharmonicity (except those given in Table II-1*a*).

According to Eq. 3.6, the wavenumber of the vibration in a diatomic molecule is given by

$$\tilde{\nu} = \frac{1}{2\pi c}\sqrt{\frac{K}{\mu}} \tag{3.6}$$

A more exact expression is given by using the wavenumber corrected for anharmonicity:

$$\omega_e = \frac{1}{2\pi c}\sqrt{\frac{K}{\mu}} \tag{3.11}$$

or

$$K = 4\pi^2 c^2 \omega_e^2 \mu \tag{3.12}$$

For HCl, $\omega_e = 2989\ \text{cm}^{-1}$ and $\mu = 0.9799$ (awu). Here, awu is the atomic weight unit. Thus, we obtain

$$\begin{aligned} K &= \frac{4(3.14)^2(3\times 10^{10})^2}{6.025\times 10^{23}}\,\omega_e^2\mu \\ &= (5.8883\times 10^{-2})\omega_e^2\mu \\ &= (5.8883\times 10^{-2})(2989)^2(0.9799) \\ &= 5.16\times 10^5 (\text{dynes/cm}) \\ &= 5.16(\text{mdyn/Å})^* \end{aligned}$$

Table I-2 lists the observed frequencies ($\tilde{\nu}$), wavenumbers corrected for anharmonicity (ω_e), reduced masses (μ), and force constants (K) for several

*10^5 (dynes/cm) = 10^5 (10^3 millidynes/10^8 Å) = 1 (mdyn/Å) = 10^2 N/m (SI unit).

TABLE I-2. Relationship Among Vibrational Frequency, Reduced Mass, and Force Constant

Molecule	Obs. $\tilde{\nu}$ (cm^{-1})	ω_e (cm^{-1})	μ (awu)	K (mdyn/Å)
H_2	4160	4395	0.5041	5.73
HD	3632	3817	0.6719	5.77
D_2	2994	3118	1.0074	5.77
HF	3962	4139	0.9573	9.65
HCl	2886	2989	0.9799	5.16
HBr	2558	2650	0.9956	4.12
HI	2233	2310	1.002	3.12
F_2	892	—	9.5023	4.45
Cl_2	546	565	17.4814	3.19
Br_2	319	323	39.958	2.46
I_2	213	215	63.466	1.76
N_2	2331	2360	7.004	22.9
CO	2145	2170	6.8584	19.0
NO	1877	1904	7.4688	15.8
O_2	1555	1580	8.000	11.8

series of diatomic molecules. In the first series, ω_e decreases in the order H_2 > HD > D_2, because μ increases in the same order, while K is almost constant (*mass effect*). In the second series, ω_e decreses in the order HF > HCl > HBr > HI, because K decreases in the same order, while μ shows little change (*force constant effect*). In the third series, ω_e decreases in the order $F_2 > Cl_2 > Br_2 > I_2$, because μ increases and K decreases in the same order. In this case, both mass effect and force constant effect are operative. In the last series, ω_e decreases in the order N_2 > CO > NO > O_2, mainly owing to the force constant effect. This may be attributed to the differences in bond order:

$$\mathrm{N{\equiv}N} > \mathrm{C^{-}{\equiv}O^{+}} \longleftrightarrow \mathrm{C{=}O} > \mathrm{N{=}O} \longleftrightarrow \mathrm{N^{-}{=}O^{+}} > \mathrm{O{=}O} \longleftrightarrow \mathrm{O^{-}{-}O^{+}}$$

More examples of similar series in diatomic molecules are found in Sec. II-1. These simple rules, obtained for a diatomic molecule, are helpful in understanding the vibrational spectra of polyatomic molecules.

Figure I-8 indicates the relationship between the force constant and the dissociation energy for the three series listed in Table I-2. In the series of hydrogen halides, the dissociation energy decreases almost linearly as the force constant decreases. Thus, the force constant may be used as a measure of the bond strength in this case. However, such a monotonic relationship does not hold for the other two series. This is not unexpected, because the force constant is a measure of the curvature of the potential well near the equilibrium position:

$$K = \left(\frac{d^2V}{dq^2}\right)_{q\to 0} \tag{3.13}$$

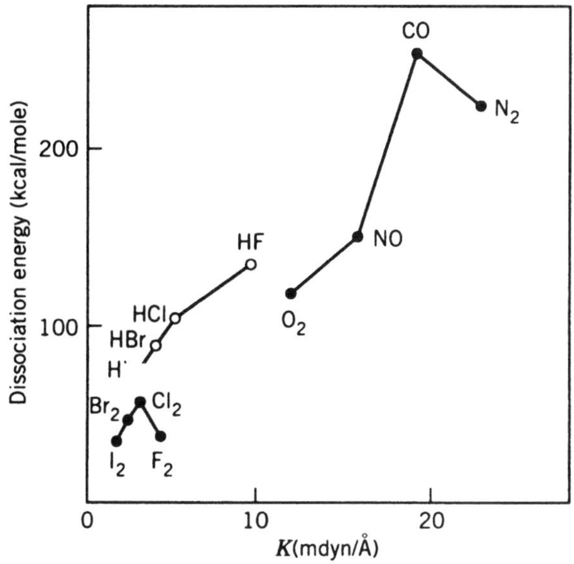

Fig. I-8. Relationships between force constants and dissociation energies in diatomic molecules.

whereas the dissociation energy D_e is given by the depth of the potential energy curve (Fig. I-7). Thus, a large force constant means sharp curvature of the potential well near the bottom, but does not necessarily indicate a deep potential well. Usually, however, a larger force constant is an indication of a stronger bond if the nature of the bond is similar in a series.

It is difficult to derive a general theoretical relationship between the force constant and the dissociation energy even for diatomic molecules.

In the case of small molecules, attempts have been made to calculate the force constants by quantum mechanical methods. The principle of the method is to express the total electronic energy of a molecule as a function of nuclear displacements near the equilibrium position and to calculate its second derivatives, $\partial^2 V/\partial q_i^2$, and so on for each displacement coordinate q_i. Thus far, *ab initio* calculations of force constants have been made for molecules such as HF, H_2O, and NH_3.[76] The force constants thus obtained are in good agreement with those calculated from the analysis of vibrational spectra. Recent progress in computer technology has made it possible to extend this approach to more complex molecules.[77]

There are several empirical relationships which relate the force constant to the bond distance. Badger's rule[78] is given by

$$K = 1.86(r - d_{ij})^{-3} \tag{3.14}$$

where r is the bond distance and d_{ij} has a fixed value for bonds between atoms of row i and j in the periodic table. Gordy's rule[79] is expressed as

$$K = aN\left(\frac{\chi_A\chi_B}{d^2}\right)^{3/4} + b \qquad (3.15)$$

where χ_A and χ_B are the electronegativities of atoms A and B constituting the bond, d is the bond distance, N is the bond order, and a and b are constants for certain broad classes of compounds. Herschbach and Laurie[80] modified Badger's rule in the form

$$r = d_{ij} + (a_{ij} - d_{ij})K^{-1/3} \qquad (3.16)$$

Here, a_{ij} and d_{ij} are constants for atoms of rows i and j in the periodic table.

Another empirical relationship is found by plotting stretching frequencies against bond distances for a series of compounds having common bonds. Such plots are highly important in estimating the bond distance from the observed frequency. Typical examples are found in the OH $\cdots$ O hydrogen-bonded compounds,[81] carbon–oxygen and carbon–nitrogen bonded compounds,[82] and molybdenum–oxygen[83] and vanadium–oxygen bonded compounds.[84] More examples are given in Sec. II-2.

I-4. NORMAL COORDINATES AND NORMAL VIBRATIONS

In diatomic molecules, the vibration of the nuclei occurs only along the line connecting two nuclei. In polyatomic molecules, however, the situation is much more complicated because all the nuclei perform their own harmonic oscillations. It can be shown, however, that any of these extremely complicated vibrations of the molecule may be represented as a superposition of a number of *normal vibrations.*

Let the displacement of each nucleus be expressed in terms of rectangular coordinate systems with the origin of each system at the equilibrium position of each nucleus. Then the kinetic energy of an N-atom molecule would be expressed as

$$T = \frac{1}{2}\sum_N m_N\left[\left(\frac{d\Delta x_N}{dt}\right)^2 + \left(\frac{d\Delta y_N}{dt}\right)^2 + \left(\frac{d\Delta z_N}{dt}\right)^2\right] \qquad (4.1)$$

If generalized coordinates such as

$$q_1 = \sqrt{m_1}\Delta x_1, \qquad q_2 = \sqrt{m_1}\Delta y_1, \qquad q_3 = \sqrt{m_1}\Delta z_1, \qquad q_4 = \sqrt{m_2}\Delta x_2, \cdots \qquad (4.2)$$

are used, the kinetic energy is simply written as

$$T = \frac{1}{2} \sum_{i}^{3N} \dot{q}_i^2 \tag{4.3}$$

The potential energy of the system is a complex function of all the coordinates involved. For small values of the displacements, it may be expanded in a Taylor's series as

$$V(q_1, q_2, \ldots, q_{3N}) = V_0 + \sum_{i}^{3N} \left(\frac{\partial V}{\partial q_i} \right)_0 q_i + \frac{1}{2} \sum_{i,j}^{3N} \left(\frac{\partial^2 V}{\partial q_i \partial q_j} \right)_0 q_i q_j + \cdots \tag{4.4}$$

where the derivatives are evaluated at $q_i = 0$, the equilibrium position. The constant term V_0 can be taken as zero if the potential energy at $q_i = 0$ is taken as a standard. The $(\partial V/\partial q_i)_0$ terms also become zero, since V must be a minimum at $q_i = 0$. Thus, V may be represented by

$$V = \frac{1}{2} \sum_{i,j}^{3N} \left(\frac{\partial^2 V}{\partial q_i \partial q_j} \right)_0 q_i q_j = \frac{1}{2} \sum_{i,j}^{3N} b_{ij} q_i q_j \tag{4.5}$$

neglecting higher-order terms.

If the potential energy given by Eq. 4.5 did not include any cross products such as $q_i q_j$, the problem could be solved directly by using Lagrange's equation:

$$\frac{d}{dt} \left(\frac{\partial T}{\partial \dot{q}_i} \right) + \frac{\partial V}{\partial q_i} = 0, \qquad i = 1, 2, \ldots, 3N \tag{4.6}$$

From Eqs. 4.3 and 4.5, Eq. 4.6 is written as

$$\ddot{q}_i + \sum_{j} b_{ij} q_j = 0, \qquad j = 1, 2, \ldots, 3N \tag{4.7}$$

If $b_{ij} = 0$ for $i \neq j$, Eq. 4.7 becomes

$$\ddot{q}_i + b_{ii} q_i = 0 \tag{4.8}$$

and the solution is given by

$$q_i = q_i^0 \sin(\sqrt{b_{ii}}t + \delta_i) \tag{4.9}$$

where q_i^0 and δ_i are the amplitude and the phase constant, respectively.

Since, in general, this simplification is not applicable, the coordinates q_i must be transformed into a set of new coordinates Q_i through the relations

$$\begin{aligned} q_1 &= \sum_i B_{1i}Q_i \\ q_2 &= \sum_i B_{2i}Q_i \\ &\vdots \\ q_k &= \sum_i B_{ki}Q_i \end{aligned} \tag{4.10}$$

The Q_i are called *normal coordinates* for the system. By appropriate choice of the coefficients B_{ki}, both the potential and the kinetic energies can be written as

$$T = \frac{1}{2}\sum_i \dot{Q}_i^2 \tag{4.11}$$

$$V = \frac{1}{2}\sum_i \lambda_i Q_i^2 \tag{4.12}$$

without any cross products.

If Eqs. 4.11 and 4.12 are combined with Lagrange's equation (4.6), there results

$$\ddot{Q}_i + \lambda_i Q_i = 0 \tag{4.13}$$

The solution of this equation is given by

$$Q_i = Q_i^0 \sin(\sqrt{\lambda_i}t + \delta_i) \tag{4.14}$$

and the frequency is

$$\nu_i = \frac{1}{2\pi} \sqrt{\lambda_i} \tag{4.15}$$

Such a vibration is called a *normal vibration.*

For the general N-atom molecule, it is obvious that the number of the normal vibrations is only $3N - 6$, since six coordinates are required to describe the translational and rotational motion of the molecule as a whole. Linear molecules have $3N - 5$ normal vibrations, as no rotational freedom exists around the molecular axis. Thus, the general form of the molecular vibration is a superposition of the $3N - 6$ (or $3N - 5$) normal vibrations given by Eq. 4.14.

The physical meaning of the normal vibration may be demonstrated in the following way. As shown in Eq. 4.10, the original displacement coordinate is related to the normal coordinate by

$$q_k = \sum_i B_{ki} Q_i \tag{4.10}$$

Since all the normal vibrations are independent of each other, consideration may be limited to a special case in which only one normal vibration, subscripted by 1, is excited (i.e., $Q_1^0 \neq 0$, $Q_2^0 = Q_3^0 = \cdots = 0$). Then, it follows from Eqs. 4.10 and 4.14 that

$$\begin{aligned} q_k = B_{k1} Q_1 &= B_{k1} Q_1^0 \sin(\sqrt{\lambda_1}\, t + \delta_1) \\ &= A_{k1} \sin(\sqrt{\lambda_1}\, t + \delta_1) \end{aligned} \tag{4.16}$$

This relation holds for all k. Thus, it is seen that the excitation of one normal vibration of the system causes vibrations, given by Eq. 4.16, of all the nuclei in the system. In other words, in the normal vibration, all the nuclei move with the same frequency and in phase.

This is true for any other normal vibration. Thus Eq. 4.16 may be written in the more general form

$$q_k = A_k \sin(\sqrt{\lambda}\, t + \delta) \tag{4.17}$$

If Eq. 4.17 is combined with Eq. 4.7, there results

$$-\lambda A_k + \sum_j b_{kj} A_j = 0 \tag{4.18}$$

This is a system of first-order simultaneous equations with respect to A. In order for all the A's to be nonzero,

$$\begin{vmatrix} b_{11}-\lambda & b_{12} & b_{13} & \cdots \\ b_{21} & b_{22}-\lambda & b_{23} & \cdots \\ b_{31} & b_{32} & b_{33}-\lambda & \cdots \\ \vdots & \vdots & \vdots & \end{vmatrix} = 0 \tag{4.19}$$

The order of this secular equation is equal to $3N$. Suppose that one root, λ_1, is found for Eq. 4.19. If it is inserted in Eq. 4.18, $A_{k1}, A_{k2}, \cdots$ are obtained for all the nuclei. The same is true for the other roots of Eq. 4.19. Thus, the most general solution may be written as a superposition of all the normal vibrations:

$$q_k = \sum_l B_{kl} Q_l^0 \sin(\sqrt{\lambda_l}\, t + \delta_l) \tag{4.20}$$

The general discussion developed above may be understood more easily if we apply it to a simple molecule such as CO_2, which is constrained to move in only one direction. If the mass and the displacement of each atom are defined as follows:

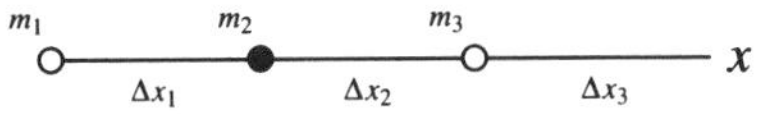

the potential energy is given by

$$V = \tfrac{1}{2}k[(\Delta x_1 - \Delta x_2)^2 + (\Delta x_2 - \Delta x_3)^2] \tag{4.21}$$

Considering that $m_1 = m_3$, we find that the kinetic energy is written as

$$T = \tfrac{1}{2}m_1(\Delta\dot{x}_1^2 + \Delta\dot{x}_3^2) + \tfrac{1}{2}m_2\Delta\dot{x}_2^2 \tag{4.22}$$

Using the generalized coordinates defined by Eq. 4.2, we may rewrite these energies as

$$2V = k\left[\left(\frac{q_1}{\sqrt{m_1}} - \frac{q_2}{\sqrt{m_2}}\right)^2 + \left(\frac{q_2}{\sqrt{m_2}} - \frac{q_3}{\sqrt{m_1}}\right)^2\right] \tag{4.23}$$

$$2T = \sum \dot{q}_i^2 \tag{4.24}$$

From comparison of Eq. 4.23 with Eq. 4.5, we obtain

$$b_{11} = \frac{k}{m_1}, \qquad b_{22} = \frac{2k}{m_2}$$
$$b_{12} = b_{21} = -\frac{k}{\sqrt{m_1 m_2}}, \quad b_{23} = b_{32} = -\frac{k}{\sqrt{m_1 m_2}}$$
$$b_{13} = b_{31} = 0, \qquad b_{33} = \frac{k}{m_1}$$

If these terms are inserted in Eq. 4.19, we obtain the following result:

$$\begin{vmatrix} \dfrac{k}{m_1} - \lambda & -\dfrac{k}{\sqrt{m_1 m_2}} & 0 \\ -\dfrac{k}{\sqrt{m_1 m_2}} & \dfrac{2k}{m_2} - \lambda & -\dfrac{k}{\sqrt{m_1 m_2}} \\ 0 & -\dfrac{k}{\sqrt{m_1 m_2}} & \dfrac{k}{m_1} - \lambda \end{vmatrix} = 0 \tag{4.25}$$

By solving this secular equation, we obtain three roots:

$$\lambda_1 = \frac{k}{m_1}, \qquad \lambda_2 = k\mu, \qquad \lambda_3 = 0$$

where

$$\mu = \frac{2m_1 + m_2}{m_1 m_2}$$

Equation 4.18 gives the following three equations:

$$-\lambda A_1 + b_{11}A_1 + b_{12}A_2 + b_{13}A_3 = 0$$
$$-\lambda A_2 + b_{21}A_1 + b_{22}A_2 + b_{23}A_3 = 0$$
$$-\lambda A_3 + b_{31}A_1 + b_{32}A_2 + b_{33}A_3 = 0$$

Using Eq. 4.17, we rewrite these as

$$(b_{11} - \lambda)q_1 + b_{12}q_2 + b_{13}q_3 = 0$$
$$b_{21}q_1 + (b_{22} - \lambda)q_2 + b_{23}q_3 = 0$$
$$b_{31}q_1 + b_{32}q_2 + (b_{33} - \lambda)q_3 = 0$$

If $\lambda_1 = k/m_1$ is inserted in the simultaneous equations above, we obtain

$$q_1 = -q_3, \qquad q_2 = 0$$

Similar calculations give

$$q_1 = q_3, \qquad q_2 = -2\sqrt{\frac{m_1}{m_2}}\, q_1 \quad \text{for} \quad \lambda_2 = k\mu$$

$$q_1 = q_3, \qquad q_2 = \sqrt{\frac{m_2}{m_1}}\, q_1 \quad \text{for} \quad \lambda_3 = 0$$

The relative displacements are depicted in the following figure:

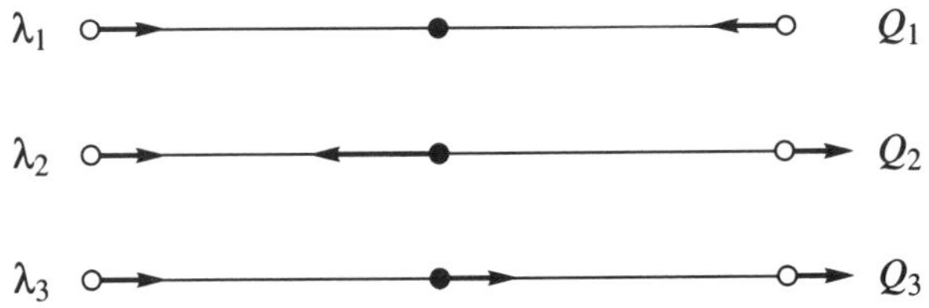

It is easy to see that λ_3 corresponds to the translational mode ($\Delta x_1 = \Delta x_2 = \Delta x_3$). The inclusion of λ_3 could be avoided if we consider the restriction that the center of gravity does not move; $m_1(\Delta x_1 + \Delta x_3) + m_2 \Delta x_2 = 0$.

The relationships between the generalized coordinates and the normal coordinates are given by Eq. 4.10. In the present case, we have

$$\begin{aligned} q_1 &= B_{11}Q_1 + B_{12}Q_2 + B_{13}Q_3 \\ q_2 &= B_{21}Q_1 + B_{22}Q_2 + B_{23}Q_3 \\ q_3 &= B_{31}Q_1 + B_{32}Q_2 + B_{33}Q_3 \end{aligned}$$

In the normal vibration whose normal coordinate is Q_1, $B_{11} : B_{21} : B_{31}$ gives the ratio of the displacements. From the previous calculation, it is obvious that $B_{11} : B_{21} : B_{31} = 1 : 0 : -1$. Similarly, $B_{12} : B_{22} : B_{32} = 1 : -2\sqrt{m_1/m_2} : 1$ gives the ratio of the displacements in the normal vibration whose normal coordinate is Q_2. Thus the mode of a normal vibration can be drawn if the normal coordinate is translated into a set of rectangular coordinates, as is shown above.

So far, we have discussed only the vibrations whose displacements occur along the molecular axis. There are, however, two other normal vibrations in which the displacements occur in the direction perpendicular to the molecular axis. They are not treated here, since the calculation is not simple. It is clear that the method described above will become more complicated as a molecule becomes larger. In this respect, the **GF** matrix method described in Sec. I-12 is important in the vibrational analysis of complex molecules.

By using the normal coordinates, the Schrödinger wave equation for the system can be written as

$$\sum_i \frac{\partial^2 \psi_n}{\partial Q_i^2} + \frac{8\pi^2}{h^2}\left(E - \frac{1}{2}\sum_i \lambda_i Q_i^2\right)\psi_n = 0 \tag{4.26}$$

Since the normal coordinates are independent of each other, it is possible to write

$$\psi_n = \psi_1(Q_1)\psi_2(Q_2)\cdots \tag{4.27}$$

and solve the simpler one-dimensional problem.

If Eq. 4.27 is substituted in Eq. 4.26, there results

$$\frac{d^2\psi_i}{dQ_i^2} + \frac{8\pi^2}{h^2}\left(E_i - \frac{1}{2}\lambda_i Q_i^2\right)\psi_i = 0 \tag{4.28}$$

where

$$E = E_1 + E_2 + \cdots$$

with

$$E_i = h\nu_i(v_i + \tfrac{1}{2})$$

$$\nu_i = \frac{1}{2\pi}\sqrt{\lambda_i} \tag{4.29}$$

I-5. SYMMETRY ELEMENTS AND POINT GROUPS[12–18]

As noted before, polyatomic molecules have $3N-6$ or, if linear, $3N-5$ normal vibrations. For any given molecule, however, only vibrations that are permitted by the selection rule for that molecule appear in the infrared and Raman spectra. Since the selection rule is determined by the symmetry of the molecule, this must first be studied.

The spatial geometrical arrangement of the nuclei constituting the molecule determines its symmetry. If a coordinate transformation (a reflection or a rotation or a combination of both) produces a configuration of the nuclei indistinguishable from the original one, this transformation is called a *symmetry operation*, and the molecule is said to have a corresponding *symmetry element.* Molecules may have the following symmetry elements.

(1) Identity *I*

This symmetry element is possessed by every molecule no matter how unsymmetrical it is, the corresponding operation being to leave the molecule unchanged. The inclusion of this element is necessitated by mathematical reasons which will be discussed in Sec. I-7.

(2) A Plane of Symmetry, σ

If reflection of a molecule with respect to some plane produces a configuration indistinguishable from the original one, the plane is called a plane of symmetry.

(3) A Center of Symmetry, *i*

If reflection at the center, that is, inversion, produces a configuration indistinguishable from the original one, the center is called a center of symmetry. This operation changes the signs of all the coordinates involved, $x_i \rightarrow -x_i \quad y_i \rightarrow -y_i \quad z_i \rightarrow -z_i$.

(4) A *p*-Fold Axis of Symmetry, C_p*

If rotation through an angle $360°/p$ about an axis produces a configuration indistinguishable from the original one, the axis is called a *p*-fold axis of symmetry, C_p. For example, a twofold axis, C_2, implies that a rotation of 180° about the axis reproduces the original configuration. A molecule may have a two-, three-, four-, five-, or sixfold, or higher axis. A linear molecule has an infinite-fold (denoted by ∞-fold) axis of symmetry, C_∞, since a rotation of $360°/\infty$, that is, an infinitely small angle, transforms the molecule into one indistinguishable from the original.

(5) A *p*-Fold Rotation–Reflection Axis, S_p*

If rotation by $360°/p$ about the axis, followed by reflection at a plane perpendicular to the axis, produces a configuration indistinguishable from the original one, the axis is called a *p*-fold rotation–reflection axis. A molecule may have a two-, three-, four-, five-, or sixfold, or higher, rotation–reflection axis. A symmetrical linear molecule has an S_∞ axis. It is easily seen that the presence of S_p always means the presence of C_p as well as σ when p is odd.

A molecule may have more than one of these symmetry elements. Combination of more and more of these elements produces systems of higher and higher symmetry. Not all combinations of symmetry elements, however, are possible.

*The notation C_p^n (or S_p^n) is used to indicate that the C_p (or S_p) operation is carried out successively n times.

For example, it is highly improbable that a molecule will have a C_3 and C_4 axis in the same direction because this requires the existence of a 12-fold axis in the molecule. It should also be noted that the presence of some symmetry elements often implies the presence of other elements. For example, if a molecule has two σ-planes at right angles to each other, the line of intersection of these two planes must be a C_2 axis. A possible combination of symmetry operations whose axes intersect at a point is called a *point group*.*

Theoretically, an infinite number of point groups exist, since there is no restriction on the order (p) of rotation axes which may exist in an isolated molecule. Practically, however, there are few molecules and ions that possess a rotation axis higher than C_6. Thus most of the compounds discussed in this book belong to the following point groups:

1. $\mathbf{C}_p$. Molecules having only a C_p and no other elements of symmetry: $\mathbf{C}_1$, $\mathbf{C}_2$, $\mathbf{C}_3$, and so on.

2. $\mathbf{C}_{ph}$. Molecules having a C_p and a σ_h perpendicular to it: $\mathbf{C}_{1h} \equiv \mathbf{C}_s$, $\mathbf{C}_{2h}$, $\mathbf{C}_{3h}$, and so on.

3. $\mathbf{C}_{pv}$. Molecules having a C_p and $p\sigma_v$ through it: $\mathbf{C}_{1v} \equiv \mathbf{C}_s$, $\mathbf{C}_{2v}$, $\mathbf{C}_{3v}$, $\mathbf{C}_{4v}, \ldots, \mathbf{C}_{\infty v}$.

4. $\mathbf{D}_p$. Molecules having a C_p and pC_2 perpendicular to the C_p and at equal angles to one another: $\mathbf{D}_1 \equiv \mathbf{C}_2$, $\mathbf{D}_2 \equiv \mathbf{V}$, $\mathbf{D}_3$, $\mathbf{D}_4$, and so on.

5. $\mathbf{D}_{ph}$. Molecules having a C_p, $p\sigma_v$ through it at angles of $360°/2p$ to one another, and a σ_h perpendicular to the C_p: $\mathbf{D}_{1h} \equiv \mathbf{C}_{2v}$, $\mathbf{D}_{2h} \equiv \mathbf{V}_h$, $\mathbf{D}_{3h}$, $\mathbf{D}_{4h}$, $\mathbf{D}_{5h}$, $\mathbf{D}_{6h}, \ldots, \mathbf{D}_{\infty h}$.

6. $\mathbf{D}_{pd}$. Molecules having a C_p, pC_2 perpendicular to it, and $p\sigma_d$ which go through the C_p and bisect the angles between two successive C_2 axes: $\mathbf{D}_{2d} \equiv \mathbf{V}_d$, $\mathbf{D}_{3d}$, $\mathbf{D}_{4d}$, $\mathbf{D}_{5d}$, and so on.

7. $\mathbf{S}_p$. Molecules having only a S_p (p even). For p odd, S_p is equivalent to $C_p \times \sigma_h$, for which other notations such as $\mathbf{C}_{3h}$ are used: $\mathbf{S}_2 \equiv \mathbf{C}_i$, $\mathbf{S}_4$, $\mathbf{S}_6$, and so on.

8. $\mathbf{T}_d$. Molecules having three mutually perpendicular C_2 axes, four C_3 axes, and a σ_d through each pair of C_3 axes: regular tetrahedral molecules.

9. $\mathbf{O}_h$. Molecules having three mutually perpendicular C_4 axes, four C_3 axes, and a center of symmetry, i: regular octahedral and cubic molecules.

10. $\mathbf{I}_h$. Molecules having six C_5 axes, ten C_3 axes, fifteen C_2 axes, fifteen σ planes, and a center of symmetry. In total, such molecules possess 120 symmetry elements. One example is an icosahedron having twenty equilateral triangular faces, which is found in the B_{12} skeleton of the $B_{12}H_{12}^{2-}$ ion (Sec. II-13). Another example is a regular dodecahedron having twelve regular pentagonal faces. Buckminsterfullerene, C_{60} (Sec. II-14), also belongs to the $\mathbf{I}_h$

*In this respect, point groups differ from space groups, which involve translations and rotations about nonintersecting axes (see Sec. I-25).

point group. It is a truncated icosahedron with twenty hexagonal and twelve pentagonal faces.

Figure I-9 illustrates the symmetry elements present in the point group, $\mathbf{D}_{4h}$. Complete listings of the symmetry elements for common point groups are found in the character tables included in Appendix 1. Figure I-10 illustrates examples of molecules belonging to some of these point groups. From the symmetry point

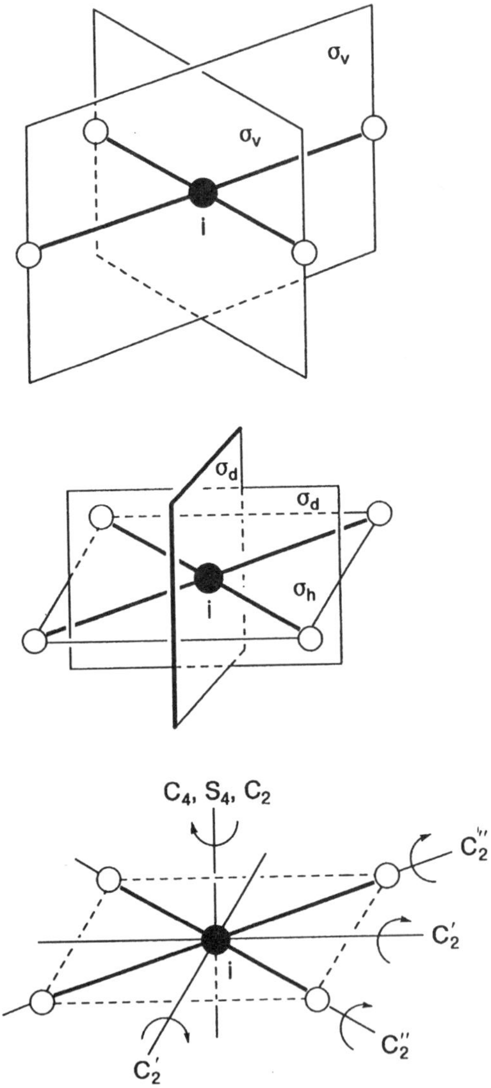

Fig. I-9. Symmetry elements in $\mathbf{D}_{4h}$ point group.

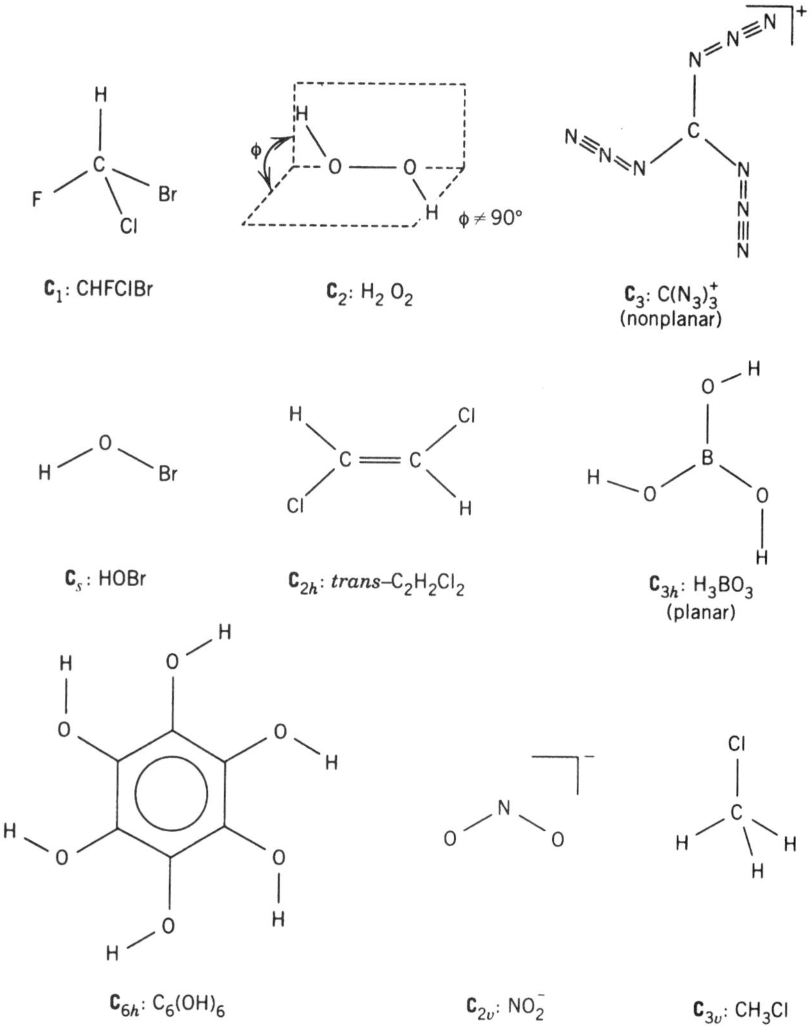

Fig. I-10. Examples of molecules belonging to some point groups.

of view, molecules belonging to the $\mathbf{C}_1$, $\mathbf{C}_2$, $\mathbf{C}_3$, $\mathbf{D}_2 \equiv \mathbf{V}$, and $\mathbf{D}_3$ groups possess only C_p axes, and are thus optically active.

I-6. SYMMETRY OF NORMAL VIBRATIONS AND SELECTION RULES

Figures I-11 and I-12 illustrate the normal modes of vibration of CO_2 and H_2O molecules, respectively. In each normal vibration, the individual nuclei carry

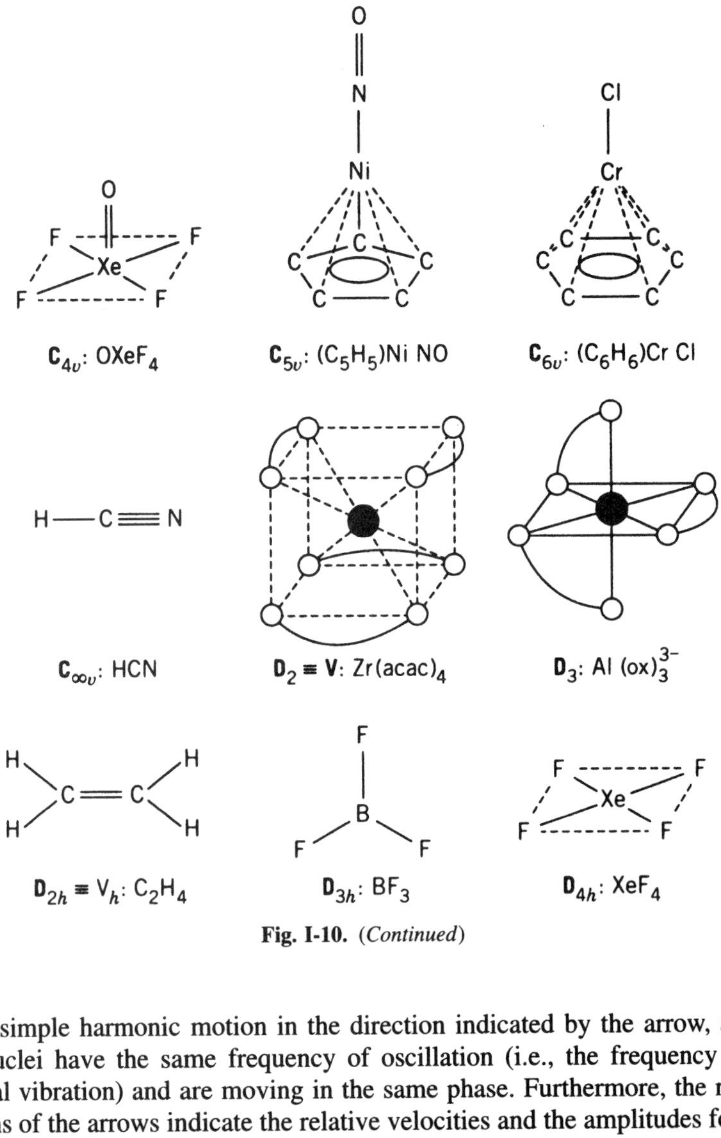

Fig. I-10. (*Continued*)

out a simple harmonic motion in the direction indicated by the arrow, and all the nuclei have the same frequency of oscillation (i.e., the frequency of the normal vibration) and are moving in the same phase. Furthermore, the relative lengths of the arrows indicate the relative velocities and the amplitudes for each nucleus.* The ν_2 vibrations in CO_2 are worth comment, since they differ from the others in that two vibrations (ν_{2a} and ν_{2b}) have exactly the same frequency. Apparently, there are an infinite number of normal vibrations of this type, which differ only in their directions perpendicular to the molecular axis. Any of them, however, can be resolved into two vibrations such as ν_{2a} and ν_{2b}, which are perpendicular to each other. In this respect, the ν_2 vibrations in CO_2 are called *doubly degenerate vibrations*. Doubly degenerate vibrations occur only when

*In this respect, all the normal modes of vibration shown in this book are only approximate.

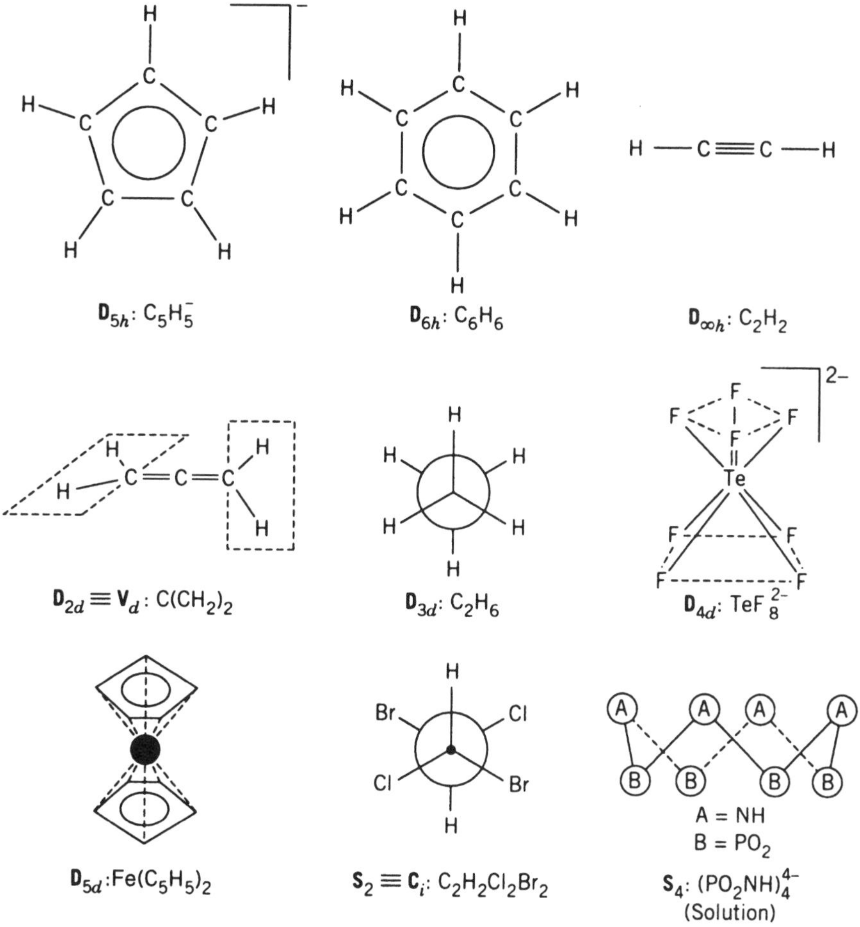

Fig. I-10. *(Continued)*

a molecule has an axis higher than twofold. *Triply degenerate vibrations* also occur in molecules having more than one C_3 axis.

To determine the symmetry of a normal vibration, it is necessary to begin by considering the kinetic and potential energies of the system. These were discussed in Sec. I-4.

$$T = \frac{1}{2} \sum_i \dot{Q}_i^2 \tag{6.1}$$

$$V = \frac{1}{2} \sum_i \lambda_i Q_i^2 \tag{6.2}$$

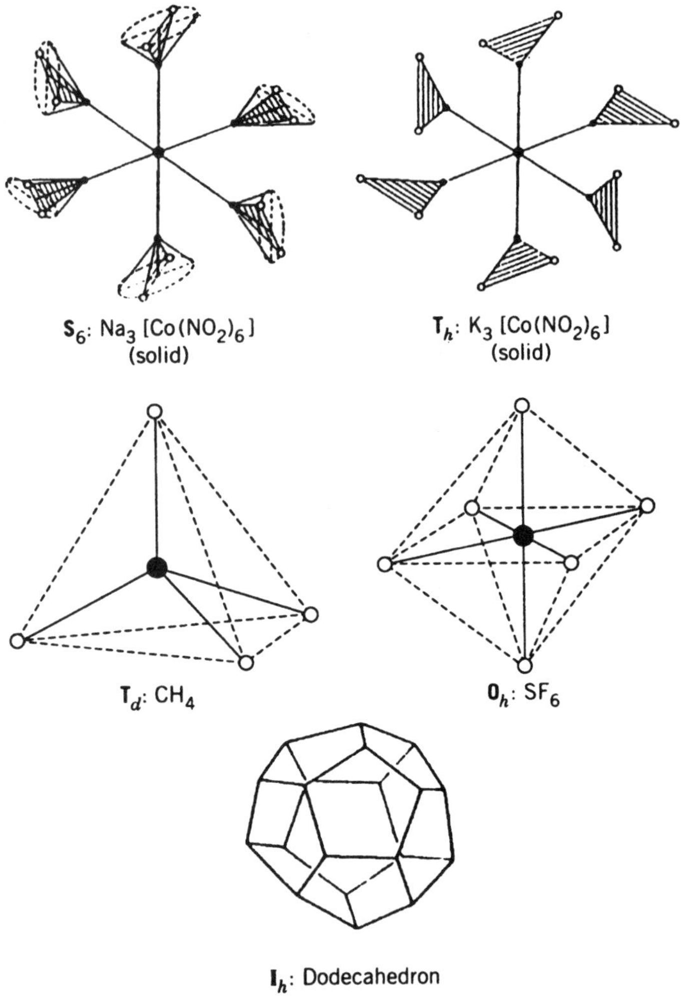

Fig. I-10. (*Continued*)

Consider a case in which a molecule performs only one normal vibration, Q_i. Then, $T = \frac{1}{2}\dot{Q}_i^2$ and $V = \frac{1}{2}\lambda_i Q_i^2$. These energies must be invariant when a symmetry operation, R, changes Q_i to RQ_i. Thus, we obtain

$$T = \tfrac{1}{2}\dot{Q}_i^2 = \tfrac{1}{2}(R\dot{Q}_i)^2$$
$$V = \tfrac{1}{2}\lambda_i Q_i^2 = \tfrac{1}{2}\lambda_i(RQ_i)^2$$

For these relations to hold, it is necessary that

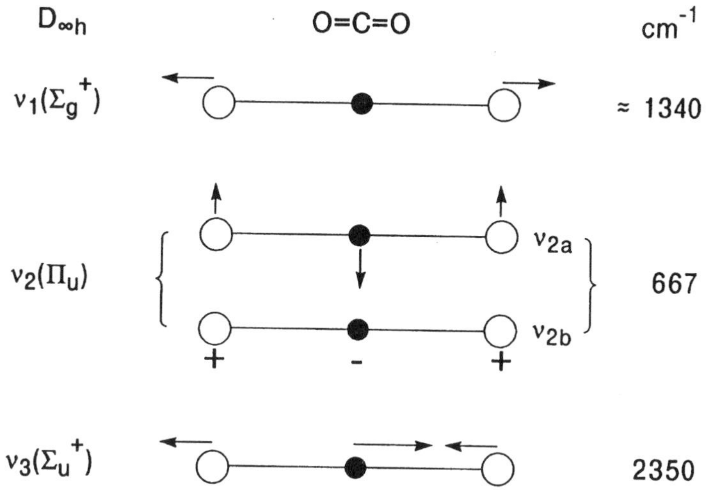

Fig. I-11. Normal modes of vibration in CO_2 (+ and – denote vibrations going upward and downward, respectively, in the direction perpendicular to the paper plane).

$$(RQ_i)^2 = Q_i^2 \quad \text{or} \quad RQ_i = \pm Q_i \tag{6.3}$$

Thus, the normal coordinate must change either into itself or into its negative. If $Q_i = RQ_i$, the vibration is said to be *symmetric*. If $Q_i = -RQ_i$, it is said to be *antisymmetric*.

If the vibration is doubly degenerate, we have

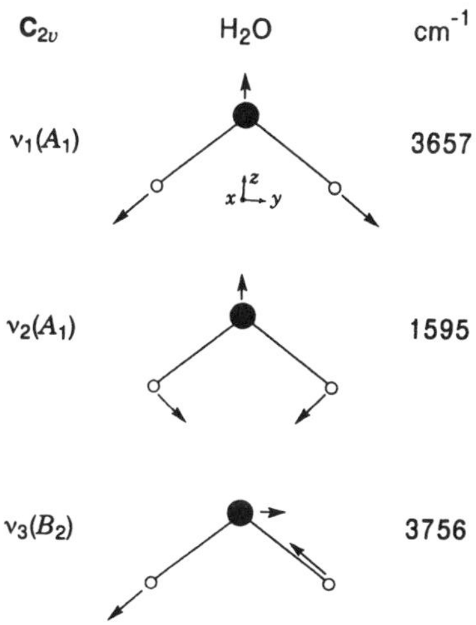

Fig. I-12. Normal modes of vibration in H_2O.

$$T = \frac{1}{2}\dot{Q}_{ia}^2 + \frac{1}{2}\dot{Q}_{ib}^2$$
$$V = \frac{1}{2}\lambda_i(Q_{ia})^2 + \frac{1}{2}\lambda_i(Q_{ib})^2$$

In this case, a relation such as

$$(RQ_{ia})^2 + (RQ_{ib})^2 = Q_{ia}^2 + Q_{ib}^2 \tag{6.4}$$

must hold. As will be shown later, such a relationship is expressed more conveniently by using a matrix form:*

$$R\begin{bmatrix} Q_{ia} \\ Q_{ib} \end{bmatrix} = \begin{bmatrix} A & B \\ C & D \end{bmatrix}\begin{bmatrix} Q_{ia} \\ Q_{ib} \end{bmatrix}$$

where the values of A, B, C, and D depend on the symmetry operation, R. In any case, the normal vibration must be either symmetric or antisymmetric or degenerate for each symmetry operation.

The symmetry properties of the normal vibrations of the H_2O molecule shown in Fig. I-12 are classified as indicated in Table I-3. Here, +1 and −1 denote symmetric and antisymmetric, respectively. In the ν_1 and ν_2 vibrations, all the symmetry properties are preserved during the vibration. Therefore they are *symmetric vibrations* and are called, in particular, *totally symmetric vibrations*. In the ν_3 vibration, however, symmetry elements such as C_2 and $\sigma_v(xz)$ are lost. Thus, it is called a *nonsymmetric vibration*. If a molecule has a number of symmetry elements, the normal vibrations are classified according to the number and the kind of symmetry elements preserved during the vibration.

To determine the activity of the vibrations in the infrared and Raman spectra, the selection rule must be applied to each normal vibration. As will be shown in Sec. I-10, rigorous selection rules can be derived quantum mechanically, and applied to individual molecules using group theory.

For small molecules, however, the IR and Raman activities may be determined by simple inspection of their normal modes. First, we consider the general rule which states that *the vibration is IR-active if the dipole moment is*

TABLE I-3.

C_{2v}	I	$C_2(z)$	$\sigma_v(xz)^a$	$\sigma_v(yz)^a$
Q_1, Q_2	+1	+1	+1	+1
Q_3	+1	−1	−1	+1

[a] σ_v = vertical plane of symmetry.

*For matrix algebra, see Appendix II.

changed during the vibration. It is obvious that the vibration of a homopolar diatomic molecule is not IR-active, whereas that of a heteropolar diatomic molecule is IR-active. In the case of CO_2, shown in Fig. I-11, it is readily seen that the ν_1 is not IR-active, whereas the ν_2 and ν_3 are IR-active. Figure I-13 illustrates the changes in dipole moment during the three normal vibrations of H_2O. It is readily seen that all the vibrations are IR-active because the dipole moment is changed as indicated.

To discuss Raman activity, we must consider the nature of polarizability introduced in Sec. I-2. When a molecule interacts with the electric field of a laser beam, its electron cloud is distorted because the positively charged nuclei are attracted toward the negative pole, and the electrons toward the positive pole, as shown in Fig. I-14. The charge separation produces an induced dipole moment (P) given by

$$P = \alpha E^* \tag{6.5}$$

where E is the strength of the electric field and α is the polarizability. By resolving P, α, and E in the x, y, and z directions, we can write

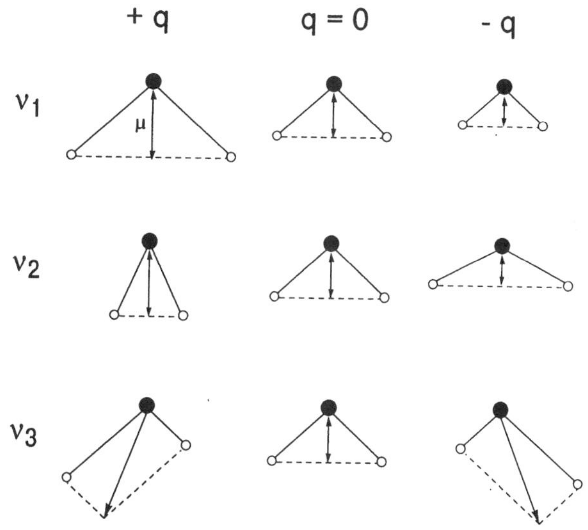

Fig. I-13. Changes in dipole moment for H_2O during each normal vibration.

*A more complete form of this equation is $P = \alpha E + \frac{1}{2}\beta E^2 \cdots$. Here, $\beta \ll \alpha$ and β is called the hyperpolarizability. The second term becomes significant only when E is large (~ 10^9 V cm^{-1}). In this case, we observe novel spectroscopic phenomena such as the hyper Raman effect, stimulated Raman effect, inverse Raman effect, and coherent anti-Stokes Raman scattering (CARS). For a discussion of nonlinear Raman spectroscopy, see Ref. 28.

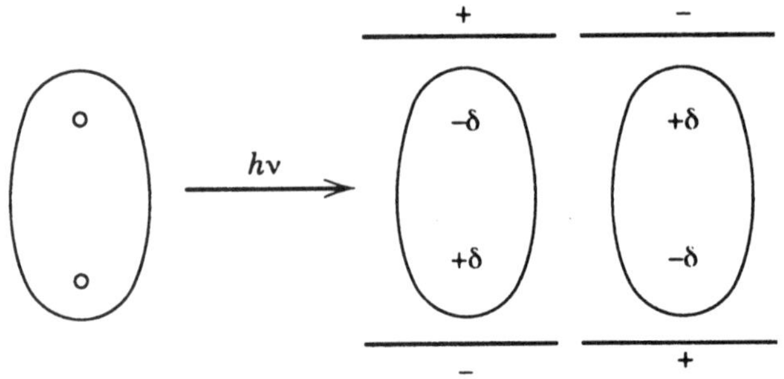

Fig. I-14. Polarization of a diatomic molecule in an electric field.

$$P_x = \alpha_x E_x, \qquad P_y = \alpha_y E_y, \qquad P_z = \alpha_z E_z \tag{6.6}$$

However, the actual relationship is more complicated since the direction of polarization may not coincide with the direction of the applied field. This is so because the direction of chemical bonds in the molecule also affects the direction of polarization. Thus, instead of Eq. 6.6, we have the relationship

$$\begin{aligned} P_x &= \alpha_{xx} E_x + \alpha_{xy} E_y + \alpha_{xz} E_z \\ P_y &= \alpha_{yx} E_x + \alpha_{yy} E_y + \alpha_{yz} E_z \\ P_z &= \alpha_{zx} E_x + \alpha_{zy} E_y + \alpha_{zz} E_z \end{aligned} \tag{6.7}$$

In matrix form, Eq. 6.7 is written as

$$\begin{bmatrix} P_x \\ P_y \\ P_z \end{bmatrix} = \begin{bmatrix} \alpha_{xx} & \alpha_{xy} & \alpha_{xz} \\ \alpha_{yx} & \alpha_{yy} & \alpha_{yz} \\ \alpha_{zx} & \alpha_{zy} & \alpha_{zz} \end{bmatrix} \begin{bmatrix} E_x \\ E_y \\ E_z \end{bmatrix} \tag{6.8}$$

and the first matrix on the right-hand side is called the *polarizability tensor.* In normal Raman scattering, the tensor is symmetric; $\alpha_{xy} = \alpha_{yx}$, $\alpha_{yz} = \alpha_{zy}$, and $\alpha_{xz} = \alpha_{zx}$. This is not so, however, in the case of resonance Raman scattering (Sec. I-22).

According to quantum mechanics, *the vibration is Raman-active if one of these six components of the polarizability changes during the vibration.* Thus, it is obvious that the vibration of a homopolar diatomic molecule is Raman-active but not IR-active, whereas the vibration of a heteropolar diatomic molecule is both IR- and Raman-active.

Changes in the polarizability tensor can be visualized if we draw a *polarizability ellipsoid* by plotting $1/\sqrt{\alpha}$ in every direction from the origin. This gives a three-dimensional surface such as shown below:

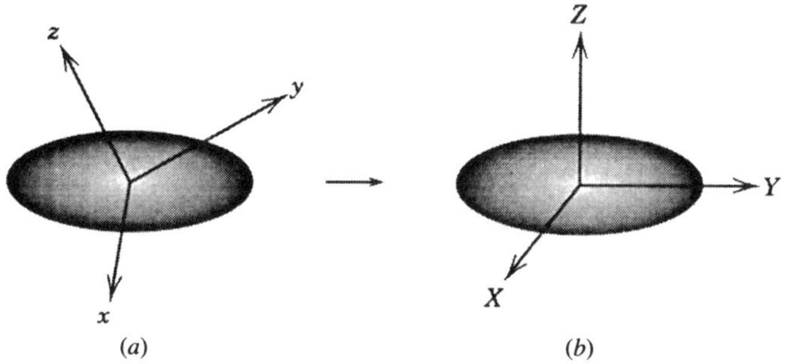

If we rotate such an ellipsoid so that its principal axes coincide with the molecular axes (X, Y, and Z), Eq. 6.8 is simplified to

$$\begin{bmatrix} P_X \\ P_Y \\ P_Z \end{bmatrix} = \begin{bmatrix} \alpha_{XX} & 0 & 0 \\ 0 & \alpha_{YY} & 0 \\ 0 & 0 & \alpha_{ZZ} \end{bmatrix} \begin{bmatrix} E_X \\ E_Y \\ E_Z \end{bmatrix} \tag{6.9}$$

Such axes are called "principal axes of polarizability." In terms of the polarizability ellipsoid, *the vibration is Raman-active if the polarizability ellipsoid changes in size, shape, or orientation during the vibration.*

As an example, Fig. I-15 illustrates the polarizability ellipsoids for the three normal vibrations of H_2O at the equilibrium ($q = 0$) and two extreme configurations ($q = \pm q$). It is readily seen that both ν_1 and ν_2 are Raman-active because the size and the shape of the ellipsoid (α_{xx}, α_{yy}, and α_{zz}) change during these vibrations. The ν_3 is also Raman-active because the orientation of the ellipsoid

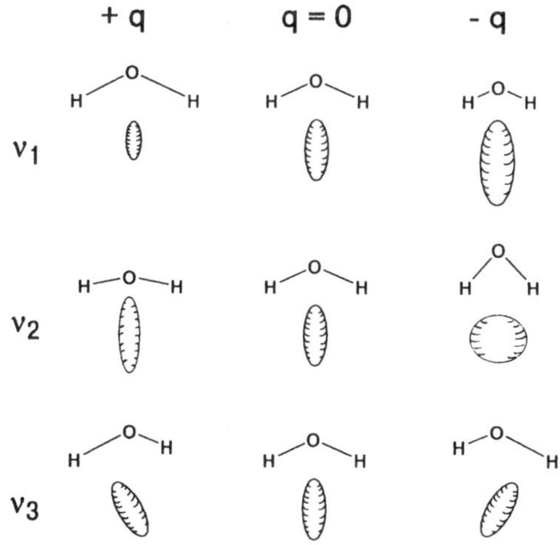

Fig. I-15. Changes in polarizability ellipsoid during normal vibrations of H_2O.

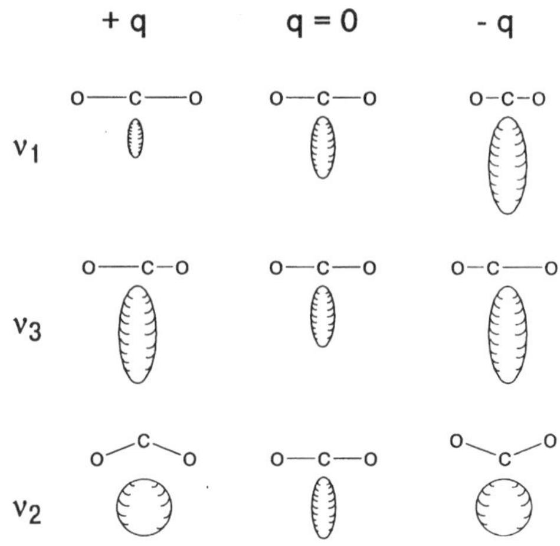

Fig. I-16. Changes in polarizability ellipsoid during normal vibrations of CO_2.

(α_{yz}) changes during the vibration. Thus, all three normal vibrations of H_2O are Raman-active. Figure I-16 illustrates the changes in polarizability ellipsoids during the normal vibrations of CO_2. It is readily seen that ν_1 is Raman-active because the size of the ellipsoid changes during the vibration (α_{xx}, α_{yy}, and α_{zz} change proportionately). Although the size and/or shape of the ellipsoid change during the ν_2 and ν_3 vibrations, they are identical in two extreme positions, as seen in Fig. I-16. If we consider a limiting case where the nuclei undergo very small displacements, there is effectively no change in the polarizability. This is illustrated in Fig. I-17. Thus, these two vibrations are not Raman-active.

It should be noted that in CO_2 the vibration symmetric with respect to the center of symmetry (ν_1) is Raman-active and not IR-active, whereas the vibrations antisymmetric with respect to the center of symmetry (ν_2 and ν_3) are IR-active but not Raman-active. In a polyatomic molecule having a center of

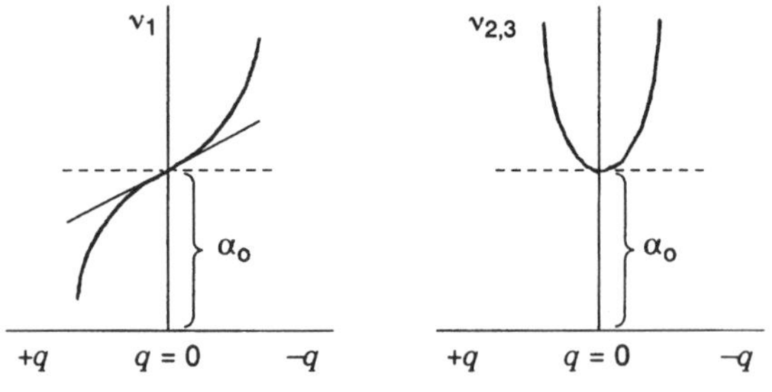

$$\left(\frac{d\alpha}{dq}\right)_0 \neq 0 \qquad \left(\frac{d\alpha}{dq}\right)_0 = 0$$

Fig. I-17. Changes in polarizability with respect to displacement coordinate during the ν_1 and $\nu_{2,3}$ vibrations in CO_2.

symmetry, the vibrations symmetric with respect to the center of symmetry (g vibrations*) are Raman-active and not IR-active, but the vibrations antisymmetric with respect to the center of symmetry (u vibrations*) are IR-active and not Raman-active. This rule is called the "mutual exclusion rule." It should be noted, however, that in polyatomic molecules having several symmetry elements in addition to the center of symmetry, the vibrations that should be active according to this rule may not necessarily be active, because of the presence of other symmetry elements. An example is seen in a square-planar XY_4-type molecule of $\mathbf{D}_{4h}$ symmetry, where the A_{2g} vibrations are not Raman-active and the A_{1u}, B_{1u}, and B_{2u} vibrations are not IR-active (see Sec. II-6).

I-7. INTRODUCTION TO GROUP THEORY[16–18]

In Sec. I-5, the symmetry and the point group allocation of a given molecule were discussed. To understand the symmetry and selection rules of normal vibrations in polyatomic molecules, however, a knowledge of group theory is required. The minimum amount of group theory needed for this purpose is given here.

Consider a pyramidal XY_3 molecule (Fig. I-18) for which the symmetry operations I, C_3^+, C_3^-, σ_1, σ_2, and σ_3 are applicable. Here, C_3^+ and C_3^- denote rotation through 120° in the clockwise and counterclockwise directions, respectively, and σ_1, σ_2, and σ_3 indicate the symmetry planes that pass through X and Y_1, X and Y_2, and X and Y_3, respectively. For simplicity, let these symme-

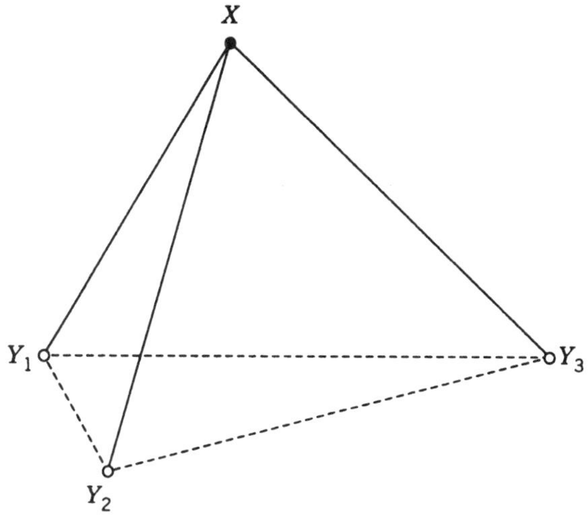

Fig. I-18. Pyramidal XY_3 molecule.

*The symbols g and u stand for *gerade* and *ungerade* (German), respectively.

try operations be denoted by I, A, B, C, D, and E, respectively. Other symmetry operations are possible, but each is equivalent to some one of the operations mentioned. For instance, a clockwise rotation through 240° is identical with operation B. It may also be shown that two successive applications of any one of these operations is equivalent to some single operation of the group mentioned. Let operation C be applied to the original figure. This interchanges Y_2 and Y_3. If operation A is applied to the resulting figure, the net result is the same as application of the single operation D to the original figure. This is written as $CA = D$. If all the possible multiplicative combinations are made, Table I-4, in which the operation applied first is written across the top, is obtained. This is called the *multiplication table* of the group.

It is seen that a group consisting of the mathematical elements (symmetry operations) I, A, B, C, D, and E satisfies the following conditions:

1. The product of any two elements in the set is another element in the set.
2. The set contains the identity operation that satisfies the relation $IP = PI = P$, where P is any element in the set.
3. The associative law holds for all the elements in the set, that is, $(CB)A = C(BA)$, for example.
4. Every element in the set has its reciprocal, X, which satisfies the relation $XP = PX = I$, where P is any element in the set. This reciprocal is usually denoted by P^{-1}.

These are necessary and sufficient conditions for a set of elements to form a *group*. It is evident that operations I, A, B, C, D, and E form a group in this sense. It should be noted that the commutative law of multiplication does not necessarily hold. For example, Table I-4 shows that $CD \neq DC$.

The six elements can be classified into three types of operations: the identity operation I, the rotations C_3^+ and C_3^-, and the reflections σ_1, σ_2, and σ_3. Each of these sets of operations is said to form a *class*. More precisely, two operations, P and Q, which satisfy the relation $X^{-1}PX = P$ or Q, where X is any operation of the group and X^{-1} is its reciprocal, are said to belong to the same class. It can easily be shown that C_3^+ and C_3^-, for example, satisfy the relation. Thus the six elements of the point group $\mathbf{C}_{3v}$ are usually abbreviated as I, $2C_3$, and $3\sigma_v$.

TABLE I-4.

	I	A	B	C	D	E
I	I	A	B	C	D	E
A	A	B	I	D	E	C
B	B	I	A	E	C	D
C	C	E	D	I	B	A
D	D	C	E	A	I	B
E	E	D	C	B	A	I

TABLE I-5.

$\mathbf{C}_{3v}$	I	A	B	C	D	E
$A_1(\Gamma_1)$	1	1	1	1	1	1
$A_2(\Gamma_2)$	1	1	1	-1	-1	-1
$E(\Gamma_3)$	$\begin{pmatrix} 1 & 0 \\ 0 & 1 \end{pmatrix}$	$\begin{pmatrix} -\frac{1}{2} & \frac{\sqrt{3}}{2} \\ -\frac{\sqrt{3}}{2} & -\frac{1}{2} \end{pmatrix}$	$\begin{pmatrix} -\frac{1}{2} & -\frac{\sqrt{3}}{2} \\ \frac{\sqrt{3}}{2} & -\frac{1}{2} \end{pmatrix}$	$\begin{pmatrix} -1 & 0 \\ 0 & 1 \end{pmatrix}$	$\begin{pmatrix} \frac{1}{2} & -\frac{\sqrt{3}}{2} \\ -\frac{\sqrt{3}}{2} & -\frac{1}{2} \end{pmatrix}$	$\begin{pmatrix} \frac{1}{2} & \frac{\sqrt{3}}{2} \\ \frac{\sqrt{3}}{2} & -\frac{1}{2} \end{pmatrix}$

The relations between the elements of the group are shown in the multiplication table (Table I-4). Such a tabulation of a group is, however, awkward to handle. The essential features of the table may be abstracted by replacing the elements by some analytical function that reproduces the multiplication table. Such an analytical expression may be composed of a simple integer, an exponential function, or a matrix. Any set of such expressions that satisfies the relations given by the multiplication table is called a *representation* of the group and is designated by Γ. The representations of the point group $\mathbf{C}_{3v}$ discussed above are indicated in Table I-5. It can easily be proved that each representation in the table satisfies the multiplication table.

In addition to the three representations in Table I-5, it is possible to write an infinite number of other representations of the group. If a set of six matrices of the type $S^{-1}R(K)S$ is chosen, where $R(K)$ is a representation of the element K given in Table I-5, $S(|S| \neq 0)$ is any matrix of the same order as R, and S^{-1} is the reciprocal of S, this set also satisfies the relations given by the multiplication table. The reason is obvious from the relation

$$S^{-1}R(K)SS^{-1}R(L)S = S^{-1}R(K)R(L)S = S^{-1}R(KL)S$$

Such a transformation is called a *similarity transformation*. Thus, it is possible to make an infinite number of representations by means of similarity transformations.

On the other hand, this statement suggests that a given representation may possibly be broken into simpler ones. If each representation of the symmetry element K is transformed into the form

$$R(K) = \left| \begin{array}{c|c|c|c} Q_1(K) & 0 & 0 & 0 \\ \hline 0 & Q_2(K) & 0 & 0 \\ \hline 0 & 0 & Q_3(K) & 0 \\ \hline 0 & 0 & 0 & Q_3(K) \end{array} \right| \tag{7.1}$$

by a similarity transformation, $Q_1(K)$, $Q_2(K)$, ... are simpler representations. In such a case, $R(K)$ is called *reducible*. If a representation cannot be simplified any further, it is said to be *irreducible*. The representations Γ_1, Γ_2, and Γ_3 in Table I-5 are all irreducible representations. It can be shown generally that the number of irreducible representations is equal to the number of classes. Thus, only three irreducible representations exist for the point group $\mathbf{C}_{3v}$. These representations are entirely independent of each other. Furthermore, the sum of the squares of the dimensions (l) of the irreducible representations of a group is always equal to the total number of the symmetry elements, namely, the *order of the group* (h). Thus

$$\sum l_i^2 = l_1^2 + l_2^2 + \cdots = h \tag{7.2}$$

In the point group $\mathbf{C}_{3v}$, it is seen that

$$1^2 + 1^2 + 2^2 = 6$$

A point group is classified into *species* according to its irreducible representations. In the point group $\mathbf{C}_{3v}$, the species having the irreducible representations Γ_1, Γ_2, and Γ_3 are called the A_1, A_2, and E species, respectively.*

The sum of the diagonal elements of a matrix is called the *character* of the matrix and is denoted by χ. It is to be noted in Table I-5 that the character of each of the elements belonging to the same class is the same. Thus, using the character, Table I-5 can be simplified to Table I-6. Such a table is called the *character table* of the point group $\mathbf{C}_{3v}$.

That the *character* of a matrix is not changed by a similarity transformation can be proved as follows. If a similarity transformation is expressed by $T = S^{-1}RS$, then

$$\chi_T = \sum_i (S^{-1}RS)_{ii} = \sum_{i,j,k} (S^{-1})_{ij}R_{jk}S_{ki} = \sum_{j,k,i} S_{ki}(S^{-1})_{ij}R_{jk}$$
$$= \sum_{j,k} \delta_{kj}R_{jk} = \sum_k R_{kk} = \chi_R$$

TABLE 1-6. The Character Table of the Point Group $\mathbf{C}_{3v}$

$\mathbf{C}_{3v}$	I	$2C_3(z)$	$3\sigma_v$
$A_1(\chi_1)$	1	1	1
$A_2(\chi_2)$	1	1	−1
$E(\chi_3)$	2	−1	0

*For the labeling of the irreducible representations (species), see Appendix 1.

where δ_{kj} is Kronecker's delta (0 for $k \neq j$ and 1 for $k = j$). Thus, any reducible representation can be reduced to its irreducible representations by a similarity transformation that leaves the character unchanged. Therefore, the character of the reducible representation, $\chi(K)$, is written as

$$\chi(K) = \sum_{m} a_m \chi_m(K) \tag{7.3}$$

where $\chi_m(K)$ is the character of $Q_m(K)$, and a_m is a positive integer that indicates the number of times $Q_m(K)$ appears in the matrix of Eq. 7.1. Hereafter the character will be used rather than the corresponding representation because a 1 : 1 correspondence exists between these two, and the former is sufficient for vibrational problems.

It is important to note that the following relation holds in Table I-6:

$$\sum_{K} \chi_i(K)\chi_j(K) = h\delta_{ij} \tag{7.4}$$

If Eq. 7.3 is multiplied by $\chi_i(K)$ on both sides, and the summation is taken over all the symmetry operations, then

$$\sum_{K} \chi(K)\chi_i(K) = \sum_{K} \sum_{m} a_m \chi_m(K)\chi_i(K)$$
$$= \sum_{m} \sum_{K} a_m \chi_m(K)\chi_i(K)$$

For a fixed m, we have

$$\sum_{K} a_m \chi_m(K)\chi_i(K) = a_m \sum_{K} \chi_m(K)\chi_i(K) = a_m h\delta_{im}$$

If we consider the sum of such a term over m, only the sum in which $m = i$ remains. Thus, we obtain

$$\sum_{K} \chi(K)\chi_m(K) = ha_m$$

or

$$a_m = \frac{1}{h} \sum_K \chi(K)\chi_m(K) \tag{7.5}$$

This formula is written more conveniently as

$$\boxed{a_m = \frac{1}{h} \sum n\chi(K)\chi_m(K)} \tag{7.6}$$

where n is the number of symmetry elements in any one class, and the summation is made over the different classes. As Sec. I-8 will show, this formula is very useful in determining the number of normal vibrations belonging to each species.*

I-8. THE NUMBER OF NORMAL VIBRATIONS FOR EACH SPECIES

As shown in Sec. I-6, the $3N - 6$ (or $3N - 5$) normal vibrations of an N-atom molecule can be classified into various species according to their symmetry properties. The number of normal vibrations in each species can be calculated by using the general equations given in Appendix III. These equations were derived from consideration of the vibrational degrees of freedom contributed by each set of identical nuclei for each symmetry species.[1] As an example, let us consider the NH_3 molecule belonging to the $\mathbf{C}_{3v}$ point group. The general equations are as follows:

$$\begin{aligned}
&A_1 \text{ species}: \quad 3m + 2m_v + m_0 - 1\\
&A_2 \text{ species}: \quad 3m + m_v - 1\\
&E \text{ species}: \quad 6m + 3m_v + m_0 - 2\\
&N \text{ (total number of atoms)} = 6m + 3m_v + m_0
\end{aligned}$$

From the definitions given in the footnotes of Appendix III, it is obvious that $m = 0$, $m_0 = 1$, and $m_v = 1$ in this case. To check these numbers, we calculate the total number of atoms from the equation for N given above. Since the result is 4, these assigned numbers are correct. Then, the number of normal vibrations in each species can be calculated by inserting these numbers in the general equations given above: 2, 0, and 2, respectively, for the A_1, A_2, and E

*Since this equation is not applicable to the infinite point groups ($\mathbf{C}_{\infty v}$ and $\mathbf{D}_{\infty h}$), several alternative approaches have been proposed (see Refs. 85 and 86).

species. Since the E species is doubly degenerate, the total number of vibrations is counted as 6, which is expected from the $3N - 6$ rule.

A more general method of finding the number of normal vibrations in each species can be developed by using group theory. The principle of the method is that all the representations are irreducible if normal coordinates are used as the basis for the representations. For example, the representations for the symmetry operations based on three normal coordinates, Q_1, Q_2, and Q_3, which correspond to the ν_1, ν_2, and ν_3 vibrations in the H_2O molecule of Fig. I-12, are as follows:

$$I\begin{bmatrix} Q_1 \\ Q_2 \\ Q_3 \end{bmatrix} = \begin{bmatrix} 1 & 0 & 0 \\ 0 & 1 & 0 \\ 0 & 0 & 1 \end{bmatrix}\begin{bmatrix} Q_1 \\ Q_2 \\ Q_3 \end{bmatrix}, \quad C_2(z)\begin{bmatrix} Q_1 \\ Q_2 \\ Q_3 \end{bmatrix} = \begin{bmatrix} 1 & 0 & 0 \\ 0 & 1 & 0 \\ 0 & 0 & -1 \end{bmatrix}\begin{bmatrix} Q_1 \\ Q_2 \\ Q_3 \end{bmatrix}$$

$$\sigma_v(xz)\begin{bmatrix} Q_1 \\ Q_2 \\ Q_3 \end{bmatrix} = \begin{bmatrix} 1 & 0 & 0 \\ 0 & 1 & 0 \\ 0 & 0 & -1 \end{bmatrix}\begin{bmatrix} Q_1 \\ Q_2 \\ Q_3 \end{bmatrix}, \quad \sigma_v(yz)\begin{bmatrix} Q_1 \\ Q_2 \\ Q_3 \end{bmatrix} = \begin{bmatrix} 1 & 0 & 0 \\ 0 & 1 & 0 \\ 0 & 0 & 1 \end{bmatrix}\begin{bmatrix} Q_1 \\ Q_2 \\ Q_3 \end{bmatrix}$$

Let a representation be written with the $3N$ rectangular coordinates of an N-atom molecule as its basis. If it is decomposed into its irreducible components, the basis for these irreducible representations must be the normal coordinates, and the number of appearances of the same irreducible representation must be equal to the number of normal vibrations belonging to the species represented by this irreducible representation. As stated previously, however, the $3N$ rectangular coordinates involve six (or five) coordinates, which correspond to the translational and rotational motions of the molecule as a whole. Therefore, the representations that have such coordinates as their basis must be subtracted from the result obtained above. Use of the character of the representation, rather than the representation itself, yields the same result.

For example, consider a pyramidal XY_3 molecule that has six normal vibrations. At first, the representations for the various symmetry operations must be written with the 12 rectangular coordinates in Fig. I-19 as their basis. Consider pure rotation C_p^+. If the clockwise rotation of the point (x, y, z) around the z axis by the angle θ brings it to the point denoted by the coordinates (x', y', z'), the relations between these two sets of coordinates are given by

$$\begin{aligned} x' &= x\cos\theta + y\sin\theta \\ y' &= -x\sin\theta + y\cos\theta \\ z' &= z \end{aligned} \tag{8.1}$$

By using matrix notation,* this can be written as

*For matrix algebra, see Appendix II.

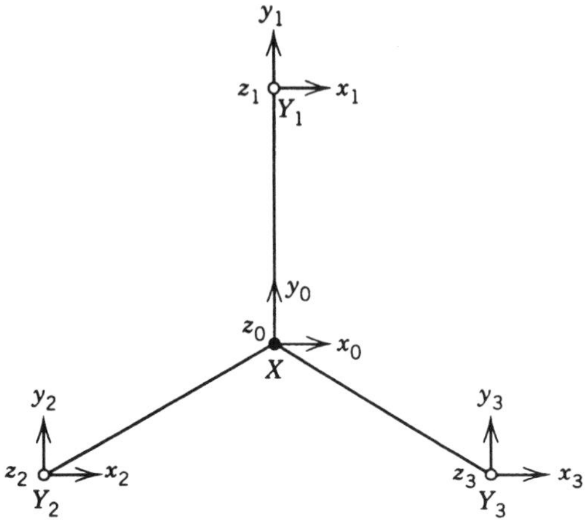

Fig. I-19. Rectangular coordinates in a pyramidal XY_3 molecule (Z axis is perpendicular to the paper plane).

$$\begin{bmatrix} x' \\ y' \\ z' \end{bmatrix} = C_\theta^+ \begin{bmatrix} x \\ y \\ z \end{bmatrix} = \begin{bmatrix} \cos\theta & \sin\theta & 0 \\ -\sin\theta & \cos\theta & 0 \\ 0 & 0 & 1 \end{bmatrix} \begin{bmatrix} x \\ y \\ z \end{bmatrix} \tag{8.2}$$

Then the character of the matrix is given by

$$\chi(C_\theta^+) = 1 + 2\cos\theta \tag{8.3}$$

The same result is obtained for $\chi(C_\theta^-)$. If this symmetry operation is applied to all the coordinates of the XY_3 molecule, the result is

$$C_\theta \begin{bmatrix} x_0 \\ y_0 \\ z_0 \\ x_1 \\ y_1 \\ z_1 \\ x_2 \\ y_2 \\ z_2 \\ x_3 \\ y_3 \\ z_3 \end{bmatrix} = \left[\begin{array}{c:c:c:c} \mathbf{A} & 0 & 0 & 0 \\ \hdashline 0 & 0 & 0 & \mathbf{A} \\ \hdashline 0 & \mathbf{A} & 0 & 0 \\ \hdashline 0 & 0 & \mathbf{A} & 0 \end{array}\right] \begin{bmatrix} x_0 \\ y_0 \\ z_0 \\ x_1 \\ y_1 \\ z_1 \\ x_2 \\ y_2 \\ z_2 \\ x_3 \\ y_3 \\ z_3 \end{bmatrix} \tag{8.4}$$

where **A** denotes the small square matrix given by Eq. 8.2. Thus, the character of this representation is simply given by Eq. 8.3. It should be noted in Eq. 8.4

that only the small matrix **A**, related to the nuclei unchanged by the symmetry operation, appears as a diagonal element. Thus, a more general form of the character of the representation for rotation around the axis by θ is

$$\boxed{\chi(R) = N_R(1 + 2\cos\theta)} \tag{8.5}$$

where N_R is the number of nuclei unchanged by the rotation. In the present case, $N_R = 1$ and $\theta = 120°$. Therefore, we obtain

$$\chi(C_3) = 0 \tag{8.6}$$

Identity (I) can be regarded as a special case of Eq. 8.5 in which $N_R = 4$ and $\theta = 0°$. The character of the representation is

$$\chi(I) = 12 \tag{8.7}$$

Pure rotation and identity are called *proper rotation.*

It is evident from Fig. I-19 that a symmetry plane such as σ_1 changes the coordinates from (x_i, y_i, z_i) to $(-x_i, y_i, z_i)$. The corresponding representation is therefore written as

$$\sigma_1 \begin{bmatrix} x \\ y \\ z \end{bmatrix} = \begin{bmatrix} -1 & 0 & 0 \\ 0 & 1 & 0 \\ 0 & 0 & 1 \end{bmatrix} \begin{bmatrix} x \\ y \\ z \end{bmatrix} \tag{8.8}$$

The result of such an operation on all the coordinates is

$$\sigma_1 \begin{bmatrix} x_0 \\ y_0 \\ z_0 \\ x_1 \\ y_1 \\ z_1 \\ x_2 \\ y_2 \\ z_2 \\ x_3 \\ y_3 \\ z_3 \end{bmatrix} = \left[\begin{array}{c:c:c:c} \mathbf{B} & 0 & 0 & 0 \\ \hdashline 0 & \mathbf{B} & 0 & 0 \\ \hdashline 0 & 0 & 0 & \mathbf{B} \\ \hdashline 0 & 0 & \mathbf{B} & 0 \end{array}\right] \begin{bmatrix} x_0 \\ y_0 \\ z_0 \\ x_1 \\ y_1 \\ z_1 \\ x_2 \\ y_2 \\ z_2 \\ x_3 \\ y_3 \\ z_3 \end{bmatrix} \tag{8.9}$$

where **B** denotes the small square matrix of Eq. 8.8. Thus, the character of this representation is calculated as $2 \times 1 = 2$. It is noted again that the matrix on the diagonal is nonzero only for the nuclei unchanged by the operation.

More generally, a reflection at a plane (σ) is regarded as $\sigma = i \times C_2$. Thus, the general form of Eq. 8.8 may be written as

$$\begin{bmatrix} -1 & 0 & 0 \\ 0 & -1 & 0 \\ 0 & 0 & -1 \end{bmatrix} \begin{bmatrix} \cos\theta & \sin\theta & 0 \\ -\sin\theta & \cos\theta & 0 \\ 0 & 0 & 1 \end{bmatrix} = \begin{bmatrix} -\cos\theta & -\sin\theta & 0 \\ \sin\theta & -\cos\theta & 0 \\ 0 & 0 & -1 \end{bmatrix}$$

Then,

$$\chi(\sigma) = -(1 + 2\cos\theta)$$

As a result, the character of the large matrix shown in Eq. 8.9 is given by

$$\boxed{\chi(R) = -N_R(1 + 2\cos\theta)} \qquad (8.10)$$

In the present case, $N_R = 2$ and $\theta = 180°$. This gives

$$\chi(\sigma_v) = 2 \qquad (8.11)$$

Symmetry operations such as i and S_p are regarded as

$$\begin{aligned} i &= i \times I, & \theta &= 0° \\ S_3 &= i \times C_6, & \theta &= 60° \\ S_4 &= i \times C_4, & \theta &= 90° \\ S_6 &= i \times C_3, & \theta &= 120° \end{aligned}$$

Therefore, the characters of these symmetry operations can be calculated by Eq. 8.10 with the values of θ defined above. Operations such as σ, i, and S_p are called *improper rotations*. Thus, the character of the representation based on 12 rectangular coordinates is as follows:

I	$2C_3$	$3\sigma_v$
12	0	2

(8.12)

To determine the number of normal vibrations belonging to each species, the $\chi(R)$ thus obtained must be resolved into the $\chi_i(R)$ of the irreducible representations of each species in Table I-6. First, however, the characters corresponding to the translational and rotational motions of the molecule must be subtracted from the result shown in Eq. 8.12.

The characters for the translational motion of the molecule in the x, y, and z directions (denoted by T_x, T_y, and T_z) are the same as those obtained in Eqs. 8.5 and 8.10. They are as follows:

$$\boxed{\chi_t(R) = \pm(1 + 2\cos\theta)} \tag{8.13}$$

where the + and − signs are for proper and improper rotations, respectively. The characters for the rotations around the x, y, and z axes (denoted by R_x, R_y, and R_z) are given by

$$\boxed{\chi_r(R) = +(1 + 2\cos\theta)} \tag{8.14}$$

for both proper and improper rotations. This is due to the fact that a rotation of the vectors in the plane perpendicular to the x, y, and z axes can be regarded as a rotation of the components of angular momentum, M_x, M_y, and M_z, about the given axes. If p_x, p_y, and p_z are the components of linear momentum in the x, y, and z directions, the following relations hold:

$$\begin{aligned} M_x &= yp_z - zp_y \\ M_y &= zp_x - xp_z \\ M_z &= xp_y - yp_x \end{aligned}$$

Since (x, y, z) and (p_x, p_y, p_z) transform as shown in Eq. 8.2, it follows that

$$C_\theta \begin{bmatrix} M_x \\ M_y \\ M_z \end{bmatrix} = \begin{bmatrix} \cos\theta & \sin\theta & 0 \\ -\sin\theta & \cos\theta & 0 \\ 0 & 0 & 1 \end{bmatrix} \begin{bmatrix} M_x \\ M_y \\ M_z \end{bmatrix}$$

Then, a similar relation holds for R_x, R_y, and R_z:

$$C_\theta \begin{bmatrix} R_x \\ R_y \\ R_z \end{bmatrix} = \begin{bmatrix} \cos\theta & \sin\theta & 0 \\ -\sin\theta & \cos\theta & 0 \\ 0 & 0 & 1 \end{bmatrix} \begin{bmatrix} R_x \\ R_y \\ R_z \end{bmatrix}$$

Thus, the characters for the proper rotations are given by Eq. 8.14. The same result is obtained for the improper rotation if the latter is regarded as $i \times$ (proper rotation). Therefore, the character for the vibration is obtained from

$$\boxed{\chi_v(R) = \chi(R) - \chi_t(R) - \chi_r(R)} \tag{8.15}$$

It is convenient to tabulate the foregoing calculations as in Table I-7. By using the formula in Eq. 7.6 and the character of the irreducible representations in Table I-6, a_m can be calculated as follows:

$$a_m(A_1) = \tfrac{1}{6}[(1)(6)(1) + (2)(0)(1) + (3)(2)(1)] = 2$$
$$a_m(A_2) = \tfrac{1}{6}[(1)(6)(1) + (2)(0)(1) + (3)(2)(-1)] = 0$$
$$a_m(E) = \tfrac{1}{6}[(1)(6)(2) + (2)(0)(-1) + (3)(2)(0)] = 2$$

and

$$\chi_v = 2\chi_{A_1} + 2\chi_E \tag{8.16}$$

In other words, the six normal vibrations of a pyramidal XY_3 molecule are classified into two A_1 and two E species.

This procedure is applicable to any molecule. As another example, a similar calculation is shown in Table I-8 for an octahedral XY_6 molecule. By use of Eq. 7.6 and the character table in Appendix I, the a_m are obtained as

TABLE I-7.

Symmetry operation	I	$2C_3$	$3\sigma_v$
Kind of rotation	Proper		Improper
θ	0°	120°	180°
$\cos\theta$	1	$-\frac{1}{2}$	-1
$1 + 2\cos\theta$	3	0	-1
N_R	4	1	2
$\chi, \pm N_R(1 + 2\cos\theta)$	12	0	2
$\chi_t, \pm(1 + 2\cos\theta)$	3	0	1
$\chi_r, +(1 + 2\cos\theta)$	3	0	-1
$\chi_v, \chi - \chi_t - \chi_r$	6	0	2

TABLE I-8.

Symmetry operation	I	$8C_3$	$6C_2$	$6C_4$	$3C_4^2 \equiv C_2''$	$S_2 \equiv i$	$6S_4$	$8S_6$	$3\sigma_h$ [a]	$6\sigma_d$ [a]
Kind of rotation	Proper					Improper				
θ	0°	120°	180°	90°	180°	0°	90°	120°	180°	180°
$\cos\theta$	1	$-\frac{1}{2}$	−1	0	−1	1	0	$-\frac{1}{2}$	−1	−1
$1+2\cos\theta$	3	0	−1	1	−1	3	1	0	−1	−1
N_R	7	1	1	3	3	1	1	1	5	3
$\chi, \pm N_R(1+2\cos\theta)$	21	0	−1	3	−3	−3	−1	0	5	3
$\chi_t, \pm(1+2\cos\theta)$	3	0	−1	1	−1	−3	−1	0	1	1
$\chi_r, +(1+2\cos\theta)$	3	0	−1	1	−1	3	1	0	−1	−1
$\chi_v, \chi-\chi_t-\chi_r$	15	0	1	1	−1	−3	−1	0	5	3

[a] σ_h = horizontal plane of symmetry; σ_d = diagonal plane of symmetry.

$$
\begin{aligned}
a_m(A_{1g}) &= \tfrac{1}{48}[(1)(15)(1) + (8)(0)(1) + (6)(1)(1) + (6)(1)(1) \\
&\quad + (3)(-1)(1) + (1)(-3)(1) + (6)(-1)(1) + (8)(0)(1) \\
&\quad + (3)(5)(1) + (6)(3)(1)] \\
&= 1 \\
a_m(A_{1u}) &= \tfrac{1}{48}[(1)(15)(1) + (8)(0)(1) + (6)(1)(1) + (6)(1)(1) \\
&\quad + (3)(-1)(1) + (1)(-3)(-1) + (6)(-1)(-1) + (8)(0)(-1) \\
&\quad + (3)(5)(-1) + (6)(3)(-1)] \\
&= 0 \\
&\;\;\vdots
\end{aligned}
$$

and therefore

$$\chi_v = \chi_{A_{1g}} + \chi_{E_g} + 2\chi_{F_{1u}} + \chi_{F_{2g}} + \chi_{F_{2u}}$$

I-9. INTERNAL COORDINATES

In Sec. I-4, the potential and the kinetic energies were expressed in terms of rectangular coordinates. If, instead, these energies are expressed in terms of *internal coordinates* such as increments of the bond length and bond angle, the corresponding force constants have clearer physical meanings than those expressed in terms of rectangular coordinates, since these force constants are characteristic of the bond stretching and the angle deformation involved. The number of internal coordinates must be equal to, or greater than, $3N-6$ (or $3N-5$), the degrees of vibrational freedom of an N-atom molecule. If more than $3N-6$ (or $3N-5$) coordinates are selected as the internal coordinates, this means that these coordinates are not independent of each other. Figure I-20 illustrates the internal coordinates for various types of molecules.

In linear XYZ (*a*), bent XY_2 (*b*), and pyramidal XY_3 (*c*) molecules, the number of internal coordinates is the same as the number of normal vibrations. In a nonplanar X_2Y_2 molecule (*d*) such as H_2O_2, the number of internal coordinates is the same as the number of vibrations if the twisting angle around the central bond ($\Delta\tau$) is considered. In a tetrahedral XY_4 molecule (*e*), however, the number of internal coordinates exceeds the number of normal vibrations by one. This is due to the fact that the six angle coordinates around the central atom are not independent of each other, that is, they must satisfy the relation

$$\Delta\alpha_{12} + \Delta\alpha_{23} + \Delta\alpha_{31} + \Delta\alpha_{14} + \Delta\alpha_{24} + \Delta\alpha_{34} = 0 \tag{9.1}$$

This is called a *redundant condition*. In a planar XY_3 molecule (*f*), the number of internal coordinates is seven when the coordinate $\Delta\theta$, which represents the deviation from planarity, is considered. Since the number of vibrations is six, one redundant condition such as

$$\Delta\alpha_{12} + \Delta\alpha_{23} + \Delta\alpha_{31} = 0 \tag{9.2}$$

must be involved. Such redundant conditions always exist for the angle coordinates around the central atom. In an octahedral XY_6 molecule (*g*), the number of internal coordinates exceeds the number of normal vibrations by three. This means that, of the 12 angle coordinates around the central atom, three redundant conditions are involved.

$$\begin{aligned} \Delta\alpha_{12} + \Delta\alpha_{26} + \Delta\alpha_{64} + \Delta\alpha_{41} &= 0 \\ \Delta\alpha_{15} + \Delta\alpha_{56} + \Delta\alpha_{63} + \Delta\alpha_{31} &= 0 \\ \Delta\alpha_{23} + \Delta\alpha_{34} + \Delta\alpha_{45} + \Delta\alpha_{52} &= 0 \end{aligned} \tag{9.3}$$

The redundant conditions are more complex in ring compounds. For example, the number of internal coordinates in a triangular X_3 molecule (*h*) exceeds the number of vibrations by three. One of these redundant conditions (A_1' species) is

$$\Delta\alpha_1 + \Delta\alpha_2 + \Delta\alpha_3 = 0 \tag{9.4}$$

The other two redundant conditions (E' species) involve bond stretching and angle deformation coordinates such as

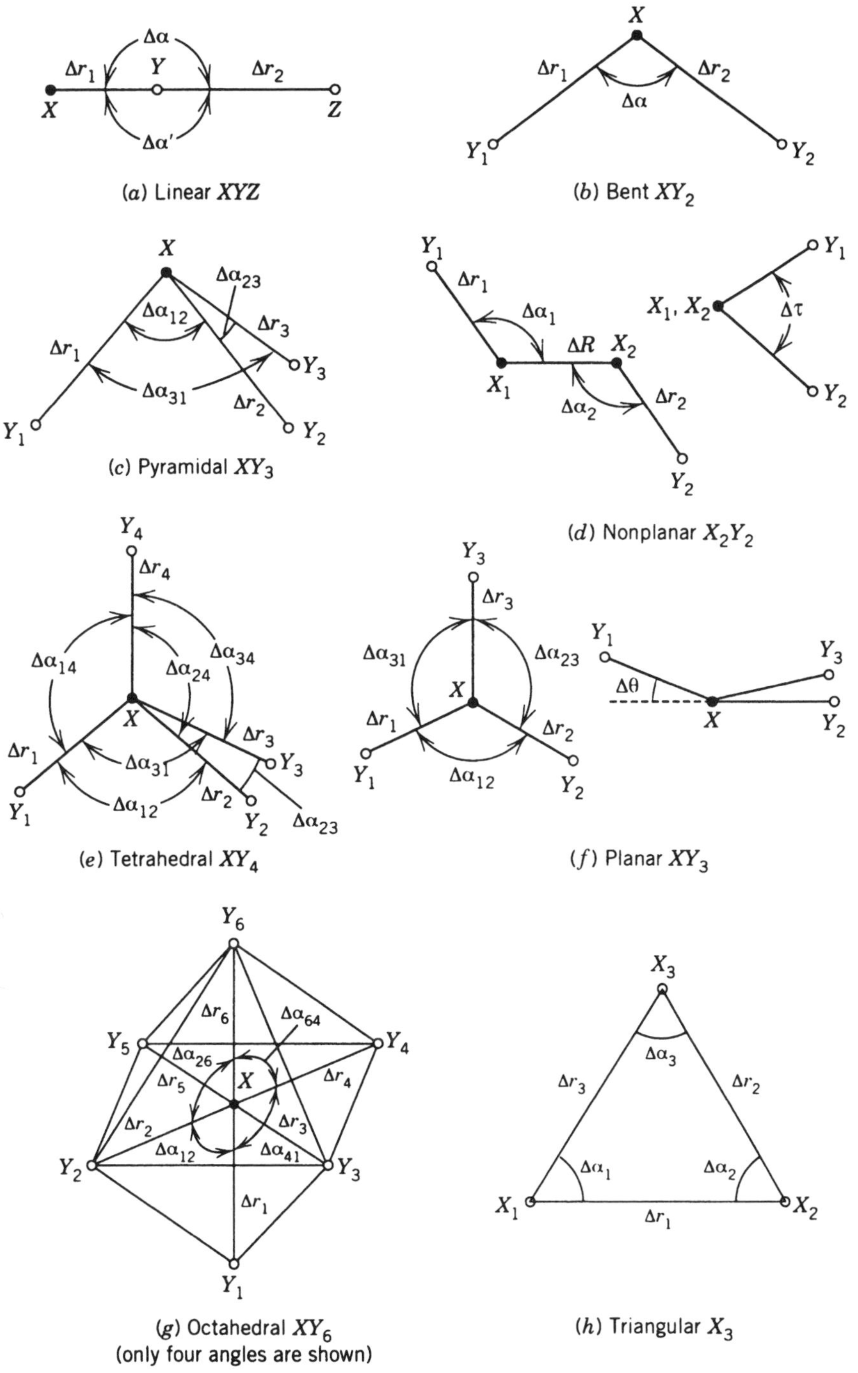

Fig. I-20. Internal coordinates for various molecules.

$$(2\Delta r_1 - \Delta r_2 - \Delta r_3) + \frac{r}{\sqrt{3}}(\Delta\alpha_1 + \Delta\alpha_2 - 2\Delta\alpha_3) = 0$$
$$(\Delta r_2 - \Delta r_3) - \frac{r}{\sqrt{3}}(\Delta\alpha_1 - \Delta\alpha_2) = 0 \tag{9.5}$$

where r is the equilibrium length of the X-X bond. The redundant conditions mentioned above can be derived by using the method described in Sec. I-12.

The procedure for finding the number of normal vibrations in each species was described in Sec. I-8. This procedure is, however, considerably simplified if internal coordinates are used. Again, consider a pyramidal XY_3 molecule. Using the internal coordinates shown in Fig. I-20c, we can write the representation for the C_3^+ operation as

$$C_3^+\begin{bmatrix}\Delta r_1\\ \Delta r_2\\ \Delta r_3\\ \Delta\alpha_{12}\\ \Delta\alpha_{23}\\ \Delta\alpha_{31}\end{bmatrix} = \begin{bmatrix}0&0&1&0&0&0\\1&0&0&0&0&0\\0&1&0&0&0&0\\0&0&0&0&0&1\\0&0&0&1&0&0\\0&0&0&0&1&0\end{bmatrix}\begin{bmatrix}\Delta r_1\\ \Delta r_2\\ \Delta r_3\\ \Delta\alpha_{12}\\ \Delta\alpha_{23}\\ \Delta\alpha_{31}\end{bmatrix} \tag{9.6}$$

Thus $\chi(C_3^+) = 0$, as does $\chi(C_3^-)$. Similarly, $\chi(I) = 6$ and $\chi(\sigma_v) = 2$. This result is exactly the same as that obtained in Table I-7 using rectangular coordinates. *When using internal coordinates, however, the character of the representation is simply given by the number of internal coordinates unchanged by each symmetry operation.*

If this procedure is made separately for stretching (Δr) and bending ($\Delta\alpha$) coordinates, it is readily seen that

$$\chi^r(R) = \chi_{A_1} + \chi_E$$
$$\chi^\alpha(R) = \chi_{A_1} + \chi_E \tag{9.7}$$

Thus, it is found that both A_1 and E species have one stretching and one bending vibration, respectively. No consideration of the translational and rotational motions is necessary if the internal coordinates are taken as the basis for the representation.

Another example, for an octahedral XY_6 molecule, is given in Table I-9. Using Eq. 7.6 and the character table in Appendix I, we find that these characters are resolved into

$$\chi^r(R) = \chi_{A_{1g}} + \chi_{E_g} + \chi_{F_{1u}} \tag{9.8}$$
$$\chi^\alpha(R) = \chi_{A_{1g}} + \chi_{E_g} + \chi_{F_{1u}} + \chi_{F_{2g}} + \chi_{F_{2u}} \tag{9.9}$$

TABLE I-9.

	I	$8C_3$	$6C_2$	$6C_4$	$3C_4^2 \equiv C_2''$	$S_2 \equiv i$	$6S_4$	$8S_6$	$3\sigma_h$	$6\sigma_d$
$\chi^r(R)$	6	0	0	2	2	0	0	0	4	2
$\chi^\alpha(R)$	12	0	2	0	0	0	0	0	4	2

Comparison of this result with that obtained in Sec. I-8 immediately suggests that three redundant conditions are included in these bending vibrations (one in A_{1g} and one in E_g). Therefore, $\chi^\alpha(R)$ for genuine vibrations becomes

$$\chi^\alpha(R) = \chi_{F_{1u}} + \chi_{F_{2g}} + \chi_{F_{2u}} \tag{9.10}$$

Thus, it is concluded that six stretching and nine bending vibrations are distributed as indicated in Eqs. 9.8 and 9.10, respectively. Although the method given above is simpler than that of Sec. I-8, caution must be exercised with respect to the bending vibrations whenever redundancy is involved. In such a case, comparison of the results obtained from both methods is useful in finding the species of redundancy.

I-10. SELECTION RULES FOR INFRARED AND RAMAN SPECTRA

According to quantum mechanics,[4,5] the selection rule for the infrared spectrum is determined by the integral:

$$[\mu]_{v'v''} = \int \psi_{v'}^*(Q_a)\mu\psi_{v''}(Q_a)\,dQ_a \tag{10.1}$$

Here, μ is the dipole moment in the electronic ground state, ψ is the vibrational eigenfunction given by Eq. 3.7, and v' and v'' are the vibrational quantum numbers before and after the transition, respectively. The activity of the normal vibration whose normal coordinate is Q_a is being determined. By resolving the dipole moment into the three components in the x, y, and z directions, we obtain the result

$$[\mu_x]_{v'v''} = \int \psi_{v}^*(Q_a)\mu_x\psi_{v''}(Q_a)\,dQ_a$$

$$[\mu_y]_{v'v''} = \int \psi_{v'}^*(Q_a)\mu_y\psi_{v''}(Q_a)\,dQ_a$$

$$[\mu_z]_{v'v''} = \int \psi_{v'}^*(Q_a)\mu_z\psi_{v''}(Q_a)\,dQ_a \tag{10.2}$$

If one of these integrals is not zero, the normal vibration associated with Q_a is infrared-active. If all the integrals are zero, the vibration is infrared-inactive.

Similar to the case of infrared spectrum, the selection rule for the Raman spectrum is determined by the integral

$$[\alpha]_{v'v''} = \int \psi^*_{v'}(Q_a)\alpha\psi_{v''}(Q_a)\,dQ_a \tag{10.3}$$

As shown in Sec. I-6, α consists of six components, α_{xx}, α_{yy}, α_{zz}, α_{xy}, α_{yz}, and α_{xz}. Thus Eq. 10.3 may be resolved into six components:

$$[\alpha_{xx}]_{v'v''} = \int \psi^*_{v'}(Q_a)\alpha_{xx}\psi_{v''}(Q_a)\,dQ_a$$

$$[\alpha_{yy}]_{v'v''} = \int \psi^*_{v'}(Q_a)\alpha_{yy}\psi_{v''}(Q_a)\,dQ_a \tag{10.4}$$

$$\cdots\cdots\cdots\cdots\cdots\cdots\cdots\cdots\cdots\cdots$$

If one of these integrals is not zero, the normal vibration associated with Q_a is Raman-active. If all the integrals are zero, the vibration is Raman-inactive. As shown below, it is possible to determine whether the integrals of Eqs. 10.2 and 10.4 are zero or nonzero from a consideration of symmetry.

(1) Selection Rules for Fundamentals. Let us consider the fundamentals in which transitions occur from $v' = 0$ to $v'' = 1$. It is evident from the form of the vibrational eigenfunction (Eq. 3.8) that $\psi_0(Q_a)$ is invariant under any symmetry operation, whereas the symmetry of $\psi_1(Q_a)$ is the same as that of Q_a. Thus, the integral does not vanish when the symmetry of μ_x, for example, is the same as that of Q_a. If the symmetry properties of μ_x and Q_a differ in even one symmetry element of the group, the integral becomes zero. In other words, for the integral to be nonzero, Q_a must belong to the same species as μ_x. *More generally, the normal vibration associated with Q_a becomes infrared-active when at least one of the components of the dipole moment belongs to the same species as Q_a.* Similar conclusions are obtained for the Raman spectrum. Namely, *the normal vibration associated with Q_a becomes Raman-active when at least one of the components of the polarizability belongs to the same species as Q_a.*

Since the species of the normal vibration can be determined by the methods described in Secs. I-8 and I-9, it is only necessary to determine the species of the components of the dipole moment and polarizability of the molecule. This can be done as follows. The components of the dipole moment, μ_x, μ_y, and μ_z, transform as do those of translational motion, T_x, T_y, and T_z, respectively. These were discussed in Sec. I-8. Thus, the character of the dipole moment is given by Eq. 8.13, which is

$$\boxed{\chi_\mu(R) = \pm(1 + 2\cos\theta)} \tag{10.5}$$

where + and − have the same meaning as before. In a pyramidal XY_3 molecule, Eq. 10.5 gives

	I	$2C_3$	$3\sigma_v$
$\chi_\mu(R)$	3	0	1

Using Eq. 7.6, we resolve this into $A_1 + E$. It is obvious that μ_z belongs to A_1. Then, μ_x and μ_y must belong to E. In fact, the pair, μ_x and μ_y, transforms as follows:

$$I\begin{bmatrix}\mu_x\\ \mu_y\end{bmatrix} = \begin{bmatrix}1 & 0\\ 0 & 1\end{bmatrix}\begin{bmatrix}\mu_x\\ \mu_y\end{bmatrix}, \qquad C_3^+\begin{bmatrix}\mu_x\\ \mu_y\end{bmatrix} = \begin{bmatrix}-\frac{1}{2} & \frac{\sqrt{3}}{2}\\ -\frac{\sqrt{3}}{2} & -\frac{1}{2}\end{bmatrix}\begin{bmatrix}\mu_x\\ \mu_y\end{bmatrix}$$

$$\chi(I) = 2, \qquad \chi(C_3^+) = -1$$

$$\sigma_1\begin{bmatrix}\mu_x\\ \mu_y\end{bmatrix} = \begin{bmatrix}-1 & 0\\ 0 & 1\end{bmatrix}\begin{bmatrix}\mu_x\\ \mu_y\end{bmatrix}$$

$$\chi(\sigma_1) = 0$$

Thus, it is found that μ_z belongs to A_1 and (μ_x, μ_y) belongs to E.

The character of the representation of the polarizability is given by

$$\boxed{\chi_\alpha(R) = 2\cos\theta(1 + 2\cos\theta)} \tag{10.6}$$

for both proper and improper rotations. This can be derived as follows. The polarizability in the x, y, and z directions is related to that in X, Y, and Z coordinates by

$$\begin{bmatrix} \alpha_{XX} & \alpha_{XY} & \alpha_{XZ} \\ \alpha_{YX} & \alpha_{YY} & \alpha_{YZ} \\ \alpha_{ZX} & \alpha_{ZY} & \alpha_{ZZ} \end{bmatrix} = \begin{bmatrix} C_{Xx} & C_{Xy} & C_{Xz} \\ C_{Yx} & C_{Yy} & C_{Yz} \\ C_{Zx} & C_{Zy} & C_{Zz} \end{bmatrix} \begin{bmatrix} \alpha_{xx} & \alpha_{xy} & \alpha_{xz} \\ \alpha_{yx} & \alpha_{yy} & \alpha_{yz} \\ \alpha_{zx} & \alpha_{zy} & \alpha_{zz} \end{bmatrix} \begin{bmatrix} C_{Xx} & C_{Yx} & C_{Zx} \\ C_{Xy} & C_{Yy} & C_{Zy} \\ C_{Xz} & C_{Yz} & C_{Zz} \end{bmatrix}$$

where C_{Xx}, C_{Xy}, and so forth denote the direction cosines between the two axes subscripted. If a rotation through θ around the Z axis superimposes the X, Y, and Z axes on the x, y, and z axes, the preceding relation becomes

$$C_\theta \begin{bmatrix} \alpha_{xx} & \alpha_{xy} & \alpha_{xz} \\ \alpha_{yx} & \alpha_{yy} & \alpha_{yz} \\ \alpha_{zx} & \alpha_{zy} & \alpha_{zz} \end{bmatrix} = \begin{bmatrix} \cos\theta & \sin\theta & 0 \\ -\sin\theta & \cos\theta & 0 \\ 0 & 0 & 1 \end{bmatrix} \begin{bmatrix} \alpha_{xx} & \alpha_{xy} & \alpha_{xz} \\ \alpha_{yx} & \alpha_{yy} & \alpha_{yz} \\ \alpha_{zx} & \alpha_{zy} & \alpha_{zz} \end{bmatrix} \begin{bmatrix} \cos\theta & -\sin\theta & 0 \\ \sin\theta & \cos\theta & 0 \\ 0 & 0 & 1 \end{bmatrix}$$

This can be written as

$$C_\theta \begin{bmatrix} \alpha_{xx} \\ \alpha_{yy} \\ \alpha_{zz} \\ \alpha_{xy} \\ \alpha_{xz} \\ \alpha_{yz} \end{bmatrix} = \begin{bmatrix} \cos^2\theta & \sin^2\theta & 0 & 2\sin\theta\cos\theta & 0 & 0 \\ \sin^2\theta & \cos^2\theta & 0 & -2\sin\theta\cos\theta & 0 & 0 \\ 0 & 0 & 1 & 0 & 0 & 0 \\ -\sin\theta\cos\theta & \sin\theta\cos\theta & 0 & 2\cos^2\theta - 1 & 0 & 0 \\ 0 & 0 & 0 & 0 & \cos\theta & \sin\theta \\ 0 & 0 & 0 & 0 & -\sin\theta & \cos\theta \end{bmatrix} \cdot \begin{bmatrix} \alpha_{xx} \\ \alpha_{yy} \\ \alpha_{zz} \\ \alpha_{xy} \\ \alpha_{xz} \\ \alpha_{yz} \end{bmatrix}$$

Thus, the character of this representation is given by Eq. 10.6. The same results are obtained for improper rotations if they are regarded as the product $i \times$ (proper rotation). For a pyramidal XY_3 molecule, Eq. 10.6 gives

	I	$2C_3$	$3\sigma_v$
$\chi_\alpha(R)$	6	0	2

Using Eq. 7.6, this is resolved into $2A_1 + 2E$. Again, it is immediately seen that* the component α_{zz} belongs to A_1, and the pair α_{zx} and α_{zy} belongs to E since

$$\begin{bmatrix} zx \\ zy \end{bmatrix} = z \begin{bmatrix} x \\ y \end{bmatrix} = A_1 \times E = E$$

It is more convenient to consider the components $\alpha_{xx} + \alpha_{yy}$ and $\alpha_{xx} - \alpha_{yy}$ than a_{xx} and a_{yy}. If a vector of unit length is considered, the relation

$$x^2 + y^2 + z^2 = 1$$

holds. Since α_{zz} belongs to A_1, $\alpha_{xx} + \alpha_{yy}$ must belong to A_1. Then, the pair $\alpha_{xx} - \alpha_{yy}$ and α_{xy} must belong to E. As a result, the character table of the point group $\mathbf{C}_{3v}$ is completed as in Table I-10. Thus, it is concluded that, in the point group $\mathbf{C}_{3v}$, both the A_1 and the E vibrations are infrared- as well as Raman-active, while the A_2 vibrations are inactive.

Complete character tables like Table I-10 have already been worked out for all the point groups. Therefore, no elaborate treatment such as that described in this section is necessary in practice. Appendix I gives complete character tables for the point groups that appear frequently in this book. From these tables, the

*The quantum-mechanical expression of the polarizability is[86a]

$$\alpha_{xx} = \frac{2}{3h} \sum_j \frac{(\mu_x)_{j0}^2 \nu_{j0}^2}{\nu_{j0}^2 - \nu^2}$$

$$\alpha_{xy} = \frac{2}{3h} \sum_j \frac{(\mu_x)_{j0}(\mu_y)_{j0}\nu_{j0}^2}{\nu_{j0}^2 - \nu^2} \quad \text{etc.}$$

Here, $(\mu_x)_{j0}$, for example, is the induced dipole moment along the x axis caused by the 0 (ground state) $\rightarrow j$ (excited state) electronic transition, ν_{j0} is the frequency of the $0 \rightarrow j$ transition, and ν is the exciting frequency. Thus, it is readily seen that the character of the polarizability components such as α_{xx} and α_{xy} is determined by considering the product of the characters of dipole moments such as μ_x and μ_y.

TABLE I-10. Character Table of the Point Group C_{3v}

C_{3v}	I	$2C_3$	$3\sigma_v$		
A_1	+1	+1	+1	μ_z	$\alpha_{xx}+\alpha_{yy}, \alpha_{zz}$
A_2	+1	+1	−1		
E	+2	−1	0	$(\mu_x, \mu_y)^a$	$(\alpha_{xz}, \alpha_{yz}),^a (a_{xx}-\alpha_{yy}, \alpha_{xy})^a$

[a] A doubly degenerate pair is represented by two terms in parentheses.

selection rules for the infrared and Raman spectra are obtained immediately: *The vibration is infrared- or Raman-active if it belongs to the same species as one of the components of the dipole moment or polarizability, respectively.* For example, the character table of the point group $\mathbf{O}_h$ signifies immediately that only the F_{1u} vibrations are infrared-active and only the A_{1g}, E_g, and F_{2g} vibrations are Raman-active, for the components of the dipole moment or the polarizability belong to these species in this point group. It is to be noted in these character tables that (1) a totally symmetric vibration is Raman-active in any point group, and (2) the infrared- and Raman-active vibrations always belong to u and g types, respectively, in point groups having a center of symmetry.

(2) Selection Rules for Combination and Overtone Bands. As stated in Sec. I-3, some combination and overtone bands appear weakly because actual vibrations are not harmonic and some of these nonfundamentals are allowed by symmetry selection rules.

Selection rules for combination bands ($\nu_i \pm \nu_j$) can be derived from the characters of the *direct products* of those of individual vibrations. Thus, we obtain:

$$\chi_{ij}(R) = \chi_i(R) \times \chi_j(R) \tag{10.7}$$

As an example, consider a molecule of $\mathbf{C}_{3v}$ symmetry. It is readily seen that

	I	$2C_3$	$3\sigma_v$
$\chi_{A_1}(R)$	1	1	1
$\times)$ $\chi_E(R)$	2	−1	0
$\chi_{A_1 \times E}(R)$	2	−1	0

$$\chi_{A_1 \times E}(R) = \chi_E(R)$$

Thus, a combination band between the A_1 and E vibrations is IR- as well as Raman-active. The activity of a combination band between two E vibrations can be determined by considering the direct product of their characters:

	I	$2C_3$	$3\sigma_v$
$\chi_E(R)$	2	−1	0
×) $\chi_E(R)$	2	−1	0
$\chi_{E\times E}(R)$	4	1	0

Using Eq. 7.6, this set of the characters can be resolved into

$$\chi_{E\times E}(R) = \chi_{A_1}(R) + \chi_{A_2}(R) + \chi_E(R)$$

Since both A_1 and E species are IR- and Raman-active, a combination band between two E vibrations is also IR- and Raman-active. It is convenient to apply the general rules of Appendix IV in determining the symmetry species of direct products.

Selection rules for overtones of nondegenerate vibrations can be obtained using the relation:

$$\chi^n(R) = [\chi(R)]^n \tag{10.8}$$

Here, $\chi^n(R)$ is the character of the $(n - 1)$th overtone ($n = 2$ for the first overtone). As an example, consider a molecule of $\mathbf{C}_{3v}$ symmetry. The character of the first overtone of the A_2 fundamental is calculated as:

	I	$2C_3$	$3\sigma_v$
$\chi_{A_2}(R)$	1	1	−1
×) $\chi_{A_2}(R)$	1	1	−1
$\chi^2_{A_2}(R)$	1	1	1

Namely, it is IR- as well as Raman-active because $\chi^2_{A_2}(R) = \chi_{A_1}(R)$. However, the second overtone of the A_2 fundamental is IR- as well as Raman-inactive because $\chi^3_{A_2}(R) = \chi^2_{A_2}(R) \times \chi_{A_2}(R) = \chi_{A_2}(R)$ (inactive). In general, odd overtones (A_1) are IR- and Raman-active, whereas even overtones (A_2) are inactive. It is obvious that all the overtones of the A_1 fundamental are IR- and Raman-active.

Selection rules for overtones of doubly degenerate vibrations (E species) are determined by:

$$\chi^n_E(R) = (1/2)[\chi^{n-1}_E(R) \cdot \chi_E(R) + \chi_E(R^n)] \tag{10.9}$$

For the first overtone, this is written as

$$\chi_E^2(R) = (1/2)[\{\chi_E(R)\}^2 + \chi_E(R^2)]$$

Here, $\chi_E(R^2)$ is the character which corresponds to the operation R performed twice successively. Thus, one obtains

$$\chi_E(I^2) = \chi_E(I) = 2$$
$$\chi_E[(C_3^+)^2] = \chi_E(C_3^-) = -1$$
$$\chi_E[(\sigma_v)^2] = \chi_E(I) = 2$$

Therefore, $\chi_E^2(R)$ can be calculated as follows:

		I	$2C_3$	$3\sigma_v$
	$\chi_E(R)$	2	−1	0
×)	$\chi_E(R)$	2	−1	0
	$\{\chi_E(R)\}^2$	4	1	0
+)	$\chi_E(R^2)$	2	−1	2
	$\{\chi_E(R)\}^2 + \chi_E(R^2)$	6	0	2
÷2)	$\chi_E^2(R)$	3	0	1

$$\chi_E^2(R) = \chi_{A_1}(R) + \chi_E(R)$$

Thus, the first overtone of the doubly degenerate vibration is IR- and Raman-active. The characters of overtones for triply degenerate vibrations are given by

$$\chi_F^n(R) = \frac{1}{3}\left\{2\chi_F(R)\chi_F^{n-1}(R) - \frac{1}{2}\chi_F^{n-2}(R)[\chi_F(R)]^2 + \frac{1}{2}\chi_F(R^2)\chi_F^{n-2}(R) + \chi_F(R^n)\right\} \quad (10.10)$$

For more details, see Refs. 3, 8, and 17.

I-11. STRUCTURE DETERMINATION

Suppose that a molecule has several probable structures, each of which belongs to a different point group. Then the number of infrared- and Raman-active fundamentals should be different for each structure. Therefore, the most probable

TABLE I-11*a*. Number of Fundamentals for Tetrahedral XeF_4

$\mathbf{T}_d$	Activity	Number of Fundamentals	XeF Stretching	FXeF Bending
A_1	R (p)	1	1	0
A_2	ia	0	0	0
E	R (dp)	1	0	1
F_1	ia	0	0	0
F_2	IR, R (dp)	2	1	1
Total	IR	2	1	1
	R	4	2	2

p, polarized; dp, depolarized (see Sec. I-20)

model can be selected by comparing the observed number of infrared- and Raman-active fundamentals with that predicted theoretically for each model.

Consider the XeF_4 molecule as an example. It may be tetrahedral or square-planar. By use of the methods described in the preceding sections, the number of infrared- or Raman-active fundamentals can be found easily for each structure. Tables I-11*a* and I-11*b* summarize the results. It is seen that the distinction of these two structures can be made by comparing the number of IR- and Raman-active FXeF bending modes; the tetrahedral structure predicts one IR and two Raman bands, whereas the square-planar structure predicts two IR and one Raman bands. In general, the XeF stretching vibrations appear above 500 cm^{-1}, whereas the FXeF bending vibrations are observed below 300 cm^{-1}. The IR and Raman spectra of XeF_4 are shown in Fig. II-17. The IR spectrum exhibits two bending bands at 291 and 123 cm^{-1}, whereas the Raman spectrum shows one bending vibration at 218 cm^{-1}. Thus, the square-planar structure is preferable to the tetrahedral structure.*

TABLE I-11*b*. Number of Fundamentals for Square-Planar XeF_4

$\mathbf{D}_{4h}$	Activity	Number of Fundamentals	XeF Stretching	FXeF Bending
A_{1g}	R (p)	1	1	0
A_{1u}	ia	0	0	0
A_{2g}	ia	0	0	0
A_{2u}	IR	1	0	1
B_{1g}	R (dp)	1	1	0
B_{1u}	ia	0	0	0
B_{2g}	R (dp)	1	0	1
B_{2u}	ia	1	0	1
E_g	R (dp)	0	0	0
E_u	IR	2	1	1
Total	IR	3	1	2
	R	3	2	1

*This conclusion may be drawn directly from observation of the mutual exclusion rule, which holds for $\mathbf{D}_{4h}$ but not for $\mathbf{T}_d$.

Another example is given by the XeF_5^- ion, which has the three possible structures shown in Fig. I-21. The results of vibrational analysis for each are summarized in Table I-12. It is seen that the numbers of IR-active vibrations are 5, 6, and 3 and those of Raman-active vibrations are 6, 9, and 3, respectively, for the $\mathbf{D}_{3h}$, $\mathbf{C}_{4v}$, and $\mathbf{D}_{5h}$ structures. Figure II-7 shows that the XeF_5^- ion exhibits three IR bands (550–400, 290, and 274 cm^{-1}) and three Raman bands (502, 423, and 377 cm^{-1}). Thus, a pentagonal planar structure is preferable to the other two structures. The somewhat unusual structures thus obtained for XeF_4 and XeF_5^- can be rationalized by the use of the valence-shell electron-pair repulsion (VSEPR) theory [Sec. II-6(3)]

This rather simple method has been used widely for the elucidation of molecular structure of inorganic and coordination compounds. In Section II, the number of IR- and Raman-active vibrations is compared for possible structures of XY_n and other molecules. Appendix V lists the number of IR- and Raman-active vibrations of MX_nY_m-type molecules. It should be noted, however, that this method does not give a clear-cut answer if the predicted numbers of infrared- and Raman-active fundamentals are similar for various probable structures. Furthermore, a practical difficulty arises in determining the number of fundamentals from the observed spectrum, since the intensities of overtone and combination bands are sometimes comparable to those of fundamentals when they appear as satellite bands of the fundamental. This is particularly true when overtone and combination bands are enhanced anomalously by *Fermi resonance* (accidental degeneracy). For example, the frequency of the first overtone of the ν_2 vibration of CO_2 (667 cm^{-1}) is very close to that of the ν_1 vibration (1337 cm^{-1}). Since these two vibrations belong to the same symmetry species (Σ_g^+), they interact with each other and give rise to two strong Raman lines at 1388 and 1286 cm^{-1}. Fermi resonances similar to the resonance observed for CO_2 may occur for a number of other molecules. It is to be noted also that the number of observed bands depends on the resolving power of the instrument used. Finally, the molecular symmetry in the isolated state is not necessarily the same as that in the crystalline state (Sec. I-26). Therefore, this method must be applied with caution to spectra obtained for compounds in the crystalline state.

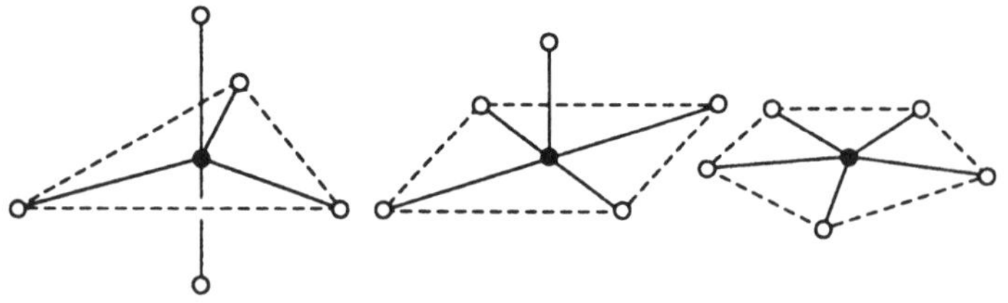

$\mathbf{D}_{3h}$ $\mathbf{V}_{4v}$ $\mathbf{D}_{5h}$

Fig. I-21. Three possible structures for the XeF_5^- ion.

TABLE I-12. Number of IR- and Raman-Active Fundamentals for Three Possible Structures of the XeF_5^- Anion

	D_{3h}	C_{4v}	D_{5h}
IR	$2A_2'' + 3E'$	$3A_1 + 3E$	$A_2'' + 2E_1'$
Raman	$2A_1' + 3E' + E''$	$3A_1 + 2B_1 + B_2 + 3E$	$A_1' + 2E_2'$

I-12. PRINCIPLE OF THE GF MATRIX METHOD*

As described in Sec. I-4, the frequency of the normal vibration is determined by the kinetic and potential energies of the system. The kinetic energy is determined by the masses of the individual atoms and their geometrical arrangement in the molecule. On the other hand, the potential energy arises from interaction between the individual atoms and is described in terms of the force constants. Since the potential energy provides valuable information about the nature of interatomic forces, it is highly desirable to obtain the force constants from the observed frequencies. This is usually done by calculating the frequencies, assuming a suitable set of force constants. If the agreement between the calculated and observed frequencies is satisfactory, this particular set of the force constants is adopted as a representation of the potential energy of the system.

To calculate the vibrational frequencies, it is necessary first to express both the potential and the kinetic energies in terms of some common coordinates (Sec. I-4). Internal coordinates (Sec. I-9) are more suitable for this purpose than rectangular coordinates, since (1) force constants expressed in terms of internal coordinates have clearer physical meanings than those expressed in terms of rectangular coordinates, and (2) a set of internal coordinates does not involve translational and rotational motion of the molecule as a whole.

Using the internal coordinates, R_i, we write the potential energy as

$$2V = \tilde{\mathbf{R}}\mathbf{F}\mathbf{R} \tag{12.1}$$

For a bent Y_1XY_2 molecule such as that in Fig. I-20*b*, **R** is a column matrix of the form

$$\mathbf{R} = \begin{bmatrix} \Delta r_1 \\ \Delta r_2 \\ \Delta\alpha \end{bmatrix}$$

$\tilde{\mathbf{R}}$ is its transpose:

*For details, see Refs. 3. The term "normal coordinate analysis" is almost synonymous with the **GF** matrix method, since most of the normal coordinate calculations are carried out by using this method.

$$\tilde{\mathbf{R}} = [\Delta r_1 \quad \Delta r_2 \quad \Delta\alpha]$$

and **F** is a matrix whose components are the force constants:

$$\mathbf{F} = \begin{bmatrix} f_{11} & f_{12} & r_1 f_{13} \\ f_{21} & f_{22} & r_2 f_{23} \\ r_1 f_{31} & r_2 f_{32} & r_1 r_2 f_{33} \end{bmatrix} \equiv \begin{bmatrix} F_{11} & F_{12} & F_{13} \\ F_{21} & F_{22} & F_{23} \\ F_{31} & F_{32} & F_{33} \end{bmatrix} \qquad (12.2)^*$$

Here r_1 and r_2 are the equilibrium lengths of the X–Y_1 and X–Y_2 bonds, respectively.

The kinetic energy is not easily expressed in terms of the same internal coordinates. Wilson[87] has shown, however, that the kinetic energy can be written as

$$2T = \tilde{\dot{\mathbf{R}}}\mathbf{G}^{-1}\dot{\mathbf{R}} \qquad (12.3)\dagger$$

where $\mathbf{G}^{-1}$ is the reciprocal of the **G** matrix, which will be defined later.

If Eqs. 12.1 and 12.3 are combined with Lagrange's equation,

$$\frac{d}{dt}\left(\frac{\partial T}{\partial \dot{R}_k}\right) + \frac{\partial V}{\partial R_k} = 0 \qquad (4.6)$$

the following secular equation, which is similar to Eq. 4.19, is obtained:

$$\begin{vmatrix} F_{11} - (G^{-1})_{11}\lambda & F_{12} - (G^{-1})_{12}\lambda & \cdots \\ F_{21} - (G^{-1})_{21}\lambda & F_{22} - (G^{-1})_{22}\lambda & \cdots \\ \vdots & \vdots & \end{vmatrix} \equiv |\mathbf{F} - \mathbf{G}^{-1}\lambda| = 0 \qquad (12.4)$$

By multiplying by the determinant of **G**

$$\begin{vmatrix} G_{11} & G_{12} & \cdots \\ G_{21} & G_{22} & \cdots \\ \vdots & \vdots & \end{vmatrix} \equiv |\mathbf{G}| \qquad (12.5)$$

*Here f_{11} and f_{22} are the stretching force constants of the X–Y_1 and X–Y_2 bonds, respectively, and f_{33} is the bending force constant of the Y_1XY_2 angle. The other symbols represent interaction force constants between stretching and stretching or between stretching and bending vibrations. To make the dimensions of all the force constants the same, f_{13} (or f_{31}), f_{23} (or f_{32}), and f_{33} are multiplied by r_1, r_2, and $r_1 r_2$, respectively.

† Appendix VI gives the derivation of Eq. 12.3.

from the left-hand side of Eq. 12.4, the following equation is obtained:

$$\begin{vmatrix} \sum G_{1t}F_{t1} - \lambda & \sum G_{1t}F_{t2} & \cdots \\ \sum G_{2t}F_{t1} & \sum G_{2t}F_{t2} - \lambda & \cdots \\ \vdots & \vdots & \end{vmatrix} \equiv |\mathbf{GF} - \mathbf{E}\lambda| = 0 \qquad (12.6)$$

Here, **E** is the unit matrix, and λ is related to the wave number $\tilde{\nu}$ by the relation $\lambda = 4\pi^2c^2\tilde{\nu}^2$.* The order of the equation is equal to the number of internal coordinates used.

The **F** matrix can be written by assuming a suitable set of force constants. If the **G** matrix is constructed by the following method, the vibrational frequencies are obtained by solving Eq. 12.6. The **G** matrix is defined as

$$\mathbf{G} = \mathbf{B}\mathbf{M}^{-1}\tilde{\mathbf{B}} \qquad (12.7)$$

Here $\mathbf{M}^{-1}$ is a diagonal matrix whose components are μ_i, where μ_i is the reciprocal of the mass of the ith atom. For a bent XY_2 molecule,

$$\mathbf{M}^{-1} = \begin{bmatrix} \mu_1 & & & & 0 \\ & \mu_1 & & & \\ & & \mu_1 & & \\ & & & \ddots & \\ 0 & & & & \mu_3 \end{bmatrix}$$

where μ_3 and μ_1 are the reciprocals of the masses of the X and Y atoms, respectively. The **B** matrix is defined as

$$\mathbf{R} = \mathbf{B}\mathbf{X} \qquad (12.8)$$

where **R** and **X** are column matrices whose components are the internal and rectangular coordinates, respectively.

To write Eq. 12.8 for a bent XY_2 molecule, we express the bond stretching (Δr_1) in terms of the x and y coordinates shown in Fig. I-22(a). It is readily seen that

$$\Delta r_1 = -(\Delta x_1)(s) - (\Delta y_1)(c) + (\Delta x_3)(s) + (\Delta y_3)(c)$$

Here, $s = \sin(\alpha/2)$, $c = \cos(\alpha/2)$, and r is the equilibrium distance between the X and Y atoms. A similar expression is obtained for Δr_2. Thus,

*Here λ should not be confused with λ_w (wavelength).

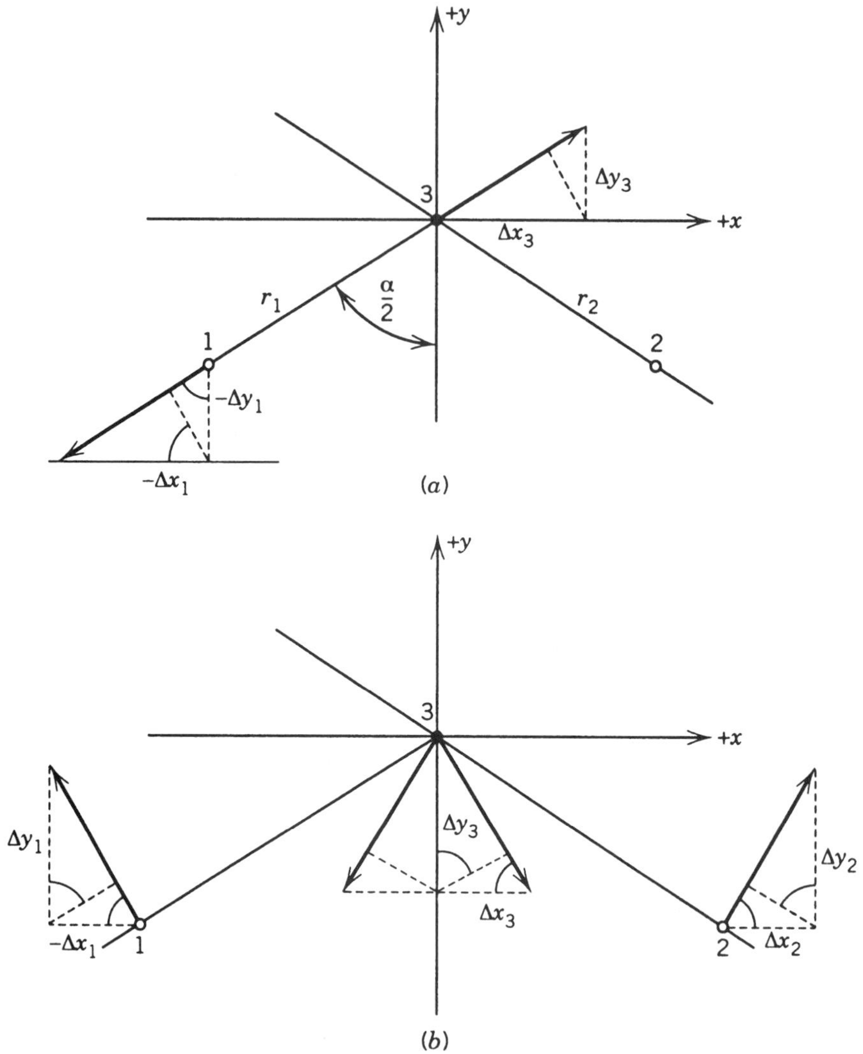

Fig. I-22. Relationship between internal and rectangular coordinates in (a) stretching and (b) bending vibrations of a bent XY_2 molecule.

$$\Delta r_2 = (\Delta x_2)(s) - (\Delta y_2)(c) - (\Delta x_3)(s) + (\Delta y_3)(c)$$

For the bond bending ($\Delta\alpha$), consider the relationships illustrated in Fig. I-22(b). It is readily seen that

$$r(\Delta\alpha) = -(\Delta x_1)(c) + (\Delta y_1)(s) + (\Delta x_2)(c) + (\Delta y_2)(s) - 2(\Delta y_3)(s)$$

If these equations are summarized in a matrix form, we obtain:

$$\begin{bmatrix} \Delta r_1 \\ \Delta r_2 \\ \Delta \alpha \end{bmatrix} = \left[\begin{array}{ccc:ccc:ccc} -s & -c & 0 & 0 & 0 & 0 & s & c & 0 \\ 0 & 0 & 0 & s & -c & 0 & -s & c & 0 \\ -c/r & s/r & 0 & c/r & s/r & 0 & 0 & -2s/r & 0 \end{array}\right] \begin{bmatrix} \Delta x_1 \\ \Delta y_1 \\ \Delta z_1 \\ --- \\ \Delta x_2 \\ \Delta y_2 \\ \Delta z_2 \\ --- \\ \Delta x_3 \\ \Delta y_3 \\ \Delta z_3 \end{bmatrix} \tag{12.9}$$

If unit vectors such as those in Fig. I-23 are considered, Eq. 12.9 can be written in a more compact form using vector notation:

$$\begin{bmatrix} \Delta r_1 \\ \Delta r_2 \\ \Delta \alpha \end{bmatrix} = \begin{bmatrix} \mathbf{e}_{31} & 0 & -\mathbf{e}_{31} \\ 0 & \mathbf{e}_{32} & -\mathbf{e}_{32} \\ \mathbf{p}_{31}/r & \mathbf{p}_{32}/r & -(\mathbf{p}_{31} + \mathbf{p}_{32})/r \end{bmatrix} \begin{bmatrix} \boldsymbol{\rho}_1 \\ \boldsymbol{\rho}_2 \\ \boldsymbol{\rho}_3 \end{bmatrix} \tag{12.10}$$

Here $\boldsymbol{\rho}_1$, $\boldsymbol{\rho}_2$, and $\boldsymbol{\rho}_3$ are the displacement vectors of atoms 1, 2, and 3, respectively. Thus Eq. 12.10 can be written simply as

$$\mathbf{R} = \mathbf{S} \cdot \boldsymbol{\rho} \tag{12.11}$$

where the dot represents the scalar product of the two vectors. Here **S** is called the **S** matrix, and its components (**S** vector) can be written according to the following formulas: (1) bond stretching,

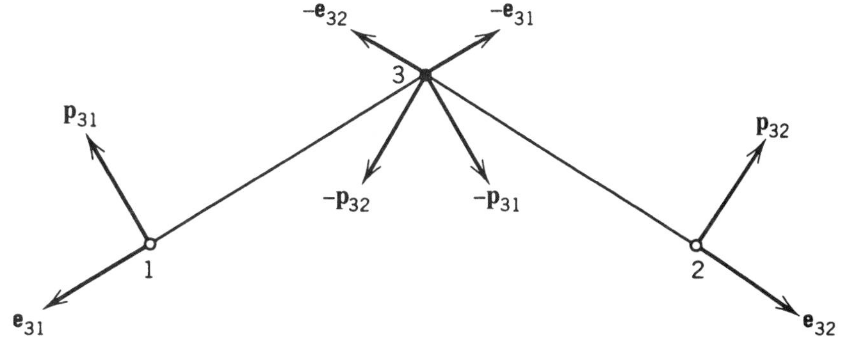

Fig. I-23. Unit vectors in a bent XY_2 molecule.

$$\Delta r_1 = \Delta_{31} = \mathbf{e}_{31} \cdot \boldsymbol{\rho}_1 - \mathbf{e}_{31} \cdot \boldsymbol{\rho}_3 \tag{12.12}$$

and (2) angle bending,

$$\Delta\alpha = \Delta\alpha_{132} = [\mathbf{p}_{31} \cdot \boldsymbol{\rho}_1 + \mathbf{p}_{32} \cdot \boldsymbol{\rho}_2 - (\mathbf{p}_{31} + \mathbf{p}_{32}) \cdot \boldsymbol{\rho}_3]/r \tag{12.13}$$

It is seen that the direction of the **S** vector is the direction in which a given displacement of the ith atom will produce the greatest increase in Δr or $\Delta\alpha$. Formulas for obtaining the **S** vectors of other internal coordinates such as those of out-of-plane ($\Delta\theta$) and torsional ($\Delta\tau$) vibrations are also available.[3]

By using the **S** matrix, Eq. 12.7 is written as

$$\mathbf{G} = \mathbf{S}\mathbf{m}^{-1}\tilde{\mathbf{S}} \tag{12.14}$$

For a bent XY_2 molecule, this becomes

$$\mathbf{G} = \begin{bmatrix} \mathbf{e}_{31} & 0 & -\mathbf{e}_{31} \\ 0 & \mathbf{e}_{32} & -\mathbf{e}_{32} \\ \mathbf{p}_{31}/r & \mathbf{p}_{32}/r & -(\mathbf{p}_{31} + \mathbf{p}_{32})/r \end{bmatrix} \begin{bmatrix} \mu_1 & 0 & 0 \\ 0 & \mu_1 & 0 \\ 0 & 0 & \mu_3 \end{bmatrix}$$

$$\times \begin{bmatrix} \mathbf{e}_{31} & 0 & \mathbf{p}_{31}/r \\ 0 & \mathbf{e}_{32} & \mathbf{p}_{32}/r \\ -\mathbf{e}_{31} & -\mathbf{e}_{32} & -(\mathbf{p}_{31} + \mathbf{p}_{32})/r \end{bmatrix}$$

$$= \begin{bmatrix} (\mu_3 + \mu_1)\mathbf{e}_{31}^2 & \mu_3 \mathbf{e}_{31} \cdot \mathbf{e}_{32} & \dfrac{\mu_1}{r}\mathbf{e}_{31} \cdot \mathbf{p}_{31} + \dfrac{\mu_3}{r}\mathbf{e}_{31} \cdot (\mathbf{p}_{31} + \mathbf{p}_{32}) \\ & (\mu_3 + \mu_1)\mathbf{e}_{32}^2 & \dfrac{\mu_1}{r}\mathbf{e}_{32} \cdot \mathbf{p}_{32} + \dfrac{\mu_3}{r}\mathbf{e}_{32} \cdot (\mathbf{p}_{31} + \mathbf{p}_{32}) \\ & & \dfrac{\mu_1}{r^2}\mathbf{p}_{31}^2 + \dfrac{\mu_1}{r^2}\mathbf{p}_{32}^2 + \dfrac{\mu_3}{r^2}(\mathbf{p}_{31} + \mathbf{p}_{32})^2 \end{bmatrix}$$

Considering

$$\mathbf{e}_{31} \cdot \mathbf{e}_{31} = \mathbf{e}_{32} \cdot \mathbf{e}_{32} = \mathbf{p}_{31} \cdot \mathbf{p}_{31} = \mathbf{p}_{32} \cdot \mathbf{p}_{32} = 1, \qquad \mathbf{e}_{31} \cdot \mathbf{p}_{31} = \mathbf{e}_{32} \cdot \mathbf{p}_{32} = 0$$

$$\mathbf{e}_{31} \cdot \mathbf{e}_{32} = \cos\alpha, \qquad \mathbf{e}_{31} \cdot \mathbf{p}_{32} = \mathbf{e}_{32} \cdot \mathbf{p}_{31} = -\sin\alpha$$

$$(\mathbf{p}_{31} + \mathbf{p}_{32})^2 = 2(1 - \cos\alpha)$$

we find that the **G** matrix is calculated as

$$\mathbf{G} = \begin{bmatrix} \mu_3 + \mu_1 & \mu_3 \cos \alpha & -\dfrac{\mu_3}{r} \sin \alpha \\ & \mu_3 + \mu_1 & -\dfrac{\mu_3}{r} \sin \alpha \\ & & \dfrac{2\mu_1}{r^2} + \dfrac{2\mu_3}{r^2} (1 - \cos \alpha) \end{bmatrix} \tag{12.15}$$

If the **G**-matrix elements obtained are written for each combination of internal coordinates, there results

$$\begin{aligned} G(\Delta r_1, \Delta r_1) &= \mu_3 + \mu_1 \\ G(\Delta r_2, \Delta r_2) &= \mu_3 + \mu_1 \\ G(\Delta r_1, \Delta r_2) &= \mu_3 \cos \alpha \\ G(\Delta \alpha, \Delta \alpha) &= \frac{2\mu_1}{r^2} + \frac{2\mu_3}{r^2} (1 - \cos \alpha) \\ G(\Delta r_1, \Delta \alpha) &= -\frac{\mu_3}{r} \sin \alpha \\ G(\Delta r_2, \Delta \alpha) &= -\frac{\mu_3}{r} \sin \alpha \end{aligned} \tag{12.16}$$

If such calculations are made for several types of molecules, it is immediately seen that the **G**-matrix elements themselves have many regularities. Decius[88] developed general formulas for writing **G**-matrix elements.* Some of them are as follows:

$$G_{rr}^2 = \mu_1 + \mu_2$$

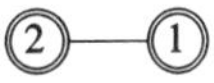

$$G_{rr}^1 = \mu_1 \cos \phi$$

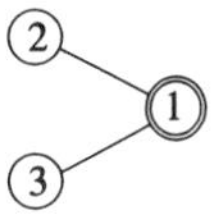

$$G_{r\phi}^2 = -\rho_{23}\mu_2 \sin \phi$$

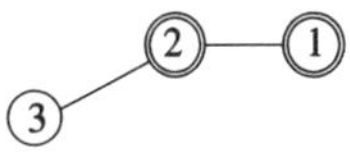

$$G_{r\phi}^1 \begin{pmatrix} 1 \\ 1 \end{pmatrix} = -(\rho_{13} \sin \phi_{213} \cos \psi_{234} + \rho_{14} \sin \phi_{214} \cos \psi_{243})\mu_1$$

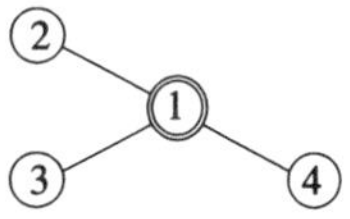

*See also Refs. 3 and 89.

$$G^3_{\phi\phi} = \rho_{12}^2 \mu_1 + \rho_{23}^2 \mu_3 + (\rho_{12}^2 + \rho_{23}^2 - 2\rho_{12}\rho_{23} \cos \phi)\mu_2$$

$$G^2_{\phi\phi}\begin{pmatrix} 1 \\ 1 \end{pmatrix} = (\rho_{12}^2 \cos \psi_{314})\mu_1 + [(\rho_{12} - \rho_{23} \cos \phi_{123} - \rho_{24} \cos \phi_{124})\rho_{12} \cos \psi_{314} + (\sin \phi_{123} \sin \phi_{124} \sin^2 \psi_{314} + \cos \phi_{324} \cos \psi_{314})\rho_{23}\rho_{24}]\mu_2$$

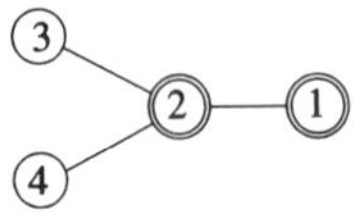

Here, the atoms surrounded by a double circle are those common to both coordinates. The symbols μ and ρ denote the reciprocals of mass and bond distance, respectively. The spherical angle $\psi_{\alpha\beta\gamma}$ in Fig. I-24 is defined as

$$\cos \psi_{\alpha\beta\gamma} = \frac{\cos \phi_{\alpha\delta\gamma} - \cos \phi_{\alpha\delta\beta} \cos \phi_{\beta\delta\gamma}}{\sin \phi_{\alpha\delta\beta} \sin \phi_{\beta\delta\gamma}} \tag{12.17}$$

The correspondence between the Decius formulas and the results obtained in Eq. 12.16 is evident.

With the Decius formulas, the **G**-matrix elements of a pyramidal XY_3 molecule have been calculated and are shown in Table I-13.

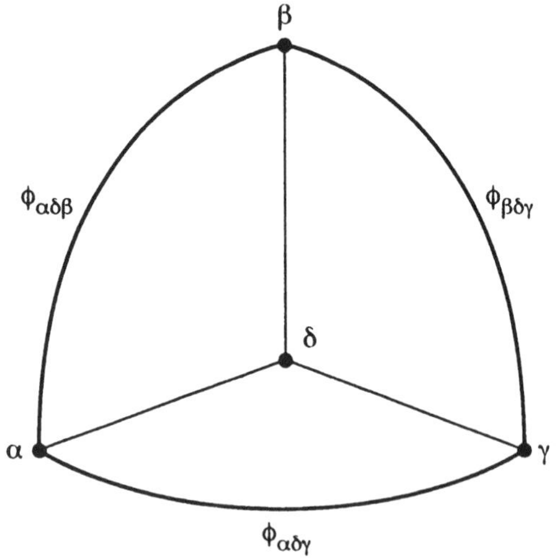

Fig. I-24. Spherical angles involving atomic positions, α, β, γ, and δ.

TABLE I-13.

	Δr_1	Δr_2	Δr_3	$\Delta\alpha_{23}$	$\Delta\alpha_{31}$	$\Delta\alpha_{12}$
Δr_1	A	B	B	C	D	D
Δr_2	—	A	B	D	C	D
Δr_3	—	—	A	D	D	C
$\Delta\alpha_{23}$	—	—	—	E	F	F
$\Delta\alpha_{31}$	—	—	—	—	E	F
$\Delta\alpha_{12}$	—	—	—	—	—	E

$A = G_{rr}^2 = \mu_X + \mu_Y$

$B = G_{rr}^1 = \mu_X \cos\alpha$

$C = G_{r\phi}^1\begin{pmatrix}1\\1\end{pmatrix} = -\dfrac{2}{r}\ \dfrac{\cos\alpha(1-\cos\alpha)\mu_X}{\sin\alpha}$

$D = G_{r\phi}^2 = -\dfrac{\mu_X}{r}\sin\alpha$

$E = G_{\phi\phi}^3 = \dfrac{2}{r^2}\,[\mu_Y + \mu_X(1-\cos\alpha)]$

$F = G_{\phi\phi}^2\begin{pmatrix}1\\1\end{pmatrix} = \dfrac{\mu_Y}{r^2}\ \dfrac{\cos\alpha}{1+\cos\alpha} + \dfrac{\mu_X}{r^2}\ \dfrac{(1+3\cos\alpha)(1-\cos\alpha)}{1+\cos\alpha}$

I-13. UTILIZATION OF SYMMETRY PROPERTIES

In view of the equivalence of the two X–Y bonds of a bent XY_2 molecule, the **F** and **G** matrices obtained in Eqs. 12.2 and 12.15 are written as

$$\mathbf{F} = \begin{bmatrix} f_{11} & f_{12} & rf_{13} \\ f_{12} & f_{11} & rf_{13} \\ rf_{13} & rf_{13} & r^2 f_{33} \end{bmatrix} \tag{13.1}$$

$$\mathbf{G} = \begin{bmatrix} \mu_3 + \mu_1 & \mu_3\cos\alpha & -\dfrac{\mu_3}{r}\sin\alpha \\ \mu_3\cos\alpha & \mu_3+\mu_1 & -\dfrac{\mu_3}{r}\sin\alpha \\ -\dfrac{\mu_3}{r}\sin\alpha & -\dfrac{\mu_3}{r}\sin\alpha & \dfrac{2\mu_1}{r^2} + \dfrac{2\mu_3}{r^2}(1-\cos\alpha) \end{bmatrix} \tag{13.2}$$

Both of these matrices are of the form

$$\begin{bmatrix} A & C & D \\ C & A & D \\ D & D & B \end{bmatrix} \tag{13.3}$$

The appearance of the same elements is evidently due to the equivalence of the two internal coordinates, Δr_1 and Δr_2. Such symmetrically equivalent sets of internal coordinates are seen in many other molecules, such as those in Fig. I-20. In these cases, it is possible to reduce the order of the **F** and **G** matrices (and hence the order of the secular equation resulting from them) by a coordinate transformation.

Let the internal coordinates R be transformed by

$$\mathbf{R}^s = \mathbf{UR} \tag{13.4}$$

Then, we obtain

$$\begin{aligned} 2T &= \tilde{\dot{\mathbf{R}}}\mathbf{G}^{-1}\dot{\mathbf{R}} = \tilde{\dot{\mathbf{R}}}^s\tilde{\mathbf{U}}^{-1}\mathbf{G}^{-1}\mathbf{U}^{-1}\dot{\mathbf{R}}^s \\ &= \tilde{\dot{\mathbf{R}}}^s\mathbf{G}_s^{-1}\dot{\mathbf{R}}^s \\ 2V &= \tilde{\mathbf{R}}\mathbf{F}\mathbf{R} = \tilde{\mathbf{R}}^s\tilde{\mathbf{U}}^{-1}\mathbf{F}\mathbf{U}^{-1}\mathbf{R}^s \\ &= \tilde{\mathbf{R}}^s\mathbf{F}_s\mathbf{R}^s \end{aligned}$$

where

$$\begin{aligned} \mathbf{G}_s^{-1} &= \tilde{\mathbf{U}}^{-1}\mathbf{G}^{-1}\mathbf{U}^{-1} \quad \text{or} \quad \mathbf{G}_s = \mathbf{UG\tilde{U}} \\ \mathbf{F}_s &= \tilde{\mathbf{U}}^{-1}\mathbf{F}\mathbf{U}^{-1} \end{aligned} \tag{13.5}$$

If **U** is an orthogonal matirx ($\mathbf{U}^{-1} = \tilde{\mathbf{U}}$), Eq. 13.5 is written as

$$\mathbf{F}_s = \mathbf{UF\tilde{U}} \quad \text{and} \quad \mathbf{G}_s = \mathbf{UG\tilde{U}} \tag{13.6}$$

Both **GF** and $\mathbf{G}_s\mathbf{F}_s$ give the same roots, since

$$\begin{aligned} |\mathbf{G}_s\mathbf{F}_s - \mathbf{E}\lambda| &= |\mathbf{UG\tilde{U}\tilde{U}}^{-1}\mathbf{FU}^{-1} - \mathbf{E}\lambda| \\ &= |\mathbf{UGFU}^{-1} - \mathbf{E}\lambda| \\ &= |\mathbf{U}||\mathbf{GF} - \mathbf{E}\lambda||\mathbf{U}^{-1}| \\ &= |\mathbf{GF} - \mathbf{E}\lambda| \end{aligned} \tag{13.7}$$

If we choose a proper **U** matrix from symmetry consideration, it is possible to factor the original **G** and **F** matrices into smaller ones. This, in turn, reduces the order of the secular equation to be solved, thus facilitating their solution. These new coordinates $\mathbf{R}^s$ are called *symmetry coordinates*.

The **U** matrix is constructed by using the equation

$$\mathbf{R}^s = N\sum_K \chi_i(K)K(\Delta r_1) \tag{13.8}$$

Here, K is a symmetry operation, and the summation is made over all symmetry operations. Also, $\chi_i(K)$ is the character of the representation to which R^s belongs. Called a generator, Δr_1 is, by symmetry operation K, transformed into $K(\Delta r_1)$, which is another coordinate of the same symmetrically equivalent set. Finally, N is a normalizing factor.

As an example, consider a bent XY_2 molecule in which Δr_1 and Δr_2 are equivalent. Using Δr_1 as a generator, we obtain

	I	$C_2(z)$	$\sigma(xz)$	$\sigma(yz)$
$K(\Delta r_1)$	Δr_1	Δr_2	Δr_2	Δr_1
$\chi_{A_1}(K)$	1	1	1	1
$\chi_{B_2}(K)$	1	−1	−1	1

Thus,

$$R^s_{A_1} = N \sum \chi_{A_1}(K) K(\Delta r_1) = 2N(\Delta r_1 + \Delta r_2)$$

$$N = \frac{1}{2\sqrt{2}} \quad \text{since } (2N)^2 + (2N)^2 = 1$$

Then

$$R^s_{A_1} = \frac{1}{\sqrt{2}}(\Delta r_1 + \Delta r_2) \tag{13.9}$$

Similarly,

$$R^s_{B_2} = \frac{1}{\sqrt{2}}(\Delta r_1 - \Delta r_2) \tag{13.10}$$

The remaining internal coordinate, $\Delta\alpha$, belongs to the A_1 species. Thus, the complete **U** matrix is written as

$$\begin{bmatrix} R^s_1(A_1) \\ R^s_2(A_1) \\ R^s_3(B_2) \end{bmatrix} = \begin{bmatrix} \frac{1}{\sqrt{2}} & \frac{1}{\sqrt{2}} & 0 \\ 0 & 0 & 1 \\ \frac{1}{\sqrt{2}} & \frac{-1}{\sqrt{2}} & 0 \end{bmatrix} \begin{bmatrix} \Delta r_1 \\ \Delta r_2 \\ \Delta\alpha \end{bmatrix} \tag{13.11}$$

If the **G** and **F** matrices of type 13.3 are transformed by relations 13.6, where **U** is given by the matrix of Eq. 13.11, they become

$$\mathbf{F_S}, \mathbf{G_S} = \left[\begin{array}{cc|c} A+C & \sqrt{2}D & 0 \\ \sqrt{2}D & B & 0 \\ \hline 0 & 0 & A-C \end{array}\right] \tag{13.12}$$

or, more explicitly,

$$\mathbf{F_S} = \left[\begin{array}{cc|c} f_{11}+f_{12} & r\sqrt{2}f_{13} & 0 \\ r\sqrt{2}f_{13} & r^2 f_{33} & 0 \\ \hline 0 & 0 & f_{11}-f_{12} \end{array}\right] \tag{13.13}$$

$$\mathbf{G_S} = \left[\begin{array}{cc|c} \mu_3(1+\cos\alpha)+\mu_1 & -\dfrac{\sqrt{2}}{r}\mu_3\sin\alpha & 0 \\ -\dfrac{\sqrt{2}}{r}\mu_3\sin\alpha & \dfrac{2\mu_1}{r^2}+\dfrac{2\mu_3}{r^2}(1-\cos\alpha) & 0 \\ \hline 0 & 0 & \mu_3(1-\cos\alpha)+\mu_1 \end{array}\right] \tag{13.14}$$

In a pyramidal XY_3 molecule (Fig. I-20*c*), Δr_1, Δr_2, and Δr_3 are the equivalent set; so are $\Delta\alpha_{23}$, $\Delta\alpha_{31}$, and $\Delta\alpha_{12}$. It is already known from Eq. 9.7 that one A_1 and one E vibration are involved both in the stretching and in the bending vibrations. Using Δr_1 as a generator, we obtain from Eq. 13.8

	I	C_3^+	C_3^-	σ_1	σ_2	σ_3
$K(\Delta r_1)$	Δr_1	Δr_2	Δr_3	Δr_1	Δr_3	Δr_2
$\chi_{A_1}(K)$	1	1	1	1	1	1
$\chi_E(K)$	2	−1	−1	0	0	0

Then, we obtain

$$R^s_{A_1} = \frac{1}{\sqrt{3}}(\Delta r_1 + \Delta r_2 + \Delta r_3) \tag{13.15}$$

$$R^s_{E_1} = \frac{1}{\sqrt{6}}(2\Delta r_1 - \Delta r_2 - \Delta r_3) \tag{13.16}$$

To find a coordinate that forms a degenerate pair with Eq. 13.16, we repeat the same procedure, using Δr_2 and Δr_3 as the generators. The results are

$$R^s_{E_2} = N(2\Delta r_2 - \Delta r_3 - \Delta r_1)$$
$$R^s_{E_3} = N(2\Delta r_3 - \Delta r_1 - \Delta r_2)$$

However, these coordinates are not orthogonal to $R^s_{E_1}$ (Eq. 13.16).

If we take a linear combination, $R^s_{E_2} + R^s_{E_3}$, we obtain Eq. 13.16. If we take $R^s_{E_2} - R^s_{E_3}$, we obtain

$$R^s_{E_4} = \frac{1}{\sqrt{2}}(\Delta r_2 - \Delta r_3) \tag{13.17}$$

Since Eqs. 13.16 and 13.17 are mutually orthogonal, these two coordinates are taken as a degenerate pair. Similar results are obtained for three angle-bending coordinates. Thus, the complete **U** matrix is written as

$$\begin{bmatrix} R^s_1(A_1) \\ R^s_2(A_1) \\ R^s_{3a}(E) \\ R^s_{4a}(E) \\ R^s_{3b}(E) \\ R^s_{4b}(E) \end{bmatrix} = \begin{bmatrix} 1/\sqrt{3} & 1/\sqrt{3} & 1/\sqrt{3} & 0 & 0 & 0 \\ 0 & 0 & 0 & 1/\sqrt{3} & 1/\sqrt{3} & 1/\sqrt{3} \\ 2/\sqrt{6} & -1/\sqrt{6} & -1/\sqrt{6} & 0 & 0 & 0 \\ 0 & 0 & 0 & 2/\sqrt{6} & -1/\sqrt{6} & -1/\sqrt{6} \\ 0 & 1/\sqrt{2} & -1/\sqrt{2} & 0 & 0 & 0 \\ 0 & 0 & 0 & 0 & 1/\sqrt{2} & -1/\sqrt{2} \end{bmatrix} \cdot \begin{bmatrix} \Delta r_1 \\ \Delta r_2 \\ \Delta r_3 \\ \Delta\alpha_{23} \\ \Delta\alpha_{31} \\ \Delta\alpha_{12} \end{bmatrix} \tag{13.18}$$

The **G** matrix of a pyramidal XY_3 molecule has already been calculated (see Table I-13). By using Eq. 13.6, the new $\mathbf{G_S}$ matrix becomes

$$\mathbf{G_S} = \left[\begin{array}{cc|cc|cc} A+2B & C+2D & & & & \\ C+2D & E+2F & \multicolumn{2}{c|}{0} & \multicolumn{2}{c}{0} \\ \hline & & A-B & C-D & & \\ \multicolumn{2}{c|}{0} & C-D & E-F & \multicolumn{2}{c}{0} \\ \hline & & & & A-B & C-D \\ \multicolumn{2}{c|}{0} & \multicolumn{2}{c|}{0} & C-D & E-F \end{array}\right] \tag{13.19}$$

Here A, B, and so forth, denote the elements in Table I-13. The **F** matrix transforms similarly. Therefore, it is necessary only to solve two quadratic equations for the A_1 and E species.

For the tetrahedral XY_4 molecule shown in Fig. I-20*e*, group theory (Secs. I-8 and I-9) predicts one A_1 and one F_2 stretching, and one E and one F_2 bending, vibration. The **U** matrix for the four stretching coordinates is

$$\begin{bmatrix} R_1^s(A_1) \\ R_{2a}^s(F_2) \\ R_{2b}^s(F_2) \\ R_{2c}^s(F_2) \end{bmatrix} = \begin{bmatrix} 1/2 & 1/2 & 1/2 & 1/2 \\ 1/\sqrt{6} & 1/\sqrt{6} & -2/\sqrt{6} & 0 \\ 1/\sqrt{12} & 1/\sqrt{12} & 1/\sqrt{12} & -3/\sqrt{12} \\ -1/\sqrt{2} & 1/\sqrt{2} & 0 & 0 \end{bmatrix} \begin{bmatrix} \Delta r_1 \\ \Delta r_2 \\ \Delta r_3 \\ \Delta r_4 \end{bmatrix} \tag{13.20}$$

whereas the **U** matrix for the six bending coordinates becomes

$$\begin{bmatrix} R_1^s(A_1) \\ R_{2a}^s(E) \\ R_{2b}^s(E) \\ R_{3a}^s(F_2) \\ R_{3b}^s(F_2) \\ R_{3c}^s(F_2) \end{bmatrix} = \begin{bmatrix} 1/\sqrt{6} & 1/\sqrt{6} & 1/\sqrt{6} & 1/\sqrt{6} & 1/\sqrt{6} & 1/\sqrt{6} \\ 2/\sqrt{12} & -1/\sqrt{12} & -1/\sqrt{12} & -1/\sqrt{12} & -1/\sqrt{12} & 2/\sqrt{12} \\ 0 & 1/2 & -1/2 & 1/2 & -1/2 & 0 \\ 2/\sqrt{12} & -1/\sqrt{12} & -1/\sqrt{12} & 1/\sqrt{12} & 1/\sqrt{12} & -2/\sqrt{12} \\ 1/\sqrt{6} & 1/\sqrt{6} & 1/\sqrt{6} & -1/\sqrt{6} & -1/\sqrt{6} & -1/\sqrt{6} \\ 0 & 1/2 & -1/2 & -1/2 & 1/2 & 0 \end{bmatrix} \cdot \begin{bmatrix} \Delta\alpha_{12} \\ \Delta\alpha_{23} \\ \Delta\alpha_{31} \\ \Delta\alpha_{14} \\ \Delta\alpha_{24} \\ \Delta\alpha_{34} \end{bmatrix} \tag{13.21}$$

The symmetry coordinate $R_1^s(A_1)$ in Eq. 13.21 represents a *redundant coordinate* (see Eq. 9.1). In such a case, a coordinate transformation reduces the order of the matrix by one, since all the **G**-matrix elements related to this coordinate become zero. Conversely, this result provides a general method of finding redundant coordinates. Suppose that the elements of the **G** matrix are calculated in terms of internal coordinates such as those in Table I-13. If a suitable combination of internal coordinates is made so that $\Sigma_j G_{ij} = 0$ (where j refers to all the equivalent internal coordinates), such a combination is a redundant coordinate. By using the **U** matrices in Eqs. 13.20 and 13.21, the problem of solving a tenth-order secular equation for a tetrahedral XY_4 molecule is reduced to that of solving two first-order (A_1 and E) and one quadratic (F_2) equation.

The normal modes of vibration of the tetrahedral XY_4 molecule are shown in Fig. II-12 of Section II. It can be seen that the normal modes, ν_1 and ν_3, correspond to the symmetry coordinates $R_1^s(A_1)$ and $R_{2b}^s(F_2)$ of Eq. 13.20, respectively, and that the normal modes, ν_2 and ν_4, correspond to the symmetry coordinates $R_{2a}^s(E)$ and $R_{3b}^s(F_2)$ of Eq. 13.21, respectively.

I-14. POTENTIAL FIELDS AND FORCE CONSTANTS

Using Eqs. 12.1 and 13.1, we write the potential energy of a bent XY_2 molecule as

$$\begin{aligned} 2V = {} & f_{11}(\Delta r_1)^2 + f_{11}(\Delta r_2)^2 + f_{33}r^2(\Delta\alpha)^2 + 2f_{12}(\Delta r_1)(\Delta r_2) \\ & + 2f_{13}r(\Delta r_1)(\Delta\alpha) + 2f_{13}r(\Delta r_2)(\Delta\alpha) \end{aligned} \tag{14.1}$$

This type of potential field is called a *generalized valence force* (GVF) field.* It consists of stretching and bending force constants, as well as the interaction force constants between them. When using such a potential field, four force constants are needed to describe the potential energy of a bent XY_2 molecule. Since only three vibrations are observed in practice, it is impossible to determine all four force constants simultaneously. One method used to circumvent this difficulty is to calculate the vibrational frequencies of isotopic molecules (e.g., D_2O and HDO for H_2O), assuming the same set of force constants.† This method is satisfactory, however, only for simple molecules. As molecules become more complex, the number of interaction force constants in the GVF field becomes too large to allow any reliable evaluation.

*A potential field consisting of stretching and bending force constants only is called *simple valence force* field.

† In addition to isotope frequency shifts, mean amplitudes of vibration, Coriolis coupling constants, centrifugal distortion constants, and so forth may be used to refine the force constants of small molecules (see Ref. 90).

In another approach, Shimanouchi[91] introduced the *Urey–Bradley force* (UBF) field, which consists of stretching and bending force constants, as well as repulsive force constants between nonbonded atoms. The general form of the potential field is given by

$$V = \sum_i \left(\frac{1}{2} K_i(\Delta r_i)^2 + K'_i r_i(\Delta r_i)\right)$$
$$+ \sum_i \left(\frac{1}{2} H_i r_{i\alpha}^2(\Delta\alpha_i)^2 + H'_i r_{i\alpha}^2(\Delta\alpha_i)\right)$$
$$+ \sum_i \left(\frac{1}{2} F_i(\Delta q_i)^2 + F'_i q_i(\Delta q_i)\right) \quad (14.2)$$

Here Δr_i, $\Delta\alpha_i$, and Δq_i are the changes in the bond lengths, bond angles, and distances between nonbonded atoms, respectively. The symbols K_i, K'_i, H_i, H'_i, and F_i, F'_i represent the stretching, bending, and repulsive force constants, respectively. Furthermore, r_i, $r_{i\alpha}$, and q_i are the values of the distances at the equilibrium positions and are inserted to make the force constants dimensionally similar. Linear force constants (K'_i, H'_i, and F'_i) must be included since Δr_i, $\Delta\alpha_i$, and Δq_i are not completely independent of each other.

Using the relation

$$q_{ij}^2 = r_i^2 + r_j^2 - 2r_i r_j \cos \alpha_{ij} \quad (14.3)$$

and considering that the first derivatives can be equated to zero in the equilibrium case, we can write the final form of the potential field as

$$V = \frac{1}{2} \sum_i \left[K_i + \sum_{j(\neq i)} (t_{ij}^2 F'_{ij} + s_{ij}^2 F_{ij}) \right] (\Delta r_i)^2$$
$$+ \frac{1}{2} \sum_{i<j} (H_{ij} - s_{ij}s_{ji}F'_{ij} + t_{ij}t_{ji}F_{ij})(\sqrt{r_i r_j}\Delta\alpha_{ij})^2$$
$$+ \sum_{i<j} (-t_{ij}t_{ji}F'_{ij} + s_{ij}s_{ji}F_{ij})(\Delta r_i)(\Delta r_j)$$

$$+\sum_{i \neq j} (t_{ij}s_{ji}F'_{ij} + t_{ji}s_{ij}F_{ij})\left(\frac{r_j}{r_i}\right)^{1/2} (\Delta r_i)(\sqrt{r_i r_j}\Delta\alpha_{ij}) \quad (14.4)^*$$

Here

$$s_{ij} = \frac{r_i - r_j \cos \alpha_{ij}}{q_{ij}}$$

$$s_{ji} = \frac{r_j - r_i \cos \alpha_{ij}}{q_{ij}}$$

$$t_{ij} = \frac{r_j \sin \alpha_{ij}}{q_{ij}}$$

$$t_{ji} = \frac{r_i \sin \alpha_{ij}}{q_{ij}} \quad (14.5)$$

In a bent XY_2 molecule, Eq. 14.4 becomes

$$\begin{aligned} V = {} & \tfrac{1}{2}(K + t^2F' + s^2F)[(\Delta r_1)^2 + (\Delta r_2)^2] + \tfrac{1}{2}(H - s^2F' + t^2F)(r\Delta\alpha)^2 \\ & + (-t^2F' + s^2F)(\Delta r_1)(\Delta r_2) + ts(F' + F)(\Delta r_1)(r\Delta\alpha) \\ & + ts(F' + F)(\Delta r_2)(r\Delta\alpha) \end{aligned} \quad (14.6)$$

where

$$s = \frac{r(1 - \cos \alpha)}{q}$$

$$t = \frac{r \sin \alpha}{q}$$

Comparing Eqs. 14.6 and 14.1, we obtain the following relations between the force constants of the generalized valence force field and those of the Urey–Bradley force field:

*In the case of tetrahedral molecules, a term

$$\sum_{i \neq j \neq k} \left(\frac{\kappa}{\sqrt{2}}\right) r_{ij}r_{ik}(r_{ij}\Delta\alpha_{ij})(r_{ik}\Delta\alpha_{ik})$$

must be added, where κ is called the internal tension.

$$
\begin{aligned}
f_{11} &= K + t^2F' + s^2F \\
r^2f_{33} &= (H - s^2F' + t^2F)r^2 \\
f_{12} &= -t^2F' + s^2F \\
rf_{13} &= ts(F' + F)r
\end{aligned}
\qquad (14.7)
$$

Although the Urey–Bradley field has four force constants, F' is usually taken as $-\frac{1}{10}F$, on the assumption that the repulsive energy between nonbonded atoms is proportional to $1/r$.[9]* Thus, only three force constants, K, H, and F, are needed to construct the **F** matrix. The *orbital valence force* (OVF) field developed by Heath and Linnett[92] is similar to the UBF field. The OVF field uses the angle ($\Delta\beta$) which represents the distortion of the bond from the axis of the bonding orbital instead of the angle between two bonds ($\Delta\alpha$).

The number of force constants in the Urey–Bradley field is, in general, much smaller than that in the generalized valence force field. In addition, the UBF field has the advantage that (1) the force constants have clearer physical meanings than those of the GVF field, and (2) they are often transferable from molecule to molecule. For example, the force constants obtained for $SiCl_4$ and $SiBr_4$ can be used for $SiCl_3Br$, $SiCl_2Br_2$, and $SiClBr_3$. Mizushima, Shimanouchi, and their co-workers[90] and Overend and Scherer[93] have given many examples that demonstrate the transferability of the force constants in the UBF field. This property of the Urey–Bradley force constants is highly useful in calculations for complex molecules. It should be mentioned, however, that ignorance of the interactions between nonneighboring stretching vibrations and between bending vibrations in the Urey–Bradley field sometimes causes difficulties in adjusting the force constants to fit the observed frequencies. In such a case, it is possible to improve the results by introducing more force constants.[93,94]

Evidently, the values of force constants depend on the force field initially assumed. Thus, a comparison of force constants between molecules should not be made unless they are obtained by using the same force field. The normal coordinate analysis developed in Secs. I-12 to I-14 has already been applied to a number of molecules of various structures. Appendix VII lists the **G** and **F** matrix elements for typical molecules.

I-15. SOLUTION OF THE SECULAR EQUATION

Once the **G** and **F** matrices are obtained, the next step is to solve the matrix secular equation:

$$|\mathbf{GF} - \mathbf{E}\lambda| = 0 \qquad (12.6)$$

*This assumption does not cause serious error in final results, since F' is small in most cases.

In diatomic molecules, $\mathbf{G} = G_{11} = 1/\mu$ and $\mathbf{F} = F_{11} = K$. Then $\lambda = G_{11}F_{11}$ and $\bar{\nu} = \sqrt{\lambda}/2\pi c = \sqrt{K/\mu}/2\pi c$ (Eq. 3.6). If the units of mass and force constant are atomic weight and mdyn/Å (or 10^5 dyn/cm), respectively,* λ is related to $\tilde{\nu}(\text{cm}^{-1})$ by

$$\tilde{\nu} = 1302.83\sqrt{\lambda}$$

or

$$\lambda = 0.58915\left(\frac{\tilde{\nu}}{1000}\right)^2 \tag{15.1}$$

As an example, for the HF molecule $\mu = 0.9573$ and $K = 9.65$ in these units. Then, from Eqs. 3.6 and 15.1, $\tilde{\nu}$ is 4139 cm^{-1}.

The **F** and **G** matrix elements of a bent XY_2 molecule are given in Eqs. 13.13 and 13.14, respectively. The secular equation for the A_1 species is quadratic:

$$|\mathbf{GF} - \mathbf{E}\lambda| = \begin{vmatrix} G_{11}F_{11} + G_{12}F_{21} - \lambda & G_{11}F_{12} + G_{12}F_{22} \\ G_{21}F_{11} + G_{22}F_{21} & G_{21}F_{12} + G_{22}F_{22} - \lambda \end{vmatrix} = 0 \tag{15.2}$$

If this is expanded into an algebraic equation, the following result is obtained:

$$\lambda^2 - (G_{11}F_{11} + G_{22}F_{22} + 2G_{12}F_{12})\lambda + (G_{11}G_{22} - G_{12}^2)(F_{11}F_{22} - F_{12}^2) = 0 \tag{15.3}$$

For the H_2O molecule,

$$\mu_1 = \mu_H = \frac{1}{1.008} = 0.99206$$

$$\mu_3 = \mu_O = \frac{1}{15.995} = 0.06252$$

$$r = 0.96(\text{Å}), \qquad \alpha = 105^\circ$$

$$\sin\alpha = \sin 105^\circ = 0.96593$$

$$\cos\alpha = \cos 105^\circ = -0.25882$$

Then, the **G** matrix elements of Eq. 13.14 are

*Although the bond distance is involved in both the **G** and **F** matrices, it is canceled during multiplication of the **G** and **F** matrix elements. Therefore, any unit can be used for the bond distance.

$$G_{11} = \mu_1 + \mu_3(1 + \cos\alpha) = 1.03840$$

$$G_{12} = -\frac{\sqrt{2}}{r}\mu_3 \sin\alpha = -0.08896$$

$$G_{22} = \frac{1}{r^2}[2\mu_1 + 2\mu_3(1 - \cos\alpha)] = 2.32370$$

If the force constants in terms of the generalized valence force field are selected as

$$f_{11} = 8.4280, \qquad f_{12} = -0.1050$$
$$f_{13} = 0.2625, \qquad f_{33} = 0.7680$$

the **F** matrix elements of Eq. 13.13 are

$$F_{11} = f_{11} + f_{12} = 8.32300$$
$$F_{12} = \sqrt{2}rf_{13} = 0.35638$$
$$F_{22} = r^2 f_{33} = 0.70779$$

Using these values, we find that Eq. 15.3 becomes

$$\lambda^2 - 10.22389\lambda + 13.86234 = 0$$

The solution of this equation gives

$$\lambda_1 = 8.61475, \qquad \lambda_2 = 1.60914$$

If these values are converted to $\tilde{\nu}$ through Eq. 15.1, we obtain

$$\tilde{\nu}_1 = 3824 \text{ cm}^{-1}, \qquad \tilde{\nu}_2 = 1653 \text{ cm}^{-1}$$

With the same set of force constants, the frequency of the B_2 vibration is calculated as

$$\lambda_3 = G_{33}F_{33} = [\mu_1 + \mu_3(1 - \cos\alpha)](f_{11} - f_{12})$$
$$= 9.13681$$
$$\tilde{\nu}_3 = 3938 \text{ cm}^{-1}$$

The observed frequencies corrected for anharmonicity are as follows: $\omega_1 = 3825$ cm^{-1}, $\omega_2 = 1654$ cm^{-1}, and $\omega_3 = 3936$ cm^{-1}.

If the secular equation is third order, it gives rise to a cubic equation:

$$
\begin{aligned}
&\lambda^3 - (G_{11}F_{11} + G_{22}F_{22} + G_{33}F_{33} + 2G_{12}F_{12} + 2G_{13}F_{13} + 2G_{23}F_{23})\lambda^2 \\
&+ \left\{ \begin{vmatrix} G_{11} & G_{12} \\ G_{21} & G_{22} \end{vmatrix} \begin{vmatrix} F_{11} & F_{12} \\ F_{21} & F_{22} \end{vmatrix} + \begin{vmatrix} G_{12} & G_{13} \\ G_{22} & G_{23} \end{vmatrix} \begin{vmatrix} F_{12} & F_{13} \\ F_{22} & F_{23} \end{vmatrix} \right. \\
&+ \begin{vmatrix} G_{11} & G_{13} \\ G_{21} & G_{23} \end{vmatrix} \begin{vmatrix} F_{11} & F_{13} \\ F_{21} & F_{23} \end{vmatrix} + \begin{vmatrix} G_{11} & G_{12} \\ G_{31} & G_{32} \end{vmatrix} \begin{vmatrix} F_{11} & F_{12} \\ F_{31} & F_{32} \end{vmatrix} \\
&+ \begin{vmatrix} G_{12} & G_{13} \\ G_{32} & G_{33} \end{vmatrix} \begin{vmatrix} F_{12} & F_{13} \\ F_{32} & F_{33} \end{vmatrix} + \begin{vmatrix} G_{11} & G_{13} \\ G_{31} & G_{33} \end{vmatrix} \begin{vmatrix} F_{11} & F_{13} \\ F_{31} & F_{33} \end{vmatrix} \\
&+ \begin{vmatrix} G_{21} & G_{22} \\ G_{31} & G_{32} \end{vmatrix} \begin{vmatrix} F_{21} & F_{22} \\ F_{31} & F_{32} \end{vmatrix} + \begin{vmatrix} G_{22} & G_{23} \\ G_{32} & G_{33} \end{vmatrix} \begin{vmatrix} F_{22} & F_{23} \\ F_{32} & F_{33} \end{vmatrix} \\
&\left. + \begin{vmatrix} G_{21} & G_{23} \\ G_{31} & G_{33} \end{vmatrix} \begin{vmatrix} F_{21} & F_{23} \\ F_{31} & F_{33} \end{vmatrix} \right\} \lambda - \begin{vmatrix} G_{11} & G_{12} & G_{13} \\ G_{21} & G_{22} & G_{23} \\ G_{31} & G_{32} & G_{33} \end{vmatrix} \begin{vmatrix} F_{11} & F_{12} & F_{13} \\ F_{21} & F_{22} & F_{23} \\ F_{31} & F_{32} & F_{33} \end{vmatrix} = 0
\end{aligned} \tag{15.4}
$$

Thus, it is possible to solve the secular equation by expanding it into an algebraic equation. If the order of the secular equation is higher than three, however, direct expansion such as that just shown becomes too cumbersome. There are several methods of calculating the coefficients of an algebraic equation using indirect expansion.[3] The use of an electronic computer greatly reduces the burden of calculation. Excellent programs written by Schachtschneider[95] and other workers are available for the vibrational analysis of polyatomic molecules.

I-16. VIBRATIONAL FREQUENCIES OF ISOTOPIC MOLECULES

As stated in Sec. I-14, the vibrational frequencies of isotopic molecules are very useful in refining a set of force constants in vibrational analysis. For large molecules, isotopic substitution is indispensable in making band assignments, since only vibrations involving the motion of the isotopic atom will be shifted by isotopic substitution.

Two important rules hold for the vibrational frequencies of isotopic molecules. The first, called the *product rule*, can be derived as follows.

Let $\lambda_1, \lambda_2, \cdots, \lambda_n$ be the roots of the secular equation $|\mathbf{GF} - \mathbf{E}\lambda| = 0$. Then,

$$\lambda_1\lambda_2 \cdots \lambda_n = |\mathbf{G}||\mathbf{F}| \tag{16.1}$$

holds for a given molecule. Since the isotopic molecule has exactly the same $|\mathbf{F}|$ as that in Eq. 16.1, a similar relation

$$\lambda_1'\lambda_2' \cdots \lambda_n' = |\mathbf{G}'||\mathbf{F}|$$

holds for this molecule. It follows that

$$\frac{\lambda_1\lambda_2 \cdots \lambda_n}{\lambda_1'\lambda_2' \cdots \lambda_n'} = \frac{|\mathbf{G}|}{|\mathbf{G}'|} \tag{16.2}$$

Since

$$\tilde{\nu} = \frac{1}{2\pi c}\sqrt{\lambda}$$

Eq. 16.2 can be written as

$$\frac{\tilde{\nu}_1\tilde{\nu}_2 \cdots \tilde{\nu}_n}{\tilde{\nu}_1'\tilde{\nu}_2' \cdots \tilde{\nu}_n'} = \sqrt{\frac{|\mathbf{G}|}{|\mathbf{G}'|}} \tag{16.3}$$

This rule has been confirmed by using pairs of molecules such as H_2O and D_2O and CH_4 and CD_4. The rule is also applicable to the product of vibrational frequencies belonging to a single symmetry species.

A more general form of Eq. 16.3 is given by the *Redlich–Teller product rule*[1]:

$$\frac{\tilde{\nu}_1\tilde{\nu}_2 \cdots \tilde{\nu}_n}{\tilde{\nu}_1'\tilde{\nu}_2' \cdots \tilde{\nu}_n'} = \sqrt{\left(\frac{m_1'}{m_1}\right)^{\alpha}\left(\frac{m_2'}{m_2}\right)^{\beta} \cdots \left(\frac{M}{M'}\right)^{t}\left(\frac{I_x}{I_x'}\right)^{\delta_x}\left(\frac{I_y}{I_y'}\right)^{\delta_y}\left(\frac{I_z}{I_z'}\right)^{\delta_z}} \tag{16.4}$$

Here $m_1, m_2, \ldots$ are the masses of the representative atoms of the various sets of equivalent nuclei (atoms represented by m, m_0, $m_{xy}, \ldots$ in the tables given in Appendix III); $\alpha, \beta, \ldots$ are the coefficients of m, m_0, $m_{xy}, \ldots$; M is the total mass of the molecule; t is the number of T_x, T_y, T_z in the symmetry type considered; I_x, I_y, I_z are the moments of inertia about the x, y, z axes, respectively, which go through the center of the mass; and δ_x, δ_y, δ_z are 1 to 0, depending on whether or not R_x, R_y, R_z belong to the symmetry type considered. A degenerate vibration is counted only once on both sides of the equation.

Another useful rule in regard to the vibrational frequencies of isotopic molecules, called the *sum rule*, can be derived as follows. It is obvious from

Eqs. 15.3 and 15.4 that

$$\lambda_1 + \lambda_2 + \cdots + \lambda_n = \sum_n \lambda = \sum_{i,j} G_{ij} F_{ij} \tag{16.5}$$

Let σ_k denote $\Sigma_{ij} G_{ij} F_{ij}$ for k different isotopic molecules, all of which have the same **F** matrix. If a suitable combination of molecules is taken, so that

$$\begin{aligned}\sigma_1 + \sigma_2 + \cdots + \sigma_k &= \left(\sum G_{ij} F_{ij}\right)_1 + \left(\sum G_{ij} F_{ij}\right)_2 + \cdots + \left(\sum G_{ij} F_{ij}\right)_k \\ &= \left[\left(\sum G_{ij}\right)_1 + \left(\sum G_{ij}\right)_2 + \cdots + \left(\sum G_{ij}\right)_k\right]\left(\sum F_{ij}\right) \\ &= 0\end{aligned}$$

then it follows that

$$\left(\sum \lambda\right)_1 + \left(\sum \lambda\right)_2 + \cdots + \left(\sum \lambda\right)_k = 0 \tag{16.6}$$

This rule has been verified for such combinations as H_2O, D_2O, and HDO, where

$$2\sigma(\text{HDO}) - \sigma(H_2O) - \sigma(D_2O) = 0$$

Such relations between the frequencies of isotopic molecules are highly useful in making band assignments.

I-17. METAL–ISOTOPE SPECTROSCOPY[96,97]

As a first approximation, vibrational spectra of coordination compounds (Section III of Part B) can be classifed into ligand vibrations which occur in the high-frequency region (4000–600 cm^{-1}) and metal–ligand vibrations which appear in the low-frequency region (below 600 cm^{-1}). The former provide information about the effect of coordination on the electronic structure of the ligand while the latter provide direct information about the structure of the coordination sphere and the nature of the metal–ligand bond. Since the main interest of coordination chemistry is the coordinate bond, it is the metal–ligand vibrations that have held the interest of inorganic vibrational spectroscopists. It is difficult, however, to make unequivocal assignments of metal–ligand vibrations since the interpretation of the low-frequency spectrum is complicated by the appearance of ligand vibrations as well as lattice vibrations in the case of solid-state spectra.

Conventional methods which have been used to assign metal–ligand vibrations are:

1. Comparison of spectra between a free ligand and its metal complex; the metal–ligand vibration should be absent in the spectrum of the free ligand. This method often fails to give a clear-cut assignment because some ligand vibrations activated by complex formation may appear in the same region as the metal–ligand vibrations.

2. The metal–ligand vibration should be metal sensitive and be shifted by changing the metal or its oxidation state. This method is applicable only when a series of metal complexes have exactly the same structure, with only the central metal being different. Also, it does not provide definitive assignments since some ligand vibrations (such as chelate ring deformations) are also metal sensitive.

3. The metal–ligand stretching vibration should appear in the same frequency region if the metal is the same and the ligands are similar. For example, the ν(Zn-N) (ν: stretching) of Zn(II) pyridine complexes are expected to be similar to those of Zn(II) α-picoline complexes. This method is applicable only when the metal–ligand vibration is known for one parent compound.

4. The metal–ligand vibration exhibits an isotope shift if the ligand is isotopically substituted. For example, the ν(Ni-N) of $[Ni(NH_3)_6]Cl_2$ at 334 cm^{-1} is shifted to 318 cm^{-1} upon deuteration of the ammonia ligands. The observed shift (16 cm^{-1}) is in good agreement with that predicted theoretically for this mode. This method was used to assign the metal–ligand vibrations of chelate compounds such as oxamido ($^{14}N/^{15}N$) and acetylacetonato($^{16}O/^{18}O$) complexes. However, isotopic substitution of the α-atom (atom directly bonded to the metal) causes shifts of not only metal–ligand vibrations but also of ligand vibrations involving the motion of the α-atom. Thus, this method alone cannot provide an unequivocal assignment of the metal–ligand vibration.

5. The frequency of a metal–ligand vibration may be predicted if the metal–ligand stretching and other force constants are known *a priori*. At present, this method is not practical since only a very limited amount of information is available on the force constants of coordination compounds.

It is obvious that none of the above methods is perfect in assigning metal–ligand vibrations. Furthermore, these methods encounter more difficulties as the structure of the complex (and hence the spectrum) becomes more complicated. Fortunately, the "metal isotope technique" which was developed in 1969 may be used to obtain reliable metal–ligand assignments.[98] Isotope pairs such as (H/D) and ($^{16}O/^{18}O$) had been used routinely by many spectroscopists. However, isotopic pairs of heavy metals such as ($^{58}Ni/^{62}Ni$) and ($^{104}Pd/^{110}Pd$) were not employed until 1969 when the first report on the assignments of the Ni-P vibrations of *trans*-$Ni(PEt_3)_2X_2$ (X = Cl and Br) was made. The delay in their use was probably due to two reasons:

1. It was thought that the magnitude of isotope shifts arising from metal isotope substitution might be too small to be of practical value.

2. Pure metal isotopes were too expensive to use routinely in the laboratory. Nakamoto and co-workers[96,97] have shown, however, that the magnitudes of metal isotope shifts are generally of the order of 2–10 cm^{-1} for stretching modes and 0–2 cm^{-1} for bending modes, and that the experimental error in measuring the frequency could be as small as ±0.2 cm^{-1} if proper precautions are taken. They have also shown that this technique is financially feasible if the compounds are prepared on a milligram scale. Normally, the vibrational spectrum of a compound can be obtained with a sample less than 10 mg. Table I-14 lists metal isotopes which are useful for vibrational studies of coordination compounds.

TABLE I-14. Some Stable Metal Isotopes[a]

Element	Inventory Form	Isotope	Natural Abundance (%)	Purity (%)
Silicon	Oxide	Si-28	92.23	>99.8
		Si-30	3.10	>94
Titanium	Oxide	Ti-46	8.0	70–96
		Ti-48	73.8	>99
Chromium	Oxide	Cr-50	4.345	>95
		Cr-52	83.79	>99.7
		Cr-53	9.50	>96
Iron	Oxide	Fe-54	5.9	>96
		Fe-56	91.72	>99.9
		Fe-57	2.1	86–93
Nickel	Metal	Ni-58	68.27	>99.9
		Ni-60	26.1	>99
		Ni-62	3.59	>96
Copper	Oxide	Cu-63	69.17	>99.8
		Cu-65	30.83	>99.6
Zinc	Oxide	Zn-64	48.6	>99.8
		Zn-66	27.9	>98
		Zn-68	18.8	97–99
Germanium	Oxide	Ge-72	27.4	>97
		Ge-74	36.5	>98
Zirconium	Oxide	Zr-90	51.45	97–99
		Zr-94	17.38	>98
Molybdenum	Metal/Oxide	Mo-92	14.84	>97
		Mo-95	15.92	>96
		Mo-97	9.55	>92
		Mo-100	9.63	>97
Ruthenium	Metal	Ru-99	12.7	>98
		Ru-102	31.6	>99
Palladium	Metal	Pd-104	11.4	>95
		Pd-108	26.46	>98
Tin	Oxide	Sn-116	14.53	>95
		Sn-120	32.59	>98
		Sn-124	5.79	>94
Barium	Carbonate	Ba-135	6.593	>93
		Ba-138	71.70	>99

[a]Available at Oak Ridge National Laboratory.

It should be noted that the central atom of a highly symmetrical molecule ($\mathbf{T}_d$, $\mathbf{O}_h$, etc.) does not move during the totally symmetric vibration. Thus, no metal–isotope shifts are expected in these cases.[99] When the central atom is coordinated by several different donor atoms, multiple isotope labeling is necessary to distinguish different coordinate bond-stretching vibrations. For example, complete assignments of bis(glycino)nickel(II) require $^{14}N/^{15}N$ and/or $^{16}O/^{18}O$ isotope shift data as well as $^{58}Ni/^{62}Ni$ isotope shift data.[100] This book contains many other applications of metal–isotope techniques. These results show that metal–isotope data are indispensable not only in assigning the metal–ligand vibrations but also in refining metal–ligand stretching force constants in normal coordinate analysis.[97] The presence of vibrational coupling between metal–ligand and other vibrations can also be detected by combining metal–isotope data with normal coordinate calculations (Sec. I-18) since both experimental and theoretical isotope shift values would be smaller when such couplings occur.

The metal–isotope techniques become more important as the molecules become larger and complex. In biological molecules such as heme proteins, structural and bonding information about the active site (iron porphyrin) can be obtained through definitive assignments of coordinate bond-stretching vibrations around the iron atom. Using resonance Raman techniques (Sec. I-22), it is possible to observe iron porphyrin and iron-axial ligand vibrations without interference from peptide chain vibrations. Thus, these vibrations can be assigned by comparing resonance Raman spectra of a natural heme protein with that of a ^{54}Fe-reconstituted heme protein. Section V of Part B includes several examples of applications of metal–isotope techniques to bioinorganic compounds [hemoglobin (^{54}Fe, ^{57}Fe), oxy-hemocyanin (^{63}Cu, ^{65}Cu), etc.]. An example of a multiple labeling is seen in the case of cytochrome P-450$_{cam}$ having the axial Fe-S linkage in addition to four Fe-N (porphyrin) bonds; Champion et al.[101] were able to assign its Fe-S stretching vibration (351 cm^{-1}) by combining ^{32}S-^{34}S and ^{54}Fe-^{56}Fe isotope shift data.

I-18. GROUP FREQUENCIES AND BAND ASSIGNMENTS

From observation of the infrared spectra of a number of compounds having a common group of atoms, it is found that, regardless of the rest of the molecule, this common group absorbs over a narrow range of frequencies, called the *group frequency*. For example, the methyl group exhibits the antisymmetric stretching [$\nu_2(CH_3)$], symmetric stretching [$\nu_s(CH_3)$], degenerate bending [$\delta_d(CH_3)$], symmetric bending [$\delta_s(CH_3)$], and rocking [$\rho_r(CH_3)$] vibrations in the ranges 3050–2950, 2970–2860, 1480–1410, 1385–1250, and 1200–800 cm^{-1}, respectively. Figure I-25 illustrates their vibrational modes together with the observed frequencies for CH_3Cl. The methylene group vibrations are also well known; the antisymmetric stretching [$\nu_a(CH_2)$], symmetric stretching [$\nu_s(CH_2)$], scissoring [$\delta(CH_2)$], wagging [$\rho_w(CH_2)$], twisting [$\rho_t(CH_2)$],

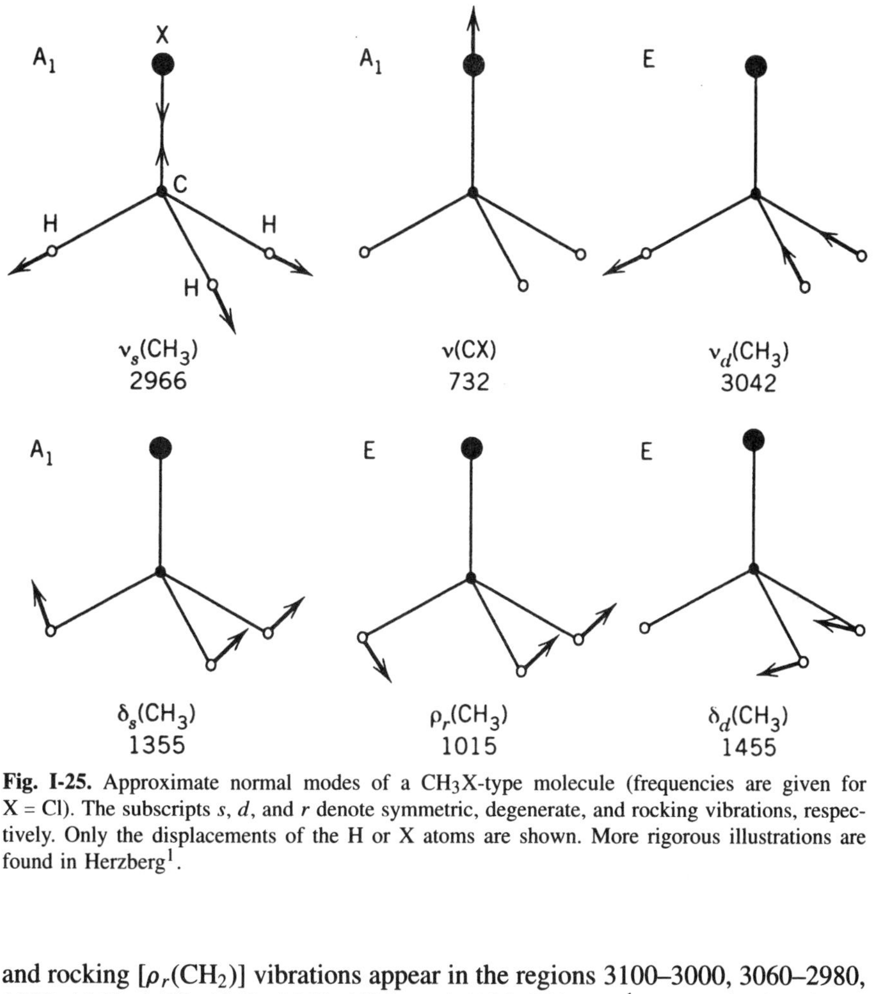

Fig. I-25. Approximate normal modes of a CH_3X-type molecule (frequencies are given for X = Cl). The subscripts *s*, *d*, and *r* denote symmetric, degenerate, and rocking vibrations, respectively. Only the displacements of the H or X atoms are shown. More rigorous illustrations are found in Herzberg[1].

and rocking [$\rho_r(CH_2)$] vibrations appear in the regions 3100–3000, 3060–2980, 1500–1350, 1400–1100, 1260–1030, and 1180–750 cm^{-1}, respectively. Figure I-26 illustrates the normal modes of these vibrations together with the observed frequencies for CH_2Cl_2. The normal modes illustrated for CH_3Cl and CH_2Cl_2 are applicable to vibrational assignments of coordination and organometallic compounds, which will be discussed in Sections III and IV of Part B, respectively. Group frequencies have been found for a number of organic and inorganic groups, and they have been summarized as *group frequency charts*,[40–42] which are highly useful in identifying the atomic groups from infrared spectra. Group frequency charts for inorganic compounds are given in Appendix VIII as well as in Figs. II-55 and II-61.

The concept of group frequency rests on the assumption that the vibrations of a particular group are relatively independent of those of the rest of the molecule. As stated in Sec. I-4, however, all the nuclei of the molecule perform their harmonic oscillations in a normal vibration. Thus, an *isolated vibration*, which the

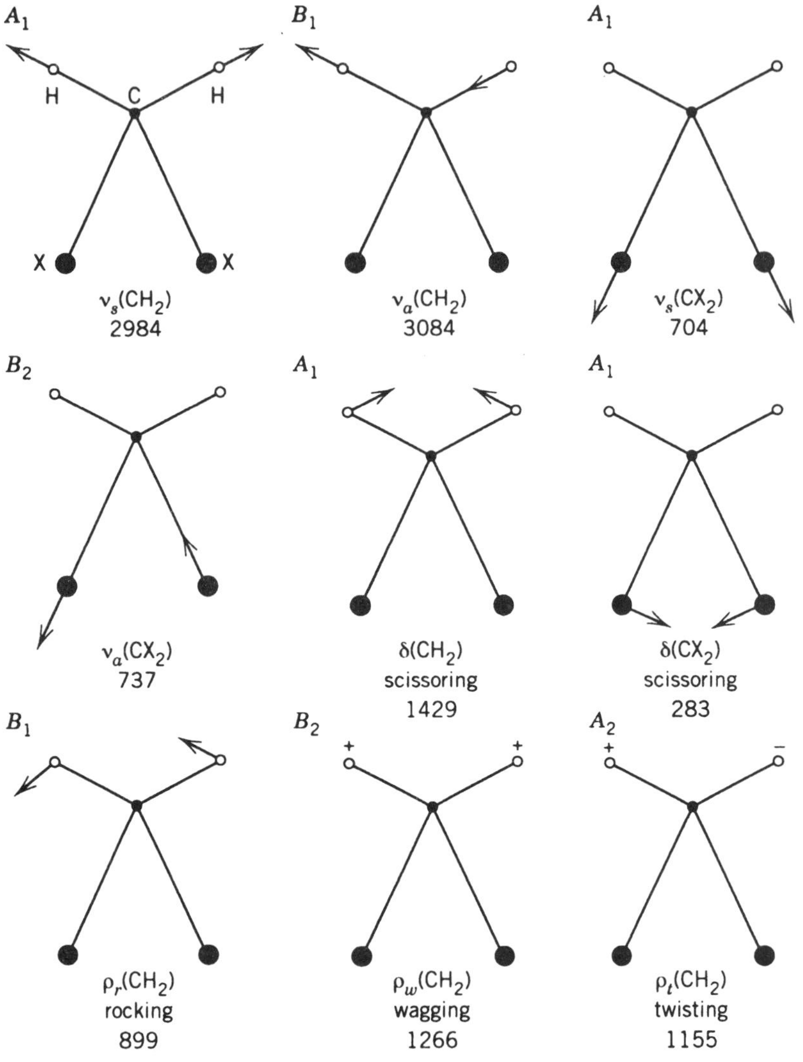

Fig. I-26. Approximate normal modes of a CH_2X_2-type molecule (frequencies are given for X = Cl). Only the displacements of the H or X atoms are shown. For more rigorous illustrations, see Herzberg[1].

group frequency would have to be, cannot be expected in polyatomic molecules. If, however, a group includes relatively light atoms such as hydrogen (OH, NH, NH_2, CH, CH_2, CH_3, etc.) or relatively heavy atoms such as the halogens (CCl, CBr, CI, etc.), as compared to other atoms in the molecule, the idea of an isolated vibration may be justified, since the amplitudes (or velocities) of the harmonic oscillation of these atoms are relatively larger or smaller than those of the other atoms in the same molecule. Vibrations of groups having multiple bonds

(C ≡C, C ≡N, C=C, C=N, C=O, etc.) may also be relatively independent of the rest of the molecule if the groups do not belong to a conjugated system.

If atoms of similar mass are connected by bonds of similar strength (force constant), the amplitude of oscillation is similar for each atom of the whole system. Therefore, it is not possible to isolate the group frequencies in a system like the following:

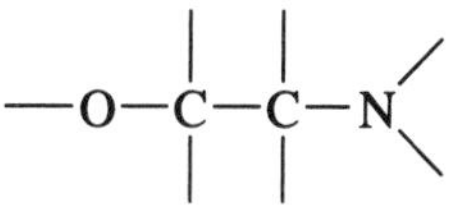

A similar situation may occur in a system in which resonance effects average out the single and multiple bonds by conjugation. Examples of this effect are seen for the metal chelate compounds discussed in Section III of Part B. When the group frequency approximation is permissible, the mode of vibration corresponding to this frequency can be inferred empirically from the band assignments obtained theoretically for simple molecules. If *coupling* between various group vibrations is serious, it is necessary to make a theoretical analysis for each individual compound, using a method like the following one.

As stated in Sec. I-4, the generalized coordinates are related to the normal coordinates by

$$q_k = \sum_i B_{ki} Q_i \tag{4.10}$$

In matrix form, this is written as

$$\mathbf{q} = \mathbf{B}_q \mathbf{Q} \tag{18.1}$$

It can be shown[3] that the internal coordinates are also related to the normal coordinates by

$$\mathbf{R} = \mathbf{L}\mathbf{Q} \tag{18.2}$$

This is written more explicitly as

$$\begin{aligned} R_1 &= l_{11}Q_1 + l_{12}Q_2 + \cdots + l_{1N}Q_N \\ R_2 &= l_{21}Q_1 + l_{22}Q_2 + \cdots + l_{2N}Q_N \\ &\vdots \qquad\qquad \vdots \\ R_i &= l_{i1}Q_1 + l_{i2}Q_2 + \cdots + l_{iN}Q_N \end{aligned} \tag{18.3}$$

In a normal vibration in which the normal coordinate Q_N changes with frequency ν_N, all the internal coordinates, R_1, $R_2, \cdots, R_i$, change with the same frequency. The amplitude of oscillation is, however, different for each internal coordinate. The relative ratio of the amplitudes of the internal coordinates in a normal vibration associated with Q_N is given by

$$l_{1N} : l_{2N} : \cdots : l_{iN} \tag{18.4}$$

If one of these elements is relatively large compared to the others, the normal vibration is said to be predominantly due to the vibration caused by the change of this coordinate.

The ratio of l's given by Eq. 18.4 can be obtained as a column matrix (or eigenvector) $\boldsymbol{l}_N$, which satisfies the relation[3,4]

$$\mathbf{GF}\boldsymbol{l}_N = \boldsymbol{l}_N \lambda_N \tag{18.5}$$

It consists of i elements, $l_{1N}, l_{2N}, \cdots, l_{iN}$, i being the number of internal coordinates, and can be calculated if the **G** and **F** matrices are known. An assembly by columns of the l elements obtained for each λ gives the relation

$$\mathbf{GFL} = \mathbf{L\Lambda} \tag{18.6}$$

where $\mathbf{\Lambda}$ is a diagonal matrix whose elements consist of λ values.

As an example, calculate the **L** matrix of the H_2O molecule, using the results obtained in Sec. I-15. The **G** and **F** matrices for the A_1 species are as follows:

$$\mathbf{G} = \begin{bmatrix} 1.03840 & -0.08896 \\ -0.08896 & 2.32370 \end{bmatrix}, \qquad \mathbf{F} = \begin{bmatrix} 8.32300 & 0.35638 \\ 0.35638 & 0.70779 \end{bmatrix}$$

with $\lambda_1 = 8.61475$ and $\lambda_2 = 1.60914$. The **GF** product becomes

$$\mathbf{GF} = \begin{bmatrix} 8.61090 & 0.30710 \\ 0.08771 & 1.61299 \end{bmatrix}$$

The **L** matrix can be calculated from Eq. 18.6:

$$\begin{bmatrix} 8.61090 & 0.30710 \\ 0.08771 & 1.61299 \end{bmatrix} \begin{bmatrix} l_{11} & l_{12} \\ l_{21} & l_{22} \end{bmatrix} = \begin{bmatrix} l_{11} & l_{12} \\ l_{21} & l_{22} \end{bmatrix} \begin{bmatrix} 8.61475 & 0 \\ 0 & 1.60914 \end{bmatrix}$$

However, this equation gives only the ratios $l_{11} : l_{21}$ and $l_{12} : l_{22}$. To determine their values, it is necessary to use the following normalization condition:

$$\mathbf{L}\tilde{\mathbf{L}} = \mathbf{G} \tag{18.7)*}$$

Then the final result is

$$\begin{bmatrix} l_{11} & l_{12} \\ l_{21} & l_{22} \end{bmatrix} = \begin{bmatrix} 1.01683 & -0.06686 \\ 0.01274 & 1.52432 \end{bmatrix}$$

This result indicates that, in the normal vibration Q_1, the relative ratio of amplitudes of two internal coordinates, R_1 (symmetric OH stretching) and R_2 (HOH bending), is 1.0168 : 0.0127. Therefore, this vibration (3824 cm^{-1}) is assigned to an almost pure OH stretching mode. The relative ratio of amplitudes for the Q_2 vibration is -0.0669 : 1.5243. Thus, this vibration is assigned to an almost pure HOH bending mode. In other cases, the l values do not provide the band assignments that are expected empirically. This occurs because the dimension of l for a stretching coordinate is different from that for a bending coordinate.

The potential energy due to a normal vibration, Q_N is written as

$$V(Q_N) = \frac{1}{2}\, Q_N^2 \sum_{ij} F_{ij}\, l_{iN}\, l_{jN} \tag{18.8)†}$$

Individual $F_{ij}l_{iN}l_{jN}$ terms in the summation represent the potential energy distribution (PED) in Q_N which gives a better measure for making band assignments.[102] In general, the value of $F_{ij}l_{iN}l_{jN}$ is large when $i = j$. Therefore, the $F_{ii}l_{iN}^2$ terms are most important in determining the distribution of the potential energy. Thus, the ratios of the $F_{ii}l_{iN}^2$ terms provide a measure of the relative contribution of each internal coordinate R_i to the normal coordinate Q_N. If any $F_{ii}l_{iN}^2$ term is exceedingly large compared with the others, the vibration is assigned to the mode associated with R_i. If $F_{ii}l_{iN}^2$ and $F_{jj}l_{jN}^2$ are relatively

*This equation can be derived as follows. According to Eq. 12.3, $2T = \tilde{\dot{\mathbf{R}}}\mathbf{G}^{-1}\dot{\mathbf{R}}$. On the other hand, Eq. 18.2 gives $\dot{\mathbf{R}} = \mathbf{L}\dot{\mathbf{Q}}$ and $\tilde{\dot{\mathbf{R}}} = \tilde{\dot{\mathbf{Q}}}\tilde{\mathbf{L}}$. Thus $2T = \tilde{\dot{\mathbf{Q}}}\tilde{\mathbf{L}}\mathbf{G}^{-1}\mathbf{L}\dot{\mathbf{Q}}$. Comparing this with $2T = \tilde{\dot{\mathbf{Q}}}\mathbf{E}\dot{\mathbf{Q}}$ (matrix form of Eq. 4.11), we obtain $\tilde{\mathbf{L}}\mathbf{G}^{-1}\mathbf{L} = \mathbf{E}$ or $\mathbf{L}\tilde{\mathbf{L}} = \mathbf{G}$.

† According to Eq. 12.1, the potential energy is written as $2V = \tilde{\mathbf{R}}\mathbf{F}\mathbf{R}$. Using Eq. 18.2, we can write this as $2V = \tilde{\mathbf{Q}}\tilde{\mathbf{L}}\mathbf{F}\mathbf{L}\mathbf{Q}$. On the other hand, Eq. 4.12 can be written as $2V = \tilde{\mathbf{Q}}\mathbf{\Lambda}\mathbf{Q}$. A comparison of these two expressions gives $\mathbf{\Lambda} = \tilde{\mathbf{L}}\mathbf{F}\mathbf{L}$. If this is written for one normal vibration whose frequency is λ_N, we have

$$\lambda_N = \sum_{ij} \tilde{l}_{Ni} F_{ij}\, l_{jN} = \sum_{ij} F_{ij}\, l_{iN}\, l_{jN}$$

Then the potential energy due to this vibration is expressed by Eq. 18.8.

large compared with the others, the vibration is assigned to a mode associated with both R_i and R_j (coupled vibration).

As an example, let us calculate the potential energy distribution for the H_2O molecule. Using the **F** and **L** matrices obtained previously, we find that the $\mathbf{\tilde{L}FL}$ matrix is calculated to be

$$\begin{bmatrix} \begin{pmatrix} l_{11}^2F_{11} & + & l_{21}^2F_{22} & + & 2l_{21}l_{11}F_{12} \\ 8.60551 & & 0.00011 & & 0.00923 \end{pmatrix} & 0 \\ 0 & \begin{pmatrix} l_{12}^2F_{11} & + & l_{22}^2F_{22} & + & 2l_{12}l_{22}F_{12} \\ 0.03721 & & 1.64459 & & -0.07264 \end{pmatrix} \end{bmatrix}$$

Then, the potential energy distribution in each normal vibration ($F_{ii}l_{iN}^2$) is given by

$$\begin{matrix} & \lambda_1 & \lambda_2 \\ \begin{matrix} R_1 \\ R_2 \end{matrix} & \multicolumn{2}{c}{\begin{bmatrix} 8.60551 & 0.03721 \\ 0.00011 & 1.64459 \end{bmatrix}} \end{matrix}$$

More conveniently, PED is expressed by calculating $(F_{ii}l_{iN}^2/\Sigma F_{ii}l_{iN}^2) \times 100$ for each coordinate:

$$\begin{matrix} & \lambda_1 & \lambda_2 \\ \begin{matrix} R_1 \\ R_2 \end{matrix} & \multicolumn{2}{c}{\begin{bmatrix} 99.99 & 2.21 \\ 0.01 & 97.79 \end{bmatrix}} \end{matrix}$$

In this case, the final results are the same whether the band assignments are based on the **L** matrix or on the potential energy distribution: Q_1 is the symmetric OH stretching and Q_2 is the HOH bending. In other cases, different results may be obtained, depending on which criterion is used for band assignments.

In the example above, no serious coupling occurs between the OH stretching and HOH bending modes of H_2O in the A_1 species. This is because the vibrational frequencies of these two modes are far apart. In other cases, however, vibrational frequencies of two or more modes in the same symmetry species are relatively close to each other. Then, the l_{iN} values for these modes become comparable.

A more rigorous method of determining the vibrational mode involves drawing the displacements of individual atoms in terms of rectangular coordinates. As in Eq. 18.2, the relationship between the rectangular and normal coordinates is given by

$$\mathbf{X} = \mathbf{L}_x\mathbf{Q} \tag{18.9}$$

The $\mathbf{L}_x$ matrix can be obtained from the relationship[103]

$$\mathbf{L}_x = \mathbf{M}^{-1}\tilde{\mathbf{B}}\mathbf{G}^{-1}\mathbf{L} \tag{18.10)*$$

The matrices on the right-hand sides have already been defined.

Three-dimensional drawings of normal modes, such as those shown in Part II, can be made from the cartesian displacement calculations obtained above. However, hand plotting of these data is laborious and complicated. Use of computer plotting programs greatly facilitates this process.[104]

I-19. INTENSITY OF INFRARED ABSORPTION[63]

The absorption of strictly monochromatic light (ν) is expressed by the Lambert–Beer law:

$$I_\nu = I_{0,\nu} e^{-\alpha_\nu p l} \tag{19.1}$$

where I_ν is the intensity of the light transmitted by a cell of length l containing a gas at pressure p, $I_{0,\nu}$ is the intensity of the incident light, and α_ν is the absorption coefficient for unit pressure. The true integrated absorption coefficient A is defined by

$$A = \int_{\text{band}} \alpha_\nu \, d\nu = \frac{1}{pl} \int_{\text{band}} \ln\left(\frac{I_{0,\nu}}{I_\nu}\right) d\nu \tag{19.2}$$

where the integration is carried over the entire frequency region of a band.

In practice, I_ν and $I_{0,\nu}$ cannot be measured accurately, since no spectrophotometers have infinite resolving power. Therefore we measure instead the apparent intensity T_ν:

$$T_\nu = \int_{\text{slit}} I(\nu) g(\nu, \nu') \, d\nu \tag{19.3}$$

where $g(\nu, \nu')$ is a function indicating the amount of light of frequency ν when the spectrophotometer reading is set at ν'. Then the apparent integrated absorption coefficient B is defined by

*By combining Eqs. 12.8 and 18.9, we have $\mathbf{R} = \mathbf{B}\mathbf{X} = \mathbf{B}\mathbf{L}_x\mathbf{Q}$. Since $\mathbf{R} = \mathbf{L}\mathbf{Q}$ (Eq. 18.2), it follows that $\mathbf{L}\mathbf{Q} = \mathbf{B}\mathbf{L}_x\mathbf{Q}$ or $\mathbf{L} = \mathbf{B}\mathbf{L}_x$. The kinetic energy is written as $2T = \tilde{\dot{\mathbf{X}}}\mathbf{M}\dot{\mathbf{X}}$. In terms of internal coordinates, it is written as $2T = \tilde{\dot{\mathbf{R}}}\mathbf{G}^{-1}\dot{\mathbf{R}} = \tilde{\dot{\mathbf{X}}}\tilde{\mathbf{B}}\mathbf{G}^{-1}\mathbf{B}\dot{\mathbf{X}}$. By comparing these two expressions, we have $\mathbf{M} = \tilde{\mathbf{B}}\mathbf{G}^{-1}\mathbf{B}$. Then we can write $\mathbf{L}_x = \mathbf{M}^{-1}\mathbf{M}\mathbf{L}_x = \mathbf{M}^{-1}\tilde{\mathbf{B}}\mathbf{G}^{-1}\mathbf{B}\mathbf{L}_x = \mathbf{M}^{-1}\tilde{\mathbf{B}}\mathbf{G}^{-1}\mathbf{L}$.

$$B = \frac{1}{pl} \int_{\text{band}} \ln \frac{\int_{\text{slit}} I_0(\nu) g(\nu, \nu')\, d\nu}{\int_{\text{slit}} I(\nu) g(\nu, \nu')\, d\nu}\, d\nu' \tag{19.4}$$

It can be shown that

$$\lim_{pl \to 0} (A - B) = 0 \tag{19.5}$$

if I_0 and α_ν are constant within the slit width used. (This condition is approximated by using a narrow slit.) In practice, we plot B/pl against pl, and extrapolate the curve to $pl \to 0$. To apply this method to gaseous molecules, it is necessary to broaden the vibrational–rotational bands by adding a high-pressure inert gas (pressure broadening).

For liquids and solutions, p and α in the preceding equations are replaced by M (molar concentration) and ε (molar absorption coefficient), respectively. However, the extrapolation method just described is not applicable, since experimental errors in determining B values become too large at low concentration or at small cell length. The true integrated absorption coefficient of a liquid can be calculated if we assume that the shape of an absorption band is represented by the Lorentz equation and that the slit function is triangular.[105]

Theoretically, the true integrated absorption coefficient A_N of the Nth normal vibration is given by[3]

$$A_N = \frac{n\pi}{3c} \left[\left(\frac{\partial \mu_x}{\partial Q_N} \right)_0^2 + \left(\frac{\partial \mu_y}{\partial Q_N} \right)_0^2 + \left(\frac{\partial \mu_z}{\partial Q_N} \right)_0^2 \right] \tag{19.6}$$

where n is the number of molecules per cubic centimeter, and c is the velocity of light. As shown by Eq. 18.2, an internal coordinate R_i is related to a set of normal coordinates by

$$R_i = \sum_N L_{iN} Q_N \tag{19.7}$$

If the additivity of the bond dipole moment is assumed, it is possible to write

$$\begin{aligned} \frac{\partial \mu}{\partial Q_N} &= \sum_i \left(\frac{\partial \mu}{\partial R_i} \right) \left(\frac{\partial R_i}{\partial Q_N} \right) \\ &= \sum_i \left(\frac{\partial \mu}{\partial R_i} \right) L_{iN} \end{aligned} \tag{19.8}$$

Then Eq. 19.6 is written as

$$A_N = \frac{n\pi}{3c}\left[\left(\sum_i \frac{\partial\mu_x}{\partial R_i} L_{iN}\right)_0^2 + \left(\sum_i \frac{\partial\mu_y}{\partial R_i} L_{iN}\right)_0^2 + \left(\sum \frac{\partial\mu_z}{\partial R_i} L_{iN}\right)_0^2\right]$$

$$= \frac{n\pi}{3c}\sum_i\left[\left(\frac{\partial\mu_x}{\partial R_i}\right)_0^2 + \left(\frac{\partial\mu_y}{\partial R_i}\right)_0^2 + \left(\frac{\partial\mu_z}{\partial R_i}\right)_0^2\right](L_{iN})^2 \qquad (19.9)$$

This equation shows that the intensity of an infrared band depends on the values of the $\partial\mu/\partial R$ terms as well as the L matrix elements.

Equation 19.9 has been applied to relatively small molecules to calculate the $\partial\mu/\partial R$ terms from the observed intensity and known L_{iN} values.[7] However, the additivity of the bond dipole moment does not strictly hold, and the results obtained are often inconsistent and conflicting. Thus far, very few studies have been made on infrared intensities of large molecules because of these difficulties.

As seen in Eq. 19.9, the IR band becomes stronger as the $\partial\mu/\partial R$ term becomes larger if the L_{iN} term does not vary significantly. In general, the more polar the bond, the larger the $\partial\mu/\partial R$ term. Thus, the IR intensities of the stretching vibrations follow the general trends:

$$\nu(\mathrm{OH}) > \nu(\mathrm{NH}) > \nu(\mathrm{CH})$$
$$\nu(\mathrm{C{=}O}) > \nu(\mathrm{C{=}N}) > \nu(\mathrm{C{=}C})$$

It is also noted that the antisymmetric stretching vibration is always stronger than the symmetric stretching vibration because the $\partial\mu/\partial R$ term is larger for the former than for the latter. This can be seen for functional groups such as $>CH_2$ and $>CCl_2$. These trends are entirely opposite to those found in Raman spectra (Sec. I-21), and make both IR and Raman spectroscopy complementary.

I-20. DEPOLARIZATION OF RAMAN LINES

As stated in Secs. I-8 and I-9, it is possible, by using group theory, to classify the normal vibration into various symmetry species. Experimentally, measurements of the infrared dichroism and polarization properties of Raman lines of an orientated crystal provide valuable information about the symmetry of normal vibrations (Sec. I-29). Here, we consider the polarization properties of Raman lines in liquids and solutions in which molecules or ions take completely random orientations.*

Suppose that we irradiate a molecule fixed at the origin of a space-fixed coordinate system with natural light from the positive-y direction, and observe the Raman scattering in the x direction as shown in Fig. I-27. The incident light vector E may be resolved into two components, E_x and E_z, of equal magnitude ($E_y = 0$). Both components give induced dipole moments, P_x, P_y, and P_z. However, only P_y and P_z contribute to the scattering along the x axis, since an oscillating dipole cannot radiate in its own direction. Then, from Eq. 6.7, we have

$$P_y = \alpha_{yx}E_x + \alpha_{yz}E_z \tag{20.1}$$

$$P_z = \alpha_{zx}E_x + \alpha_{zz}E_z \tag{20.2}$$

The intensity of the scattered light is proportional to the sum of squares of the individual $\alpha_{ij}E_j$ terms. Thus, the ratio of the intensities in the y and z directions is

$$\rho_n = \frac{I_y}{I_z} = \frac{\alpha_{yx}^2 E_x^2 + \alpha_{yz}^2 E_z^2}{\alpha_{zx}^2 E_x^2 + \alpha_{zz}^2 E_z^2} \tag{20.3}$$

where ρ_n is called the *depolarization ratio for natural light* (n).

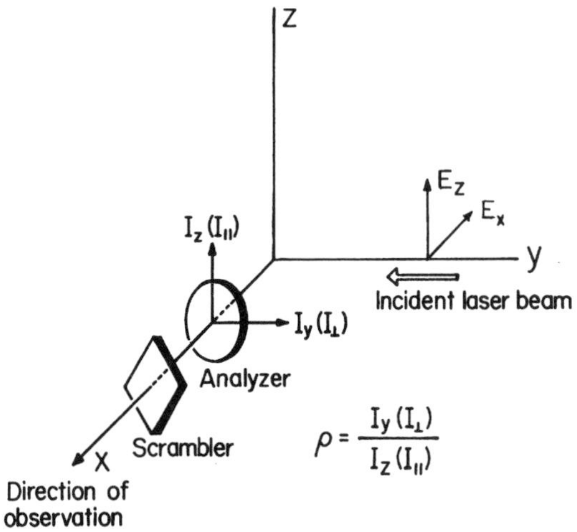

Fig. I-27. Experimental configuration for measuring depolarization ratios. The scrambler is placed after the analyzer because the monochromator gratings show different efficiencies for ⊥ and ∥ directions.

*It is possible to obtain approximate depolarization ratios of fine powders where the molecules or ions take pseudorandom orientations (see Ref. 106).

In a homogeneous liquid or gas, the molecules are randomly oriented, and we must consider the polarizability components averaged over all molecular orientations. The results are expressed in terms of two quantities: $\overline{\alpha}$ (*mean value*) and γ (*anisotropy*):

$$\overline{\alpha} = \tfrac{1}{3}(\alpha_{xx} + \alpha_{yy} + \alpha_{zz}) \tag{20.4}$$

$$\gamma^2 = \tfrac{1}{2}[(\alpha_{xx} - \alpha_{yy})^2 + (\alpha_{yy} - \alpha_{zz})^2 + (\alpha_{zz} - \alpha_{xx})^2 + 6(\alpha_{xy}^2 + \alpha_{yz}^2 + \alpha_{zx}^2)] \tag{20.5}$$

These two quantities are invariant to any coordinate transformation. It can be shown[3] that the average values of the squares of α_{ij} are

$$\overline{(\alpha_{xx})^2} = \overline{(\alpha_{yy})^2} = \overline{(\alpha_{zz})^2} = \tfrac{1}{45}[45(\overline{\alpha})^2 + 4\gamma^2] \tag{20.6}$$

$$\overline{(\alpha_{xy})^2} = \overline{(\alpha_{yz})^2} = \overline{(\alpha_{zx})^2} = \tfrac{1}{15}\gamma^2 \tag{20.7}$$

Since $E_x = E_z = E$, Eq. 20.3 can be written as

$$\rho_n = \frac{I_y}{I_z} = \frac{6\gamma^2}{45(\overline{\alpha})^2 + 7\gamma^2} \tag{20.8}$$

The total intensity I_n is given by

$$I_n = I_y + I_z = \text{const}\,\{\tfrac{1}{45}[45(\overline{\alpha})^2 + 13\gamma^2]\}E^2 \tag{20.9}$$

If the incident light is plane polarized (e.g. laser beam), with its electric vector in the z direction ($E_x = 0$), Eq. 20.8 becomes

$$\rho_p = \frac{I_y}{I_z} = \frac{3\gamma^2}{45(\overline{\alpha})^2 + 4\gamma^2} \tag{20.10}$$

where ρ_p is the *depolarization ratio for polarized light* (p). In this case, the total intensity is given by

$$I_p = I_y + I_z = \text{const}\,\{\tfrac{1}{45}[45(\overline{\alpha})^2 + 7\gamma^2]\}E^2 \tag{20.11}$$

The symmetry property of a normal vibration can be determined by measuring the depolarization ratio. From an inspection of character tables (Appendix I), it is obvious that $\overline{\alpha}$ is nonzero only for totally symmetric vibrations. Then, Eq. 20.8 gives $0 \leq \rho_n < \frac{6}{7}$, and the Raman lines are said to be *polarized*. For all nontotally symmetric vibrations, $\overline{\alpha}$ is zero, and $\rho_n = \frac{6}{7}$. Then, the Raman lines are said to be *depolarized*. If the exciting line is plane polarized, these criteria must be changed according to Eq. 20.10. Thus, $0 \leq \rho_p < \frac{3}{4}$ for totally symmetric vibrations, and $\rho_p = \frac{3}{4}$ for nontotally symmetric vibrations. Figure I-28 shows the Raman spectra of CCl_4 (500–150 cm^{-1}) in two directions of polarization obtained with the 488 nm excitation. The three bands at 459, 314, and 218 cm^{-1} give ρ_p values of approximately 0.02, 0.75, and 0.75, respectively. Thus, it is concluded that the 459 cm^{-1} band is polarized (A_1), whereas the two bands at 314 (F_2) and 218 (E) cm^{-1} are depolarized.

As stated in Sec. I-6, the polarizability tensors are symmetric in normal Raman scattering. If the exciting frequency approaches that of an electronic

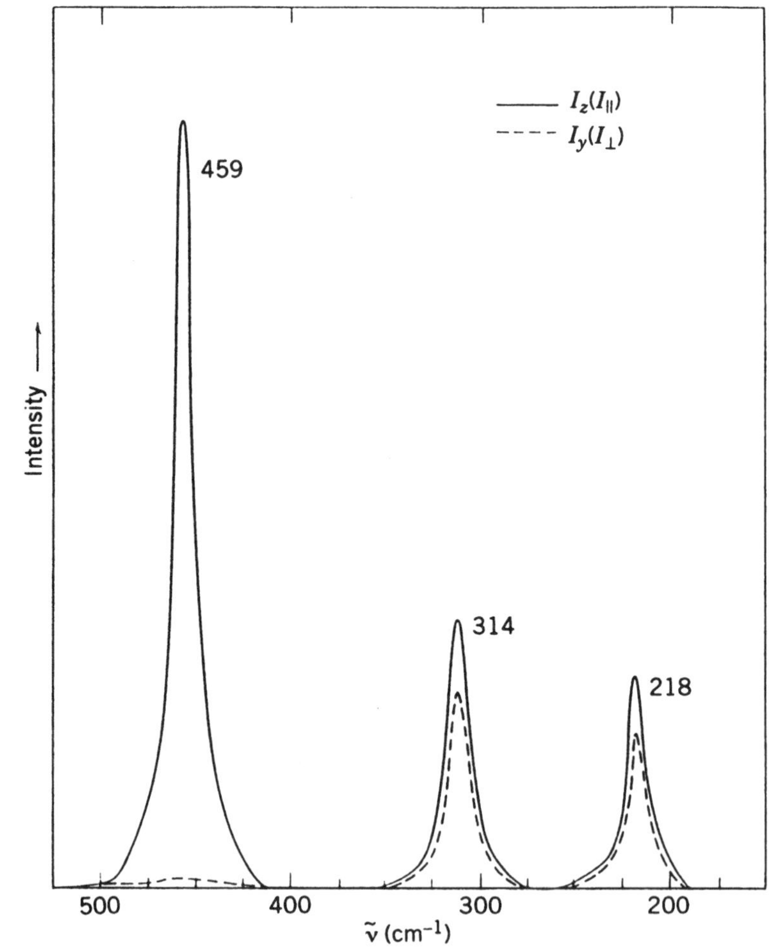

Fig. I-28. Raman spectra of CCl_4 in two directions of polarization (488 nm excitation).

absorption, some scattering tensors become antisymmetric,* and resonance Raman scattering can occur (Sec. I-22). In this case, Eq. 20.10 must be written in a more general form.[107]

$$\rho_p = \frac{3g^s + 5g^a}{10g^o + 4g^s} \tag{20.12}$$

where

$$\begin{aligned} g^o &= 3(\overline{\alpha})^2 \\ g^s &= \tfrac{1}{3}\,[(\alpha_{xx} - \alpha_{yy})^2 + (\alpha_{yy} - \alpha_{zz})^2 + (\alpha_{zz} - \alpha_{xx})^2] \\ &\quad + \tfrac{1}{2}\,[(\alpha_{xy} + \alpha_{yx})^2 + (\alpha_{xz} + \alpha_{zx})^2 + (\alpha_{yz} + \alpha_{zy})^2] \\ g^a &= \tfrac{1}{2}\,[(\alpha_{xy} - \alpha_{yx})^2 + (\alpha_{xz} - \alpha_{zx})^2 + (\alpha_{yz} - \alpha_{zy})^2] \end{aligned} \tag{20.13}$$

If we define

$$\gamma_s^2 = \tfrac{3}{2}\,g^s \quad \text{and} \quad \gamma_{as}^2 = \tfrac{3}{2}\,g^a \tag{20.14}$$

Eq. 20.12 can be written as

$$\rho_p = \frac{3\gamma_s^2 + 5\gamma_{as}^2}{45(\overline{\alpha})^2 + 4\gamma_s^2} \tag{20.15}$$

In normal Raman scattering, $\gamma_s^2 = \gamma^2$ and $\gamma_{as}^2 = 0$. Then Eq. 20.15 is reduced to Eq. 20.10.

The symmetry properties of resonance Raman lines can be predicted on the basis of Eq. 20.15. For totally symmetric vibrations, $\overline{\alpha} \neq 0$ and $\gamma_{as} = 0$. Then, Eq. 20.15 gives $0 \leq \rho_p < \frac{3}{4}$. Nontotally symmetric vibrations ($\overline{\alpha} = 0$) are classified into two types: those that have symmetric scattering tensors, and those that have antisymmetric scattering tensors. If the tensor is symmetric, $\gamma_{as} = 0$ and $\gamma_s \neq 0$. Then Eq. 20.15 gives $\rho_p = \frac{3}{4}$ (depolarized). If the tensor is antisymmetric, $\gamma_{as} \neq 0$ and $\gamma_s = 0$. Then, Eq. 20.15 gives $\rho_p = \infty$ (*inverse polarization*). In the case of the $\mathbf{D}_{4h}$ point group, the B_{1g} and B_{2g} representations belong to the former type, whereas the A_{2g} representations belongs to the latter.[108] As will be shown in Sec. I-22, Spiro and Strekas[107] observed for the first time A_{2g} vibrations which exhibit ρ_p values $\frac{3}{4} < \rho_p < \infty$. For these vibrations, the term "anomalous polarization (ap)" is used instead of "inverse polarization (ip)."[109]

*A tensor is called antisymmetric if $\alpha_{xx} = \alpha_{yy} = \alpha_{zz} = 0$ and $\alpha_{xy} = -\alpha_{yx}, \alpha_{yz} = -\alpha_{zy}$, and $\alpha_{zx} = -\alpha_{xz}$.

I-21. INTENSITY OF RAMAN SCATTERING

According to the quantum mechanical theory of light scattering, the intensity per unit solid angle of scattered light arising from a transition between states m and n is given by

$$I_{n \leftarrow m} = \text{const}\,(\nu_0 \pm \nu_{mn})^4 \sum_{\rho\sigma} |(P_{\rho\sigma})_{mn}|^2 \tag{21.1}$$

where

$$(P_{\rho\sigma})_{mn} = (\alpha_{\rho\sigma})_{mn} E = \frac{1}{h} \sum_r \left[\frac{(M_\rho)_{rn}(M_\sigma)_{mr}}{\nu_{rm} - \nu_0} + \frac{(M_\rho)_{mr}(M_\sigma)_{rn}}{\nu_{rn} + \nu_0} \right] E \tag{21.2}$$

Here ν_0 is the frequency of the incident light: ν_{rm}, ν_{rn}, and ν_{mn} are the frequencies corresponding to the energy differences between subscripted states; terms of the type $(M_\sigma)_{mr}$ are the cartesian components of transition moments such as $\int \Psi_r^* \mu_\sigma \Psi_m \, d\tau$; and E is the electric vector of the incident light. It should be noted here that the states denoted by m, n, and r represent vibronic states $\psi_g(\xi, Q)\phi_i^g(Q)$, $\psi_g(\xi, Q)\phi_j^g(Q)$, and $\psi_e(\xi, Q)\phi_v^e(Q)$, respectively, where ψ_g and ψ_e are electronic ground- and excited-state wave functions, respectively, and ϕ_i^g, ϕ_j^g, and ϕ_v^e are vibrational functions. The symbols ξ and Q represent the electronic and nuclear coordinates, respectively. These notations are illustrated in Fig. I-29. Finally, σ and ρ denote x, y, and z components.

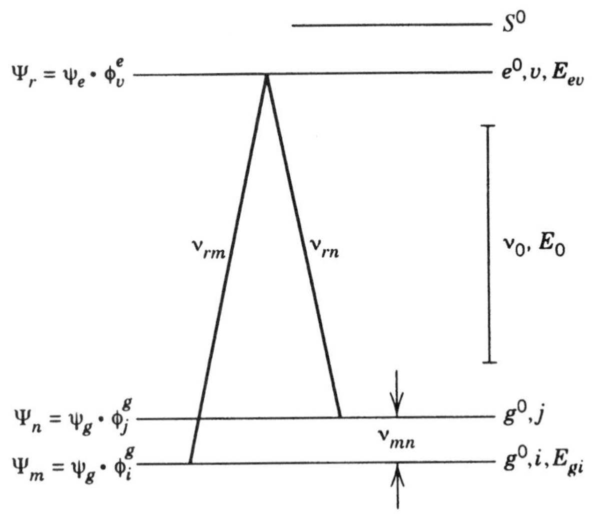

Fig. I-29. Energy-level diagram for Raman scattering.

Since the electric dipole operator acts only on the electronic wave functions, the $(\alpha_{\rho\sigma})_{mn}$ term in Eq. 21.2 can be written in the form[27]

$$(\alpha_{\rho\sigma})_{mn} = \frac{1}{h} \int \phi_j^g (\alpha_{\rho\sigma})_{gg} \phi_i^g \, dQ \tag{21.3}$$

where

$$(\alpha_{\rho\sigma})_{gg} = \sum_e \left(\frac{\int \psi_g^* \mu_\sigma \psi_e \, d\tau \cdot \int \psi_e^* \mu_\rho \psi_g \, d\tau}{\bar{\nu}_{eg} - \nu_0} + \frac{\int \psi_g^* \mu_\rho \psi_e \, d\tau \cdot \int \psi_e^* \mu_\sigma \psi_g \, d\tau}{\bar{\nu}_{eg} + \nu_0} \right)$$

Here $\bar{\nu}_{eg}$ corresponds to the energy of a pure electronic transition between the ground and excited states.

To discuss the Raman scattering, we expand the $(\alpha_{\rho\sigma})_{gg}$ term as a Taylor series with respect to the normal coordinate Q:

$$(\alpha_{\rho\sigma})_{gg} = (\alpha_{\rho\sigma})_{gg}^0 + \left[\frac{\partial (\alpha_{\rho\sigma})_{gg}}{\partial Q} \right]_0 Q + \cdots \tag{21.4}$$

Then, we write Eq. 21.3 as

$$(\alpha_{\rho\sigma})_{mn} = \frac{1}{h} (\alpha_{\rho\sigma})_{gg}^0 \int \phi_j^g \phi_i^g \, dQ + \frac{1}{h} \left[\frac{\partial (\alpha_{\rho\sigma})_{gg}}{\partial Q} \right]_0 \int \phi_j^g Q \phi_i^g \, dQ \tag{21.5}$$

The first term on the right-hand side is zero unless $i = j$. This term is responsible for Rayleigh scattering. The second term determines the activity of fundamental vibrations in Raman scattering; it vanishes for a harmonic oscillator unless $j = i \pm 1$. (Sec. I-10).

If we consider a Stokes transition, $\upsilon \rightarrow \upsilon + 1$, Eq. 21.5 is written as[3]

$$(\alpha_{\rho\sigma})_{\upsilon, \upsilon+1} = \frac{1}{h} \left[\frac{\partial (\alpha_{\rho\sigma})_{gg}}{\partial Q} \right]_0 \sqrt{\frac{(\upsilon + 1) h}{8\pi^2 \mu \nu}} \tag{21.6}$$

where μ and ν are the reduced mass and the Stokes frequency. Then Eq. 21.1 is written as

$$I = \text{const}\,(\nu_0 - \nu)^4 \frac{E^2}{h^2} \left[\frac{\partial(\alpha_{\rho\sigma})_{gg}}{\partial Q} \right]_0^2 \frac{(v+1)h}{8\pi^2\mu\nu} \tag{21.7}$$

In Sec. I-20, we derived a classical equation for Raman intensity:

$$I_n = \text{const} \left(\frac{\partial \alpha}{\partial Q} \right)^2 E^2$$

$$= \text{const} \left\{ \frac{1}{45} [45(\overline{\alpha})^2 + 13\gamma^2] \right\} E^2 \tag{20.9}$$

By replacing the $\partial\alpha/\partial Q$ term of Eq. 21.7 with the term enclosed in braces in Eq. 20.9, we obtain

$$I_n = \text{const}\,(\nu_0 - \nu)^4 \frac{(v+1)}{8\pi^2\mu\nu} \frac{E^2}{h^2} \left\{ \frac{1}{45} [45(\overline{\alpha})^2 + 13\gamma^2] \right\} \tag{21.8}$$

At room temperature, most of the scattering molecules are in the $v = 0$ state, but some are in higher vibrational states. Using the Maxwell–Bolztmann distribution law, we find that the fraction of molecules f_v with vibrational quantum number v is given by

$$f_v = \frac{e^{-[v+(1/2)]h\nu/kT}}{\sum_v e^{-[v+(1/2)]h\nu/kT}} \tag{21.9}$$

Then the total intensity is proportional to $\Sigma_v f_v(v + 1)$, which is equal to $(1 - e^{-h\nu/kT})^{-1}$ (see Ref. 24). Hence we can rewrite Eq. 21.8 in the form

$$I_n = KI_0 \frac{(\nu_0 - \nu)^4}{\mu\nu(1 - e^{-h\nu/kT})} [45(\overline{\alpha})^2 + 13\gamma^2] \tag{21.10}$$

Here I_0 is the incident light intensity which is proportional to E^2, and K summarizes all other constant terms.

If the incident light is polarized, the form of Eq. 21.10 is slightly modified:

$$I_p = KI_0 \frac{(\nu_0 - \nu)^4}{\mu\nu(1 - e^{-h\nu/kT})} [45(\overline{\alpha})^2 + 7\gamma^2] \tag{21.11}$$

As shown in Sec. I-20, the degree of depolarization ρ_p is

$$\rho_p = \frac{3\gamma^2}{45(\overline{\alpha})^2 + 4\gamma^2} \quad \text{or} \quad \gamma^2 = \frac{45(\overline{\alpha})^2 \rho_p}{3 - 4\rho_p} \tag{21.12}$$

Since $\rho_p = \frac{3}{4}$ for nontotally symmetric vibrations, Eq. 21.12 holds only for totally symmetric vibrations. Then Eq. 21.11 is written as

$$I_p = K' I_0 \frac{(\nu_0 - \nu)^4}{\mu\nu(1 - e^{-h\nu/kT})} \left(\frac{1 + \rho_p}{3 - 4\rho_p} \right) (\overline{\alpha})^2 \tag{21.13}$$

In the case of a solution, the intensity is proportional to the molar concentration C. Then Eq. 21.13 is written as

$$I_p = K'' I_0 \frac{C(\nu_0 - \nu)^4}{\mu\nu(1 - e^{-h\nu/kT})} \left(\frac{1 + \rho_p}{3 - 4\rho_p} \right) (\overline{\alpha})^2 \tag{21.14}$$

If we compare the intensities of totally symmetric vibrations (A_1 mode) of two tetrahedral XY_4-type molecules, the intensity ratio is given by

$$\frac{I_1}{I_2} = \frac{C_1}{C_2} \left(\frac{\tilde{\nu}_0 - \tilde{\nu}_1}{\tilde{\nu}_0 - \tilde{\nu}_2} \right)^4 \frac{\tilde{\nu}_2 \mu_2}{\tilde{\nu}_1 \mu_1} \frac{(1 - e^{-hc\tilde{\nu}_2/kT})(\overline{\alpha}_1)^2}{(1 - e^{-hc\tilde{\nu}_1/kT})(\overline{\alpha}_2)^2} \tag{21.15}$$

In this case, the ρ_p term drops out, since $\gamma^2 = 0$ for isotropic molecules such as tetrahedral XY_4 and octahedral XY_6 types. By using CCl_4 as the standard, it is possible to determine the relative value of the $\partial\alpha/\partial Q$ term, which provides information about the degree of covalency and the bond order.[24]

As seen above, the intensity of Raman scattering increases as the $(\partial\alpha/\partial Q)_0$ term becomes larger. The following general rules may reflect the trends in the $(\partial\alpha/\partial Q)_0$ values: (1) Stretching vibrations produce stronger Raman bands than bending vibrations. (2) Symmetric vibrations produce stronger Raman bands than antisymmetric vibrations. Totally symmetric "breathing" vibrations produce the most intense Raman bands. (3) Stretching vibrations of covalent bonds produce stronger Raman bands than that of ionic bonds. Among the covalent bonds, Raman intensities increase as the bond order increases. Thus, the ratio of relative intensities of the C≡C, C=C, and C—C stretching vibrations is approximately 3 : 2 : 1. (4) Bonds involving heavier atoms produce stronger Raman bands than those involving lighter atoms. For example, the ν(S—S) is stronger than the ν(C—C). Coordination compounds containing heavy metals and sulfur ligands are ideal for Raman studies.

I-22. PRINCIPLE OF RESONANCE RAMAN SPECTROSCOPY

In normal Raman spectroscopy, the exciting frequency lies in the region where the compound has no electronic absorption band. In resonance Raman spectroscopy, the exciting frequency falls within the electronic band (Sec. I-2). In the gaseous phase, this tends to cause resonance fluorescence since the rotational–vibrational levels are discrete. In the liquid and solid states, however, these levels are no longer discrete because of molecular collisions and/or intermolecular interactions. If such a broad vibronic bands is excited, it tends to give resonance Raman rather than resonance fluorescence spectra.[110,111]

Resonance Raman spectroscopy is particularly suited to the study of biological macromolecules such as heme proteins because only a dilute solution (biological condition) is needed to observe the spectrum and only vibrations localized within the chromophoric group are enhanced when the exciting frequency approaches that of the relevant chromophore. This *selectivity* is highly important in studying the theoretical relationship between the electronic transition and the vibrations to be resonance enhanced.

The origin of resonance Raman enhancement is explained in terms of Eq. 21.2. In normal Raman spectroscopy, ν_0 is chosen in the region that is far from the electronic absorption. Then, $\nu_{rm} \gg \nu_0$, and $\alpha_{\rho\sigma}$ is independent of the exciting frequency ν_0. In resonance Raman spectroscopy, the denominator, $\nu_{rm} - \nu_0$, becomes very small as ν_0 approaches ν_{rm}. Thus, the first term in the square brackets of Eq. 21.2 dominates all other terms and results in striking enhancement of Raman lines. However, Eq. 21.2 cannot account for the selectivity of resonance Raman enhancement since it is not specific about the states of the molecule. Albrecht[112] derived a more specific equation for the initial and final states of resonance Raman scattering by introducing the Herzberg–Teller expansion of electronic wave functions into the Kramers–Heisenberg dispersion formula. The results are as follows:

$$(\alpha_{\rho\sigma})_{gi,gj} = A + B + C \tag{22.1}$$

$$A = \sum_{e\neq g}' \sum_{v} \left[\frac{(g^0|R_\sigma|e^0)(e^0|R_\rho|g^0)}{E_{ev} - E_{gi} - E_0} + (\text{nonresonance term}) \right] \langle i|v\rangle\langle v|j\rangle \tag{22.2}$$

$$B = \sum_{e\neq g}' \sum_{v} \sum_{s\neq e}' \sum_{a} \left\{ \left[\frac{(g^0|R_\sigma|e^0)(e^0|h_a|s^0)(s^0|R_\rho|g^0)}{E_{ev} - E_{gi} - E_0} + (\text{nonresonance term}) \right] \times \frac{\langle i|v\rangle\langle v|Q_a|j\rangle}{E_e^0 - E_s^0} + \left[\frac{(g^0|R_\sigma|s^0)(s^0|h_a|e^0)(e^0|R_\rho|g^0)}{E_{ev} - E_{gi} - E_0} + \right. \right.$$
$$\left. \left. + (\text{nonresonance term}) \right] \times \frac{\langle i|Q_a|v\rangle\langle v|j\rangle}{E_e^0 - E_s^0} \right\} \tag{22.3}$$

$$C = \sum_{e \neq g}' \sum_{t \neq g}' \sum_{v} \sum_{a} \left\{ \left[\frac{(e^0|R_\rho|g^0)(g^0|h_a|t^0)(t^0|R_\sigma|e^0)}{E_{ev} - E_{gi} - E_0} + \text{(nonresonance term)} \right] \right.$$

$$\times \frac{\langle i|v\rangle\langle v|Q_a|j\rangle}{E_g^0 - E_t^0} + \left[\frac{(e^0|R_\rho|t^0)(t^0|h_a|g^0)(g^0|R_\sigma|e^0)}{E_{ev} - E_{gi} - E_0} \right.$$

$$\left. \left. + \text{(nonresonance term)} \right] \times \frac{\langle i|Q_a|v\rangle\langle v|j\rangle}{E_g^0 - E_t^0} \right\} \qquad (22.4)$$

The notations g, i, j, e, and v were explained in Sec. I-21. Other notations are as follows: s, another excited electronic state; h_a, the vibronic coupling operator $\partial\mathcal{H}/\partial Q_a$, $\mathcal{H}$ and Q_a being the electronic Hamiltonian and the ath normal coordinate of the electronic ground state, respectively; E_{gi} and E_{ev}, the energies of states gi and ev, respectively; $|g^0)$, $|e^0)$, and $|s^0)$, the electronic wave functions for the equilibrium nuclear positions of the ground and excited states; E_e^0 and E_s^0, the corresponding energies of the electronic states, e^0 and s^0, respectively; and E_0, the energy of the exciting light. The nonresonance terms are similar to the preceding terms except that the denominator is $(E_{ev} - E_{gj} + E_0)$ instead of $(E_{ev} - E_{gi} - E_0)$ and that R_σ and R_p (cartesian components of the transition moment) in the numerator are interchanged. These terms can be neglected under the strict resonance condition since the resonance terms become very large. The C term is usually neglected because its components are denominated by $E_g^0 - E_t^0$, where t refers to an excited state which is much higher in energy than the first excited state. These notations are shown on the right-hand side of Fig. I-29. In more rigorous expressions, the damping factor, $i\Gamma$, must be added to the denominators such as $E_{ev} - E_{gi} - E_0$ so that the resonance term does not become infinity under rigorous resonance conditions.[112]

For the A term to be nonzero, the $g \leftrightarrow e$ electronic transition must be allowed, and the product of the integrals $\langle i|v\rangle\langle v|j\rangle$ (Franck–Condon Factor) must be nonzero. The latter condition is satisfied if the equilibrium position is shifted by a transition ($\Delta \neq 0$ as shown in Fig. I-30). Totally symmetric vibrations tend to be resonance-enhanced via the A term since they tend to shift the equilibrium position upon electronic excitation. If the equilibrium position is not shifted by the $g \leftrightarrow e$ transition, the A term becomes zero since either one of the integrals, $\langle i|v\rangle$ or $\langle v|j\rangle$, becomes zero. This situation tends to occur for nontotally symmetric vibrations.

In contrast to the A term, the B term resonance requires at least one more electronic excited state(s), which must be mixed with the e state via normal vibration, Q_a. Namely, the integral $(e^0|h_a|s^0)$ must be nonzero. The B-term resonance is significant only when these two excited states are closely located so that the denominator, $E_e^0 - E_s^0$, is small. Other requirements are that the $g \leftrightarrow e$ and $g \leftrightarrow s$ transitions should be allowed and that the integrals $\langle i|v\rangle$ and $\langle v|Q_a|j\rangle$ should not be zero. The B term can cause resonance enhancement of nonto-

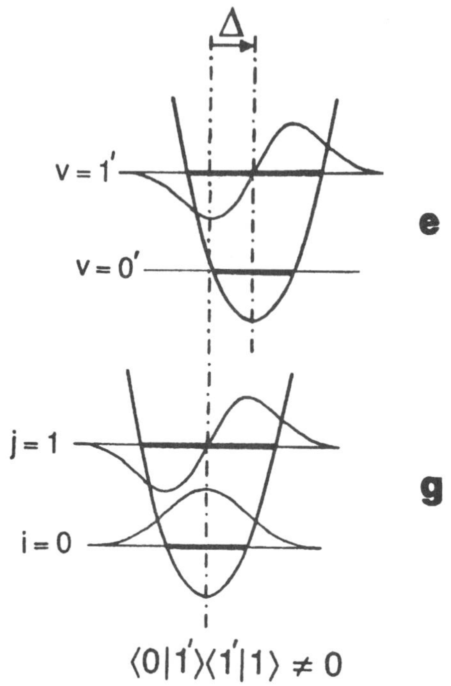

Fig. I-30. Shift of equilibrium position caused by totally symmetric vibration.

tally symmetric as well as totally symmetric vibrations, although it is mainly responsible for resonance enhancement of the former.

As seen in Eqs. 22.2 and 22.3, both A and B terms involve summation over v, which is the vibrational quantum number at the electronic excited state. Since the harmonic oscillator selection rule ($\Delta v = \pm 1$) does not hold for a large v, overtones and combination bands may appear under resonance conditions. In fact, series of these nonfundamental vibrations have been observed in the case of A-term resonance (Sec. I-23).

In Sec. I-20, we have shown that the depolarization ratio for totally symmetric vibrations is in the range $0 \leq \rho_p < \frac{3}{4}$. For the A term to be nonzero, however, the term $(g^0|R_\sigma|e^0)(e^0|R_\rho|g^0)$ must be nonzero. Since g^0 is totally symmetric, this condition is realized only when both $\chi(R_\sigma)\chi(e^0)$ and $\chi(R_\rho)\chi(e^0)$ belong to the totally symmetric species. Thus, $\chi(R_\sigma) = \chi(R_\rho)$, and only one of the diagonal terms of the polarizability tensor (α_{xx}, α_{yy}, or α_{zz}) becomes nonzero. Then, it is readily seen from Eq. 20.15 that $\rho_p = \frac{1}{3}$. This rule has been confirmed by many observations. If the electronic excited state is degenerate, two of the diagonal terms must be nonzero. In this case, $\rho_p = \frac{1}{8}$. These rules hold for any point group with at least one C_3 or higher axis, with the exception of the cubic groups in which the cartesian coordinates transform as a single irreducible degenerate representation.[109]

I-23. RESONANCE RAMAN SPECTRA

Because of the advantages mentioned in the preceding section, resonance Raman (RR) spectroscopy has been applied to vibrational studies of a number of inorganic as well as organic compounds. It is currently possible to cover the whole range of electronic transitions continuously by using excitation lines from a variety of lasers and Raman shifters.[32] In particular, the availability of excitation lines in the UV region has made it possible to carry out UV resonance Raman (UVRR) spectroscopy.[113]

In RR spectroscopy, the excitation line is chosen inside the electronic absorption band. This condition may cause thermal decomposition of the sample by local heating. To minimize thermal decomposition, several techniques have been developed. These include the rotating sample technique, the rotating (or oscillating) laser beam technique, and their combinations with low-temperature techniques.[30,32] It is always desirable to keep low laser power so that thermal decomposition is minimal. This will also minimize photodecomposition, which occurs depending on laser lines in some compounds.

An example of the *A*-term resonance is given by I_2, which has an absorption band near 500 nm and its fundamental at ~210 cm^{-1}. Figure I-31 shows the RR

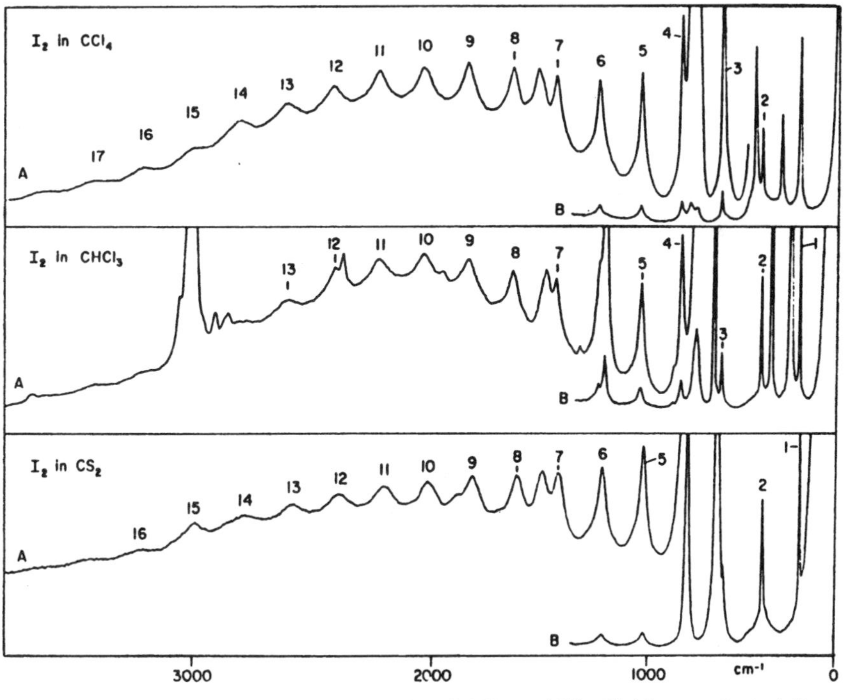

Fig. I-31. Resonance Raman spectra of I_2 in CCl_4, $CHCl_3$, and CS_2 (514.5 nm excitation). Reproduced with permission from Kiefer and Bernstein[114].

spectrum of I_2 in solution obtained by Kiefer and Bernstein[114] (excitation at 514.5 nm). It shows a series of overtones up to the seventeenth. The vibrational energy of an anharmonic oscillator including the cubic term is expressed as:

$$E_v = hc\omega_e(v + \tfrac{1}{2}) - hc\chi_e\omega_e(v + \tfrac{1}{2})^2 + hcy_e\omega_e(v + \tfrac{1}{2})^3 \cdots$$

The anharmonicity constants, $\chi_e\omega_e$ and $y_e\omega_e$ can be determined by plotting the $\tilde{\nu}(\text{obs})/v$ against the vibrational quantum number. v, as shown in Fig. I-32. The straight line gives $\chi_e\omega_e$ and the deviation from the straight line (at higher v) gives $y_e\omega_e$. The results (CCl_4 solution) in cm^{-1} are:

$$\omega_e = 212.59 \pm 0.08, \qquad \chi_e\omega_e = 0.62 \pm 0.03, \qquad y_e\omega_e = -0.005 \pm 0.002$$

As stated in the preceding section, the depolarization ratio (ρ_p) is expected to be closed to $\frac{1}{3}$ for totally symmetric vibrations. However, this value increases as the change in vibrational quantum number (Δv) increases. For example, it is 0.48 for the tenth overtone of I_2 in CCl_4.[109]

The second example is given by TiI_4, which has the absorption maximum near 514 nm. Figure I-33 shows the RR spectrum of solid TiI_4 obtained by Clark and Mitchell[115] with 514.5 nm excitation. In this case, a series of overtones up to the twelfth has been observed. Using these data, they obtained the following constants (cm^{-1}):

$$\omega_e = 161.0 \pm 0.2 \quad \text{and} \quad X_{11} = 0.11 \pm 0.03$$

Here, X_{11} is the anharmonicity constant corresponding to $\chi_e\omega_e$ of a diatomic molecule (Sec. I-3).

Theoretical treatments of Raman intensities of these overtone series under

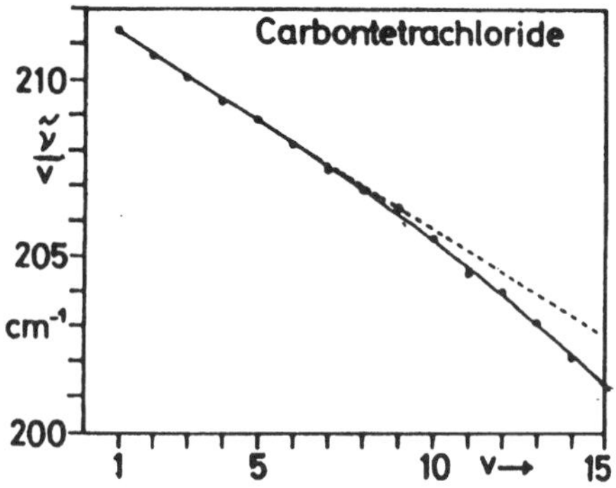

Fig. I-32. Plot of $\tilde{\nu}/v$ vs. v for I_2 in CCl_4. Reproduced with permission from Kiefer and Bernstein[114].

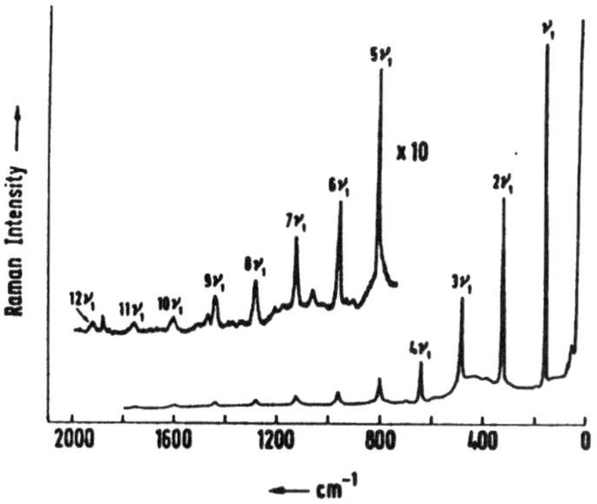

Fig. I-33. Resonance Raman spectrum of solid TiI_4 (514.5 nm excitation). Reproduced with permission from Clark and Mitchell[115].

rigorous resonance conditions have been made by Nafie et al.[116] There are many other examples of RR spectra of small molecules and ions which exhibit *A*-term resonance. Section II-6 includes references on RR spectra of the CrO_4^{2-} and MnO_4^- ions. Kiefer[110] reviewed RR studies of small molecules and ions.

Hirakawa and Tsuboi[117] noted that among the totally symmetric modes, the mode that leads to the excited-state configuration is most strongly resonance enhanced. For example, the NH_3 molecule is pyramidal in the ground state and planar in the excited state (216.8 nm above the ground state). Then, the symmetric bending mode near 950 cm^{-1} is enhanced 10 times more than the symmetric stretching mode near 3300 cm^{-1} when the excitation wavelength is changed from 514.5 to 351.1 nm.

A typical example of the *B*-term resonance is given by metalloporphyrins and heme proteins. As shown in Fig. I-34, Ni(OEP) (OEP:octaethylporphyrin) exhibits two electronic transitions referred to as the Q_0 (or α) and **B** (or Soret) bands along with a vibronic side band (Q_1 or β) in the 350–600 nm region. According to MO (molecular orbital) calculations on the porphyrin core of $\mathbf{D}_{4h}$ symmetry, Q_0 and **B** transitions result from strong interaction between the $a_{1u}(\pi) \rightarrow e_g(\pi^*)$ and $a_{2u}(\pi) \rightarrow e_g(\pi^*)$ transitions which have similar energies and the same excited-state symmetry (E_u). The transition dipoles add up for the strong **B** transition and nearly cancel out for the weak Q_0 transition. This is an ideal situation for *B*-term resonance. According to Eq. 22.3, any normal modes which give nonzero values for the integral $(e^0|h_a|s^0)$ are enhanced via the *B* term. These vibrations must belong to one of the symmetry species given by $E_u \times E_u$, that is, $A_{1g} + B_{1g} + B_{2g} + A_{2g}$.*

Figure I-35 shows the RR spectra of Ni(OEP) with excitation near the **B**, Q_1, and Q_0 transitions as obtained by Spiro and co-workers.[118] As discussed in

*For symmetry species of the direct products, see Appendix IV. Also note that the A_{1g} vibrations are not effective in vibronic mixing.

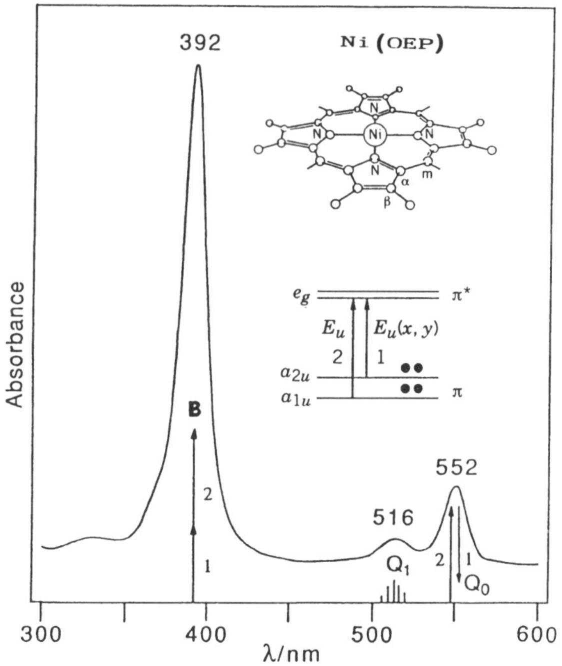

Fig. I-34. Structure, energy-level diagram, and electronic spectrum of Ni(OEP). Reproduced with permission from Spiro et al.[118].

Sec. I-20, polarization properties are expected to be A_{1g} (p, polarized), B_{1g} and B_{2g} (dp, depolarized), and A_{2g} (ap, anomalous polarization). These polarization properties, together with normal coordinate analysis, were used to make complete vibrational assignments of Ni(OEP).[119] It is seen that the spectra obtained by excitation near the Q_0 band (bottom traces) are dominated by the dp bands (B_{1g} and B_{2g} species), while those obtained by excitation between the Q_0 and Q_1 bands (middle traces) are dominated by the ap bands (A_{2g} species). The spectra obtained by excitation near the **B** band (top traces) are dominated by the p bands (A_{1g} species). In this case, the *A*-term resonance is much more important than the *B*-term resonance because of the very strong absorption strength of the **B** band.

As will be shown in Sec. II-8, anomalously polarized bands have also been observed in the RR spectra of the $IrCl_6^{2-}$ ion, although their origin is different from the vibronic coupling mentioned above.[120,121]

The majority of compounds studied thus far exhibit the *A*-term rather than the *B*-term resonance. A more complete study of resonance Raman spectra involves the observation of *excitation profiles* (Raman intensity plotted as a function of the excitation frequency for each mode), and the simulation of observed excitation profiles based on theoretical treatments of resonance Raman scattering[122].

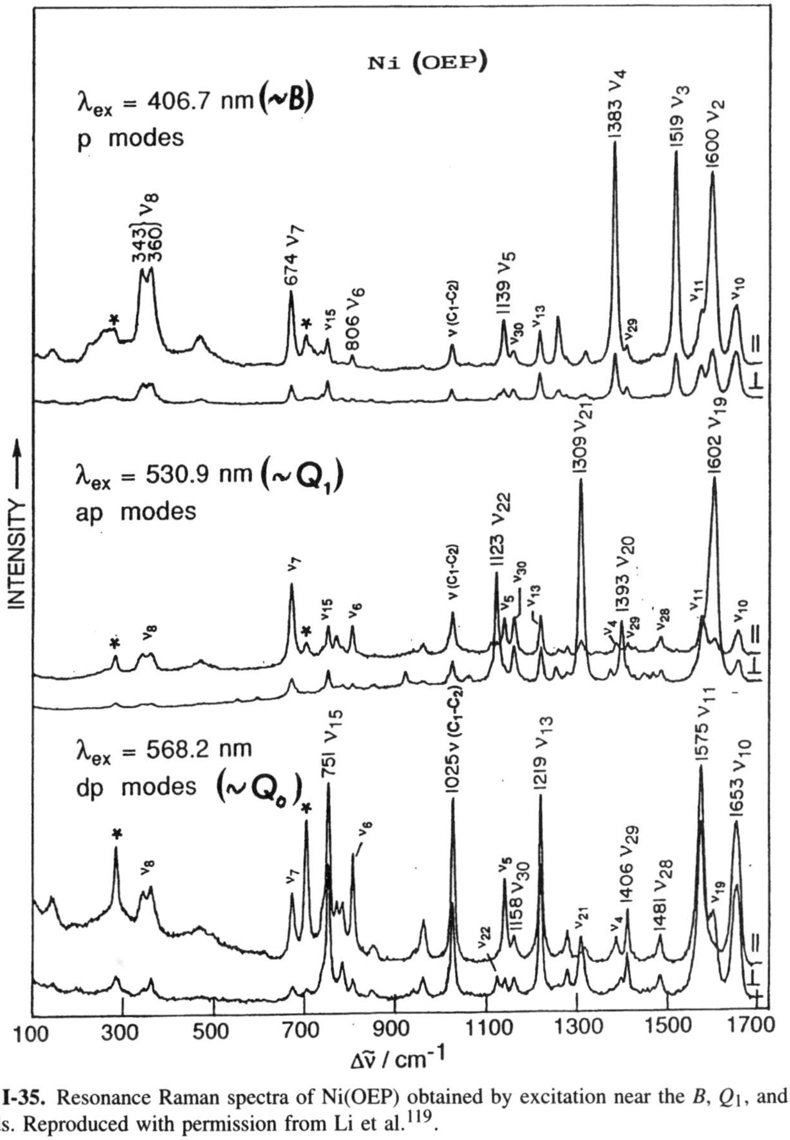

Fig. I-35. Resonance Raman spectra of Ni(OEP) obtained by excitation near the B, Q_1, and Q_0 bands. Reproduced with permission from Li et al.[119].

I-24. VIBRATIONAL SPECTRA IN GASEOUS PHASE[123] AND INERT GAS MATRICES[52,53]

As distinct from molecules in condensed phases, those in the gaseous phase are free from intermolecular interactions. If the molecules are relatively small, vibrational spectra of gases exhibit rotational fine structure (see Fig. I-3) from which moments of inertia and hence internuclear distances and bond angles can be calculated.[1,2] Furthermore, detailed analysis of rotational fine structure provides information about the magnitude of rotation–vibration interaction (Coriolis coupling), centrifugal distortion, anharmonicity, and even nuclear spin statistics in some cases. In the past, infrared spectroscopy was the main tool in measuring gas-phase vibrational spectra. Recently, Raman spectroscopy has been playing a significant role because of the development of powerful laser sources and high-resolution spectrophotometers. For example, Clark and Rippon[124] measured gas-phase Raman spectra of Group IV tetrahalides, and calculated the Coriolis coupling constants from the observed band contours. For gas-phase Raman spectra of other inorganic compounds, see Ref. 123. Unfortunately, the majority of inorganic and coordination compounds exist as solids at room temperature. Although some of these compounds can be vaporized at high temperatures without decomposition, it is rather difficult to measure their spectra by the conventional method. Furthermore, high-temperature spectra are difficult to interpret because of the increased importance of rotational and vibrational hot bands.

In 1954, Pimentel and his co-workers[125] developed the *matrix isolation technique* to study the infrared spectra of unstable and stable species. In this method, solute molecules and inert gas molecules such as Ar and N_2 are mixed at a ratio of 1 : 500 or greater and deposited on an IR window such as a CsI crystal cooled to 10–15 K. Since the solute molecules trapped in an inert gas matrix are completely isolated from each other, the matrix isolation spectrum is similar to the gas-phase spectrum; no crystal field splittings and no lattice modes are observed. However, the former spectrum is simpler than the latter because, except for a few small hydride molecules, no rotational transitions are observed because of the rigidity of the matrix environment at low temperatures. The lack of rotational structure and intermolecular interactions results in a sharpening of the solute band so that even very closely located metal isotope peaks can be resolved in a matrix environment. This technique is also applicable to a compound which is not volatile at room temperature. For example, matrix isolation spectra of metal halides can be measured by vaporizing these compounds at high temperatures in a Knudsen cell and co-condensing their vapors with inert gas molecules on a cold window.[126] The recent development of closed-cycle helium refrigerators greatly facilitated the experimental technique. The matrix isolation technique has now been applied to a number of inorganic and coordination compounds to obtain structural and bonding information. Some important applications of this technique are described below.

(1) Vibrational Spectra of Radicals

Highly reactive radicals can be produced *in situ* in inert gas matrices by photolysis and other techniques. Since these radicals are stabilized in matrix environments, their spectra can be measured by routine spectroscopic techniques. For example, the spectrum of the HOO radical[127] was obtained by measuring the spectrum of the photolysis product of a mixture of HI and O_2 in an Ar matrix at ~4 K. Section II lists the vibrational frequencies of many other radicals, such as CN, OF, and HSi, obtained by similar methods.

(2) Vibrational Spectra of High-Temperature Species

Alkali halide vapors produced at high temperatures consist mainly of monomers and dimers. The vibrational spectra of these salts at high temperatures are difficult to measure and difficult to interpret because of the presence of hot bands. The matrix isolation technique utilizing a Knudsen cell has solved this problem. The vibrational frequencies of some of these high-temperature species are listed in Section II.

(3) Isotope Splittings

As stated before, it is possible to observe individual peaks due to heavy metal isotopes in inert gas matrices since the bands are extremely sharp (half-band width, 1.5–1.0 cm^{-1}) under these conditions. Figure I-36*a* shows the infrared

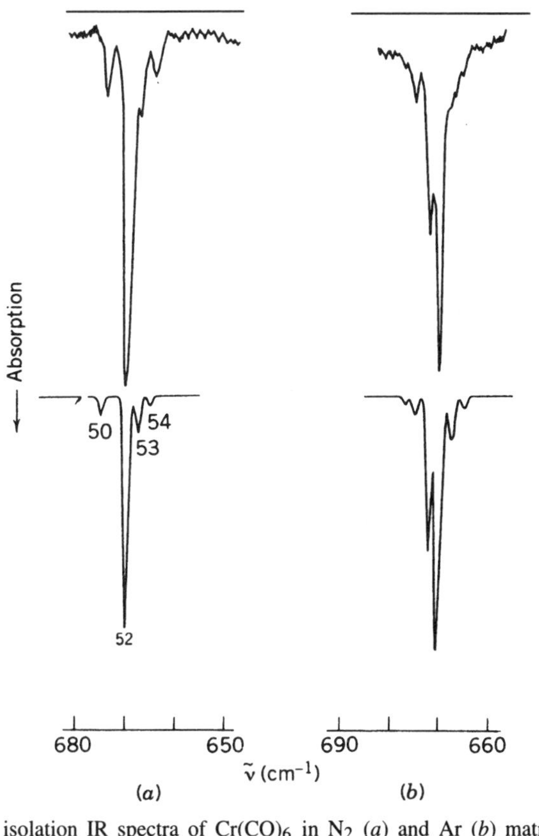

Fig. I-36. Matrix isolation IR spectra of $Cr(CO)_6$ in N_2 (*a*) and Ar (*b*) matrices. The bottom spectra were obtained by computer simulation.[128]

spectrum of the ν_7 band (coupled vibration between CrC stretching and CrCO bending modes) of $Cr(CO)_6$ in a N_2 matrix.[128] The bottom curve shows a computer simulation using the measured isotope shift of 2.5 cm^{-1} per atomic mass unit, a 1.2 cm^{-1} half-band width, and the percentages of natural abundance of Cr isotopes: ^{50}Cr (4.31%), ^{52}Cr (83.76%), ^{53}Cr (9.55%), and ^{54}Cr (2.38%). The isotope splittings of $SnCl_4$ and $GeCl_4$ in Ar matrices are discussed in Sec. II-6. These isotope frequencies are highly important in refining force constants in normal coordinate analysis.

(4) Chemical Synthesis

The matrix co-condensation technique can be used to synthesize a number of unstable and transient coordination compounds. For example, a series of nickel carbonyls of the type $Ni(CO)_x$, where $x = 1$, 2, 3, and 4, have been synthesized by allowing metal vapor to react with CO diluted in Ar on a cold window and warming the matrix carefully. Figure I-37 shows the result obtained

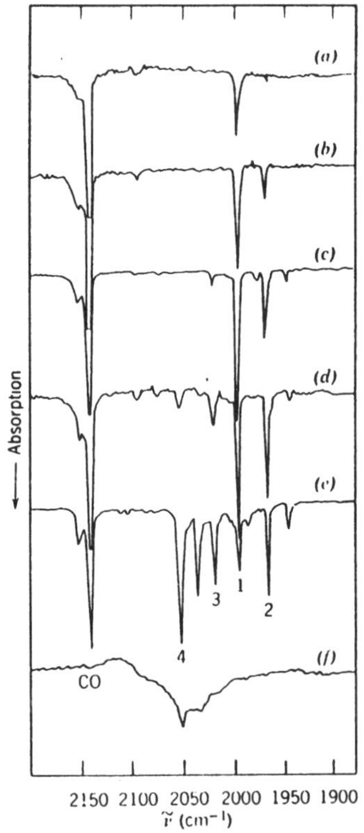

Fig. I-37. Infrared spectra of Ni atoms deposited in a 500:1 Ar/CO matrix and subsequent annealing: (*a*) original, (*b*) 17 K, (*c*) 18 K, (*d*) 19 K, (*e*) 26 K, and (*f*) 35 K (temperatures are relative). The arabic numerals refer to the relative rate of growth and disappearance of the bands and hence to *n* in $Ni(CO)_n$.[129]

by DeKock.[129] The structures of $Ni(CO)_2$ and $Ni(CO)_3$ were concluded to be linear and trigonal-planar, respectively, since these compounds exhibit only one CO stretching band in the infrared. Similar methods have been applied to the synthesis of a number of coordination compounds ML_x, where M is Pt, Pd, Ni, and so on, and L is CO, N_2, O_2, PF_3, and so on.[53] More detailed discussions of individual compounds will be given in Section III of Part B.

(5) Matrix Effect

The vibrational frequencies of matrix-isolated molecules give slight shifts when the matrix gas is changed. This result suggests the presence of weak interaction between the solute and matrix molecules. In some cases, the spectra are complicated by the presence of more than one trapping site. For example, the infrared spectrum of $Cr(CO)_6$ in an Ar matrix (Fig. I-36*b*) is markedly different from that in a N_2 matrix (Fig. I-36*a*).[128] The former spectrum can be interpreted by assuming two different sites in an Ar matrix. The computer-simulation spectrum (bottom curve) was obtained by assuming that these two sites are populated in a 2 : 1 ratio, the frequency separation of the corresponding peaks being 2 cm^{-1}. Thus, it is always desirable to obtain the matrix isolation spectra in several different environments.

I-25. SYMMETRY IN CRYSTALS

The symmetry elements and point groups of molecules and ions in the free state have been discussed in Sec. I-5. For molecules and ions in crystals, however, it is necessary to consider some additional symmetry operations which characterize translational symmetries in the lattice. Addition of these translational operations results in the formation of the space groups which can be used to classify the symmetry of molecules and ions in crystals.

(1) 32 Crystallographic Point Groups As discussed in Sec. I-5, molecular symmetry can be described by point groups which are derived from the combinations of symmetry operations, I (identity), C_p (p-fold axis of symmetry), σ (plane of symmetry), i (center of symmetry), and S_p (p-fold rotation–reflection axis), where p may range from 1 to ∞. In the case of crystal symmetry, p is limited to 1, 2, 3, 4, and 6 because of space-filling requirement in a crystal lattice. As a result, only the 32 crystallographic point groups listed in Table I-15 are possible. Here, Hermann–Mauguin (H–M) as well as Schönflies (S) notations are shown. In the H-M notation, 1, 2, 3, 4, and 6 denote the principal axis of rotation, C_p, where p is 1, 2, 3, 4, and 6, respectively, and $\bar{1}$, $\bar{2}$, $\bar{3}$, $\bar{4}$, and $\bar{6}$ denote a combination of respective rotation about an axis and inversion through a point on the axis. Thus, $\bar{1}$ means the center of inversion (i), $\bar{2}$ indicates σ on a mirror plane (m), and $\bar{3}$, $\bar{4}$, and $\bar{6}$ correspond to S_6, S_4, and S_3, respectively. For example, the point group $\mathbf{C}_{2h}$ (I, C_2, α, and i) can be written

TABLE I-15. Seven Crystallographic Systems and Thirty-two Crystallographic Point Groups

System	Axes and Angles	32 Crystallographic Point Groups						
Triclinic	*a* *b* *c* α β γ	1 C_1	$\bar{1}$ C_i					
Monoclinic	*a* *b* *c* 90° β 90°	2 C_2	*m* or $\bar{2}$ C_s	2/*m* C_{2h}				
Orthorhombic	*a* *b* *c* 90° 90° 90°				222 D_2	*mm*2 C_{2v}	*mmm* D_{2h}	
Tetragonal	*a* *a* *c* 90° 90° 90°	4 C_4	$\bar{4}$ S_4	4/*m* C_{4h}	422 D_4	4*mm* C_{4v}	$\bar{4}2m$ D_{2d}	4/*mmm* D_{4h}
Trigonal (rhombohedral)	*aaa* / ααα or *a* *a* *c* / 90° 90° 120°	3 C_3	$\bar{3}$ C_{3i}		32 D_3	3*m* C_{3v}	$\bar{3}m$ D_{3d}	
Hexagonal	*a* *a* *c* 90° 90° 120°	6 C_6	$\bar{6}$ C_{3h}	6/*m* C_{6h}	622 D_6	6*mm* C_{6v}	$\bar{6}m2$ D_{3h}	6/*mmm* D_{6h}
Cubic	*a* *a* *a* 90° 90° 90°	23 T	*m*3 T_h		432 O	$\bar{4}3m$ T_d	*m*3*m* O_h	

as $2m\bar{1}$, which can be simplified to $2/m$ where the slash means the C_2 axis is perpendicular to *m*. As shown in Table I-15, these 32 crystallographic point groups which describe the symmetry of the unit cell can be classified into seven crystallographic systems according to the most simple choice of reference axes.

(2) Symmetry of the Crystal Lattice A crystal has a lattice structure which is a repetition of identical units. Bravais has shown that only the 14 different types of lattices shown in Fig. I-38 are possible. These Bravais lattices are also classified into the seven crystallographic systems mentioned above. Here, the lattice points are regarded as "points" or "small, indentical, perfect spheres," thus representing the highest symmetry possible. In actual crystals, these points are occupied by unsymmetrical but identical molecules or ions so that their high symmetry is lowered. In Fig. I-38, P, C (or A or B), F, I, and R indicate primitive, base-centered, face-centered, body-centered, and rhombohedral lattices, respectively. The number of lattice points (LP) in each cell is summarized in Table I-16. For crystal structures designated by P, crystallographic unit cells

TABLE I-16. Primitive and Centered Lattices

Type of Crystal Structure	Number of Lattice Points (LP)
Primitive (P)	1
Base-centered (A, B, or C)	2
Body-centered (I)	2
Rhombohedral (R)	3 or 1[a]
Face-centered (F)	4

[a] The crystallographic unit cell may have been divided by three.

are identical with the Bravais unit cells. For those designated by other letters, crystallographic unit cells contain 2, 3, or 4 Bravais cells. In these cases, the number of molecules in the crystallographic unit cell must be divided by the corresponding LP since only the consideration of one Bravais cell is sufficient to interpret the vibrational spectra of crystals.

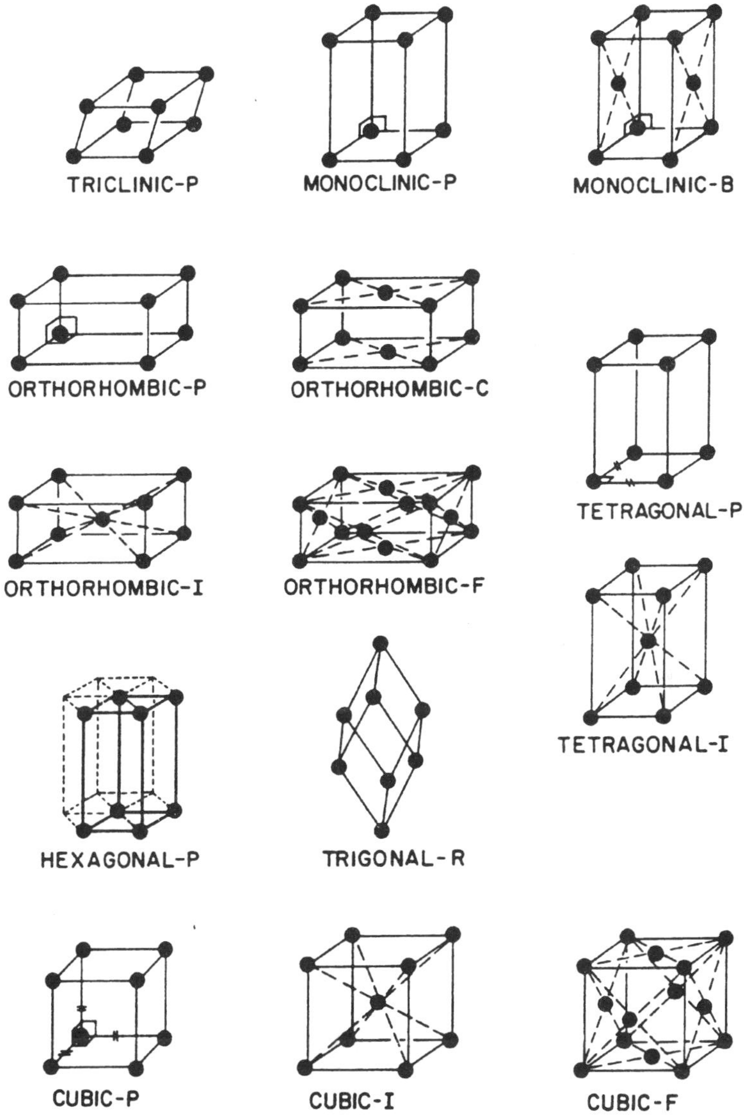

Fig. I-38. The 14 Bravais lattices belonging to the seven crystallographic systems. Reproduced with permission from Burns and Glazer[129a].

(3) Space Groups As stated above, the 32 point groups describe the symmetry of the unit cells, and the 14 Bravais lattices describe all possible arrangements of crystal lattices. To describe space symmetry, however, symmetry operations mentioned earlier are not enough. We must add three new symmetry operations.

Translation A translational operation along one lattice direction gives an identical point.

Screw Axis Rotation of a lattice point about an axis followed by translation gives an identical point. Namely, rotation by $2\pi/p$ followed by a translation of $q/p(q = 1, 2, \ldots, n-1)$ in the direction of the axis is written as p_q. For example, 2_1 indicates twofold rotation followed by a translation of $(1/2)a$, $(1/2)b$, or $(1/2)c$, as shown in Fig. I-39A. If the axis is threefold, the translation must be $1/3$ or $2/3$, and the symbols are 3_1 and 3_2, respectively.

Glide Plane Reflection in a plane followed by translation by $(1/2)a$, $(1/2)b$, $(1/2)c$, as shown in Fig. I-39B. These planes are indicated by a, b, and c, respectively. n is used for diagonal translation of $(b+c)/2$, $(c+a)/2$, or $(a+b)/2$.

Addition of these symmetry operations results in 230 different combinations which are called "space groups." Table I-17 shows the distribution of space groups among the seven crystal systems.[129b] Some of these space groups are rarely found in actual crystals. About a half of crystals belong to 13 space groups of the monoclinic system. Space groups can be described either by using Hermann–Mauguin (H–M) or Schönflies (S) notations. For example, P2/m ($\equiv C_{2h}^1$) indicates a primitive lattice with m perpendicular to 2. The first symbol (capital letter) refers to the Bravais lattice (P, A, B, C, F, I and R) and the next

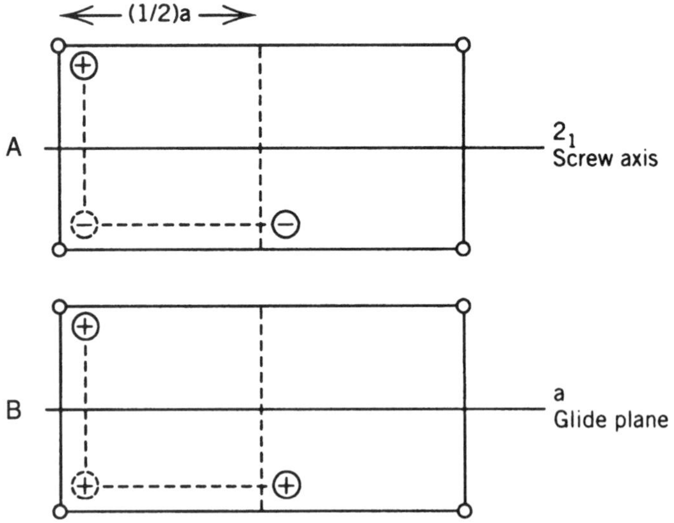

Fig. I-39. Screw axis and glide plane.

TABLE I-17. The 230 Space Groups*

	Point Groups		Space groups						
	Schfl.	H.-M.							
Triclinic system	C_1	1	$P1$						
	C_i	$\bar{1}$	$P\bar{1}$						
Monoclinic system	$C_2{}^{(1-3)}$	2	$P2$	$P2_1$	$C2$				
	$C_S{}^{(1-4)}$	m	Pm	Pc	Cm	Cc			
	$C_{2h}{}^{(1-6)}$	$2/m$	$P2/m$	$P2_1/m$	$C2/m$	$P2/c$	$P2_1/c$	$C2/c$	
Orthorhombic system	$D_2{}^{(1-9)}$	222	$P222$ $I222$	$P222_1$ $I2_12_12_1$	$P2_12_12$	$P2_12_12_1$	$C222_1$	$C222$	$F222$
	$C_{2v}{}^{(1-22)}$	$mm2$	$Pmm2$ $Pba2$ $Abm2$ $Ima2$	$Pmc2_1$ $Pna2_1$ $Ama2$	$Pcc2$ $Pnn2$ $Aba2$	$Pma2$ $Cmm2$ $Fmm2$	$Pca2_1$ $Cmc2_1$ $Fdd2$	$Pnc2$ $Ccc2$ $Imm2$	$Pmn2_1$ $Amm2$ $Iba2$
	$D_{2h}{}^{(1-28)}$	mmm	$Pmmm$ $Pcca$ $Pbca$ $Ccca$	$Pnnn$ $Pbam$ $Pnma$ $Fmmm$	$Pccm$ $Pccn$ $Cmcm$ $Fddd$	$Pban$ $Pbcm$ $Cmca$ $Immm$	$Pmma$ $Pnnm$ $Cmmm$ $Ibam$	$Pnna$ $Pmmn$ $Cccm$ $Ibca$	$Pmna$ $Pbcn$ $Cmma$ $Imma$
Tetragonal system	$C_4{}^{(1-6)}$	4	$P4$	$P4_1$	$P4_2$	$P4_3$	$I4$	$I4_1$	
	$S_4{}^{(1-2)}$	$\bar{4}$	$P\bar{4}$	$I\bar{4}$					
	$C_{4h}{}^{(1-6)}$	$4/m$	$P4/m$	$P4_2/m$	$P4/n$	$P4_2/n$	$I4/m$	$I4_1/a$	

TABLE I-17. (*Continued*)

	Point groups		Space groups						
	Schfl.	H.-M.							
Tetragonal system	$D_4^{(1-10)}$	422	$P422$	$P42_12$	$P4_122$	$P4_12_12$	$P4_222$	$P4_22_12$	$P4_322$
			$P4_32_12$	$I422$	$I4_122$				
	$C_{4v}^{(1-12)}$	$4mm$	$P4mm$	$P4bm$	$P4_2cm$	$P4_2nm$	$P4cc$	$P4nc$	$P4_2mc$
			$P4_2bc$	$I4mm$	$I4cm$	$I4_1md$	$I4_1cd$		
	$D_{2d}^{(1-12)}$	$\bar{4}2m$	$P\bar{4}2m$	$P\bar{4}2c$	$P\bar{4}2_1m$	$P\bar{4}2_1c$	$P\bar{4}m2$	$P\bar{4}c2$	$P\bar{4}b2$
			$P\bar{4}n2$	$I\bar{4}m2$	$I\bar{4}c2$	$I\bar{4}2m$	$I\bar{4}2d$		
	$D_{4h}^{(1-20)}$	$4/mmm$	$P4/mmm$	$P4/mcc$	$P4/nbm$	$P4/nnc$	$P4/mbm$	$P4/mnc$	$P4/nmm$
			$P4/ncc$	$P4_2/mmc$	$P4_2/mcm$	$P4_2/nbc$	$P4_2/nnm$	$P4_2/mbc$	$P4_2/mnm$
			$P4_2/ncm$	$P4_2/ncm$	$I4/mmm$	$I4/mcm$	$I4_1/amd$	$I4_1/acd$	
Trigonal system	$C_3^{(1-4)}$	3	$P3$	$P3_1$	$P3_2$	$R3$			
	$C_{3i}^{(1-2)}$	$\bar{3}$	$P\bar{3}$	$R\bar{3}$					
	$D_3^{(1-7)}$	32	$P312$	$P321$	$P3_112$	$P3_121$	$P3_212$	$P3_221$	$R32$
	$C_{3v}^{(1-6)}$	$3m$	$P3m1$	$P31m$	$P3c1$	$P31c$	$R3m$	$R3c$	
	$D_{3d}^{(1-6)}$	$\bar{3}m$	$P\bar{3}1m$	$P\bar{3}1c$	$P\bar{3}m1$	$P\bar{3}c1$	$R\bar{3}m$	$R\bar{3}c$	

Hexagonal system	$C_6^{(1-6)}$	6	$P6$	$P6_1$	$P6_5$	$P6_2$	$P6_4$	$P6_3$	
	$C_{3h}^{(1)}$	$\bar{6}$	$P\bar{6}$						
	$C_{6h}^{(1-2)}$	$6/m$	$P6/m$	$P6_3/m$					
	$D_6^{(1-6)}$	622	$P622$	$P6_122$	$P6_522$	$P6_222$	$P6_422$	$P6_322$	
	$C_{6v}^{(1-4)}$	$6mm$	$P6mm$	$P6cc$	$P6_3cm$	$P6_3mc$			
	$D_{3h}^{(1-4)}$	$\bar{6}m2$	$P\bar{6}m2$	$P\bar{6}c2$	$P\bar{6}2m$	$P\bar{6}2c$			
	$D_{6h}^{(1-4)}$	$6/mmm$	$P6/mmm$	$P6/mcc$	$P6_3/mcm$	$P6_3/mmc$			
Cubic system	$T^{(1-5)}$	23	$P23$	$F23$	$I23$	$P2_13$	$I2_13$		
	$T_h^{(1-7)}$	$m3$	$Pm3$	$Pn3$	$Fm3$	$Fd3$	$Im3$	$Pa3$	$Ia3$
	$O^{(1-8)}$	432	$P432$ $I4_132$	$P4_232$	$F432$	$F4_132$	$I432$	$P4_332$	$P4_132$
	$T_d^{(1-6)}$	$\bar{4}3m$	$P\bar{4}3m$	$F\bar{4}3m$	$I\bar{4}3m$	$P\bar{4}3n$	$F\bar{4}3c$	$I\bar{4}3d$	
	$O_h^{(1-10)}$	$m3m$	$Pm3m$ $Fd3c$	$Pn3n$ $Im3m$	$Pm3n$ $Ia3d$	$Pn3m$	$Fm3m$	$Fm3c$	$Fd3m$

*Reproduced with permission from Robertson[129b].

symbol indicates fundamental symmetry elements which lie along the special direction in the crystal. In the case of the monoclinic system, it is the twofold axis. Thus, $P2_1/m$ ($\equiv C_{2h}^2$) is a primitive cell with m perpendicular to 2_1. These symbols indicate only the minimum symmetry elements that are necessary to distinguish the space groups. The differences among space groups can be found by comparing the "unit cell maps" given in the "International Tables for X-Ray Crystallography."[130]

I-26. VIBRATIONAL ANALYSIS OF CRYSTALS[56–60]

Because of intermolecular interactions, the symmetry of a molecule is generally lower in the crystalline state than in the gaseous (isolated) state. This change in symmetry may split the degenerate vibrations and activate infrared- (or Raman-) inactive vibrations. Secondly, intermolecular (interionic) interactions in the crystal lattice cause frequency shifts relative to the gaseous state. Finally, the spectra obtained in the crystalline state exhibit *lattice modes*—vibrations due to translatory and rotatory motions of a molecule in the crystalline lattice. Although their frequencies are usually lower than 300 cm^{-1}, they may appear in the high-frequency region as the combination bands with internal modes (see Fig. II-13, for example). Thus, the vibrational spectra of crystals must be interpreted with caution, especially in the low-frequency region.

To analyze the spectra of crystals, it is necessary to carry out a site-group or factor-group analysis, as described in the following subsection.

(1) Subgroups and Correlation Tables

For any point group, there are "subgroups" which consist of some, but not all, symmetry elements of the "parent group." For example, the character table of the point group, $\mathbf{C}_{3v}$ (Table I-6) contains I, C_3, and σ_v as the symmetry elements. Then, the subgroups of $\mathbf{C}_{3v}$ are $\mathbf{C}_3$, consisting of I and C_3, and $\mathbf{C}_s$, consisting of I and σ_v only. The relationship between the symmetry species in the "parent group" and those in "subgroups" is given in a "correlation table." Such correlation tables have already been worked out for all common point groups and are listed in Appendix IX. As an example, the correlation table for $\mathbf{C}_{3v}$ is shown below:

$\mathbf{C}_{3v}$	$\mathbf{C}_3$	$\mathbf{C}_s$
A_1	A	A'
A_2	A	A''
E	E	$A' + A''$

A pyramidal XY_3 molecule such as NH_3 belongs to the point group $\mathbf{C}_{3v}$. If one of the Y atoms is replaced by a Z atom, the resulting XY_2Z molecule belongs to the point group $\mathbf{C}_s$. As a result, the doubly degenerate vibration (E) splits into two vibrations ($A' + A''$). These correlation tables are highly important in predicting the effect of lowering of symmetry on molecular vibrations. Section II includes a number of examples of symmetry lowering by substitution of an atomic group by another group.

(2) Site Group Analysis

According to Halford,[131] the vibrations of a molecule in the crystalline state are governed by a new selection rule derived from *site symmetry*—a local symmetry around the center of gravity of a molecule in a unit cell. The site symmetry can be found by using the following two conditions: (1) the site group must be a subgroup of both the space group of the crystal and the molecular point group of the isolated molecule, and (2) the number of equivalent sites must be equal to the number of molecules in the unit cell. Halford derived a complete table that lists possible site symmetries and the number of equivalent sites for 230 space groups. Suppose that the space group of the crystal, the number of molecules in the unit cell (Z), and the point group of the isolated molecule are known. Then, the site symmetry can be found from the "Table of Site Symmetry for the 230 Space Groups" which was originally derived by Halford. Appendix X gives its modified version by Ferraro and Ziomek.[17] In general, the site symmetry is lower than the molecular symmetry in an isolated state. In some cases, it may be difficult to make an unambiguous choice of site symmetry by the method cited above. Then, Wyckoff's tables on crystallographic data[132] must be consulted.

The vibrational spectra of calcite and aragonite crystals are markedly different, although both have the same composition (Sec. II-4). This result can be explained if we consider the difference in site symmetry of the CO_3^{2-} ion between these crystals. According to X-ray analysis, the space group of calcite is $\mathbf{D}_{3d}^6$ and Z is two, Appendix X gives

$$\mathbf{D}_3(2), \quad \mathbf{C}_{3i}(2), \quad \mathbf{C}_3(4), \quad \mathbf{C}_i(6), \quad \mathbf{C}_2(6), \quad \mathbf{C}_1(12)$$

as possible site symmetries for space group $\mathbf{D}_{3d}^6$ (the number in front of point group notation indicates the number of distinct sets of sites, and that in parenthesis denotes the number of equivalent sites for each distinct set). Rule 2 eliminates all but $\mathbf{D}_3(2)$ and $\mathbf{C}_{3i}(2)$. Rule 1 eliminates the latter since $\mathbf{C}_{3i}$ is not a subgroup of $\mathbf{D}_{3h}$. Thus, the site symmetry of the CO_3^{2-} ion in calcite must be $\mathbf{D}_3$. On the other hand, the space group of aragonite is $\mathbf{D}_{2h}^{16}$ and Z is four. Appendix X gives

$$2\mathbf{C}_i(4), \quad \mathbf{C}_s(4), \quad \mathbf{C}_1(8)$$

Since $\mathbf{C}_i$ is not a subgroup of $\mathbf{D}_{3h}$, the site symmetry of the CO_3^{2-} ion in aragonite must be $\mathbf{C}_s$. Thus, the $\mathbf{D}_{3h}$ symmetry of the CO_3^{2-} ion in an isolated state

TABLE I-18. Correlation Table for D_{3h}, D_3, C_{2v}, and C_s

Point Group	ν_1	ν_2	ν_3	ν_4
D_{3h}	A_1'(R)	A_2''(I)	E'(I, R)	E'(I, R)
D_3	A_1(R)	A_2(I)	E(I, R)	E(I, R)
C_{2v}	A_1(I, R)	B_1(I, R)	A_1(I, R) + B_2(I, R)	A_1(I, R) + B_2(I, R)
C_s	A'(I, R)	A''(I, R)	A'(I, R) + A'(I, R)	A'(I, R) + A'(I, R)

is lowered to $\mathbf{D}_3$ in calcite and to $\mathbf{C}_s$ in aragonite. Then, the selection rules are changed as shown in Table I-18.

There is no change in the selection rule in going from the free CO_3^{2-} ion to calcite. In aragonite, however, ν_1 becomes infrared active, and ν_3 and ν_4 each split into two bands. The observed spectra of calcite and aragonite are in good agreement with these predictions (see Table II-4*b*).

(3) Factor Group Analysis

A more complete analysis including lattice modes can be made by the method of factor group analysis developed by Bhagavantam and Venkatarayudu.[133] In this method, we consider all the normal vibrations for an entire Bravais cell. Figure I-40 illustrates the Bravais cell of calcite, which consists

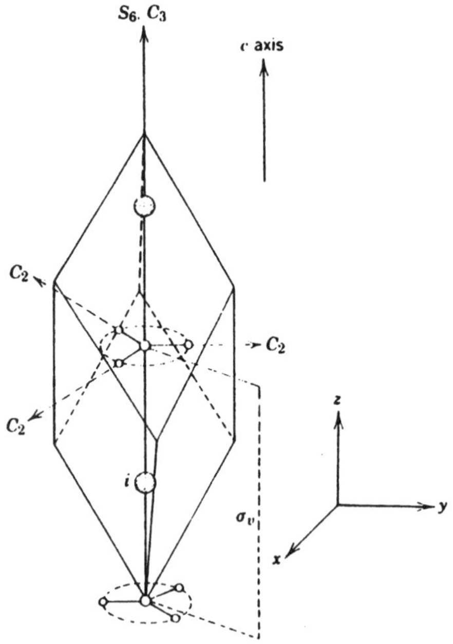

Fig. I-40. The Bravais cell of calcite.

of the following symmetry elements: I, $2S_6$, $2S_6^2 \equiv 2C_3$, $S_6^3 \equiv i$, $3C_2$, and $3\sigma_v$ (glide plane). These elements are exactly the same as those of the point group $\mathbf{D}_{3d}$, although the last element is a glide plane rather than a plane of symmetry in a single molecule.

As mentioned in Sec. I-25, it is possible to derive the 230 space groups by combining operations possessed by the 32 crystallographic point groups with operations such as pure translation, screw rotation (translation + rotation), and glide plane reflection (translation + reflection). If we regard the translations that carry a point in a unit cell into the equivalent point in another cell as identity, we define the 230 factor groups that are the subgroups of the corresponding space groups. In the case of calcite, the factor group consists of the symmetry elements described above, and is denoted by the same notation as that used for the space group ($\mathbf{D}_{3d}^6$). The site group discussed previously is a subgroup of a factor group.

Since the Bravais cell contains 10 atoms, it has $3 \times 10 - 3 = 27$ normal vibrations, excluding three translational motions of the cell as a whole.* These 27 vibrations can be classified into various symmetry species of the factor group $\mathbf{D}_{3d}^6$, using a procedure similar to that described in Sec. I-8 for internal vibrations. First, we calculate the characters of representations corresponding to the entire freedom possessed by the Bravais cell [$\chi_R(N)$], translational motions of the whole cell [$\chi_R(T)$], translatory lattice modes [$\chi_R(T')$], rotatory lattice modes [$\chi_R(R')$], and internal modes [$\chi_R(n)$], using the equations given in Table I-19. Then, each of these characters is resolved into the symmetry species of the point group, $\mathbf{D}_{3d}$. The final results show that three internal modes (A_{2u} and two E_u), three translatory modes (A_{2u} and two E_u), and two rotatory modes (A_{2u} and E_u) are infrared active, and three internal modes (A_{1g} and two E_g), one translatory mode (E_g), and one rotatory mode (E_g) are Raman active. As will be shown in the following subsections, these predictions are in perfect agreement with the observed spectra.

I-27. THE CORRELATION METHOD

In the preceding section, we described the application of factor group analysis to the calcite crystal. However, the correlation method developed largely by Fateley et al.[19] is simpler and gives the same results. In this method, intramolecular and lattice vibrations are classified under point groups of molecular symmetry, site symmetry, and factor group symmetry, and correlations are made using the correlation tables given in Appendix IX. In the following, we demonstrate its utility using calcite as an example. For more detailed discussions and applications, the reader should consult the books by Fateley et al.[19] and Ferraro and Ziomek.[17]

*These three motions give acoustical modes (Sec. I-28).

TABLE I-19. Factor Group Analysis of Calcite Crystal

$\mathbf{D}_{3d}^{6}$	I	$2S_6$	$2C_3 \equiv 2S_6^2$	$i \equiv S_6^3$	$3C_2$	$3\sigma_v$	Number of Vibrations N	T	T'	R'	n		
A_{1g}	1	1	1	1	1	1	1	0	0	0	1		$\alpha_{xx}+\alpha_{yy}, \alpha_{zz}$
A_{1u}	1	−1	1	−1	1	−1	2	0	1	0	1		
A_{2g}	1	1	1	1	−1	−1	3	0	1	1	1		
A_{2u}	1	−1	1	−1	−1	1	4	1	1	1	1	T_z	
E_g	2	−1	−1	2	0	0	4	0	1	1	2		$(\alpha_{xx}-\alpha_{yy}, \alpha_{xy}), (\alpha_{xz}, \alpha_{yx})$
E_u	2	1	−1	−2	0	0	6	1	2	1	2	(T_x, T_y)	
$N_R(p)$	10	2	4	2	4	0	↑	↑	↑	↑	↑		
$N_R(s)$	4	2	4	2	2	0							
$N_R(s-v)$	2	0	2	0	2	0							
$\chi_R(N)$	30	0	0	−6	−4	0							
$\chi_R(T)$	3	0	0	−3	−1	1							
$\chi_R(T')$	9	0	0	−3	−1	−1							
$\chi_R(R')$	6	0	0	0	−2	0							
$\chi_R(n)$	12	0	0	0	0	0							

p, total number of atoms in the Bravais cell.
s, total number of molecules (ions) in the Bravais cell.
v, total number of monoatomic molecules (ions) in the Bravais cell.
$N_R(p)$, number of atoms unchanged by symmetry operation R.
$N_R(s)$, number of molecules (ions) whose center of gravity is unchanged by symmetry operation R.
$N_R(s-v)$, $N_R(s)$ minus number of monoatomic molecules (ions) unchanged by symmetry operation R.
$\chi_R(N) = N_R(p)[\pm(1+2\cos\theta)]$, character of representation for entire freedom possessed by the Bravais cell.
$\chi_R(T) = \pm(1+2\cos\theta)$, character of representation for translational motions of the whole Bravais cell.
$\chi_R(T') = \{N_R(s)-1\}\{\pm(1+2\cos\theta)\}$ character of representation for translatory lattice modes.
$\chi_R(R') = N_R(s-v)\{\pm(1+2\cos\theta)\}$, character of representation for rotary lattice modes.
$\chi_R(n) = \chi_R(N) - \chi_R(T) - \chi_R(T') - \chi_R(R')$, character of representation for internal modes.
Note that + and − signs are for proper and improper rotations, respectively. The symbol θ should be taken as defined in Sec. I-8.

(1) The CO_3^{2-} Ion in the Free State As mentioned in Sec. I-4, normal vibrations of a molecule can be described in terms of translational motions of the individual atoms along the x, y, and z axes. Thus, the normal modes of an N-atom molecule can be expressed by using $3N$ translational motions. Furthermore, the symmetry species of these normal modes must correlate with those of the translational motions of an atom located at a particular site within the molecule. Since the latter are known from the molecular structure, the former may be determined directly by using the correlation table.

As an example, consider a planar CO_3^{2-} ion for which the x, y, and z axes are chosen as shown in Fig. I-41. It is readily seen that the C atom is situated at the $\mathbf{D}_{3h}$ site, whereas the O atom is situated at a site where the local symmetry is only $\mathbf{C}_{2v}$ (I, C_2, σ_h, and σ_v). The translational motions of the C atom under $\mathbf{D}_{3h}$ symmetry are: $T_z(A_2'')$ and $(T_x, T_y)(E')$. The translational motions of the O atom under $\mathbf{C}_{2v}$ symmetry are: $T_x(A_1)$, $T_y(B_2)$, and $T_z(B_1)$. In Table I-20, the symmetry species of these translational motions are connected by arrows to those of the whole ion using the correlation table. Then, the number of times a particular symmetry species occurs in the total representation is given by the number of arrows which terminate on that species. The number of normal vibrations of the CO_3^{2-} ion in each symmetry species is given by subtracting those of the translational and rotational motions of the whole ion from the total representation. The results are shown at the bottom of Table I-20.

Thus, the CO_3^{2-} ion exhibits four normal vibrations; $\nu_1(A_1'$, Raman-active), $\nu_2(A_2''$, IR-active), $\nu_3(E'$, IR- and Raman-active), and $\nu_4(E'$, IR- and Raman-active). The normal modes of these four vibrations are shown in Fig. II-8.

In Sec. I-8, we derived the general method to classify normal vibrations into symmetry species based on group theory. As seen above, this procedure is greatly simplified if we use the correlation method.

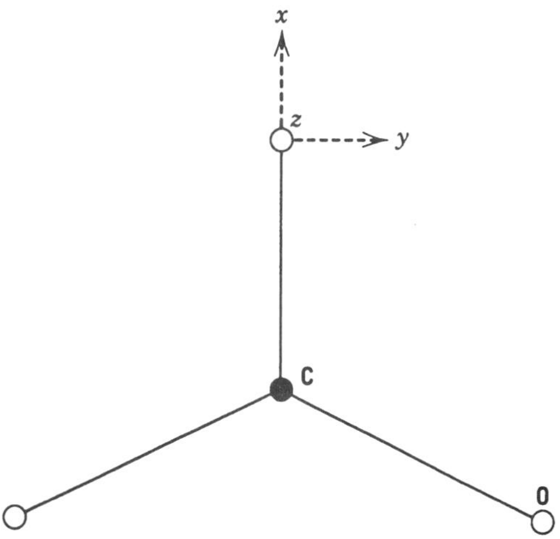

Fig. I-41. The x, y, and z axes chosen for planar CO_3^{2-} ion.

TABLE I-20. Correlation Method for the CO_3^{2-} Ion in the Free State

C Atom $\mathbf{D}_{3h}$	CO_3^{2-} Ion $\mathbf{D}_{3h}$	O Atom $\mathbf{C}_{2v}$
	A_1'	
	$A_2'(R_z)$	$A_1(T_X)$
	A_1''	
$A_2''(T_z)$	$A_2''(T_z)$	$B_2(T_y)$
$E'(T_X, T_y)$	$E'(T_X, T_y)$	$B_1(T_z)$
	$E''(R_X, R_y)$	

$\chi_{(\text{total})} = A_1' + A_2' + 2A_2'' + 3E' + E''$

$\chi_{(\text{trans})} = A_2'' + E'$

$\chi_{(\text{rot})} = A_2' + E''$

$\chi_{(\text{vib})} = A_1' + A_2'' + 2E'$

(2) Intramolecular Vibrations in the CO_3^{2-} Ion in Calcite Based on the same principle as that used above, we can classify the normal vibrations of the CO_3^{2-} ion in the calcite lattice by using the correlation method. As discussed in Sec. I-26, calcite belongs to the space group $\mathbf{D}_{3d}^6$, and Z is two. From the Bravais cell shown in Fig. I-40, it is readily seen that the CO_3^{2-} ion is at the $\mathbf{D}_3$ site, whereas the Ca^{2+} ion is at the $\mathbf{C}_{3i}$ ($\equiv \mathbf{S}_6$) site. Table I-21 shows the correlations among the molecular symmetry ($\mathbf{D}_{3h}$), site symmetry ($\mathbf{D}_3$), and factor group symmetry ($\mathbf{D}_{3d}$). Under the "vib" column, we list the number of vibrations belonging to each species of $\mathbf{D}_{3h}$ symmetry which was obtained in (1). Throughout the present correlations, the number of degrees of vibrational freedom for all the doubly degenerate species must be multiplied by 2 since there are two E' modes (ν_3 and ν_4). Thus, it is 4 for the E' species. The number under "f" indicates "vib" times Z (=2), which is the number of vibrations of the unit cell. In going from $\mathbf{D}_{3h}$ to $\mathbf{D}_3$, no changes occur except for changes in notations of symmetry species. Column C indicates the degeneracy, and column a_m shows the degree of vibrational freedom contributed by the corresponding molecular symmetry species. Finally, the species under $\mathbf{D}_3$ symmetry are connected to those of the factor group by the arrows using the correlation table. The last column, a_ξ, is the degree of vibrational freedom contributed by the corresponding site symmetry species to the factor group species. It should be noted that the number of degrees of vibrational freedom must be 12 throughout the described correlations above. Such bookkeeping must be carried out for every correlation.

(3) Lattice Vibrations in Calcite Table I-22 shows the correlation diagram for lattice vibrations of the CO_3^{2-} ion. The variables "t" and "f" denote the degrees of translational freedom of the CO_3^{2-} ion for each ion and for the Bravais cell, respectively. The same result is obtained for the rotatory lattice vibrations. Table I-23 shows the correlation diagram for translatory lattice vibrations

TABLE I-21. Correlation Among Molecular Symmetry, Site Symmetry, and Factor Group Symmetry for Intramolecular Vibrations of CO_3^{2-} Ion in Calcite

f	Vib.	Molecular Symmetry ($\mathbf{D}_{3h}$)	Site Symmetry ($\mathbf{D}_3$)	C a_m	Factor Group Symmetry ($\mathbf{D}_{3d}$)	C a_ζ
					A_{1g}	1×1
2	1	A_1' (ν_1) →	A_1	1×2		
					A_{1u}	1×1
					E_g	2×2
8	4	E' (ν_3, ν_4) →	E	2×4		
					E_u	2×2
					A_{2g}	1×1
2	1	A_2'' (ν_2) →	A_2	1×2		
					A_{2u}	1×1

$\chi(\text{intra}) = A_{1g} + A_{1u} + A_{2g} + A_{2u} + 2E_g + 2E_u$

TABLE I-22. Correlation Between Site Symmetry and Factor Group Symmetry for Lattice Vibrations of CO_3^{2-} Ion in Calcite

f	t	Site Symmetry ($\mathbf{D}_3$)	Factor Group Symmetry ($\mathbf{D}_{3d}$)	C	a_ζ
			$E_g(R_x, R_y)$	2	1
4	2	$E(T_x, T_y)$ (R_x, R_y)			
			$E_u(T_x, T_y)$	2	1
			$A_{2g}(R_z)$	1	1
2	1	$A_2(T_z)$ (R_z)			
			$A_{2u}(T_z)$	1	1

$\chi(\text{trans}, CO_3^{2-}) = E_g + E_u + A_{2g} + A_{2u}$

$\chi(\text{rot}, CO_3^{2-}) = E_g + E_u + A_{2g} + A_{2u}$

TABLE I-23. Correlation Between Site Symmetry and Factor Group Symmetry for Lattice Vibrations of the Ca^{2+} Ion in Calcite

f	t	Site Symmetry ($\mathbf{C}_{3i}$)	Factor Group Symmetry ($\mathbf{D}_{3d}$)	C	a_ζ
			A_{1u}	1	1
2	1	$A_u(T_z)$			
			$A_{2u}(T_z)$	1	1
4	2	$E_u(T_x, T_y)$ →	$E_u(T_x, T_y)$	2	2

$\chi(\text{trans}, Ca^{2+}) = A_{1u} + A_{2u} + 2E_u$

$\chi(\text{acoustical}) = A_{2u} + E_u$

$$\chi(\text{trans, total}) = \chi(\text{trans}, CO_3^{2-}) + \chi(\text{trans}, Ca^{2+}) - \chi(\text{acoustical}) = A_{1u} + A_{2g} + A_{2u} + E_g + 2E_u$$

TABLE I-24. Distribution of Normal Vibrations of Calcite as Obtained by the Correlation Method

Symmetry Species of Factor Group ($\mathbf{D}_{3d}$)	Translatory Lattice	Acoustical	Rotatory Lattice	Intramolecular
A_{1g} (Raman)	0	0	0	1
A_{1u}	1	0	0	1
A_{2g}	1	0	1	1
A_{2u} (IR)	1	1	1	1
E_g (Raman)	1	0	1	2
E_u (IR)	2	1	1	2
Total	9	3	6	12

of the Ca^{2+} ion. No rotatory lattice vibrations exist for single-atom ions such as the Ca^{2+} ion. The distribution of all the translatory lattice vibrations can be obtained by subtracting χ (acoustical) from χ (trans, CO_3^{2-}) + χ (trans, Ca^{2+}).

(4) Summary Table I-24 summarizes the results obtained in Tables I-21 to I-23. The total number of vibrations including the acoustical modes should be 30 since the Bravais cell contains 10 atoms. These results are in complete agreement with that obtained by factor group analysis (Table I-19).

I-28. LATTICE VIBRATIONS

Consider a lattice consisting of two alternate layers of atoms; atoms 1 of mass M_1 lie on one set of planes and atoms 2 of mass M_2 lie on another set of planes, as shown in Fig. I-42. For example, such an arrangement is found in the 111 plane of the NaCl crystal. Let a denote the distance between atoms 1 and 2. Then the length of the primitive cell is $2a$. We consider waves that propagate in the direction shown by the arrow, and assume that each plane interacts only with its neighboring planes. If the force constants (f) are identical between all these neighboring planes, we have

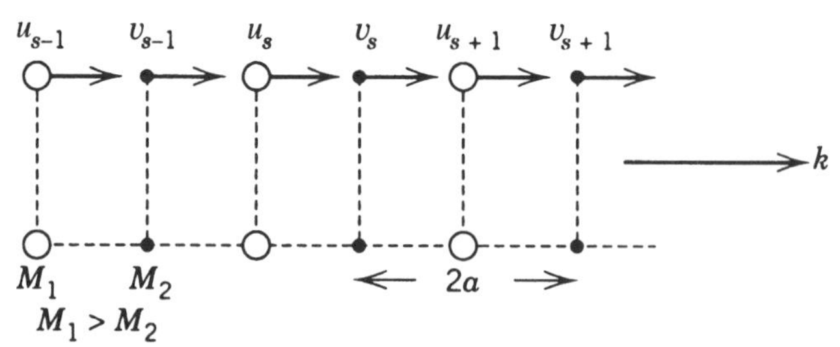

Fig. I-42. Displacements of atoms 1 and 2 in a diatomic lattice.

$$M_1 \frac{d^2 u_s}{dt^2} = f(v_s + v_{s-1} - 2u_s) \tag{28.1}$$

$$M_2 \frac{d^2 v_s}{dt^2} = f(u_{s+1} + u_s - 2v_s) \tag{28.2}$$

where u_s and v_s denote the displacements of atoms 1 and 2 in the cell indexed by s.

The solutions for these equations take the form of a traveling wave having different amplitudes u and v. Thus, we obtain

$$u_s = u \exp[i(\omega t + 2ska)] \tag{28.3}$$

$$v_s = v \exp[i(\omega t + 2ska)] \tag{28.4}$$

Here ω is angular frequency, $2\pi\nu$ (sec^{-1}). If these are substituted in Eqs. 28.1 and 28.2, respectively, we obtain

$$[M_1\omega^2 - 2f]u + f[1 + \exp(-2ika)]v = 0 \tag{28.5}$$

$$f[1 + \exp(2ika)]u + [M_2\omega^2 - 2f]v = 0 \tag{28.6}$$

These homogeneous linear equations have a nontrivial solution if the following determinant is zero:

$$\begin{vmatrix} M_1\omega^2 - 2f & f[1 + \exp(-2ika)] \\ f[1 + \exp(2ika)] & M_2\omega^2 - 2f \end{vmatrix} = 0 \tag{28.7}$$

By solving this equation, we obtain

$$\omega^2 = f\left(\frac{1}{M_1} + \frac{1}{M_2}\right) + f\left[\left(\frac{1}{M_1} + \frac{1}{M_2}\right)^2 - \frac{4\sin^2 ka}{M_1M_2}\right]^{1/2}$$

(optical branch) (28.8)

$$\omega^2 = f\left(\frac{1}{M_1} + \frac{1}{M_2}\right) - f\left[\left(\frac{1}{M_1} + \frac{1}{M_2}\right)^2 - \frac{4\sin^2 ka}{M_1M_2}\right]^{1/2}$$

(acoustical branch) (28.9)

The term k in Eqs. 28.3–28.9 is called the wavevector and indicates the phase difference between equivalent atoms in each unit cell. In the case of a one-dimentional lattice, $|\mathbf{k}| = k$. Thus, we use k rather than $\mathbf{k}$ in this case, and k can take any value between $-\pi/2a$ and $+\pi/2a$. This region is called the first Brillouin zone. Figure I-43 shows a plot of ω versus k for the positive half of the first Brillouin zone. There are two values for each ω which constitute the "optical" and "acoustical" branches in the dispersion curves.

At the center of the first Brillouin zone ($k = 0$), we have

$$\omega = 0 \quad \text{and} \quad u = v \qquad \text{(acoustical)} \tag{28.10}$$

$$\omega = \sqrt{2f\left(\frac{1}{M_1} + \frac{1}{M_2}\right)} \quad \text{and} \quad \frac{u}{v} = -\frac{M_2}{M_1} \quad \text{or} \quad M_1 u + M_2 v = 0$$

$$\text{(optical)} \tag{28.11}$$

At the end of the first Brillouin zone ($k = \pi/2a$), we have

$$\omega = \sqrt{\frac{2f}{M_2}} \quad \text{and} \quad u = 0 \qquad \text{(optical)} \tag{28.12}$$

$$\omega = \sqrt{\frac{2f}{M_1}} \quad \text{and} \quad v = 0 \qquad \text{(acoustical)} \tag{28.13}$$

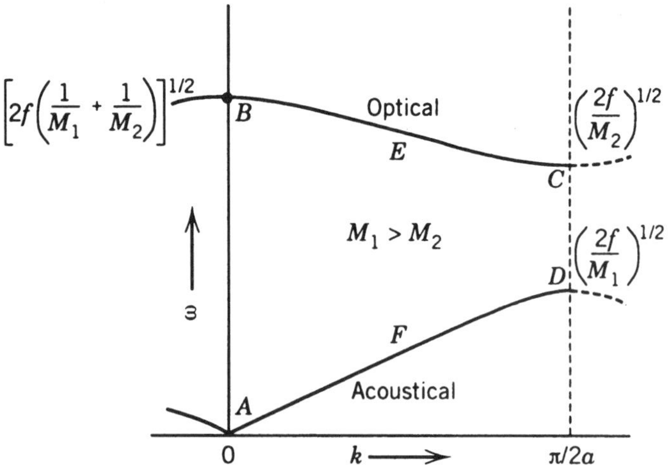

Fig. I-43. Dispersion curves for lattice vibrations.

The points A, B, C, and D in Fig. I-43 correspond to Eqs. 28.10, 28.11, 28.12, and 28.13, respectively.

To observe lattice vibrations in IR spectra, the momentum of the IR photon must be equal to that of the phonon.* The momentum of the photon (P) is given by

$$P = \frac{h}{\lambda} = \frac{h/2\pi}{\lambda/2\pi} = \hbar Q \tag{28.14}$$

where $Q = 2\pi/\lambda$. On the other hand, the momentum of the phonon is given by $\hbar k$.[134] Thus, the following relationship must hold:

$$\hbar Q = \hbar k \tag{28.15}$$

Since lattice vibrations are observed in the low-frequency region ($\lambda \cong 10^{-3}$ cm), $Q = 2\pi/\lambda \cong 10^3$ cm^{-1}. The k value at the end of the Brillouin zone is $k = \pi/2a \cong 10^8$ cm^{-1}. Thus, the k value which corresponds to the IR photon for lattice vibration is much smaller than the k value at the end of the Brillouin zone. This result indicates that optical transitions we observe in IR spectra occur practically at $k \cong 0$. A similar conclusion can be derived for Raman spectra of lattice vibrations.

Figure I-44 shows the modes of lattice vibrations corresponding to various points on the dispersion curve shown in Fig. I-43. At point A (acoustical mode), all atoms move in the same direction (translational motion of the whole lattice) and its frequency is zero. This is seen in Fig. I-44*a*. In a three-dimensional lattice, there are three such modes. Thus, we subtract 3 from our calculations in factor group analysis (Sec. I-26).

On the other hand, at point B (optical branch), the two atoms move in opposite directions, but the center of gravity of the unit cell remains unshifted (Fig. I-44*b*). Furthermore, the equivalent atom in each lattice moves in phase. If the two atoms carry opposite electrical charges, such a motion produces an oscillating dipole moment which can interact with incident IR radiation. Thus, it is possible to observe it optically. It should be noted that the frequency of a diatomic molecule in the free state is $\omega = \sqrt{f/\mu}$, whereas that of a diatomic lattice is $\omega = \sqrt{2f/\mu}$ (μ = reduced mass).

At point C (optical branch), the lighter atoms are moving back and forth against each other while the heavier atoms are fixed (Fig. I-44*c*). At point D (acoustical branch), the situation is opposite to that of point C. The vibrational modes in the middle of the Brillouin zone (points E and F) are shown in Figs. I-44*e* and *f*, respectively.

Thus far, we have discussed the lattice vibrations of a one-dimensional chain. The treatment of the three-dimensional lattice is basically the same, although

*The lattice vibration causes elastic waves in crystals. The quantum of the lattice vibrational energy is called a "phonon," in analogy with the photon of the electromagnetic wave.

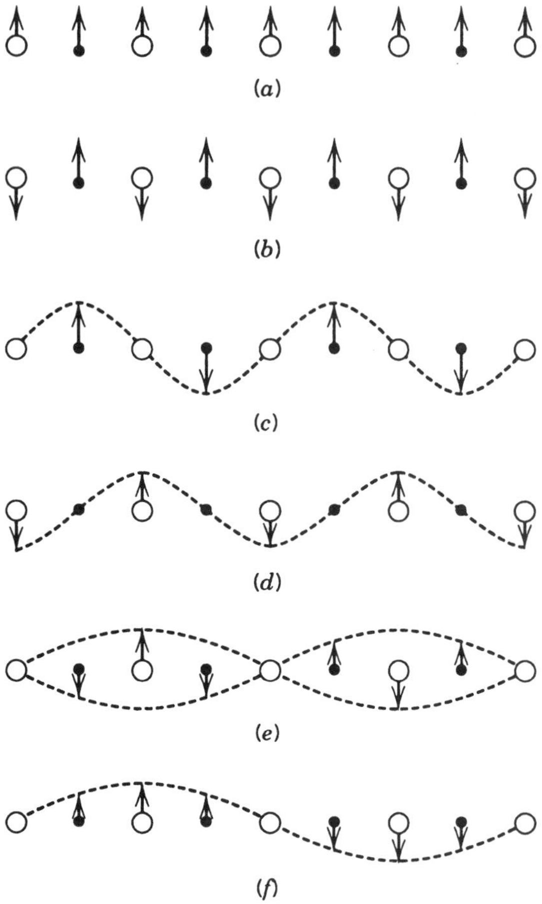

Fig. I-44. Wave motions corresponding to various points on the dispersion curves.

more complicated.[56] If the primitive cell contains σ molecules, each of which consists of N atoms, there are 3 acoustical modes and $3N\sigma - 3$ optical modes. The latter is grouped into $(3N - 6)\sigma$ internal modes and $6\sigma - 3$ lattice modes. The general forms of the dispersion curves of such a crystal are shown in Fig. I-45. In this book, our interest is focused on vibrational analysis of $3N\sigma - 3$ optical modes at $\mathbf{k} \cong 0$. Examples are found in diamond and graphite [Sec. II-14(3)] and quartz [Sec. II-15(2)].

I-29. POLARIZED SPECTRA OF SINGLE CRYSTALS

In the preceding section, the 30 normal vibrations of the Bravais cell of calcite crystal have been classified into symmetry species of the factor group $\mathbf{D}_{3d}$. The

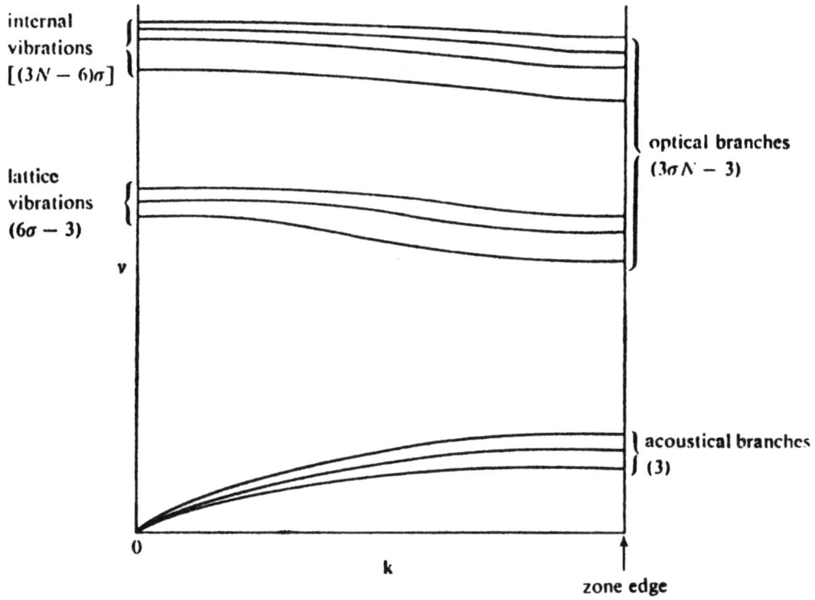

Fig. I-45. General form of dispersion curves for a molecular crystal. Reproduced with permission from Turrell[56].

results (Tables I-19 and I-24) show that three intramolecular ($A_{2u} + 2E_u$), three translatory lattice ($A_{2u} + 2E_u$) and two rotatory lattice ($A_{2u} + E_u$) vibrations are IR-active, whereas three intramolecular ($A_{1g} + 2E_g$), one translatory lattice (E_g), and one rotatory lattice (E_g) are Raman-active. In order to classify the observed bands into these symmetry species, it is desirable to measure infrared dichroism and polarized Raman spectra using single crystals of calcite.

(1) Infrared Dichroism

Suppose that we irradiate a single crystal of calcite with polarized infrared radiation whose electric vector vibrates along the *c* axis (*z* direction) in Fig. I-40. Then the infrared spectrum shown by the solid curve of Fig. I-46 is obtained.[135] According to Table I-19, only the A_{2u} vibrations are activated under

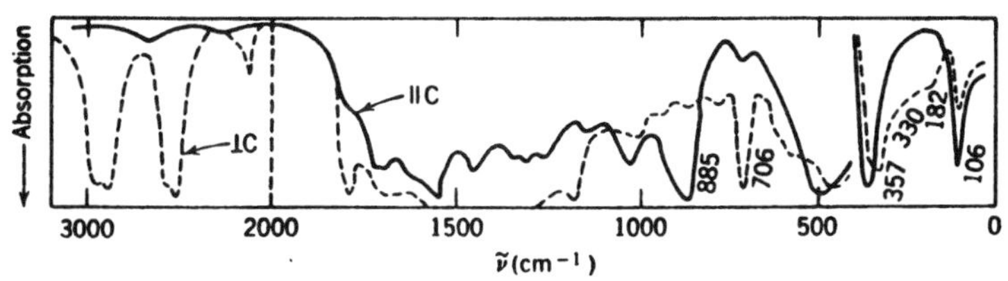

Fig. I-46. Infrared dichroism of calcite.[135]

such conditions. Thus, the three bands observed at 885(υ), 357(t), and 106(r) cm^{-1} are assigned to the A_{2u} species. The spectrum shown by the dotted curve is obtained if the direction of polarization is perpendicular to the c axis (x, y plane). In this case, only the E_u vibrations should be infrared-active. Therefore the five bands observed at 1484(υ), 706(υ), 330(t), 182(t), and 106(r) cm^{-1} are assigned to the E_u species. Here, υ, t, and r denote intramolecular, translatory lattice, and rotatory lattice modes, respectively.

(2) Polarized Raman Spectra

Polarized Raman spectra provide more information about the symmetry properties of normal vibrations than do polarized infrared spectra.[136] Again consider a single crystal of calcite. According to Table I-19, the A_{1g} vibrations become Raman active if any one of the polarizability components, α_{xx}, α_{yy}, and α_{zz}, is changed. Suppose that we irradiate a calcite crystal from the y direction, using polarized radiation whose electric vector vibrates parallel to the z axis (see Fig. I-47), and observe the Raman scattering in the x direction with its polarization in the z direction. This condition is abbreviated as $y(zz)x$. In this case, Eq. 6.7 is simplified to $P_z = \alpha_{zz}E_z$ because $E_x = E_y = 0$ and $P_x = P_y = 0$. Since α_{zz} belongs to the A_{1g} species, only the A_{1g} vibrations are observed under this condition. Figure I-48c illustrates the Raman spectrum obtained with this condition. Thus, the strong Raman line at 1088 cm^{-1}(υ) is assigned to the A_{1g} species. Both the A_{1g} and E_g vibrations are observed if the $z(xx)y$ condition is used. The Raman spectrum (Fig. I-48a) shows that five Raman lines [1088(υ), 714(υ), 283(r), 156(t), and 1434(υ) cm^{-1} (not shown)] are observed under this

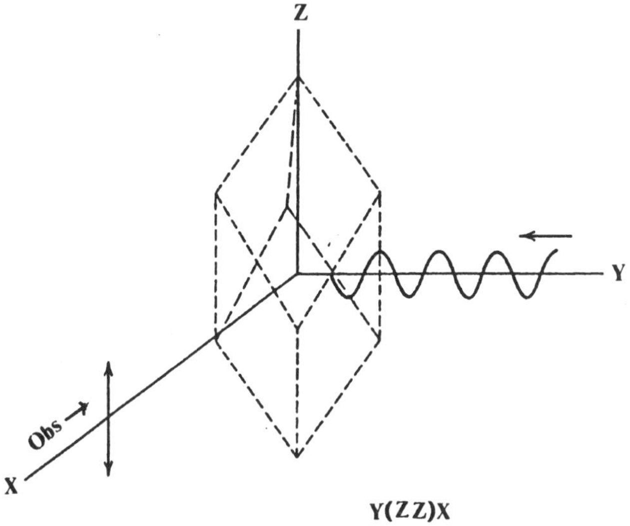

Fig. I-47. Schematic representation of experimental condition, $y(zz)x$.

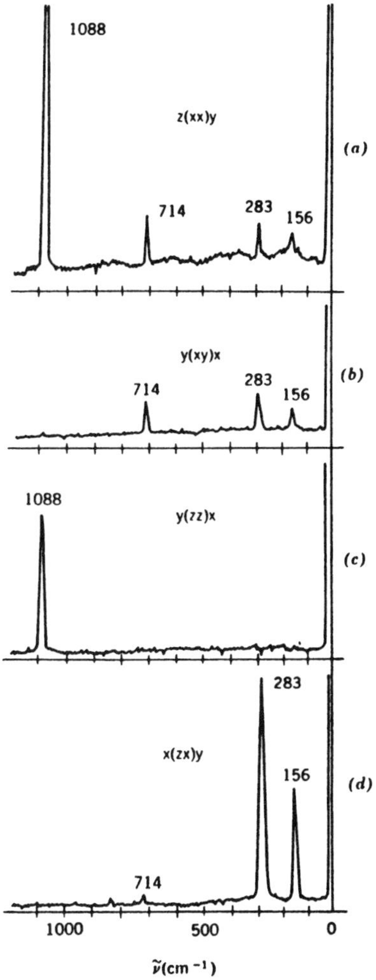

Fig. I-48. Polarized Raman spectra of calcite.[136]

condition. Since the 1088 cm^{-1} line belongs to the A_{1g} species, the remaining four must belong to the E_g species. These assignments can also be confirmed by measuring Raman spectra using the *y*(*xy*)*x* and *x*(*zx*)*y* conditions (Fig. I-48*b* and *d*).

(3) Normal Coordinate Analysis on the Bravais Cell

In the discussion above, we have assigned several bands in the same symmetry species to the *υ*-, *t*-, and *r*-types. In general, the intramolecular (*υ*) vibrations appear above 400 cm^{-1}, whereas the lattice vibrations appear below 400 cm^{-1}. However, more complete assignments can only be made via normal coordinate analysis on the entire Bravais cell.[137] Such calculations have been made by Nakagawa and Walter[138] on crystals of alkali–metal nitrates which are isomorphous with calcite. These workers employed four intramolecular and seven intermolecular force constants. The latter are in the range of 0.12–0.00 mdyn/Å. Figure I-49 illustrates the vibrational modes of the 18 (27 if *E* modes are counted as 2) optically active vibrations together with the corresponding frequencies of calcite.

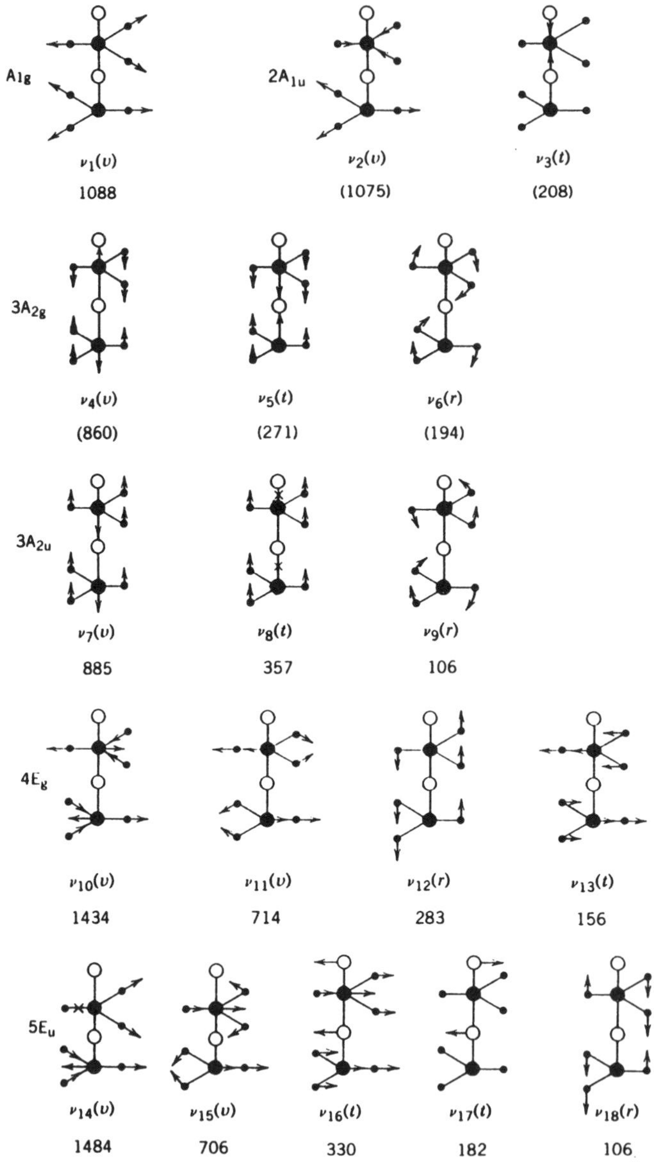

Fig. I-49. Vibrational modes of calcite. The observed and calculated (in parentheses) are listed under each mode.[138]

I-30. VIBRATIONAL ANALYSIS OF CERAMIC SUPERCONDUCTORS

In 1987, Wu et al.[139] synthesized a ceramic superconductor of the composition, $YBa_2Cu_3O_{7-\delta}$, whose superconducting critical temperature (T_c) was above the boiling point of liquid nitrogen (77 K). Since then, IR and Raman spectra of this and related compounds have been studied extensively, and the results are reviewed by Ferraro and Maroni.[140,141] Here, we limit our discussion to the Raman spectra of the above superconductor and their significance in studying oxygen deficiency and the structural changes resulting from it.

The superconductor, $YBa_2Cu_3O_{7-\delta}$ (abbreviated as the 123 conductor), can be obtained by baking a mixture of Y_2O_3, and $BaCO_3$, and CuO in a proper ratio. The product is normally a mixture of an orthorhombic form ($0 < \delta < 1$) which is superconducting and a tetragonal form ($\delta = 1$) which is an insulator. The T_c increases as δ approaches 0.

Figure I-50 shows the Bravais unit cells of the orthorhombic (Pmmm = D_{2h}^1) and tetragonal (P4/mmm = D_{4h}^1) forms.[142] The former is a distorted, oxygen-deficient form of perovskite. The orthorhombic unit cell contains 13 atoms, and their possible site symmetries can be found from the tables of site symmetries (Appendix X) as:

$$8\mathbf{D}_{2h}(1), \qquad 12\mathbf{C}_{2v}(2), \qquad 6\mathbf{C}_s(4), \qquad \mathbf{C}_1(8)$$

It is seen in Fig. I-50 that the three atoms, Y, O(1), and Cu(1), are at the $\mathbf{D}_{2h}$ sites. These atoms contribute $3B_{1u} + 3B_{2u} + 3B_{3u}$ vibrations since T_x, T_y, and T_z

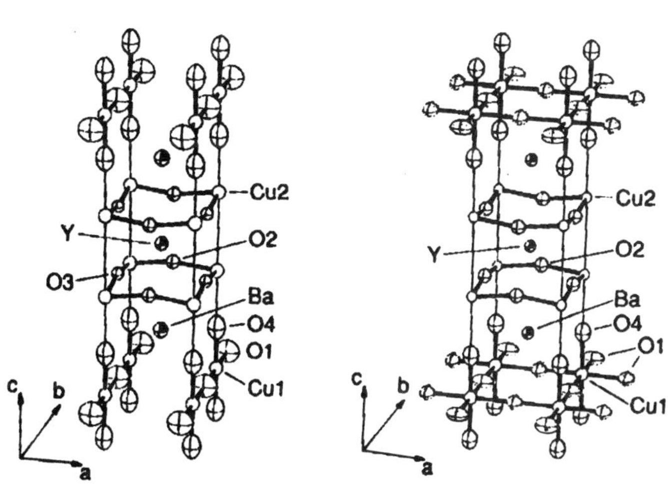

Fig. I-50. The Bravais unit cells of the orthorhombic and tetragonal forms of $YBa_2Cu_3O_{7-\delta}$. Reproduced with permission from Schuller and Jorgensen[142].

TABLE I-25. Correlation Between Site Symmetry ($\mathbf{C}_{2v}$) and Factor Group Symmetry ($\mathbf{D}_{2h}$) for the Orthorhombic Form of the 123 Conductor[a]

f	t	Site Symmetry ($\mathbf{C}_{2v}$), $C_2(z)$	Factor Group Symmetry ($\mathbf{D}_{2h}$)	a_ζ
2	1	$A_1(T_z)$	A_g	1
			$B_{1u}(T_z)$	1
2	1	$B_1(T_x, R_y)$	$B_{2g}(R_y)$	1
			$B_{3u}(T_x)$	1
2	1	$B_2(T_y, R_x)$	$B_{3g}(R_x)$	1
			$B_{2u}(T_y)$	1

[a] For f, t, and a, see Sec. I-27. The complete correlation table is found in Appendix IX.

belong to the B_{3u}, B_{2u}, and B_{1u} species, respectively, in the $\mathbf{D}_{2h}$ point group. On the other hand, the 10 atoms, 2Ba, 2Cu(2), 2O(2), 2O(3), and 2O(4) are at the $\mathbf{C}_{2v}$ sites. As shown in Table I-25, each pair of these atoms possess six degrees of vibrational freedom ($2A_1+2B_1+2B_2$) which are split into $A_g+B_{2g}+B_{3g}+B_{1u}+B_{2u}+B_{3u}$ under $\mathbf{D}_{2h}$ symmetry. The number of optical modes at $k \cong 0$ in each species can be obtained by subtracting three acoustical modes ($B_{1u}+B_{2u}+B_{3u}$) from the above counting. Table I-26 summarizes the results. It is seen that the orthorhombic unit cell has 15 Raman-active modes ($5A_g$+$5B_{2g}$+$5B_{3g}$) and 21 IR-active modes ($7B_{1u}+7B_{2u}+7B_{3u}$). It should be noted that the mutual exclusion rule holds in this case since the $\mathbf{D}_{2h}$ point group has a center of symmetry. Although the same results can be obtained by using factor group analysis,[143] the correlation method is simpler and straightforward.

Similar calculations on the tetragonal unit cell ($YBa_2Cu_3O_6$) show that 10

TABLE I-26. Vibrational Analysis for Orthorhombic Form of $YBa_2Cu_3O_{7-\delta}$ Using the Correlation Method

$\mathbf{D}_{2h}$	$\mathbf{D}_{2h}$ Y, O(1), Cu(1)	$\mathbf{C}_{2v}$ Ba, Cu(2), O(2), O(3), O(4)	Acoustical Vib.	Total Optical Vib.	Activity
A_g	0	5	0	5	Raman
B_{1g}	0	0	0	0	Raman
B_{2g}	0	5	0	5	Raman
B_{3g}	0	5	0	5	Raman
A_u	0	0	0	0	Inactive
B_{1u}	3	5	1	7	IR
B_{2u}	3	5	1	7	IR
B_{3u}	3	5	1	7	IR
Total	9	30	3	36	

vibrations ($4A_{1g}+B_{1g}+5E_g$) are Raman-active while 11 vibrations ($5A_{2u}+6E_u$) are IR-active under $\mathbf{D}_{4h}$ symmetry.

Figure I-51 shows the Raman spectrum of a polished sintered pellet of the 123 conductor ($\delta = 0.3$) obtained by Ferraro et al.[144] The five A_g modes appear strongly, and the most probable assignments for these bands are.[145,146]

492 cm^{-1} Axial motion of the O(4) atoms

445 cm^{-1} O(2)–Cu(2)–O(3) bending with the two O atoms moving in phase

336 cm^{-1} O(2)–Cu(2)–O(3) bending with the two O atoms moving out of phase

145 cm^{-1} Axial motion of the Cu(2) atoms

112 cm^{-1} Axial motion of the Ba atoms

These values fluctuate within ±10 cm^{-1} depending upon variations in oxygen stoichiometry and crystalline disorder. The results of normal coordinate analysis[147] indicate considerable mixing among the vibrations represented by the cartesian coordinates of individual atoms.

If the 123 conductor is prepared in pure oxygen and the oxygen is removed quantitatively by heating the sample in argon, a series of samples having $\delta = 0$,

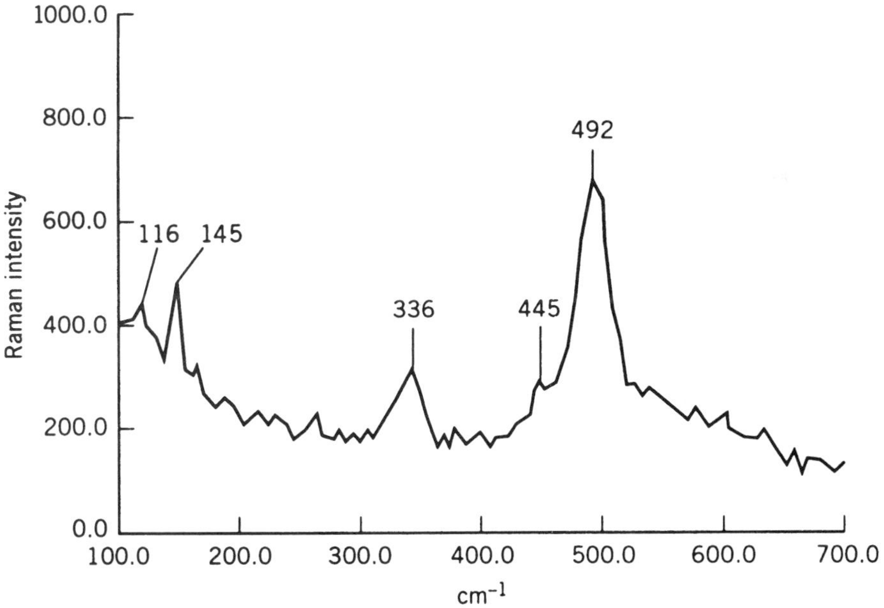

Fig. I-51. Backscattered Raman spectrum of a polished sintered pellet of $YBa_2Cu_3O_{7-\delta}$. Reproduced with permission from Ferraro and Maroni[144].

0.2, 0.5, and 0.7 can be obtained. The T_c values of these samples were found to be 94, 77, 50, and 20 K, respectively. Thomsen et al.[148] prepared a series of such samples, and measured their Raman spectra. Figure I-52 shows a plot of vibrational frequencies of the four A_g modes mentioned above against δ values. It is seen that the two modes at 502 and 154 cm^{-1} are softened and the two modes at 438 and 334 cm^{-1} are hardened as the oxygen is removed from the sample. These results seem to suggest that the T_c of the 123 conductor is related to oxygen deficiency in the O(2)–Cu(2)–O(3) layer. The effect of changing the size of the cation (Y)[149] and isotopic substitution ($^{16}O/^{18}O$)[150] on these Raman bands has also been studied.

Thus far, IR studies on the 123 conductor have been hindered by the difficulties in obtaining IR spectra from highly opaque samples and in making reliable band assignments.[140,141]

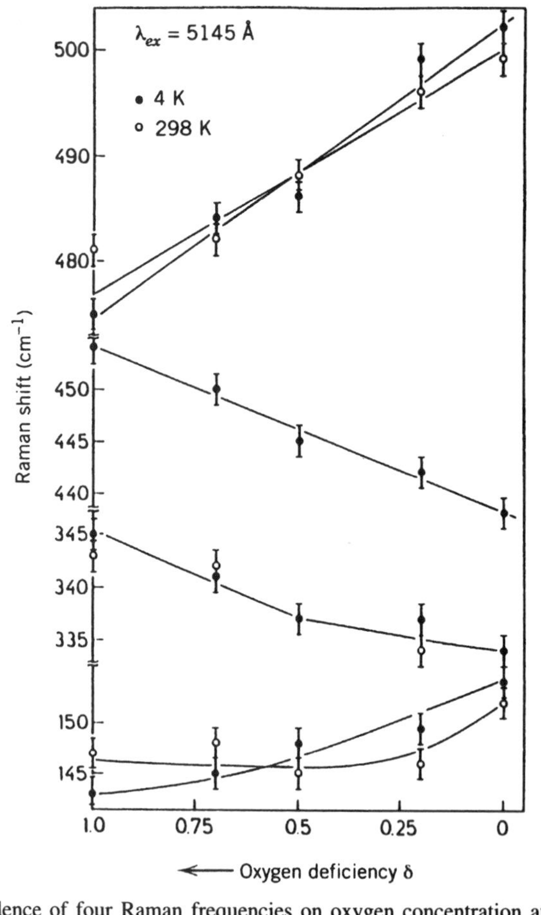

Fig. I-52. Dependence of four Raman frequencies on oxygen concentration at 4 K and 298 K. Reproduced with permission from Thomsen et al.[148].

GENERAL READINGS

THEORY OF MOLECULAR VIBRATIONS

1. G. Herzberg, *Molecular Spectra and Molecular Structure.* Vol. II: *Infrared and Raman Spectra of Polyatomic Molecules*, Van Nostrand, Princeton, NJ, 1945.
2. G. Herzberg, *Molecular Spectra and Molecular Structure.* Vol. I: *Spectra of Diatomic Molecules*, Van Nostrand, Princeton, NJ, 1950.
3. E. B. Wilson, J. C. Decius, and P. C. Cross, *Molecular Vibrations*, McGraw-Hill, New York, 1955.
4. G. W. King, *Spectroscopy and Molecular Structure*, Holt, Rinehart, and Winston, New York, 1964.
5. C. J. H. Schutte, *The Theory of Molecular Spectroscopy.* Vol. I: *The Quantum Mechanics and Group Theory of Vibrating and Rotating Molecules*, North Holland, Amsterdam, 1976.
6. P. Gans, *Vibrating Molecules*, Chapman and Hall, London, 1971.
7. D. Steele, *Theory of Vibrational Spectroscopy*, Saunders, London, 1971.
8. L. A. Woodward, *Introduction to the Theory of Molecular Vibrations and Vibrational Spectroscopy*, Oxford University Press, London, 1972.
9. C. N. Banwell, *Fundamentals of Molecular Spectroscopy*, 2nd ed., McGraw-Hill, London, 1972.
10. S. J. Cyvin, *Molecular Structures and Vibrations*, Elsevier, Amsterdam, 1972.
11. A. Fadini, *Molekülkraftkonstanten*, Steinkopff-Verlag, Darmstadt, 1976.

SYMMETRY AND GROUP THEORY

12. M. Orchin and H. H. Jaffé, *Symmetry, Orbitals and Spectra*, Wiley, New York, 1971.
13. D. C. Harris and M. D. Bertolucci, *Symmetry and Spectroscopy*, Oxford University Press, New York, 1978.
14. P. R. Bunker, *Molecular Symmetry and Spectroscopy*, Academic Press, New York, 1979.
15. S. F. A. Kettle, *Symmetry and Structure*, Wiley, New York, 1985.
16. L. H. Hall, *Group Theory and Symmetry in Chemistry*, McGraw-Hill, New York, 1969.
17. J. R. Ferraro and J. S. Ziomek, *Introductory Group Theory and its Application to Molecular Structure*, 2nd ed., Plenum Press, New York, 1975.
18. F. A. Cotton, *Chemical Application of Group Theory*, 3rd ed., Wiley-Interscience, New York, 1990.

CORRELATION METHOD

19. W. G. Fateley, F. R. Dollish, N. T. McDevitt, and F. F. Bentley, *Infrared and Raman Selection Rules for Molecular and Lattice Vibrations: The Correlation Method*, Wiley-Interscience, New York, 1972.

FOURIER-TRANSFORM INFRARED SPECTROSCOPY

20. R. J. Bell, *Introductory Fourier Transform Spectroscopy*, Academic Press, New York, 1972.
21. J. R. Ferraro and L. J. Basile, eds., *Fourier Transform Infrared Spectroscopy*, Vol. I, Academic Press, New York, 1978 to present.
22. P. G. Griffiths and J. A. de Haseth, *Fourier Transform Infrared Spectrometry*, Wiley, New York, 1986.
23. J. R. Ferraro and K. Krishnan, eds., *Practical Fourier Transform Infrared Spectroscopy*, Academic Press, San Diego, CA, 1990.

RAMAN SPECTROSCOPY

24. H. A. Szymanski, ed., *Raman Spectroscopy: Theory and Practice*, Vol. I, Plenum Press, New York, 1967, Vol. 2, 1970.
25. T. R. Gilson and P. J. Hendra, *Laser Raman Spectroscopy*, Wiley, New York, 1970.
26. M. C. Tobin, *Laser Raman Spectroscopy*, Wiley-Interscience, New York, 1971.
27. J. A. Koningstein, *Introduction to the Theory of the Raman Effect*, D. Reidel, Dordrecht (Holland), 1973.
28. D. A. Long, *Raman Spectroscopy*, McGraw-Hill, New York, 1977.
29. J. G. Grasselli, M. K. Snavely, and B. J. Bulkin, *Chemical Applications of Raman Spectroscopy*, Wiley, New York, 1981.
30. D. P. Strommen and K. Nakamoto, *Laboratory Raman Spectroscopy*, Wiley, New York, 1984.
31. J. G. Grasselli and B. J. Bulkin, eds., *Analytical Raman Spectroscopy*, Wiley, New York, 1991.
32. J. R. Ferraro and K. Nakamoto, *Introductory Raman Spectroscopy*, Academic Press, San Diego, CA, 1994.

VIBRATIONAL SPECTRA OF INORGANIC, COORDINATION, AND ORGANOMETALLIC COMPOUNDS

33. H. Siebert, *Anwendungen der Schwingungsspektroskopie in der Anorganischen Chemie*, Springer-Verlag, Berlin, 1966.
34. D. M. Adams, *Metal-Ligand and Related Vibrations*, Edward Arnold, London, 1967.

35. J. R. Ferraro, *Low-frequency Vibrations of Inorganic and Coordination Compounds*, Plenum Press, New York, 1971.

36. L. H. Jones, *Inorganic Spectroscopy*, Vol. I, Marcel Dekker, New York, 1971.

37. R. A. Nyquist and R. O. Kagel, *Infrared Spectra of Inorganic Compounds*, Academic Press, New York, 1971.

38. S. D. Ross, *Inorganic Infrared and Raman Spectra*, McGraw-Hill, New York, 1972.

39. E. Maslowsky, Jr., *Vibrational Spectra of Organometalalic Compounds*, Wiley, New York, 1976.

VIBRATIONAL SPECTRA OF ORGANIC COMPOUNDS

40. L. J. Bellamy, *The Infrared Spectra of Complex Molecules*, 3rd ed., Vol. I, Chapman and Hall, London, 1975; Vol. 2, 1980.

41. N. B. Colthup, L. H. Daly, and S. E. Wiberley, *Introduction to Infrared and Raman Spectroscopy*, 3rd ed., Academic Press, San Diego, CA, 1990.

42. D. Lin-Vien, N. B. Colthup, W. G. Fateley, and J. G. Grasselli, *The Handbook of Infrared and Raman Characteristic Frequencies of Organic Molecules*, Academic Press, San Diego, CA, 1991.

VIBRATIONAL SPECTRA OF BIOLOGICAL COMPOUNDS

43. F. S. Parker, *Application of Infrared, Raman and Resonance Raman Spectroscopy in Biochemistry*, Plenum Press, New York, 1983.

44. A. T. Tu, *Raman Spectroscopy in Biology*, Wiley, New York, 1982.

45. P. R. Carey, *Biochemical Applications of Raman and Resonance Raman Spectroscopies*, Academic Press, New York, 1982.

46. T. G. Spiro, ed., *Biological Applications of Raman Spectroscopy*, Vols. 1–3, Wiley, New York, 1987–1988.

VIBRATIONAL SPECTRA OF ADSORBED SPECIES

47. L. H. Little, *Infrared Spectra of Adsorbed Species*, Academic Press, New York, 1967.

48. M. L. Hair, *Infrared Spectroscopy in Surface Chemistry*, Marcel Dekker, New York, 1967.

49. A. T. Bell and M. L. Hair, eds., "*Vibrational Spectroscopies for Adsorbed Species*," American Chemical Society, Washington, DC, 1980.

50. J. T. Yates, Jr. and T. E. Madey, eds., *Vibrational Spectroscopy of Molecules on Surfaces*, Plenum Press, New York, 1987.

LOW-TEMPERATURE AND MATRIX ISOLATION SPECTROSCOPY

51. B. Meyer, *Low-Temperature Spectroscopy*, Elsevier, Amsterdam, 1971.

52. H. E. Hallam, ed., *Vibrational Spectroscopy of Trapped Species*, Wiley, New York, 1973.

53. M. Moskovits and G. A. Ozin, ed., *Cryochemistry*, Wiley, New York, 1976.

TIME-RESOLVED SPECTROSCOPY

54. G. H. Atkinson, *Time-Resolved Vibrational Spectroscopy*, Academic Press, New York, 1983.

HIGH-PRESSURE SPECTROSCOPY

55. J. R. Ferraro, *Vibrational Spectroscopy at High External Pressures—The Diamond Anvil Cell*, Academic Press, New York, 1984.

VIBRATIONAL SPECTRA OF CRYSTALS AND MINERALS

56. G. Turrell, *Infrared and Raman Spectra of Crystals*, Academic Press, New York, 1972.

57. J. C. Decius and R. M. Hexter, *Molecular Vibrations in Crystals*, McGraw-Hill, New York, 1977.

58. V. C. Farmer, *The Infrared Spectra of Minerals*, Mineralogical Society, London, 1974.

59. J. A. Gadsden, *Infrared Spectra of Minerals and Related Inorganic Compounds*, Butterworth, London, 1975.

60. C. Karr, ed., *Infrared and Raman Spectroscopy of Lunar and Terrestrial Minerals*, Academic Press, New York, 1975.

61. R. J. P. Lyon, *Minerals in the Infrared ... A Critical Bibliography*, Stanford Research Institute, Menlo Park, CA, 1962.

62. K. Omori, "Infrared Absorption Spectra of Some Essential Minerals," *Sci. Rep., Tohoku Univ., Ser. 3*, **1**(1), p. 102 (1961).

VIBRATIONAL INTENSITIES

63. W. B. Person and G. Zerbi, eds., *Vibrational Intensities in Infrared and Raman Spectroscopy*, Elsevier, Amsterdam, 1982.

ADVANCES SERIES

64. *Spectroscopic Properties of Inorganic and Organometallic Compounds*, Vol. 1 to present, The Chemical Society, London.
65. *Molecular Spectroscopy—Specialist Periodical Reports*, Vol. 1 to present, The Chemical Society, London.
66. R. J. H. Clark and R. E. Hester, eds., *Advances in Infrared and Raman Spectroscopy*, Vol. 1 to present, Heyden, London.
67. J. Durig, ed., *Vibrational Spectra and Structure*, Vol. 1 to present, Elsevier, Amsterdam.
68. C. B. Moore, ed., *Chemical and Biochemical Applications of Lasers*, Vol. 1 to present, Academic Press, New York.
69. *Structure and Bonding*, Vol. 1 to present, Springer-Verlag, New York.

INDEX AND COLLECTION OF SPECTRAL DATA

70. N. N. Greenwood, E. J. F. Ross, and B. P. Straughan, *Index of Vibrational Spectra of Inorganic and Organometallic Compounds*, Vols. 1–3. Butterworth, London, 1972–1977.
71. Sadtler's IR Handbook of Inorganic Chemicals. Bio-Rad Laboratories, Sadtler Division, Philadelphia, PA.
72. B. Schrader, *Raman/IR Atlas of Organic Compounds*, VCH, New York, 1989.
73. T. Shimanouchi, Tables of Molecular Vibrational Frequencies, Consolidated Volume, *Natl. Stand. Ref. Data Ser. (U.S., Natl. Bur. Stand.)*, **39,** June (1972); also see *J. Phys. Chem. Ref. Data*, **1,** 189 (1972); **2,** 121, 225 (1973); **3,** 269 (1974).

REFERENCES

74. W. Holzer, W. F. Murphy, and H. J. Bernstein, *J. Chem. Phys.*, **52,** 399 (1970).
75. C. F. Shaw, III, *J. Chem. Educ.*, **58,** 343 (1981).
76. P. Pulay, *Mol. Phys.*, **17,** 197 (1969); **18,** 473 (1970); P. Pulay and W. Meyer, *J. Mol. Spectrosc.*, **40,** 59 (1971).
77. Y. Nishimura, M. Tsuboi, S. Kato, and K. Morokuma, *J. Am. Chem. Soc.*, **103,** 1354 (1981).
78. R. M. Badger, *J. Chem. Phys.*, **2,** 128 (1934).
79. W. Gordy, *J. Chem. Phys.*, **14,** 305 (1946).
80. D. R. Herschbach and V. W. Laurie, *J. Chem. Phys.*, **35,** 458 (1961).
81. K. Nakamoto, M. Margoshes, and R. E. Rundle, *J. Am. Chem. Soc.*, **77,** 6480 (1955).
82. E. M. Layton, Jr., R. D. Cross, and V. A. Fassel, *J. Chem. Phys.*, **25,** 135 (1956).
83. F. D. Hardcastle and I. E. Wachs, *J. Raman Spectrosc.*, **21,** 683 (1990).
84. F. D. Hardcastle and I. E. Wachs, *J. Phys. Chem.*, **95,** 5031 (1991).

85. D. P. Strommen and E. R. Lippincott, *J. Chem. Educ.*, **49,** 341 (1972).
86. G. C. Lie, *J. Chem. Educ.*, **56,** 636 (1979).
86a. H. Eyring, J. Walter, and G. E. Kimball, *Quantum Chemistry*, 5th ed., Wiley, New York, 1949, p. 121.
87. E. B. Wilson, *J. Chem. Phys.*, **7,** 1047 (1939); **9,** 76 (1941).
88. J. C. Decius, *J. Chem. Phys.*, **16,** 1025 (1948).
89. T. Shimanouchi, *J. Chem. Phys.*, **25,** 660 (1956).
90. T. Shimanouchi, "The Molecular Force Field," in D. Henderson, ed., *Physical Chemistry: An Advanced Treatise*, Vol. 4, Academic Press, New York, 1970.
91. T. Shimanouchi, *J. Chem. Phys.*, **17,** 245, 734, 848 (1949).
92. D. F. Heath and J. W. Linnett, *Trans. Faraday Soc.*, **44,** 556, 873, 878, 884 (1948); **45,** 264 (1949).
93. J. Overend and J. R. Scherer, *J. Chem. Phys.*, **32,** 1289, 1296, 1720 (1960); **33,** 446 (1960); **34,** 547 (1961); **36,** 3308 (1962).
94. T. Shimanouchi, *Pure Appl. Chem.*, **7,** 131 (1963).
95. J. H. Schachtschneider, "Vibrational Analysis of Polyatomic Molecules," Pts. V and VI, Tech. Rept. 231-64 and 53-65, Shell Development Co., Emeryville, CA, 1964 and 1965.
96. K. Nakamoto, *Angew. Chem.*, **11,** 666 (1972).
97. N. Mohan, K. Nakamoto, and A. Müller, "The Metal Isotope Effect on Molecular Vibrations," in R. J. H. Clark and R. E. Hester, eds., *Advances in Infrared and Raman Spectroscopy*, Vol. 1, Heyden, London, 1976.
98. K. Nakamoto, K. Shobatake, and B. Hutchinson, *Chem. Commun.*, 1451 (1969).
99. J. Takemoto and K. Nakamoto, *Chem. Commun.*, 1017 (1970).
100. J. R. Kincaid and K. Nakamoto, *Spectrochim. Acta*, **32A,** 277 (1976).
101. P. M. Champion, B. R. Stallard, G. C. Wagner, and I. C. Gunsalus, *J. Am. Chem. Soc.*, **104,** 5469 (1982).
102. Y. Morino and K. Kuchitsu, *J. Chem. Phys.*, **20,** 1809 (1952).
103. B. L. Crawford and W. H. Fletcher, *J. Chem. Phys.*, **19,** 141 (1951).
104. P. LaBonville and J. M. Williams, *Appl. Spectrosc.*, **25,** 672 (1971).
105. D. A. Ramsey, *J. Am. Chem. Soc.*, **74,** 72 (1952).
106. D. P. Strommen and K. Nakamoto, *Appl. Spectrosc.*, **37,** 436 (1983).
107. T. G. Spiro and T. C. Strekas, *Proc. Natl. Acad. Sci., U.S.A.*, **69,** 2622 (1972).
108. W. M. McClain, *J. Chem. Phys.*, **55,** 2789 (1971).
109. D. P. Strommen, *J. Chem. Educ.*, **69,** 803 (1992).
110. W. Kiefer, *Appl. Spectrosc.*, **28,** 115 (1974).
111. M. Mingardi and W. Siebrand, *J. Chem. Phys.*, **62,** 1074 (1975).
112. A. C. Albrecht, *J. Chem. Phys.*, **34,** 1476 (1961).
113. S. A. Asher, *Anal. Chem.*, **65,** 59A, 201A (1993).
114. W. Kiefer and H. J. Bernstein, *J. Raman Spectrosc.*, **1,** 417 (1973).
115. R. J. H. Clark and P. D. Mitchell, *J. Am. Chem. Soc.*, **95,** 8300 (1973).
116. L. A. Nafie, P. Stein, and W. L. Peticolas, *Chem. Phys. Lett.*, **12,** 131 (1971).
117. Y. Hirakawa and M. Tsuboi, *Science*, **188,** 359 (1975).

118. T. G. Spiro, R. S. Czernuszewicz, and X.-Y. Li, *Coord. Chem. Rev.*, **100,** 541 (1990).

119. X.-Y. Li, R. S. Czernuszewicz, J. R. Kincaid, P. Stein, and T. G. Spiro, *J. Phys. Chem.*, **94,** 47 (1990).

120. H. Hamaguchi, I. Harada, and T. Shimanouchi, *Chem. Phys. Lett.*, **32,** 103 (1975).

121. H. Hamaguchi, *J. Chem. Phys.*, **69,** 569 (1978).

122. F. Inagaki, M. Tasumi, and T. Miyazawa, *J. Mol. Spectrosc.*, **50,** 286 (1974).

123. G. A. Ozin, "Single Crystal and Gas Phase Raman Spectroscopy in Inorganic Chemistry," in S. J. Lippard, ed., *Progress in Inorganic Chemistry*, Vol. 14, Wiley-Interscience, New York, 1971.

124. R. J. H. Clark and D. M. Rippon, *J. Mol. Spectrosc.*, **44,** 479 (1972).

125. E. Whittle, D. A. Dows, and G. C. Pimentel, *J. Chem. Phys.*, **22,** 1943 (1954).

126. M. J. Linevsky, *J. Chem. Phys.*, **34,** 587 (1961).

127. D. E. Milligan and M. E. Jacox, *J. Chem. Phys.*, **38,** 2627 (1963).

128. D. Tevault and K. Nakamoto, *Inorg. Chem.*, **14,** 2371 (1975).

129. R. L. DeKock, *Inorg. Chem.*, **10,** 1205 (1971).

129a. G. Burns and A. M. Glazer, *Space Groups for Solid State Scientists*, Academic Press, New York, 1978, p. 81.

129b. J. M. Robertson, *Organic Crystals and Molecules*, Cornell University Press, Ithaca, NY, 1953, p. 44.

130. *International Tables for X-Ray Crystallography*, Knoch Press, Birmingham, England, 1952.

131. R. S. Halford, *J. Chem. Phys.*, **14,** 8 (1946).

132. R. W. C. Wyckoff, *Crystal Structures*, Vols. I and II, Wiley-Interscience, New York, 1964.

133. S. Bhagavantam and T. Venkatarayudu, *Proc. Indian Acad. Sci.*, **9A,** 224 (1939); *Theory of Groups and Its Application to Physical Problems*, Andhra University, Waltair, India, 1951.

134. C. Kittel, *Introduction to Solid State Physics*, 5th ed., Wiley, New York, 1976, p. 118.

135. M. Tsuboi, *Infrared Absorption Spectra*, Vol. 6, Nankodo, Tokyo, 1958, p. 41.

136. S. P. Porto, J. A. Giordmaine, and T. C. Damen, *Phys. Rev.*, **147,** 608 (1966).

137. T. Shimanouchi, M. Tsuboi, and T. Miyazawa, *J. Chem. Phys.*, **35,** 1597 (1961).

138. I. Nakagawa and J. L. Walter, *J. Chem. Phys.*, **51,** 1389 (1969).

139. M. K. Wu, J. R. Ashburn, C J. Torng, P. H. Hor, R. L. Meng, L. Gao, Z. J. Huang, Y. Q. Wang, and C. W. Chu, *Phys. Rev. Lett.*, **55,** 908 (1987).

140. J. R. Ferraro and V. A. Maroni, *Appl. Spectrosc.*, **44,** 351 (1990).

141. V. A. Maroni and J. R. Ferraro, *Practical Fourier Transform Infrared Spectroscopy*, Academic Press, San Diego, CA, 1990, p. 1.

142. I. K. Schuller and J. D. Jorgensen, *Mater. Res. Soc. Bull.*, **14**(1), 27 (1989).

143. J. Hanuza, J. Klamut, R. Horyń, and B. Jeżowska-Trzebiatowska, *J. Mol. Struct.*, **193,** 57 (1989).

144. J. R. Ferraro and V. A. Maroni, private communication.

145. M. Stavola, D. M. Krol, L. F. Schneemeyer, S. A. Sunshine, J. V. Waszczak, and S. G. Kosinski, *Phys. Rev. B: Condens. Matter* [3], **39,** 287 (1989).

146. Y. Morioka, A. Tokiwa, M. Kikuchi, and Y. Syono, *Solid State Commun.*, **67,** 267 (1988).

147. F. E. Bates and J. E. Eldridge, *Solid State Commun.*, **64,** 1435 (1987); F. E. Bates, *Phys. Rev.*, **B39,** 322 (1989).

148. C. Thomsen, R. Liu, M. Bauer, A. Wittlin, L. Genzel, M. Cardona, E. Schönherr, W. Bauhofer, and W. König, *Solid State Commun.*, **65,** 55 (1988).

149. M. Cardona, R. Liu, C. Thomsen, M. Bauer, L. Genzel, W. König, A. Wittlin, U. Amador, M. Barahona, F. Fernandez, C. Otero, and R. Saez, *Solid State Commun.*, **65,** 71 (1988).

150. B. Batlogg, R. J. Cava, A. Jarayaman, R. B. van Dover, G. A. Kourouklis, S. Sunshine, D. W. Murphy, L. W. Rupp, H. S. Chen, A. White, K. T. Short, A. M. Mujsce, and E. A. Rietman, *Phys. Rev. Lett.*, **58,** 2333 (1987).

II

APPLICATIONS IN INORGANIC CHEMISTRY

II-1. DIATOMIC MOLECULES

As shown in Sec. I-3, diatomic molecules have only one vibration along the chemical bond; its frequency is given by

$$\tilde{\nu} = \frac{1}{2\pi c}\sqrt{\frac{K}{\mu}}$$

where K is the force constant, μ the reduced mass, and c the velocity of light. In homopolar XX molecules ($\mathbf{D}_{\infty h}$), the vibration is not infrared-active but is Raman-active, whereas it is both infrared- and Raman-active in heteropolar XY molecules ($\mathbf{C}_{\infty v}$).

Table II-1*a* lists a number of diatomic molecules and ions for which frequencies corrected for anharmonicity (ω_e) and anharmonicity constants ($x_e\omega_e$) are known. The force constants can be calculated directly from these ω_e values. The ω_e values cover a wide range of frequencies, the highest being that of H_2 (4395 cm^{-1}) and the lowest being that of Cs_2 (42 cm^{-1}). Correspondingly, the $x_e\omega_e$ value is largest for H_2 (117.91 cm^{-1}) and smallest for Cs_2 (0.08 cm^{-1}). Extensive studies on hydrides show that their frequencies span from 4395 cm^{-1} (H_2) to 891 cm^{-1} (HCs). Table II-1*a* also contains many isotopic frequencies involving light as well as heavy atoms.

The effects of changing the mass and/or force constant in a series of diatomic molecules have been discussed in Sec. I-3. Similar series are found in Table II-1a. For example, we find that (all in units of cm^{-1})

Table II-1*a*. Harmonic Frequencies and Anharmonicity Constants of Diatomic Molecules and Ions (cm^{-1})[a, b]

Molecule	ω_e	$\chi_e\omega_e$	Refs.	Molecule	ω_e	$\chi_e\omega_e$	Refs.
H H	4395.2	117.91	1	H ^{88}Sr	1206.89	17.03	10
(H H)$^+$	2297	62	1	D ^{88}Sr	858.85	8.64	11
H D	3817.1	94.96	1	H ^{138}Ba	1168.43	14.61	12
D D	3118.5	64.10	1	D Ba	829.84	7.37	11
H ^{6}Li	1420.12	23.66	2, 3	H Mn	1546.85	27.60	13
H ^{7}Li	1405.51	23.18	2, 3	H Co	(1890)	—	1
D ^{6}Li	1074.33	13.54	2	H Ni	2000	40	14
D ^{7}Li	1054.94	13.05	2	H Cu	1940.4	37	14
H ^{23}Na	1171.76	19.52	4, 5	H ^{107}Ag	1759.67	33.93	15
H ^{39}K	985.0	14.65	1	D ^{109}Ag	1250.68	17.21	16
H ^{85}Rb	937.10	14.28	6	H ^{197}Au	2305.0	43.12	1
H ^{87}Rb	936.98	14.27	6	H ^{64}Zn	1615.72	59.62	17
H Cs	891.25	12.82	7	D ^{64}Zn	1147.36	28.72	11
H ^{9}Be	2058.6	35.5	1	H ^{110}Cd	1461.13	61.99	17
H ^{24}Mg	1495.26	31.64	8	D ^{116}Cd	1032.04	29.29	11
D ^{24}Mg	1078.14	16.15	8	H Hg	1387.1	83.01	1
H Ca	1298.40	19.18	9	H ^{11}B	(2366)	(49)	1
H ^{27}Al	1682.4	29.11	14, 18	H ^{16}O	3735.2	82.81	1
H ^{69}Ga	1603.96	28.42	19	D ^{16}O	2720.9	44.2	1,31
D ^{71}Ga	1142.77	14.42	16	(D O)$^-$	2723.5	49.72	31
H ^{115}In	1475.43	25.16	20	(H ^{32}S)$^-$	2647.1	53.28	32
D ^{115}In	1048.60	12.70	16	H ^{19}F	4138.55	90.07	1
H ^{205}Tl	1391.27	23.10	21	D ^{19}F	2998.3	45.71	1
D ^{205}Tl	987.04	11.67	16	H ^{35}Cl	2991.0	52.85	33
H ^{12}C	2860.4	64.11	22	D ^{35}Cl	2145.2	27.18	33
D ^{12}C	2101.0	34.7	1	H Br	2649.7	45.21	1
H ^{28}Si	2042.5	35.67	23	H I	2309.5	39.73	1
H Ge	1831.85	32.86	24	(H ^{132}Xe)$^+$	2270.0	41.33	34
H ^{120}Sn	1655.49	28.83	25, 26	^{7}Li ^{7}Li	351.4	2.59	1
H ^{208}Pb	1560.53	28.79	27	^{7}Li F	906.2	7.90	35
H ^{14}N	(3300)	—	1	^{7}Li Cl	641.1	4.2	35
(H ^{14}N)$^-$	3191.5	85.6	28	^{7}Li Br	563.2	3.53	35
H ^{31}P	(2380)	—	1	^{7}Li I	498.2	3.39	35
H ^{209}Bi	1699.52	31.93	29, 30	^{23}Na ^{23}Na	159.2	0.73	1
^{23}Na K	123.3	0.40	1	Cs Cl	209	0.75	37
^{23}Na Rb	106.6	0.46	1	^{133}Cs Br	(194)	(2.0)	1
^{23}Na ^{19}F	536.1	3.83	36	^{133}Cs ^{127}I	142	(1.2)	1
^{23}Na Cl	366	2.05	37	^{9}Be ^{19}F	1265.5	9.12	1
^{23}Na Br	302	1.50	37	^{9}Be ^{35}Cl	846.65	5.11	1
^{23}Na I	258	1.08	37	^{9}Be ^{16}O	1487.3	11.83	1
^{39}K ^{39}K	92.6	0.35	1	^{24}Mg ^{19}F	717.6	3.84	1
K F	426	2.4	38	^{24}Mg ^{35}Cl	465.4	2.05	1
K Cl	281	1.30	37	^{24}Mg ^{79}Br	373.8	1.34	1
K Br	213	0.80	37	Mg ^{127}I	[312]	—	1
K I	212	0.70	1	^{24}Mg ^{16}O	785.1	5.18	1
^{85}Rb ^{85}Rb	57.3	0.96	1	Mg S	525.2	2.93	1
Rb ^{133}Cs	49.4	—	1	^{40}Ca ^{19}F	587.1	2.74	1
^{85}Rb ^{19}F	373.44	1.90	38, 39	Ca ^{35}Cl	369.8	1.31	1

Table II-1a. (Continued)

Molecule	ω_e	$\chi_e\omega_e$	Refs.	Molecule	ω_e	$\chi_e\omega_e$	Refs.
Rb Cl	288	0.92	37	Ca ^{79}Br	285.3	0.86	1
^{133}Cs ^{133}Cs	42.0	0.08	1	CA ^{127}I	242.0	0.64	1
^{133}Cs ^{19}F	352.62	1.63	39	Ca ^{16}O	732.03	4.83	40
Sr ^{19}F	500.1	2.21	1	^{48}Ti ^{35}Cl	456.4	6.3	1
Sr ^{35}Cl	302.3	0.95	1	^{48}Ti ^{16}O	1008.4	4.61	1
Sr ^{79}Br	216.5	0.51	1	^{90}Zr ^{16}O	936.6	3.45	1
Sr ^{127}I	173.9	0.42	1	V ^{16}O	1012.7	4.9	1
Sr ^{16}O	653.5	4.0	1	Cr ^{16}O	898.8	6.5	1
Ba ^{19}F	468.9	1.79	1	Mn Mn	68.1	1.05	42
^{138}Ba ^{35}Cl	279.3	0.89	1	^{55}Mn F	618.8	3.01	1
Ba ^{79}Br	193.8	0.42	1	^{55}Mn ^{16}Cl	384.9	1.4	1
^{139}Ba O	669.73	2.02	41	^{55}Mn Br	289.7	0.9	1
^{45}Sc ^{16}O	971.6	3.95	1	^{55}Mn ^{16}O	840.7	4.89	1
^{89}Y ^{16}O	852.5	2.45	1	Fe ^{35}Cl	406.6	1.2	1
^{129}La ^{16}O	811.6	2.23	1	Fe ^{16}O	880	5	1
Ce ^{16}O	865.0	2.99	1	Co Cl	421.2	0.74	1
^{141}Pr ^{16}O	818.9	1.20	1	Ni Cl	419.2	1.04	1
Gd ^{16}O	841.0	3.70	1	^{63}Cu ^{19}F	622.7	3.95	1
Lu ^{16}O	841.7	4.07	1	^{63}Cu ^{35}Cl	416.9	1.57	1
Yb Cl	293.6	1.23	1	^{63}Cu ^{79}Br	314.1	0.87	1
^{63}Cu ^{127}I	264.8	0.71	1	^{202}Hg ^{81}Br	186.39	0.98	1
Cu ^{16}O	628	3	1	Hg ^{127}I	125.65	1.09	1
^{107}Ag ^{35}Cl	343.6	1.16	1	Hg Tl	26.9	0.69	1
^{109}Ag ^{81}Br	247.7	0.68	1	^{11}B ^{11}B	1051.3	9.4	1
^{107}Ag ^{127}I	206.2	0.43	1	^{11}B ^{19}F	1399.8	11.3	1
Ag ^{16}O	493.21	4.10	1	^{11}B ^{35}Cl	839.1	5.11	1
^{197}Au ^{35}Cl	382.8	1.30	1	^{11}B Br	587.0	4.86	43
Zn ^{19}F	(630)	(3.5)	1	^{11}B ^{14}N	1514.6	12.3	1
Zn ^{35}Cl	390.56	1.55	1	^{11}B ^{16}O	1885.4	11.77	1
Zn Br	(220)	—	1	^{27}Al ^{19}F	814.58	8.1	1
^{64}Zn ^{127}I	223.40	0.75	1	^{27}Al ^{35}Cl	481.4	2.03	44
Cd ^{19}F	(535)	—	1	^{27}Al ^{79}Br	378.02	1.28	1
Cd ^{35}Cl	330.55	1.2	1	^{27}Al ^{127}I	316.1	1.0	1
Cd Br	230.0	0.50	1	^{27}Al ^{16}O	978.23	7.12	1
Cd ^{127}I	178.56	0.63	1	^{69}Ga F	622.37	3.30	45
Hg ^{19}F	490.8	4.05	1	^{69}Ga ^{35}Cl	365.0	1.11	1
Hg ^{35}Cl	292.60	1.60	1	^{69}Ga ^{81}Br	263.0	0.81	1
^{69}Ga ^{127}I	216.4	0.5	1	^{12}C ^{14}N	2068.7	13.14	1
GaAs(gr)	215	3	46	^{12}C ^{31}P	1239.75	6.86	1
GaAs(ex)	152	2.89	46	^{12}C ^{16}O(gr)	2170.23	13.46	1
Ga ^{16}O	767.7	6.34	1	C O(ex)	1743.76	14.57	50
^{115}In F	535.36	2.67	47	$(^{12}$C ^{16}O$)^+$	2214.24	15.16	1
^{115}In ^{35}Cl	317.41	1.01	1	^{12}C ^{32}S	1285.1	6.5	1
^{115}In ^{81}Br	221.0	0.65	1	^{12}C Se	1036.0	4.8	1
^{115}In ^{127}I	177.1	0.4	1	Si Si	(750)	—	1
In ^{16}O	703.1	3.71	1	(Si F)$^+$	1050.47	4.95	51
Tl ^{19}F	475.0	1.89	1	^{28}Si ^{19}F	856.7	4.7	1
Tl ^{35}Cl	287.5	1.24	1	^{28}Si ^{35}Cl	535.4	2.20	1

Table II-1*a*. (*Continued*)

Molecule	ω_e	$\chi_e\omega_e$	Refs.	Molecule	ω_e	$\chi_e\omega_e$	Refs.
Tl ^{81}Br	192.18	0.39	1	Si Br	425.4	1.52	1
Tl ^{127}I	150	—	1	^{28}Si ^{14}N	1151.7	6.56	1
^{12}C ^{12}C	1641.4	11.67	1	^{28}Si ^{16}O	1242.0	6.05	1
(C F)$^+$	1792.76	13.23	48	^{28}Si ^{32}S	749.5	2.56	1
(C Cl)$^+$	1177.7	6.65	49	^{28}Si Se	580.0	1.78	1
^{12}C ^{35}Cl	846	1.0	1	^{28}Si Te	481.2	1.30	1
(^{74}Ge F)$^+$	815.60	3.22	52	Pb ^{35}Cl	303.8	0.88	1
Ge F	666.5	3.15	53	Pb ^{79}Br	207.55	0.50	1
^{74}Ge ^{35}Cl	407.6	1.36	1	Pb ^{127}I	160.53	0.25	1
Ge Br	296.6	0.9	1	Pb ^{16}O	721.8	3.70	1
^{74}Ge ^{16}O	985.7	4.30	1	^{208}Pb ^{32}S	428.1	1.20	1
^{74}Ge ^{32}S	575.8	1.80	1	Pb Se	277.6	0.51	1
^{74}Ge ^{80}Se	406.8	1.2	1	Pb Te	211.8	0.12	1
^{74}Ge ^{130}Te	323.4	1.0	1	^{14}N ^{14}N	2359.6	14.46	1
Sn ^{19}F	582.9	2.69	1	(^{14}N ^{14}N)$^+$	2207.2	16.14	1
Sn ^{35}Cl	352.9	1.06	1	^{14}N ^{16}O	1903.9	13.97	1
Sn Br	247.7	0.62	1	^{14}N S	1220.0	7.75	1
Sn ^{16}O	822.4	3.73	1	^{14}N Br	693	5.0	1
^{120}Sn ^{32}S	487.23	1.35	54	^{14}N ^{31}P	1337.2	6.98	1
Sn Se	331.2	0.74	1	^{14}N ^{75}As	1068.0	5.36	1
Sn Te	259.5	0.50	1	^{14}N Sb	942.0	5.6	1
Pb Pb	256.5	2.96	1	^{31}P ^{31}P	780.4	2.80	1
Pb ^{19}F	507.2	2.30	1	P F	846.7	4.49	55
P Cl	551.4	2.23	56	^{209}Bi ^{127}I	163.9	0.31	1
^{31}P ^{16}O	1230.66	6.52	1	Bi O	688.4	4.8	59
P S	739.1	2.79	57	^{16}O ^{16}O	1580.4	12.07	1
^{75}As ^{75}As	429.4	1.12	1	(O O)$^+$	1876.4	16.53	1
(^{75}As ^{75}As)$^+$	(314.8)	(1.25)	1	O F	1053.45	10.23	60
(As ^{35}Cl)$^+$	527.7	1.74	58	^{16}O Cl	(780)	—	1
^{75}As ^{16}O	967.4	5.3	1	^{16}O Br	713	7	1
Sb Sb	269.9	0.59	1	^{16}O I	(687)	(5)	1
Sb ^{209}Bi	220.0	0.50	1	^{16}O ^{32}S	1123.7	6.12	1
Sb ^{19}F	641.2	2.77	1	^{32}S ^{32}S	725.7	2.85	1
Sb ^{35}Cl	369.0	0.92	1	Se ^{16}O	907.1	4.61	1
Sb ^{14}N	942.05	5.6	1	^{80}Se ^{80}Se	391.8	1.06	1
Sb ^{16}O	817.2	5.30	1	Te ^{16}O	796.0	3.50	1
^{209}Bi ^{209}Bi	172.7	0.32	1	Te Te	251	0.55	1
^{209}Bi ^{19}F	510.8	2.05	1	^{19}F ^{19}F	[892.1]	—	1
^{209}Bi ^{35}Cl	308.0	0.96	1	^{19}F ^{35}Cl	783.5	4.95	61
^{209}Bi ^{79}Br	209.3	0.47	1	^{19}F Br	662.3	3.80	61
^{35}Cl ^{35}Cl	564.9	4.0	1	I F	610.26	3.14	64, 65
(^{35}Cl ^{35}Cl)$^+$	645.3	2.90	1	^{127}I ^{35}Cl	384.2	1.47	1
Cl Br	442.5	1.5	62	^{127}I ^{79}Br	268.4	0.78	1
^{79}Br ^{79}Br	325.43	1.10	63	^{127}I ^{127}I	214.6	0.61	1
^{79}Br ^{81}Br	323.2	1.07	1				

[a] The compounds are listed in the order of the periodic table of the first atom: IA, IAA, . . . , IB, IIB, ...
[b] Values in parentheses are not certain, and those in square brackets are observed frequencies without anharmonicity correction. The ω_e and $\chi_e\omega_e$ values are rounded off at two decimals. The abbreviations (gr) and (ex) indicate the ground and excited states, respectively.

Li_2 (351.4) > Na_2 (159.2) > K_2 (92.6) > Rb_2 (57.3) > Cs_2 (42.0)

HBe (2058.6) > HMg (1495.3) > HCa (1298.4) > HSr (1206.9) > HBa (1168.4)

Across the periodic table, we find that

HB (~ 2366) < HC (2860.4) < HN (~ 3300) < HO (3735.2) < HF (4138.6)

The frequency decreases upon removal of bonding electrons:

N_2 (2359.6) > N_2^+ (2207.2), As_2 (429.4) > As_2^+ (314.8)

while the opposite trend is found when antibonding electrons are removed:

CO^+ (2214.2) > CO (2170.2), SiF^+ (1050.5) > SiF (856.7),

GeF^+ (815.6) > GeF (666.5)

Similar trends are found in Table II-1b. Table II-1a also includes two examples for which vibrational frequencies were reported for both ground and excited electronic states:

CO ($X^1\Sigma^+$, ground state), 2170.2 > CO ($^3\Pi$, excited state), 1743.8
GaAs ($X^3\Sigma^-$, ground state), 215 > GaAs ($^3\Pi_{0+}$, excited state), 152

In inert gas matrices, a variety of aggregates of diatomic molecules are produced, depending upon the experimental conditions employed. For example, the Raman spectra of H_2, HD, and D_2 in Ar matrices suggest the formation of aggregates of more than six molecules after annealing above 35 K.[66] The vibrational spectra and structures of the O_4^+, O_4^-, and N_4^+ ions will be discussed later. The nitric oxide dimer, $(NO)_2$, in Ar matrices exists as the *cis*-form (1862 and 1768 cm^{-1}) or the *trans*-form (1740 cm^{-1}), with the ν(NO) shown in the brackets.[67] The IR spectrum of the *cis*-dimer in the gaseous state has been assigned.[68]

Table II-1*b* lists the observed frequencies of homopolar diatomic molecules, ions, and radicals. Combining these data with those given in Table II-1*a*, the following frequency trends are immediately obvious:

$O_2^+ > O_2 > O_2^- > O_2^{2-}$, $S_2 > S_2^-$, $Cl_2^+ > Cl_2 > Cl_2^-$,

$Br_2^+ > Br_2$, $I_2^+ > I_2 > I_2^-$

As mentioned earlier, these orders can be explained in terms of simple M.O.

TABLE II-1*b*. Observed Frequencies of Homopolar Diatomic Molecules, Ions, and Radicals (cm^{-1})

Species	State[a]	$\tilde{\nu}$	Refs.	Species	State[a]	$\tilde{\nu}$	Refs.
T_2[b]	Liquid	2458	69	Cl_2[c]	Mat	546	77
Sn_2	Mat	188	70	Cl_2^-[c]	Mat	247	78
Pb_2	Mat	112	70	Br_2^+	Sol'n	360	79
P_2	Gas	775	71	Br_2	Gas	319	80
As_2	Gas	421	71	I_2^+	Sol'n	238	81
O_2^-	Mat	1097	72	I_2	Gas	213	80
O_2^{2-}	Solid	794, 738	73	I_2^-	Mat	115	82
S_2	Mat	718	74	Ag_2	Mat	194	83
S_2^-	Solid[d]	623	75	Zn_2	Mat	80	84
Se_2^-	Solid[d]	349	75	Cd_2	Mat	58	84
F_2^-	Mat	475	76				

[a]Mat = inert gas matrix.
[b]T = ^3H (tritium).
[c]Cl = ^{35}Cl.
[d]Doped in KCl.

theory. Table II-1*c* shows the relationship among various molecular parameters in the $O_2^+ > O_2 > O_2^- > O_2^{2-}$ series. It is seen that, as the bond order decreases, the bond distance increases, the bond energy decreases, and the vibrational frequency decreases. Although the force constant is not rigorously related to the bond energy (Sec. I-3), there is an approximate linear relationship between these parameters. Thus, these frequency trends provide valuable information about the nature of diatomic ligands in coordination compounds (see Section III of Part B). The O_2^+ ion was found in compounds such as $O_2^+[MF_6]^-$ (M = As, Sb, etc.) and $O_2^+[M_2F_{11}]^-$ (M = Nb, Ta, etc.),[86] whereas the O_2^- ion was observed in a triangular MO_2 type complex formed by the reaction of alkali metals with O_2 in an Ar matrix.[87] The $\nu(O_2)$ of simple metal superoxides ($M^+ O_2^-$) and peroxides ($M^{2+} O_2^{2-}$) have been measured.[88] Linear relationships are noted between $\nu(S_2)$ and S—S bond distances for sulfur compounds.[89]

TABLE II-1*c*. Relationship Between Various Molecular Parameters in the O_2 and Its Ions

Species	Bond Order	Bond Distance (Å)[a]	Bond Energy (Kcal/mole)[a]	$\nu(O_2)$ (cm^{-1})	Force Constant (mdyn/Å)
O_2^+	2.5	1.123	149.4	1876	16.59
O_2	2.0	1.207	117.2	1580	11.76
O_2^-	1.5	1.280	—	1094[b]	5.67
O_2^{2-}	1.0	1.49	48.8	791/736[b,c]	2.76

[a]Ref. 85.
[b]Na salt.
[c]Solid-state splitting.

In an Ar matrix, $K^{+}(O_4)^{-}$ takes the trans structure:

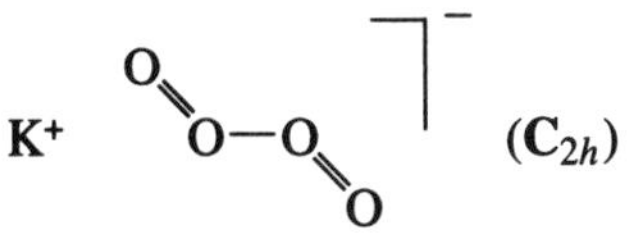

and exhibits ν(O=O) (A_g), ν(O=O) (B_u), ν(O—O) (A_g), and δ(O=O—O) at 1291.5, 993, 305, and 131 cm^{-1}, respectively.[90,91] The structure of the O_4 molecule is predicted to be either quasi-square $\mathbf{D}_{4h}$[92] or triangular $\mathbf{D}_{3h}$ (similar to the CO_3^{2-} and NO_3^{-} ions).[93] The N_4^{+} ion exhibits an IR band at 2237.6 cm^{-1}, and its structure has been found to be linear and centrosymmetric.[94]

Table II-1*d* lists vibrational frequencies of heteropolar diatomic molecules, ions, and radicals. It also gives isotopic frequencies involving (^{1}H, ^{2}D, ^{3}T), (^{6}Li, ^{7}Li), (^{12}C, ^{13}C), (^{32}S, ^{34}S), and (^{35}Cl, ^{39}Cl). Charge effects on vibrational

Table II-1*d*. Observed Frequencies of Heteropolar Diatomic Molecules, Ions, and Radicals (cm^{-1})[a]

Species	State[b]	$\tilde{\nu}$	Refs.	Species	State	$\tilde{\nu}$	Refs.
H Al	Mat	1593	95	(H ^{32}S)$^{+}$	Gas	2547.2	107
H Si	Mat	1967	96	(D ^{32}S)+	Gas	1829.1	107
	Gas	1971.04	97	H F	Gas	3961.42	108
H N	Mat	3131.6	98	H Cr	Mat	1548	109
H As	Gas	2077.00	99	D Cr	Mat	1112	109
H O	Mat	3452.3	100	H Ni	Mat	1906	94
		3428.2		H Cu	Mat	1882	94
	Mat	3548.2	101	^{7}Li O	Mat	745	110, 111
D O	Mat	2616.1	101	Li F	Mat	885	112
(H O)$^{-}$	Solid	3637.4	102	Li Cl	Mat	575	113
	Gas	3555.59	103	^{6}Li ^{35}Cl	Gas	686.1	114
(H O)$^{+}$	Gas	2956.37	104	^{7}Li ^{35}Cl	Gas	643.0	114
(T O)$^{-}$	Solid	2225	105	^{6}Li ^{37}Cl	Gas	683.3	114
H S	Mat	2540.8	106	^{7}Li ^{37}Cl	Gas	640.1	114
	Gas	2591	106	Li Br	Mat	510	113
D S	Mat	1847.8	106	Li I	Mat	433	113
	Gas	1885	106	Na F	Mat	515	112
Na ^{35}Cl	Gas	364.7	115, 116	U N	Mat	995	129
Mg F	Mat	738	117	U O	Mat	820	130
^{26}Mg O	Mat	815.4	118	Pu N	Mat	855.7	131
Mg O	Gas	774.74	119	Pu O	Mat	~820	132
^{40}Ca O	Mat	707.0	120	Zn O	Mat	802	133
Ti O	Mat	1005	121	Cd O	Mat	719	133
Ti N	Mat	1037	122	B F	Gas	1378.7	134
Zr O	Mat	975	121	Al O	Mat	975	135
Nb N	Mat	1002.5	123	Al F	Mat	793	136

Table II-1d. (Continued)

Species	State[b]	$\tilde{\nu}$	Refs.	Species	State	$\tilde{\nu}$	Refs.
Nb O	Mat	968	123	Tl F	Mat	441	137
Ta O	Mat	1020	124	Tl Cl	Mat	261	137
W O	Mat	1050.9	125	^{205}Tl ^{35}Cl	Mat	284.7	138
^{56}Fe O	Mat	873.1	126	Tl Br	Mat	179	137
Ru O	Mat	834.4	127	Tl I	Mat	143	137
Co O	Mat	846.4	126	C F	Mat	1279	139
^{58}Ni O	Mat	825.7	126	C O	Mat	2138.4	140
Th N	Mat	934.6	128	C S	Mat	1274	141
(C N)	Sol'n	2080	142	Pb ^{80}Se	Mat	275.1	148
C N	Mat	2046	143	$(N O)^+$	Solid	2273	152
^{29}Si N	Gas	1972.80	144	N O	Mat	1880	153
^{28}Si O	Mat	1225.9	145	$(N O)^-$	Mat	1358 ~ 1374	154
^{28}Si ^{32}S	Gas	749.65	146	$(N O)^{2-}$	Mat	886	155
^{28}Si ^{34}S	Gas	739.30	146	P N	Mat	1323	156
^{74}Ge O	Mat	973.3	147	S O	Mat	1101.4	157
^{74}Ge S	Mat	566.6	148	$(S Se)^-$	Solid	464	158
^{74}Ge^{80}Se	Mat	397.9	148	F N	Mat	1117	159
^{74}Ge Te	Mat	317.67	148	F O	Mat	1028.7	160
Sn O	Mat	816.1	149	Cl N	Mat	818.5, 825	159
^{120}Sn S	Mat	480.5	148	Cl O	Mat	850.7	161
^{120}Sn ^{80}Se	Mat	325.2	148	Br N	Mat	691	159
Pb O	Mat	718.4	150	Br O	Mat	729.9	162
Pb S	Mat	423.1	148	S Cl	Mat	617	163
^{208}Pb ^{32}S	Gas	426.6	151	S Br	Mat	518	163
S I	Mat	443	163	^{79}Br ^{35}Cl	Gas	444.3	167
$(Cl F)^+$	Sol'n	819	164	I N	Mat	590	168
^{79}Br O	Gas	723.4	165	I F	Gas	610.2	169
$(Br O)^-$	Sol'n	618	166	I Cl	Gas	381	170
^{79}Br F	Gas	669.9	167	I Br	Gas	265	170

[a]The species are arranged in the order of the perodic table: IA, IIA, ... VIIA, IB, IIB, ... VIIIB.
[b]Mat = inert gas matrix.

frequencies are seen for

$$(HO)^- > HO > (HO)^+, \qquad HS > (HS)^+, \qquad (NO)^+ > NO > (NO)^- > (NO)^{2-}$$

Reaction of Li atoms with NO in Ar matrices yields an ionic species. $Li^+ (NO)^-$, which exhibits the $\nu(NO^-)$ and $\nu(Li^+{-}NO^-)$ at 1352 and 651 cm^{-1}, respectively, in IR spectra.[155] Monoxo groups such as M(IV)=O (M = V, Cr, Mn, etc.) and nitrido groups such as M(V)≡N (M = Cr, Mn, Fe, etc.) exhibit their stretching vibrations in the 1050–750 cm^{-1} region. Vibrational spectra of coordination compounds containing these groups are discussed in Part B.

Alkali halide vapor consists mainly of monomers and dimers. As stated in Sec. I-24, Linevsky[171] developed a technique to isolate these monomers and dimers, produced at high temperature, in inert gas matrices. This technique has been used extensively to study the infrared spectra and structures of a number of inorganic salts. Some references on metal halide dimers are: $(LiF)_2$,[172] $(LiCl)_2$,[113,173] $(LiBr)_2$,[113,173] and $(NaX)_2$ (X = F, Cl, Br, and I).[174] These dimers are known to be cyclic-planar ($\mathbf{D}_{2h}$). On the other hand, $(TlX)_2$(X = F and Cl) are linear and symmetrical ($\mathbf{D}_{\infty h}$, X—Tl—Tl—X).[137] Such structures have been well known for $(HgX)_2$ (X = Cl, Br, and I).[175] The dimer $(LiO)_2$ is also cyclic-planar.[110] However, $(NH)_2$ is *trans*-planar ($\mathbf{C}_{2h}$) in a N_2 matrix[176] but takes the *cis* structure upon complex formation with the $Cr(CO)_5$ group.[177]

Hydrogen halides polymerize in the condensed phases; hydrogen fluoride polymerizes even in the gaseous phase.[178] The HX stretching bands are shifted markedly to lower frequencies by polymerization. For example, the monomer frequencies of HF (3962 cm^{-1}), HCl (2886 cm^{-1}), HBr (2558 cm^{-1}), and HI (2230 cm^{-1}) in the gaseous phase are lowered to 3420–3060,[179] 2746–2704,[180] 2438–2404,[180] and 2120 cm^{-1},[180] respectively, in the solid phase.

The IR spectra of $(HF)_2$. (HF) (DF) and $(DF)_2$ in Ar matrices have been assigned based on the structure[181]

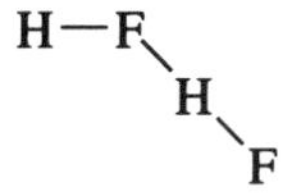

This work has been extended to higher polymers, $(HF)_3$ and $(HF)_4$.[181a] Recently, IR spectra of van der Waals complexes between inert gases and hydrogen halides have been obtained. In Ne—HF[182] and Ne—HCl,[183] the ν(HF) and ν(HCl) are shifted by +0.4722 and +0.3024 cm^{-1}, respectively, from their values in the free state. For analogous Ar complexes, see Refs. 184–188.

Halogens form molecular compounds with organic solvents. For example, the band at 213 cm^{-1} of gaseous I_2 is shifted to 201 cm^{-1} in benzene solution,[189] and the band at 381.5 cm^{-1} of gaseous ICl is shifted to 275 cm^{-1} in pyridine solution.[190] The Raman spectra of I_2, Br_2, and ICl have been studied in many solvents.[191] These frequencies are much lower than the corresponding gas-phase frequencies because of charge-transfer interaction with solvent molecules. The resonance Raman spectrum of I_2 is discussed in Sec. I-23.

The hydroxyl ion $[OH]^-$ is characterized by a sharp band at 3700–3500 cm^{-1}. For example, $LiOH{\cdot}H_2O$ exhibits a sharp OH stretching band at 3574 cm^{-1} and a broad OH_2 stretching band in the 3200–2800 cm^{-1} region.[192] The cyanide ion $[CN]^-$ exhibits a relatively sharp band in the 2250–2050 cm^{-1} region. The CN stretching bands are at 2080 cm^{-1} for ionic cyanides such as Na[CN] and K[CN][193] and at 2170–2250 cm^{-1} for covalent cyanides such as Cu[CN] and Au[CN], in which two metals are bridged by the CN groups.[194] The vibrational frequencies of cyanogen, N≡C—C≡N, have been determined and its harmonic frequencies and anharmonicity constants calculated using ^{13}C and ^{15}N isotope data.[195]

II-2. TRIATOMIC MOLECULES

The three normal modes of linear X_3-($\mathbf{D}_{\infty h}$) and YXY-($\mathbf{D}_{\infty h}$) type molecules were shown in Fig. I-11; ν_1 is Raman-active but not infrared-active, whereas ν_2 and ν_3 are infrared-active but not Raman-active (mutual exclusion rule). However, all three vibrations become infrared- as well as Raman-active in linear XYZ-type molecules ($\mathbf{C}_{\infty v}$), shown in Fig. II-1. The three normal modes of bent X_3-($\mathbf{C}_{2v}$) and YXY-($\mathbf{C}_{2v}$) type molecules were also shown in Fig. I-12. In this case, all three vibrations are both infrared- and Raman-active. The same holds for bent XXY- and XYZ-($\mathbf{C}_s$) type molecules. Section I-12 describes the procedure for normal coordinate analysis of the bent YXY-type molecule. In the following, vibrational frequencies of a number of triatomic molecules are listed for each class of compounds.

Table II-2*a* lists the vibrational frequencies of XY_2-type halides. Most of these data were obtained in inert gas matrices. It is seen that the majority of compounds follow the trend $\nu_3 > \nu_1$, and that exceptions occur for some bent molecules. Although the structures of these halides are classified either as linear or bent, it should be noted that the bond angles of the latter range from 95° to 170°.[230] Thus, some bent molecules are almost linear. In principle, the distinction between linear and bent structures can be made by IR/Raman selection rules. For example, the ν_1 is IR-active for the bent structure but IR-inactive for the linear structure. However, such a simple criterion may lead to conflicting results.[233]

The bond angle (α) of the YXY molecule can be calculated if the ν_3' of the YX′Y molecule is available. Here, X′ is an isotope of X. Using the G and F elements given in Appendix VII, the following equation can readily be derived:

$$\left(\frac{\tilde{\nu}_3'}{\tilde{\nu}_3}\right)^2 = \frac{M_X}{M_{X'}}\left[\frac{M_{X'} + M_Y(1 - \cos\alpha)}{M_X + M_Y(1 - \cos\alpha)}\right]$$

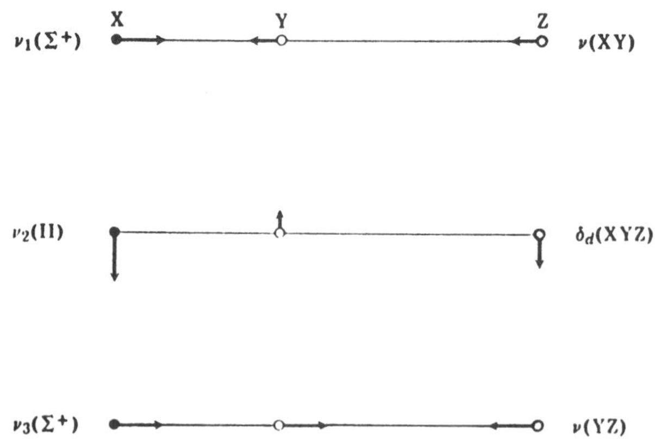

Fig. II-1. Normal modes of vibration of linear XYZ molecules.

Table II-2a. Vibrational Frequencies of XY_2-Type Metal Halides (cm^{-1})

Compound[a]	Structure	ν_1	ν_2	ν_3	Refs.
BeF_2	Linear	(680)	345	1555	196
$BeCl_2$	Linear	(390)	250	1135	196
$BeBr_2$	Linear	(230)	220	1010	196
BeI_2	Linear	(160)	(175)	873	196
MgF_2	Linear	550	249	842	197
$MgCl_2$	Linear	327	93	601	197
$MgBr_2$	Linear	198	82	497	197
MgI_2	Linear	148	56	445	197
$^{40}CaF_2$	Bent	484.8	163.4	553.7	198
$^{40}Ca^{35}Cl_2$	Linear	—	63.6	402.3	199
CaI_2 (g)	Linear	—	43.3	292.3	199a
$^{86}SrF_2$	Bent	441.5	82.0	443.4	198
$^{88}Sr^{35}Cl_2$	Bent	269.3	43.7	299.5	199
BaF_2	Bent	389.6	(64)	413.2	198
$BaCl_2$	Bent	255.2	—	260.0	199
$^{69}GaCl_2$	Bent	373.0	—	415.1	200
$InCl_2^-$	Bent	328	177	291	201
CF_2	Bent	1102	668	1222	202
$C^{35}Cl_2$	Bent	719.5	—	745.7	203
CBr_2	Bent	595.0	—	640.5	204
SiF_2	Bent	851.5	(345)	864.6	205
$^{28}Si^{35}Cl_2$	Bent	513	—	502	206
$^{28}SiBr_2$	Bent	402.6	—	399.5	207
GeF_2	Bent	692	263	663	208
$GeCl_2$	Bent	398	—	373	209
SnF_2	Bent	592.7	197	570.9	210
$SnCl_2$	Bent	354	(120)	334	209
$SnBr_2$	Bent	237	84	223	211
PbF_2	Bent	531.2	165	507.2	210
$PbCl_2$	Bent	297	—	321	209
$PbBr_2$ (g)	Bent	200	64	—	212
NF_2	Bent	1069.6	573.4	930.7	213
PF_2	Bent	834.0	—	843.5	214
PCl_2	Bent	452.0	—	524.8	215
PBr_2	Bent	369.0	—	410.0	215
$O^{35}Cl_2$	Bent	630.7	296.4	670.8	216
$^{32}SF_2$	Bent	838	355	817	217, 217a
SCl_2	Bent	518	208	526	218
SBr_2	Bent	405	—	418	218
SI_2	Bent	368	—	376	218
$SeCl_2$ (g)	Bent	415	153	377	219
$TeCl_2$ (g)	Bent	377	125	—	219
KrF_2 (g)	Linear	449	233	596, 580	220
XeF_2	Linear	497	213.2	555	221
$Xe^{35}Cl_2$	Linear	—	—	314.1	222
$^{63}CuF_2$	Bent	—	183.0	743.9	223
$[CuCl_2]^-$(s)	Linear	302	108	407	223
$[CuBr_2]^-$(s)	Linear	194	81	326	223
$[CuI_2]^-$(s)	Linear	148	65	279	224

Table II-2a. (*Continued*)

Compound[a]	Structure	ν_1	ν_2	ν_3	Refs.
$[AgCl_2]^-$(s)	Linear	268	88	333	224
$[AgBr_2]^-$(s)	Linear	170	61	253	224
$[AgI_2]^-$(s)	Linear	133	49	215	224
$[AuCl_2]^-$(s)	Linear	329	120, 112	350	225
$[AuBr_2]^-$(s)	Linear	209	79, 75	254	225
$[AuI_2]^-$(s)	Linear	158	67, 59	210	225
ZnF_2	Linear	596	150	754	226
$ZnCl_2$	Linear	352	103, 100	503	227
$ZnBr_2$	Linear	223	71	404	227
ZnI_2 (g)	Linear	163	67.6	337.5	199a, 227
CdF_2	Linear	555	121	660	226
$CdCl_2$	Linear	(327)	88	419	228
$CdBr_2$	Linear	(205)	62	319	228
CdI_2	Linear	(149)	(50)	269	228
HgF_2	Linear	568	170	642	226
$HgCl_2$	Linear	(348)	107	405	228
$HgBr_2$	Linear	(219)	73	294	228
HgI_2	Linear	(158)	63	237	228
TiF_2	Bent	665	~180	766	229
VF_2	Bent	—	—	733.2	230
CrF_2	Linear	668	125	749	231
$CrCl_2$	Linear	—	—	493.5	232
MnF_2	Linear	—	124.8	700.1	230
$MnCl_2$	Linear	—	83	476.8	232
MnI_2 (g)	Linear	—	54.5	324.2	199a
FeF_2	Linear	—	141.0	731.3	230
$FeCl_2$	Linear	—	88	493.2	232
CoF_2	Bent	—	151.0	723.5	230
$CoCl_2$	Linear	—	94.5	493.4	232
NiF_2	Bent	—	139.7	779.6	230
$NiCl_2$	Linear	(350)	85	520.6	232
$NiBr_2$	Linear	—	69	414.2	232

[a] All data were obtained in inert gas matrices except those for which the physical state is indicated as g (gas), l (liquid), or s (solid).

Here *M* denotes the mass of the atom subscripted. Figure II-2 shows the spectra of NiF_2 in Ne and Ar matrices obtained by Hastie et al.[234] Using the ν_3 values obtained for isotopic NiF_2, the F—Ni—F angle was calculated to be 154–167°. The bond angle depends upon the nature of the matrix gas employed. For example, $NiCl_2$ is linear in Ar matrices, but bent (~130°) in N_2 matrices.[235] Thus, structural data obtained in gas matrices must be interpreted with caution. As another example, the bond angle of CaF_2 increases from 139° to 143° to 156° upon changing the matrix gas from Kr to Ne to N_2.[236]

The structure of the dimeric species $(MX_2)_2$ is known to be cyclic-planar:

($\mathbf{D}_{2h}$)[237,238]

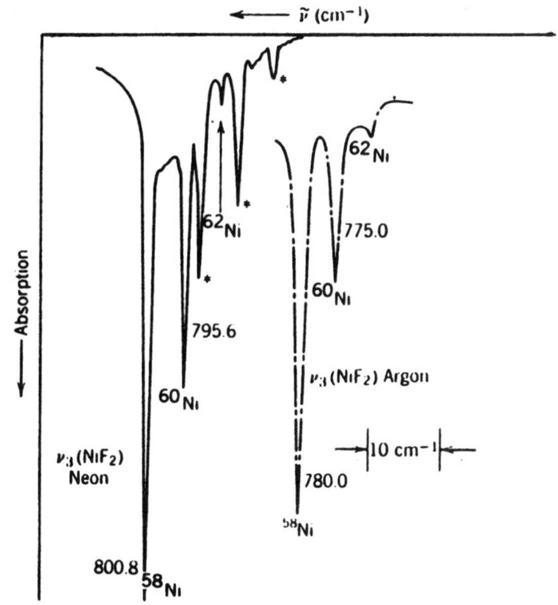

Fig. II-2. IR spectra (ν_3) of NiF_2 in Ne and Ar matrices. Matrix splitting, indicated by asterisks, is present in the Ne matrix spectrum (reproduced with permission from Ref. 234).

although an exception is reported for $(GeF_2)_2$:[239]

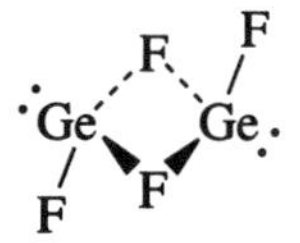

(C_{2h})

(Nonplanar)

Mixed fluorides such as $LiNaF_2$ and $CaSrF_4$ take planar ring stuctures in inert gas matrices:[240]

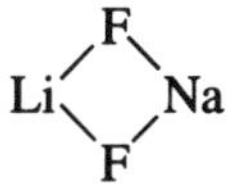

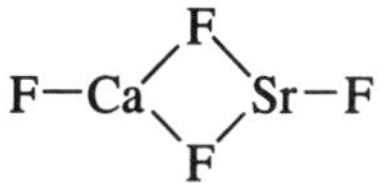

Table II-2*b* lists the vibrational frequencies of triatomic oxides, sulfides, selenides, and related compounds. Most of these data were obtained in inert gas matrices.

Although the NO_2^- (nitrite) ion is bent, the NO_2^+ (nitronium) ion is linear in most compounds. Solid N_2O_5 consists of the NO_2^+ and NO_3^- ions, and the

Table II-2*b*. Vibrational Frequencies of YXY- and XXY-Type Oxides and Related Compounds (cm^{-1})

Compound[a]	Structure	ν_1[b]	ν_2	ν_3[b]	Refs.
$^6Li^6LiO$	Linear	—	118.	1028.5	110
O^6LiO	Bent	—	243.4	730	111
ONaO	Bent	1080.0	390.7	332.8	241
OBO	Linear	—	—	1276	242
AlOAl	Bent	715	(120)	994	243
GaOGa	Bent	472	—	809.4	244
$O^{12}CO$ (g)	Linear	(1337)[c]	667	2349	245, 246
$O^{13}CO$	Linear	—	(649)	2284	246
$[O\ C\ O]^+$	Linear	1280	623	1469	247, 248
$[O\ C\ O]^-$	Linear	—	—	1658.3	248
SCS (g)	Linear	658	397	1533	249
SeCSe	Linear	369.1	313.1	1301.9	250
$O^{120}SnO$	Bent	—	—	863.1	251
$[O\ N\ O]^-$	Bent	1327	806	1286	252
O N O	Bent	1325	752	1610	253, 254
$[O\ N\ O]^+$ (s)	Linear	1396	570	2360	255
NNO	Linear	2223.8	588.7	1284.9	256
$[O_3]^-$	Bent	1016	600.9	802.3	257
O_3	Bent	1134.9	716.0	1089.2	258
S_3 (g)	Bent	575	256	656	259
$[S_3]^-$ (l)	Bent	535	235.5	571	260
$[S_3]^{2-}$ (s)	Bent	458	277	476	260a
OSO	Bent	1147	517	1351	261
SSO (g)	Bent	679.1	382	1166.5	262
$[O\ S\ O]^-$	Bent	985.1	495.5	1041.9	263
$O^{80}SeO$	Bent	992.0	372.5	965.6	264
OTeO	Bent	831.7	294	848.3	265
O O H (g)	Bent	1391.8	3436.2	1097.6	266
O O D (g)	Bent	1020.1	2549.2	1121.5	266
F O O	Bent	~1500	586.4	376.0	267
F O F	Bent	925/915	461	821	268
ClOCl(s)	Bent	630.7	296.4	670.8	269
O O Cl	Bent	1477.8	432.4	214.9	270, 270a
$[O\ Cl\ O]^-$(s)	Bent	790	400	840	271
O Cl O	Bent	944.8	448.7	1107.6	270, 272
$[O\ Cl\ O]^+$(s)	Bent	1055	520	1295/1230	273
Br O Br(s)	Bent	504	197	587	274, 275
$[O\ Br\ O]^+$	Bent	865	375	932	276
$[O\ Br\ O]^-$	Bent	710	320	—	271
OCeO	Bent	757.0	—	736.8	277
OTbO	Bent	758.7	—	718.8	277
OThO	Bent	787.4	—	735.3	277
O Ru O	Bent	926	—	902.2	278
OPrO	Bent	—	—	730.4	277
$O^{48}TiO$	Bent	—	—	934.8	279
OTaO	Bent	971	—	912	280
OWO	Bent	992	—	928	281
OUO	Linear	(765.4)	—	776.1	282

Table II-2*b*. (*Continued*)

Compound[a]	Structure	ν_1[b]	ν_2	ν_3[b]	Refs.
$[O\ U\ O]^{2+}$(s)	Linear	880	140	950	283
O Pu O	Linear	—	—	794.3	284
$[P\ Be\ P]^{4-}$(s)	Linear	379	340	860	285
$[As\ Be\ As]^{4-}$(s)	Linear	222	310	775	285
$[Sb\ Be\ Sb]^{4-}$(s)	Linear	—	255	700	285

[a] All data were obtained in inert gas matrices except those for which the physical state is indicated as g (gas), l (liquid), or s (solid).
[b] For XXY molecules, ν_1 and ν_3 are XX and XY stretching modes, respectively.
[c] Fermi resonance with $2\nu_2$ (see Sec. 1-11).

former is slightly bent, as indicated by the appearance of a strong δ(ONO) band at 534 cm^{-1} in the Raman spectrum.[286] The O_3 molecule is bent (bond angle, ~117°),[258] so is the O_3^- ion (~108°) in Ar matrices.[287,288] The S_3 molecule, S_3^- and S_3^{2-} ions, are all bent (103° ~115°), and the frequencies of all three fundamentals decrease progressively in this order. A T-shaped structure of $\mathbf{C}_{2v}$ symmetry has been proposed for SiC_2 prepared in an Ar matrix at 8 K.[289]

Metal dioxides produced by sputtering techniques[290] in inert gas matrices take linear or bent O—M—O structures. Metal dinitrides produced by the same technique also take linear N—M—N (M = U[291] and Pu[292]) or bent N—M—N (M = Th[293]) structures. These structures are markedly different from those of molecular oxygen and nitrogen complexes of various metals produced by the conventional matrix cocondensation technique (Section III of Part B).

The NpO_2^{2+}, PuO_2^{2+}, and AmO_2^{2+} ions are similar to the UO_2^{2+} ion in that they are linear and symmetrical. They exhibit the ν_3 bands at 969, 962, and 939 cm^{-1}, respectively, in IR spectra.[294] The relationship between the U═O stretching force constant and the U═O distance was derived by Jones.[295] According to Bartlett and Cooney,[296] the U═O distance R (in pm = 10^{-2} Å) is given by

$$R = 9141(\nu_3)^{-2/3} + 80.4$$
$$R = 10650(\nu_1)^{-2/3} + 57.5$$

where ν is in units of cm^{-1}. The relationship between the UO_2 stretching frequency and the U═O distance has also been studied by other investigators.[297–299] McGlynn et al.[300] noted that, in a series of $K_xUO_2L_y(NO_3)_2$-type compounds, the U═O stretching frequency decreases as L is changed in the order of the spectrochemical series:

$$CN^- > en > NH_3 > NCS^- > ONO^- > py > H_2O > F^- > NO_3^-$$

Vibrational spectra of coordination compounds containing dioxo (O═M═O) groups are discussed in Section III of Part B.

Table II-2*c* lists the vibrational frequencies of triatomic interhalogeno compounds. The resonance Raman spectrum of the I_3^- ion gives a series of overtones of the ν_1 vibration.[327,328] The resonance Raman spectra of the I_2Br^- and IBr_2^- ions and their complexes with amylose have been studied.[329] The same table also lists the vibrational frequencies of XHY-type (X, Y: halogens) compounds. All these species are linear except the $ClHCl^-$ ion, which was found to be bent in an inelastic neutron scattering (INS) and Raman spectral study.[321]

Ault and co-workers have utilized salt–molecule reactions to produce a number of novel triatomic and other anions in isolated environments.[330] For example, the linear symmetric $[FHF]^-$ ion was produced in Ar matrices via codeposition of CsF vapor with HF gas diluted in Ar.[318]

$$Cs^+F^- + HF \longrightarrow Cs^+[F—H—F]$$

The reaction of CsF vapor with HCl gas diluted in Ar yields a mixture of two types of the $[FHCl]^-$ anions; in Type I, the H atom is not equidistant from the F and Cl atoms, and its ν_3 [mainly $\nu(HF)$] is observed near 2500 cm^{-1}, whereas in Type II, the H atom is symmetrically located and its ν_3 is at 933 cm^{-1}.[331]

$$Cs^+F^- + HCl \begin{cases} \nearrow Cs^+[F—H\cdots Cl]^- \quad \text{(Type I)} \\ \searrow Cs^+[F—H—Cl]^- \quad \text{(Type II)} \end{cases}$$

The yield of each type is determined by the location of the cation in the ion pair, and the Type I/Type II ratio varies, depending on the cation used; smaller alkali metal cations favor Type II while larger cations prefer Type I. These classifications were made originally by Evans and Lo who observed crystalline HCl_2^- salts of Types I and II depending upon the cation employed.[331a]

Table II-2*d* lists the vibrational frequencies of bent XH_2-type molecules. The XH stretching frequencies are lower and the XH_2 bending frequencies are higher in condensed phases than in the vapor phase because of hydrogen bonding in the former. This trend is also seen for H_2O and H_2S dissolved in organic solvents such as pyridine and dioxane.[352,353]

In both liquid water and ice, H_2O molecules interact extensively via O—H···O bonds. However, there are marked differences between the two phases. In the latter, H_2O molecules are tetrahedrally hydrogen-bonded, and this local structure is repeated throughout the crystal. In liquid water, however, the O—H···O bond distance and angle vary locally, and the bond is sometimes broken. Thus, its vibrations cannot be described simply by using the three normal modes of the isolated H_2O molecule. According to Walrafen et al.,[354] an isosbestic point exists at 3403 cm^{-1} in the Raman spectrum of liquid water obtained as a function of temperature, and the bands above and below this frequency are mainly due to non-hydrogen-bonded and hydrogen-bonded species,

Table II-2c. Vibrational Frequencies of Triatomic Halogeno Compounds (cm^{-1})

Compound[a]	Structure[b]	ν_1[c]	ν_2	ν_3[c]	Refs.	Compound[a]	Structure[b]	ν_1[c]	ν_2	ν_3[c]	Refs.
$FClF^+$ (s)	b	807	387	830	301	$BrII^-$ (s)	l	117	84	168	315
FClF (m)	b	500	242	578, 570	302	$BrII^+$ (s)	b	258	—	198	312
$FClF^-$ (s)	l	510, 478	—	636	303	I_3^- (sl)	l	114	52	145	310
$FClCl^+$ (s)	b	744	299, 293	535, 528	304	HFH^+ (s)	b	2970	1680	3080	316
$FBrF^+$ (s)	b	706	362	702	305	FHF^- (s)	l	596, 603	1233	1450	317
$FBrF^-$ (s)	l	442	198	596, 562	306	FHF^- (m)	l	—	1217	1364	318
$ClFCl^+$ (s)	b	~529	293, 258	586, 593	307	FHF^- (g)	l	583.1	1286.0	1331.2	319
Cl_3^+ (s)	b	505.8	231.5	515.7	308	$ClHF^-$ (s)	l	275	863, 823	2710	320
Cl_3^- (m)	b	253	—	340	309	$ClHCl^-$ (s)	b	199	660, 602	1670	321
$ClBrCl^-$ (sl)	l	278	~135	225	310	$ClHBr^-$ (s)	l	—	508	1705	322
$ClICl^+$ (s)	b	371	147	364	311	$ClHI^-$ (s)	l	—	485	2200	322
$ClICl^-$ (sl)	l	269	127	226	310	$BrHF^-$ (s)	l	220	740	2900	320
$ClII^+$ (s)	b	360	(126)	197	312	$BrHBr^-$ (s)	l	126	1038	1420	323
Br_3^+ (s)	b	293	124	297	313	$BrHBr^-$ (m)	l	168	—	670	324
Br_3^- (sl)	l	164	53	191	310	IHF^- (s)	l	180	635	3145	320
Br_3 (m)	l	190	—	—	314	IHI (m)	l	(120.7)	—	682.1	325
$BrIBr^+$ (s)	l	256	124	256	312	F_3^- (m)	l	461	—	550	326

[a] m = inert gas matrix; sl = solution; s = solid.
[b] b = bent; l = linear.
[c] For XYZ type, ν_1 and ν_3 correspond to ν(XY) and ν(YZ), respectively.

Table II-2*d*. Vibrational Frequencies of Bent XH_2-Type Molecules (cm^{-1})

Molecule	State	ν_1	ν_2	ν_3	Refs.
TiH_2	Matrix	1483.2	496	1435.5	332
VH_2	Matrix	1532.4	529	1508.3	332
CrH_2	Matrix	1651.3	—	1614.9	332a
MoH_2	Matrix	1752.7	—	1727.4	332a
AlH_2	Matrix	1766	760	1799	333
SiH_2	Matrix	1964.4	994.8	1973.3	334
GeH_2	Matrix	1887	920	1864	335
NH_2	Matrix	—	1499	3220	336
	Surface	3290	1610	3380	337
$[NH_2]^-$	Solid	3270	1556	3323	338
$[PH_2]^-$	Solid	2230	1080	2160	339
$[OH_2]^+$	Matrix	3182.7	1401.7	3219.5	340, 340a
$^{16}OH_2$	Gas	3657	1595	3756	341
	Liquid	3450	1640	3615	342
	Solid	3400	1620	3220	343
$^{16}OHD^a$	Gas	2727	1402	3707	341
	Solid	2416	1490	3275	343
$^{16}OD_2$	Gas	2671	1178	2788	341
	Solid	2495, 2336	1210	2432	343
$^{18}OD_2$	Gas	2657	1169	2764	344
$OT_2{}^b$	Gas	—	996	2370	345
	Solid	1988	1023	2104	346
SH_2	Gas	2615	1183	2627	347
	Matrix	2619.5	—	2632.6	348
	Solid	2532, 2523	1186, 1171	2544	349
SD_2	Gas	1892	934	2000	350
	Solid	1843, 1835	857, 847	1854	349
SeH_2	Gas	2345	1034	2358	351
SeD_2	Gas	1687	741	1697	351

^{a}Here ν_1 and ν_3 denote ν(OD) and ν(OH), respectively.
^{b}T = ^{3}H (tritium).

respectively. In addition, liquid water exhibits librational and restricted translational modes which correspond to rotational and translational motions of the isolated molecule, respectively. The librations yield a broad contour at 1000–330 cm^{-1}, while the restricted translations appear at ca. 170 and 60 cm^{-1}.[355] For more details, see the review by Walrafen.[356]

Hornig et al.[343] assigned the spectrum of ice I as shown in Table II-2*d*. They also assigned some librational and translational modes, Bertie and Whalley[357] studied the vibrational spectra of ice in other phases, and found the spectra to be consistent with reported crystal structures in each phase. For crystal water and aquo complexes, see Sec. III-7 of Part B. The vibrational frequencies of H_2O, D_2O, HDO, and their dimers in argon matrices have been reported.[358]

Table II-2*e* lists the vibrational frequencies of linear XYZ-type molecules.

Table II-2*e*. Vibrational Frequencies of Linear XYZ and X_3 Molecules (cm^{-1})

XYZ	State	ν_1(XY)	$\nu_2(\delta)$	ν_3(YZ)	Refs.
HCN	Gas	3311	712	2097	359
	Matrix	3306	721	—	360
DCN	Gas	2630	569	1925	359
TCN	Gas	2460	513	1724	361
FCN	Gas	1077	449	2290	362
ClCN	Gas	714, 784[b]	380	2219	363
	Matrix	718	384, 387[a]	—	364
BrCN	Gas	574	342.5	2200	363
	Matrix	575	349, 351[a]	—	364
ICN	Gas	485	304	2188	365
NaCN	Matrix	382	170	2044	366
HNC	Matrix	3620	477	2029	367
	Gas	3692.7	464.2	2023.9	368
^{6}LiNC	Matrix	722.9	121.7	2080.5	369
$H^{11}BO$	Matrix	(2849)	754	1817	370
$[HNN]^+$	Gas	3234.0	686.8	2257.9	371
CCO	Matrix	1074	381	1978	372
SCO	Gas	859	520	2062	373
$[HCS]^+$	Gas	3141.7	766.5	—	374
SCSe	Gas	1435	(355)	506	375
SCTe	Sol'n	1347	(377)	423	375
$[CNO]^-$	Solid	2096	471	1106	376
$[NNF]^+$	Solid	2371	391	1057	377
NCN	Matrix	—	423	1475	378
NNO	Gas	2277	596.5	1300.3	379
NNC	Matrix	1235	394	2824	380
NCO	Matrix	1275	487	1922	381
$[NCO]^-$	Sol'n	2155	630	1282, 1202[c]	382
$[NNN]^-$	Solid	1344	645	2041	383
$[NNN]^+$	Matrix	1287	472.7	1657.5	384
$[NC^{32}S]^-$	Solid	2053	486, 471[a]	748	385
$[NCSe]^-$	Solid	2070	424, 416[a]	558	386
$[NCTe]^-$	Solid	2073	366	450	387
PNO	Matrix	865.2	—	1754.7	388

[a]Splitting due to matrix or solid state effect.
[b]Fermi resonance between ν_1 and $2\nu_2$.
[c]Fermi resonance between ν_3 and $2\nu_2$.

Some of these vibrations split into two because of Fermi resonance, matrix effects, or crystal-field effects. Vibrational spectra of coordination compounds containing pseudohalide ions such as NCS^-, NCO^-, and N_3^- are discussed in Sec. III-16 of Part B.

Table II-2*f* lists the vibrational frequencies of bent XYZ-type molecules. The spectra of most of these compounds were measured in inert gas matrices. As an example, Fig. II-3 shows the matrix-isolation IR spectra of four isotopic species of F—Cl—O obtained by Andrews et al.[410] These species were

Table II-2f. Vibrational Frequencies of Bent XYZ-Type Molecules (cm^{-1})

XYZ	State	ν_1(XY)	$\nu_2(\delta)$	ν_3(YZ)	Refs.
HCO	Matrix	2483	1087	1863	389
$H^{12}CF$	Matrix	—	1405	1181.5	139
HNO	Matrix	3450	1110	1563	390
$H^{14}NF$	Matrix	—	1432	1000	159
HOO	Matrix	3414	1389	1101	391
FOH	Matrix	884	1393	3483	392
ClOO	Matrix	407	373	1441	393
BrOO	Matrix	—	—	1487	394
HOF	Matrix	3537.1	1359.0	886.0	395
HOCl	Matrix	3578	1239	728	396
HOBr	Matrix	3589	1164	626	396
HOI	Matrix	3597	1103	575	397
$H^{72}GeBr$	Matrix	1858	701	283	398
NOF	Matrix	1886.2	734.9	492.2	399
NSF	Gas	1372	366	640	400
NSCl	Gas	1325	273	414	401
ONF	Gas	1843.5	765.8	519.9	402
$ON^{35}Cl$	Gas	1799.7	595.8	331.9	403
ONBr	Gas	1799.0	542.0	266.4	404
ONI	Matrix	1809	470	216	405
OPF	Matrix	1292.2	416.0	811.4	406
OPCl	Gas	1218	—	780	407
OCF	Matrix	1855	626	1018	408
OCCl	Matrix	1880	281	570	409
$O^{35}ClF$	Matrix	1038.0	315.2	593.5	410
SPF	Matrix	720.2	313.6	791.4	411
SPBr	Matrix	712	—	372	412
$^{35}ClCF$	Matrix	742	—	1146	413
$^{35}Cl^{32}SN$	Matrix	1327.3	403.8	267.4	414
ClSnBr	Matrix	328	—	228	415
ClPbBr	Matrix	295	—	200	415
$Br^{32}SN$	Matrix	1312.9	346.1	226.2	414

produced via the photolysis of Ar matrix samples containing F—Cl (^{35}Cl and ^{37}Cl in natural abundance) and an isotopically scrambled O_3 ($^{16}O_3$, $^{16,18}O_3$ and $^{18}O_3$).

$$F{-}Cl + O_3 \longrightarrow F{-}Cl{-}O + O_2$$

The results of normal coordinate analyses show that the bands at 1038–990, 593–587, and 315–307 cm^{-1} are almost pure ν(ClO), ν(FCl), and δ(FClO), respectively. The ν(ClO) of FClO is much higher than that of diatomic ClO (780 cm^{-1}) because the F atom removes the electron density from the antibonding orbital of the Cl—O group.

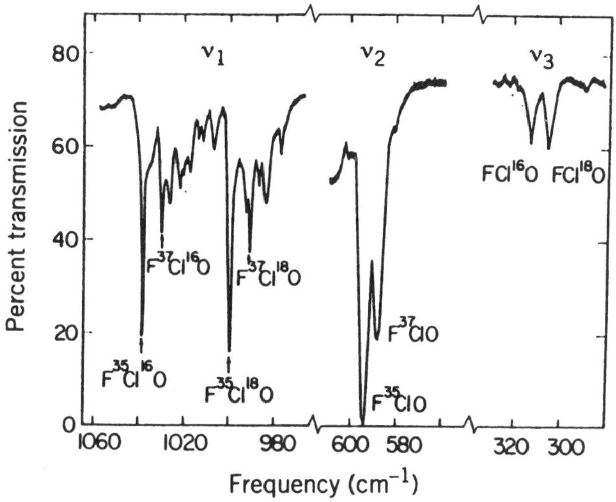

Fig. II-3. IR spectra of four isotopic species of FClO in Ar matrices (reproduced with permission from Ref. 410).

II-3. PYRAMIDAL FOUR-ATOM MOLECULES

(1) XY_3 Molecules (C_{3v})

The four normal modes of vibrations of a pyramidal XY_3 molecule are shown in Fig. II-4. All four vibrations are both infrared- and Raman-active. The G and F matrix elements of the pyramidal XY_3 molecule are given in Appendix VII.

Table II-3*a* lists the fundamental frequencies of XH_3-type molecules. Several bands marked by an asterisk are split into two by *inversion doubling*. As is shown in Fig. II-5, two configurations of the XH_3 molecule are equally probable. If the potential barrier between them is small, the molecule may resonate between the two structures. As a result, each vibrational level splits into two levels (positive and negative).[416] Transitions between levels of different sign are allowed in the infrared spectrum, whereas those between levels of the same sign are allowed in the Raman spectrum. The transition between the two levels at $v = 0$ is also observed in the microwave region ($\tilde{\nu} = 0.79$ cm^{-1}). If the

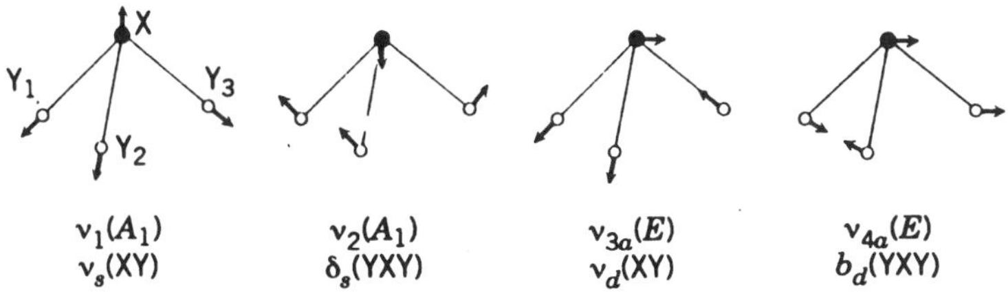

Fig. II-4. Normal modes of vibration of pyramidal XY_3 molecules.

Table II-3a. Vibrational Frequencies of Pyramidal XH_3 Molecules (cm^{-1})

Molecule	State	$\nu_1(A_1)$	$\nu_2(A_1)$	$\nu_3(E)$	$\nu_4(E)$	Refs.
NH_3	Gas	3335.9, 3337.5 [a]	931.6, 968.1 [a]	3414	1627.5	416
	Solid	3223	1060	3378	1646	417
ND_3	Gas	2419	748.6, 749.0 [a]	2555	1191.0	416
	Solid	2318	815	2500	1196	417
$^{15}NH_3$	Gas	3335	926, 961 [a]	3335	1625	418
NT_3	Gas	2016	647	2163	1000	419
PH_3	Gas	2327	990, 992 [a]	2421	1121	420
PD_3	Gas	1694	730	(1698)	806	420
AsH_3	Gas	2122	906	2185	1005	420
AsD_3	Gas	1534	660	—	714	420
SbH_3	Gas	1891	782	1894	831	421
SbD_3	Gas	1359	561	1362	593	421
$[OH_3]^+SbCl_6^-$	Sol'n	3560	1095	3510	1600	422
$[OH_3]^+ClO_4^-$	Solid	3285	1175	3100	1577	423
$[OH_3]^+NO_3^-$	Solid	2780	1135	2780	1680	424
$[OH_3]^+$	Liquid SO_2	3385	—	3470, 3400	1700, 1635	425
$[OD_3]^+$	Liquid SO_2	2490	—	2660, 2580	1255	425
$[SeH_3]^+$	Solid	2302	936	2320	1057	425a
$[CH_3]^-Na^+$	Matrix	2760	1092	2805	1384	426
SiH_3	Gas	—	727.9	—	—	427
$[GeH_3]^-$	Liquid NH_3	1740	809	—	886	428

[a]Inversion doubling.

potential barrier is sufficiently high, and if the three Y groups are not identical, optical isomers may be anticipated.

As is seen in Table II-3*a*, ν_1 and ν_3 overlap or are close in most compounds. The presence of the hydronium (H_3O^+) ion in hydrated acids has been confirmed by observing its characteristic frequencies. For example, it was shown from infrared spectra that $H_2PtCl_6 \cdot 2H_2O$ should be formulated as $(H_3O)_2[PtCl_6]$.[429] For normal coordinate analysis of pyramidal XH_3 molecules, see Refs. 430–432.

Reaction of CH_3I with Na (or K) atoms in N_2 matrices at 20K[426] produces an ion-pair complex CH_3^- Na^+ (or CH_3^- K^+), of $\mathbf{C}_{3v}$ symmetry. The ν(Na—CH_3) and ν(K—CH_3) vibrations were observed at 298 and 280 cm^{-1}, respectively. Vibrational spectra of the $(NH_3)_2$ dimer in Ar matrices show that two NH_3 molecules are bonded via a very weak hydrogen bond.[433]

Table II-3*b* lists the vibrational frequencies of pyramidal XY_3 halogeno compounds. Clark and Rippon[449] have measured the Raman spectra of a number

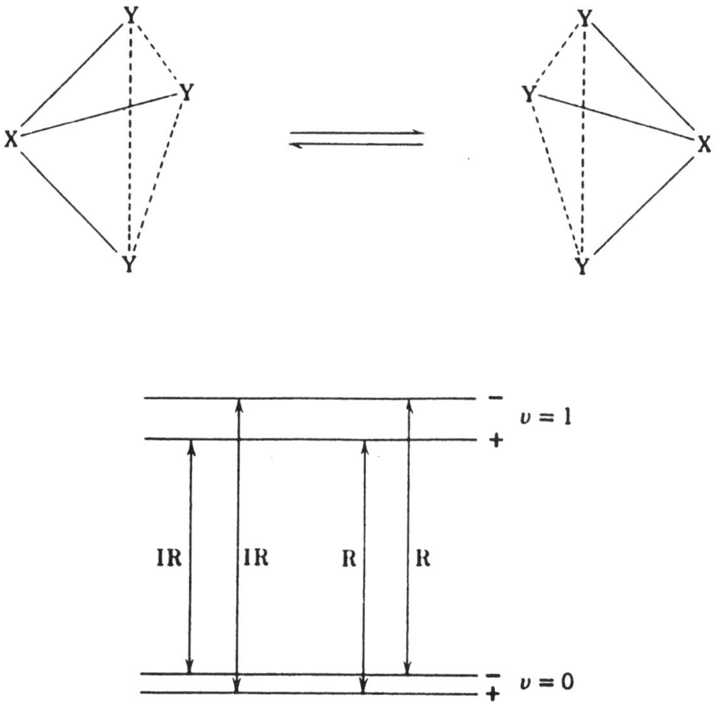

Fig. II-5. Inversion doubling of XY_3-type molecules.

of these compounds in the gaseous phase. The compounds show a $\nu_1 > \nu_3$ and $\nu_2 > \nu_4$ trend, whereas the opposite trend holds for the neutral XH_3 molecules listed in Table II-3*a*. In some cases, the two stretching frequencies (ν_1 and ν_3) are too close to be distinguished empirically. This is also true for the two bending bands (ν_2 and ν_4).

Figure II-6 shows the Raman spectra of the SnX_3^- ions (X = Cl, Br, and I) in solution obtained by Taylor.[442] In crystalline $[As(Ph)_4][SnX_3]$, the symmetry of the SnX_3^- ion is lowered to $\mathbf{C}_s$ so that ν_3 and ν_4 each split into two bands.[440] Normal coordinate analyses on the SnX_3^- ion (X = F, Cl, Br, and I)[440] and Group VB trihalides[455] have been carried out.

Table II-3*c* lists the vibrational frequencies of pyramidal XO_3-type compounds. Rocchiciolli[469] has measured the infrared spectra of a number of sulfites, selenites, chlorates, and bromates. Dasent and Waddington[470] also measured the infrared spectra of metal iodates, and suggested that extra bands at 480–420 cm^{-1} may be due to the metal–oxygen vibrations. Again, the ν_3 and ν_4 vibrations may split into two bands because of lowering of symmetry in the crystalline state. Although $\nu_2 > \nu_4$ holds in all cases, the order of two stretching frequencies (ν_1 and ν_3) depends on the nature of the central metal. It is to be noted that the SO_3^- and SO_3^{2-} anions are pyramidal while the SO_3 molecule

Table II-3*b*. Vibrational Frequencies of Pyramidal XY_3 Halogeno Compounds (cm^{-1})

Molecule	State	ν_1	ν_2	ν_3	ν_4	Refs.
$[InCl_3]^-$	Solid	252	102	185	97	434
$[InBr_3]^-$	Solid	177	74	149	46	434
$[InI_3]^-$	Solid	136	78	110	40	434
CF_3	Matrix	1084	703	1250	600, 500	435
$^{28}SiF_3$	Matrix	832	406	954	290	436
$^{28}SiCl_3$	Matrix	470.2	—	582.0	—	437
$GeCl_3$	Matrix	388	—	362	—	438
$[GeCl_3]^-$	Solid	303	—	285	—	439
$[SnF_3]^-$	Solid	520	280	477	224	440
	Matrix	509	256	454	—	441
$[SnCl_3]^-$	Sol'n	297	130	256	104	442
$[SnBr_3]^-$	Sol'n	205	82	180	65	442
$[SnI_3]^-$	Sol'n	162	61	148	48	442
$[PbF_3]^-$	Matrix	456	—	405	—	443
$[PbCl_3]^-$	Sol'n	249	—	—	—	444
$[PbBr_3]^-$	Sol'n	176	—	164	58	444
$[PbI_3]^-$	Sol'n	137	—	127	30 ~ 45	444
NF_3	Gas	1035	649	910	500	445
NCl_3	Sol'n	535	347	637	254	446
$^{14}NCl_3$	Matrix	554.2	365.2	644	263	447
NI_3	Solid	279	146	354	90	448
PF_3	Gas	893.2	486.5	858.4	345.6	449
PCl_3	Gas	515.0	258.3	504.0	186.0	449
PBr_3	Gas	390.0	159.9	384.4	112.8	449
PI_3	Solid	303	111	325	79	450
AsF_3	Gas	738.5	336.8	698.8	262.0	449
$AsCl_3$	Gas	416.5	192.5	391.0	150.2	449
$AsBr_3$	Melt	272	128	287	99	451
AsI_3	Gas	212.0	89.6	(201)	63.9	449
SbF_3	Matrix	654	259	624	—	452
$SbCl_3$	Gas	380.7	150.8	358.9	121.8	449
$SbBr_3$	Sol'n	254	101	245	81	451
SbI_3	Gas	186.5	74.0	(147)	54.3	449
$BiCl_3$	Gas	342	123	322	107	453
$BiBr_3$	Gas	220	77	214	63	454
BiI_3	Solid	145.5	90.2	115.2	71.0	455
$[SF_3]^+$	Melt	943	690	922	356	456
$[SCl_3]^+$	Solid	498	276	533, 521	215, 208	457
$[SBr_3]^+$	Solid	375	175	429, 414	128	458
$[SeF_3]^+$	Melt	781	381	743	275	456
$[SeCl_3]^+$	Sol'n	430	206	415	172	459
$[SeBr_3]^+$	Sol'n	291	138	298	108	451
$[TeCl_3]^+$	Sol'n	362	186	347	(150)	460
$[TeBr_3]^+$	Sol'n	265	112	266	92	451
$[MnBr_3]^-$	Solid	280	110	150	80	461
$FeCl_3$	Matrix	363.0	68.7	460.2	113.8	462
YCl_3	Gas	378	78	359	58.6	462a

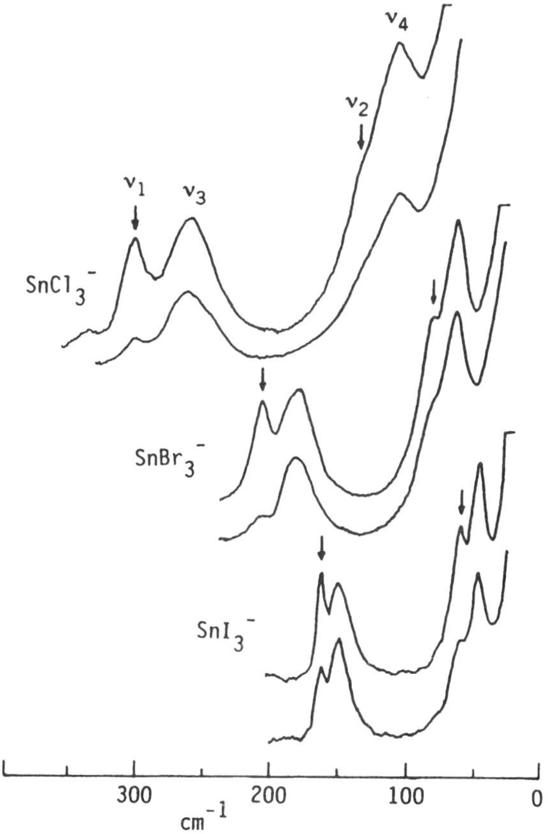

Fig. II-6. Raman spectra of the SnX_3^- ions in diethyl ether extracts from SnX_2 in HX solutions. The arrows denote polarized bands (reproduced with permission from Ref. 442).

Table II-3c. Vibrational Frequencies of Pyramidal XO_3 Molecules (cm^{-1})

Molecule	State	$\nu_1(A_1)$	$\nu_2(A_1)$	$\nu_3(E)$	$\nu_4(E)$	Refs.
$[SO_3]^{2-}$	Sol'n	967	620	933	469	463
$[SO_3]^-$	Matrix	1000	604	1175	511	464
$[SeO_3]^{2-}$	Sol'n	807	432	737	374	465
$[TeO_3]^{2-}$	Sol'n	758	364	703	326	465
$[ClO_3]^-$	Sol'n	933	608	977	477	466
	Solid	939	614	971	489	467
$[BrO_3]^-$	Sol'n	805	418	805	358	466
	Solid	810	428	790	361	467
$[IO_3]^-$	Sol'n	805	358	775	320	466
	Solid	796	348	745	306	467
XeO_3	Sol'n	780	344	833	317	468

is planar (Table II-4b). Figure II-7 shows the IR spectra of $KClO_3$ and KIO_3 obtained in the crystalline state.

(2) $ZXY_2(C_s)$ and ZXYW (C_1) Molecules

Substitution of one of the Y atoms of a pyramidal XY_3 molecule by a Z atom lowers the symmetry from $\mathbf{C}_{3v}$ to $\mathbf{C}_s$. Then the degenerate vibrations split into two bands, and all six vibrations become infrared- and Raman-active. The relationship between $\mathbf{C}_{3v}$ and $\mathbf{C}_s$ is shown in Table II-3*d*. Table II-3*e* lists the vibrational frequencies of pyramidal ZXY_2 molecules.

Vibrational spectra of the $[SCl_nF_{3-n}]^+$ ions ($n = 0 \sim 3$)[498] and PH_nF_{3-n} ($n = 0 \sim 3$)[499] have been assigned. The $[XeO_2F]^+$ ion takes a pseudo-tetrahedral structure owing to the presence of a sterically active lone pair electron of the Xe atom.[497]

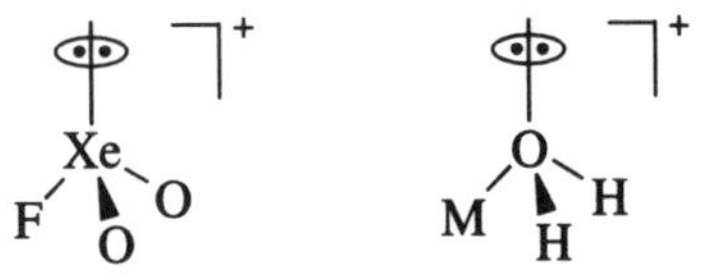

Matrix cocondensation reactions of alkali halide molecules (M^+X^-) with H_2O diluted in Ar produce pyramidal $[MOH_2]^+$ ions which exhibit the $\nu(OH_2)$ and $\delta(OH_2)$ in the 3300–3100 and 700–400 cm^{-1} regions, respectively.[500] These ions may serve as simple models of aquo complexes.

The ZXYW-type molecule belongs to the $\mathbf{C}_1$ point group, and all six vibrations are infrared- and Raman-active. The vibrational spectra of OSClBr[501] and $[XSnYZ]^-$ (X, Y, Z: a halogen) have been reported.[502] For example, the Raman spectrum of the $[ClSnBrI]^-$ ion in the solid state exhibits the $\nu(SnCl)$, ν(Sn-Br), and $\nu(SnI)$ at 275, 193, and 155 cm^{-1}, respectively.

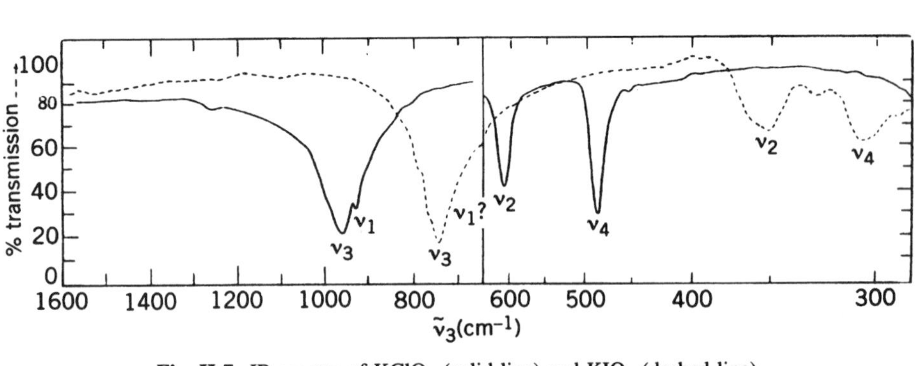

Fig. II-7. IR spectra of $KClO_3$ (solid line) and KIO_3 (dashed line).

Table II-3*d*. Relationship Between C_{3v} and C_s

C_{3v}	$\nu_1(A_1)$	$\nu_2(A_1)$	$\nu_3(E)$		$\nu_4(E)$	
XY_3	$\nu_s(XY)$	$\delta_s(YXY)$	$\nu_d(XY)$		$\delta_d(YXY)$	
	↓	↓	↙	↘	↙	↘
C_s	$\nu_1(A')$	$\nu_3(A')$	$\nu_2(A')$	$\nu_5(A'')$	$\nu_4(A')$	$\nu_6(A'')$
ZXY_2	$\nu_s(XZ)$	$\delta_s(YXZ)$	$\nu_s(XY)$	$\nu_a(XY)$	$\delta_s(YXY)$	$\delta_a(YXZ)$

Table II-3*e*. Vibrational Frequencies of Pyramidal ZXY_2-Type Metal Halides (cm^{-1})

$Z{-}X{<}^{Y}_{Y}$	$\nu_1(A')$ $\nu(XZ)$	$\nu_2(A')$ $\nu_s(XZ)$	$\nu_3(A')$ $\delta_s(YXZ)$	$\nu_4(A')$ $\delta(YXY)$	$\nu_5(A'')$ $\nu_a(XY)$	$\nu_6(A'')$ $\delta_a(YXZ)$	Refs.
HNF_2	3193	972	500	1307	888	1424	471
$HNCl_2$	3279	687	—	1002	695	1295	472
FNH_2	891	3234	1233	1564	3346	1241	473, 474
$ClNH_2$	679 674	3297	1056	1534	3374	1063	473
$ClNF_2$	692	918	552	366	842	382	475
FPH_2	795	2304	934	1090	2310	—	476
$FAsH_2$	649	2108	842	984	2117	—	477
$ClPF_2$	545	864	411	(302)	852	260	478
$BrPF_2$	459	858	244	391	849	215	478
$FPCl_2$	838	525	328	203	525	267	478
$FPBr_2$	824	398	258	123	423	221	478
IPF_2	375	851	198	413	846	204	479
$[FOH_2]^+$	865	3386	1067	1630	3225	1261	480
$[BrOF_2]^+$	1062	655	365	290	630	315	481
OSF_2	1333	808	530	(410)	748	390	482
$OSCl_2$	1251	492	194	344	455	284	483, 484
$OSBr_2$	1121	405	120	267	379	223	485
$[FSO_2]^-$	450	1098	248	600	1150	270	486
$[FSH_2]^+$	853	2500	987	1181	2447	1020	486a
$[ClSO_2]^-$	214	1120	172	526	1312	(103)	487
$[BrSO_2]^-$	203	1117	115	530	1308	—	487
$[ISO_2]^-$	184	1112	(55)	530	1300	—	487
$[FSeO_2]^-$	~430	903	283	324	888	283	488
$OSeF_2$	1049	667	362	282	637	253	489
$OSeCl_2$	995	388	161	279	347	255	490
$OSe(OH)_2$ [a]	831	702	430	336	690	364	491, 492
$FClO_2$	602	1097	398	533	1253	351	493
$FBrO_2$	506	908	305	394	953	271	494
$[OClF_2]^+$	1333	731	513	384	695	404	495, 496
$[FXeO_2]^+$	594	873	281	321	931	249	497

[a] The OH group was assumed to be a single atom.

II-4. PLANAR FOUR-ATOM MOLECULES

(1) XY_3 Molecules (D_{3h})

The four normal modes of vibration of planar XY_3 molecules are shown in Fig. II-8; ν_2, ν_3, and ν_4 are infrared-active, and ν_1, ν_3, and ν_4 are Raman-active. This case should be contrasted with pyramidal XY_3 molecules, for which all four vibrations are both infrared- and Raman-active. Appendix VII lists the G and F matrix elements of the planar XY_3 molecule.

Table II-4*a* lists the vibrational frequencies of planar XY_3 molecules. As stated above, pyramidal and planar structures can be distinguished based on the difference in selection rules. Thus, the pyramidal structure should exhibit two stretching bands (A_1 and E) while the planar structure should show only one stretching band (E') in IR spectra. Andrews and Pimentel[517] prepared the methyl radical via the reaction of Li atoms with CH_3Br or CH_3I in Ar matrices. It exhibits two IR bands at 1383 and 730.3 cm^{-1} which were assigned to the in-plane and out-of-plane bending modes, respectively, of the planar CH_3 radical based on normal coordinate calculations.

The IR spectrum and normal coordinate analysis have been reported for $AlCl_3 \cdot NH_3$ (staggered, C_{3v}).[523] Vibrational spectra of dimeric species such as Al_2F_6 and Al_2Cl_6 are discussed in Sec. II-10. Normal coordinate calculations on planar XY_3 molecules have been carried out by many investigators.[524–529]

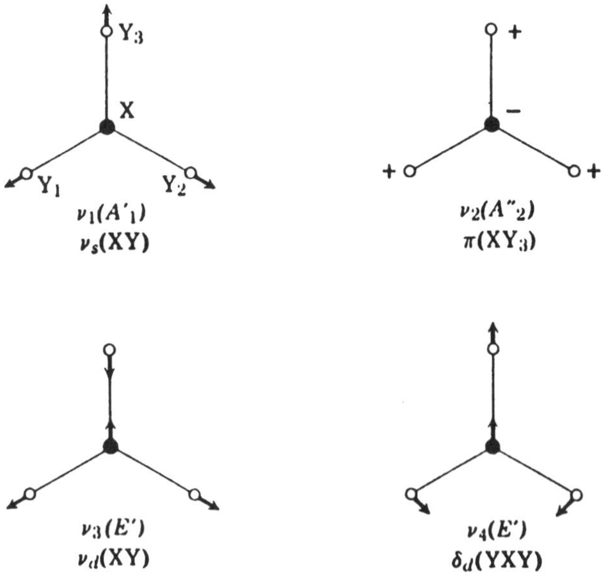

Fig. II-8. Normal modes of vibration of planar XY_3 molecules.

Table II-4*a*. Vibrational Frequencies of Planar XY_3 Molecules (cm^{-1})

Molecule	State[a]	$\nu_1(A_1')$	$\nu_2(A_2'')$	$\nu_3(E')$	$\nu_4(E')$	Refs.
$^{10}BH_3$	Mat	(2623)	1132	2820	1610	503
$^{10}BF_3$	Gas	888	719.3	1505.8	481.1	504, 505
$^{11}BCl_3$	Liquid	472.7	—	950.7	253.7	506, 507
$^{11}BBr_3$	Liquid	278	374	802	150	506, 508
$^{11}BI_3$	Liquid	192	—	691.8	101.0	506, 508
AlH_3	Mat	—	697.8	1882.8	783.4	508a
AlF_3	Mat	—	286.2	909.4	276.9	509
$AlCl_3$	Mat	393.5	—	618.8	150	510
$AlBr_3$	Gas	228	107	450–500	93	511
AlI_3	Gas	156	77	370–410	64	511
$[AlSb_3^-]^{6-}$	Solid	132	181	293, 318	—	512
$GaCl_3$	Mat	386.2	—	469.3	132	510
$GaBr_3$	Gas	219, 237	95	—	84	510, 513
GaI_3	Gas	147	63	275	50	510, 513
$[GaSb_3]^{6-}$	Solid	128	110	192, 210	—	512
$InCl_3$	Mat	359.0	—	400.5	119	510, 514
$InBr_3$	Gas	212	74	280	62	511
InI_3	Gas	151	56	200–230	44	511
$[InAs_3]^{6-}$	Solid	171	99	197, 215	—	515
$TlBr_3$	Sol'n	190	125	220	51	516
CH_3	Mat	—	730.3	—	1383	517
$[CdCl_3]^-$	Sol'n	265	—	287	90	518
$[CdBr_3]^-$	Sol'n	168	—	184	58	518
$[CdI_3]^-$	Sol'n	124	—	161	51	518
$[HgCl_3]^-$	Solid	273	113	263	100	519
$[PS_3]^-$	Melt	476	540	695	242	520
PrF_3	Mat	526, 542	86	458	99	521, 522

[a]Mat: inert gas matrix.

Table II-4*b* lists the vibrational frequencies of planar XO_3-type compounds. Figure II-9 shows the Raman spectra of KNO_3 in the crystalline state and in aqueous solution. As discussed in Sec. I-26, the spectra of calcite and aragonite are markedly different because of the difference in crystal structure. Recent normal coordinate calculations[545] on the CO_3 radical indicate a *trans*-$\mathbf{C}_s$ structure (A) rather than a three-membered ring $\mathbf{C}_{2v}$ structure (B) suggested originally.[546]

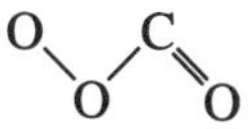

(A)

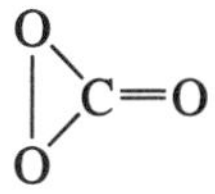

(B)

Crystals of $LiNO_3$, $NaNO_3$, and KNO_3 take the calcite structure (Sec. I-29). Nakagawa and Walter[537] carried out normal coordinate analyses on the whole Bravais lattices of these crystals. The spectra of anhydrous metal nitrates such as $Zn(NO_3)_2$[547] and $UO_2(NO_3)_2$[548] can be interpreted in terms of $\mathbf{C}_{2v}$ symmetry

Table II-4*b*. Vibrational Frequencies of Planar XO_3 and Related Compounds in the Crystalline State (cm^{-1})

Compound		$\nu_1(A_1')$	$\nu_2(A_2'')$	$\nu_3(E')$	$\nu_4(E')$	Refs.
$La[^{10}BO_3]$	IR	939	740.5	1330.0	606.2	530, 531
$H_3[BO_3]$	IR	1060	668, 648	1490–1428	545	532
$Ca[CO_3]$ (calcite)	IR	—	879	1429–1492	706	533
	R	1087	—	1432	714	533
$Ca[CO_3]$ (aragonite)	IR	1080	866	1504, 1492	711, 706	533
	R	1084	852	1460	704	533
$Ba[CS_3]$	R	510	516[b]	920	314	534, 535
$Ba[CSe_3]$	IR	290	420	802	185	535, 536
$Na[NO_3]$	IR	—	831	1405	692	537
	R	1068	—	1385	724	537
$K[NO_3]$	IR	—	828	1370	695	537
	R	1049	—	1390	716	537
NO_3	IR	1060	762	1480	380	538–540
SO_3[a]	IR	(1068)	495	1391	529	541, 542
	R	1065	497.5	1390	530.2	543
SeO_3[a]	R	923	—	1219	305	544

[a]Gaseous state.
[b]Infrared.

since the NO_3 group is covalently bonded to the metal. Raman spectra of metal nitrates in the molten state[549,550] indicate that the degeneracy of the ν_3 vibration is lost and the Raman-inactive ν_2 vibration appears. Apparently, the $\mathbf{D}_{3h}$ selection rule is violated because of cation–anion interaction. The infrared spectrum of monomeric lithium nitrate ($LiNO_3$) in an inert gas matrix shows a large splitting of the ν_3 vibration (240 cm^{-1}) due to distortion of the NO_3 group.[551] Like CO_3^{2-} and NO_3^- ions, CS_3^{2-} and CSe_3^{2-} ions act as chelating ligands. For normal coordinate analyses of planar XO_3 molecules, see Refs. 552 and 553.

Some XY_3-type halides take the unusual T-shaped structure of $\mathbf{C}_{2v}$ symmetry shown below. This geometry is derived from a trigonal-bipyramidal structure in which two equatorial positions are occupied by two lone-pair electrons. Typical examples are ClF_3 and BrF_3. With the equatorial Y atom represented as Y′,

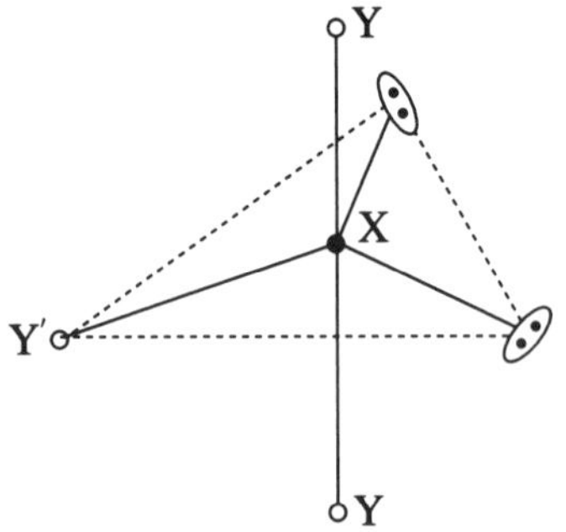

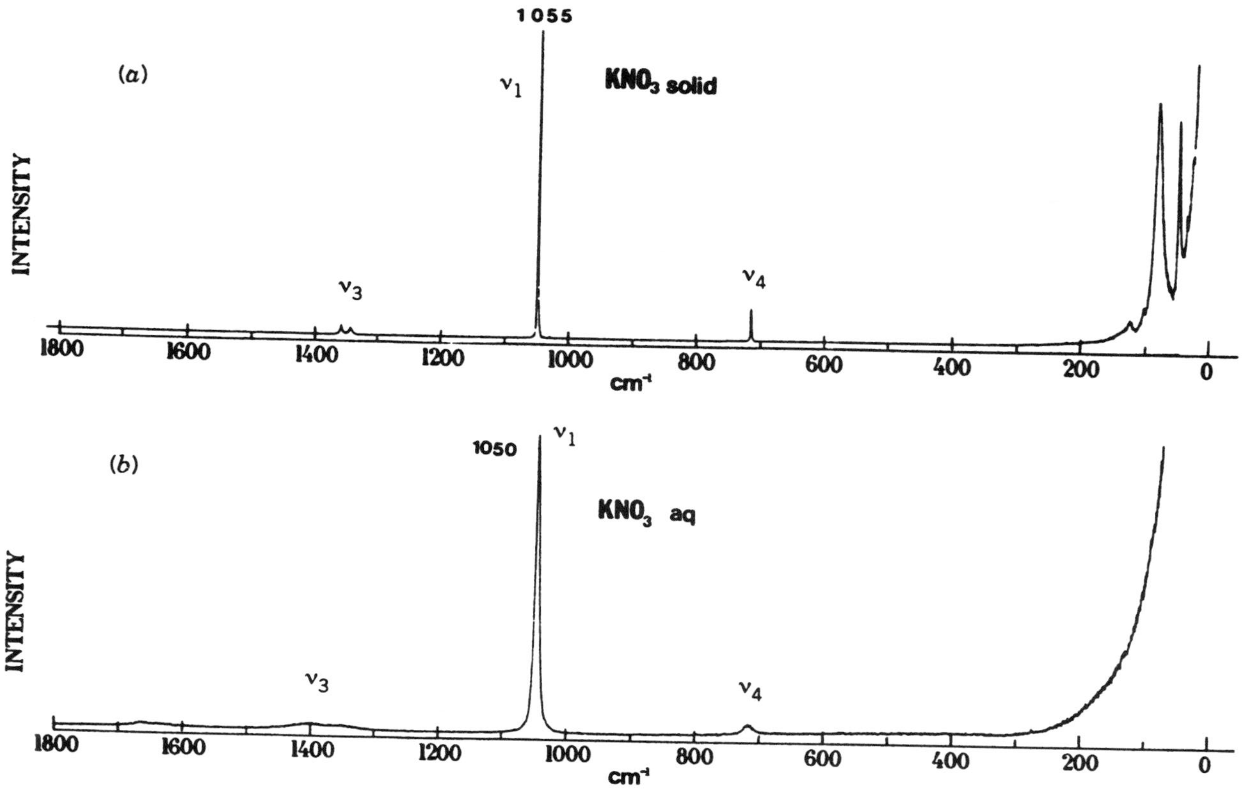

Fig. II-9. Raman spectra of KNO_3 (*a*) in the solid state and (*b*) in aqueous solution.

the following assignments have been made for these molecules:[554] $\nu(XY')$, A_1, 754 and 672; $\nu(XY)$, B_1, 683.2 and 597; $\nu(XY)$, A_1, 523 and 547; δ, A_1, 328 and 235; δ, B_1, 431 and 347; π, B_2, 332 and 251.5 cm^{-1} (for each mode, the former value is for ClF$_3$ and the latter is for BrF$_3$). The Raman spectrum of XeF$_4$ in SbF$_5$ exhibits two strong polarized bands at 643 and 584 cm^{-1}, which were assigned to the T-shaped XeF$_3^+$ ion in [XeF$_3$][SbF$_6$].[555] XeOF$_2$ takes a similar T structure.[556] In an inert gas matrix UO$_3$ gives an infrared spectrum consistent with the T-shaped structure.[557]

(2) ZXY$_2$ (C$_{2v}$) and ZXYW(C$_s$) Molecules

If one of the Y atoms of a planar XY$_3$ molecule is replaced by a Z atom, the symmetry is lowered to C$_{2v}$. If two of the Y atoms are replaced by two different atoms, W and Z, the symmetry is lowered to C$_s$. As a result, the selection rules are changed, as already shown in Table I-18. In both cases, all six vibrations become active in infrared and Raman spectra. Table II-4c lists the vibrational frequencies of planar ZXY$_2$ and ZXYW molecules.

Silanone produced by the silane–ozone photochemical reaction in Ar matrices gives rise to a band at 1202 cm^{-1} in the IR spectrum which is assigned to the $\nu(\text{Si}=\text{O})$.[579]

$$\text{SiH}_4 \; + \; \text{O}_3 \quad \longrightarrow \quad \begin{matrix} \text{H} \\ \diagdown \\ \quad \text{Si}=\text{O} \\ \diagup \\ \text{H} \end{matrix}$$

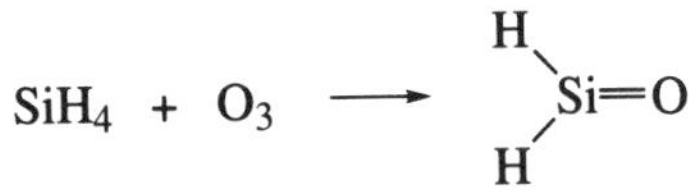

The frequencies listed for the formate and acetate ions were obtained in aqueous solution. Vibrational spectra of metal salts of these ions are discussed in Sec. III-8 of Part B. Although not listed in this table, the IR spectra of binary mixed halides of boron[580] and aluminum[581] have also been reported.

II-5. OTHER FOUR-ATOM MOLECULES

(1) X$_2$Y$_2$ Molecules

Molecules like O$_2$H$_2$ take the nonplanar C$_2$ structure (twisted about the O—O bond by ca. 90°), whereas N$_2$F$_2$ and [N$_2$O$_2$]$^{2-}$ exist in two forms: *trans*-planar (C$_{2h}$) and *cis*-planar (C$_{2v}$). Figure II-10 shows the six normal modes of vibration for these three structures. The selection rules for the C$_{2v}$ and C$_2$ structures are different only in the ν_6 vibration, which is infrared-inactive and Raman depolarized in the planar model but infrared-active and Raman polarized in the nonplanar model.

Table II-5a lists the vibrational frequencies of X$_2$Y$_2$-type molecules. In N$_2$ matrices, (NO)$_2$ exists in three forms; *cis*, *trans*, and another form of uncertain

Table II-4c. Vibrational Frequencies of Planar ZXY_2 and ZXYW Molecules (cm^{-1})

$XY_3(\mathbf{D}_{3h})$	$\nu_1(A_1')$ ν_s(XY)	$\nu_2(A_2'')$ $\pi(XY_3)$	$\nu_3(E')$ ν_d(XY)		$\nu_4(E')$ δ_d(YXY)		
$ZXY_2(\mathbf{C}_{2v})$	$\nu_1(A_1)$ ν(XZ)	$\nu_6(B_1)$ $\pi(ZXY_2)$	$\nu_2(A_1)$ ν_s(XY)	$\nu_4(B_2)$ ν_a(XY)	$\nu_3(A_1)$ δ_s(ZXY)	$\nu_5(B_2)$ δ_a(ZXY)	
ZXYW($\mathbf{C}_s$)	$\nu_1(A')$ ν(XZ)	$\nu_6(A'')$ π	$\nu_2(A')$ ν(XY)	$\nu_4(A')$ ν(XW)	$\nu_3(A')$ δ(ZXY)	$\nu_5(A')$ δ(ZXW)	Refs.
H—BCl_2	2616	786	735	1091	—	895	558
Cl—GaH_2	406.9	620	1964.6	1978.1	731.4	510.1	558a
H—$GaCl_2$	2015.3	464.3	414.3	437.3	130	607.5	558b
[F—$CO_2]^-$	883	—	1316	1749	—	—	559
[(HO)—$CO_2]^{-,a}$	960	835	1338	1697	712	579	560
[H—$CO_2]^-$	2803	1069	1351	1585	760	1383	561
[(CH_3)—$CO_2]^{-,a}$	926	621	1413	1556	650	471	561
O=CF_2	1930	767.4	965.6	1243.7	582.9	619.9	562, 563
O=CCl_2	1827	580	569	849	285	440	564
O=CBr_2	1828	512	425	757	181	350	564
O=CClF	1868	667	776	1095	501	415	564
O=CBrCl	1828	547	517	806	240	372	564
O=CBrF	1874	620	721	1068	398	335	564
O=CHF	1837	—	2981	1065	1343	663	565
S=CH_2	1063	993	2970	~3028	1550	1437	566
S=CF_2	1368	622	787	1189	526	417	567
S=CCl_2	1120	472	~500	812	296	~302	568
Se=CF_2	1280	560	710	1208	432	352	569
[Se—$CS_2]^-$	442	485	433	925	284	265	570
S=$SiCl_2$	739.1	—	—	609.9	—	186	571
S=$GeCl_2$	580	—	404	440	—	—	572
(HO)—NO_2 ^{a}	902	767	1311	1697	660	597	573
F—NO_2	~555	740	1308	1800	810	~555	574
Cl—NO_2	370	652	1318	1670	787	411	574
[O=$NF_2]^+$	1862	715	897	1163	569	647	575, 576
[O=$NCl_2]^+$	1648	—	628	735	220	420	577
Br—PO_2	363.2	—	1146.5	1440.1	513.5	230	578

[a]Only the ZXY_2 skeletal vibrations are listed.

structure. The *cis*-form is most stable, and its ν_s and ν_a are at 1870 and 1776 cm^{-1}, respectively.[593] The NO reacts with Lewis acids such as BF_3 to form a red species at 77 K. Ohlsen and Laane[594] measured its resonance Raman spectrum and concluded that it is an asymmetrical $(NO)_2$ dimer having a *cis*-planar structure:

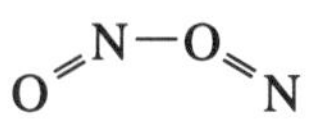

A possible mechanism for the formation of such a dimer in the presence of

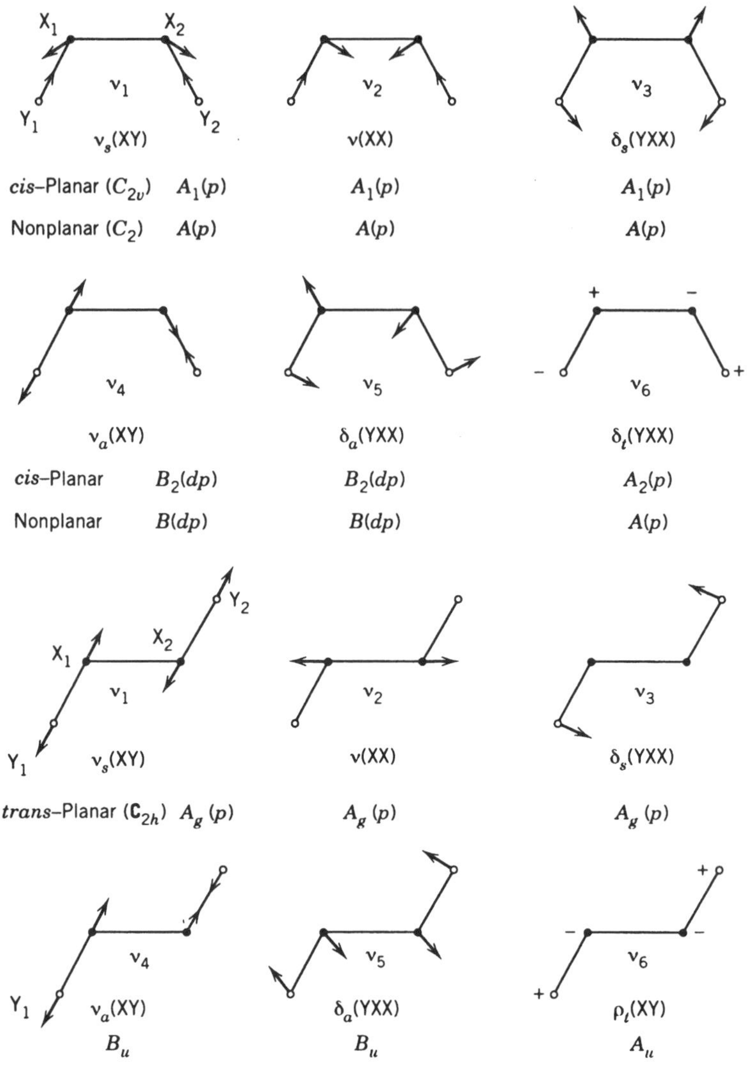

Fig. II-10. Normal modes of vibration of nonlinear X_2Y_2 molecules (*p*: polarized; *dp*: depolarized).[567]

Lewis acids has been proposed. Although the $N_2O_2^{2-}$ ion takes a *cis* or *trans* structure in the solid state, the $N_2O_2^{2-}$ ion produced in Ar matrices exhibits an IR band at 1205 cm^{-1} and its structure was suggested to be[595]

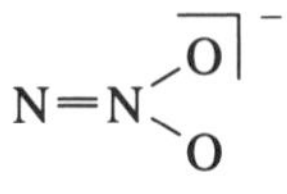

Table II-5a. Vibrational Frequencies of X_2Y_2 Molecules (cm^{-1})[a]

		ν_1	ν_2	ν_3	ν_4	ν_5	ν_6	
trans-X_2Y_2	C_{2h}	A_g	A_g	A_g	B_u	B_u	A_u	
cis-X_2Y_2	C_{2v}	A_1	A_1	A_1	B_2	B_2	A_2	
Twisted X_2Y_2	C_2	A	A	A	B	B	A	
Assignment		$\nu_s(XY)$	$\nu(XX)$	$\delta_s(YXX)$	$\nu_a(XY)$	$\delta_a(YXX)$	π	Refs.
cis-$[N_2O_2]^{2-}$	Solid	830	1314	584	1047	330	—	582
trans-$[N_2O_2]^{2-}$	Solid	(1419)	(1121)	(696)	1031	371	492	583, 584
cis-N_2F_2	Liquid	896	1525	341	952	737	(550)	585
trans-N_2F_2	Liquid	1010	1522	600	990	423	364	586
O_2H_2	Gas	3607	1394	864	3608	1266	317	587
O_2D_2	Gas	2669	1029	867	2661	947	230	587
O_2F_2	Matrix	611	1290	366	624	459	(202)	588
S_2H_2	Liquid	2509	509	(868)[b]	2557[c]	882	—	589
S_2F_2	Gas	717	615	320	681	301	183	590
S_2Cl_2	Gas	466	546	202	457	240	92	591
S_2Br_2	Liquid	365	529	172	351	200	66	592
Se_2Cl_2	Liquid	367	288	130	367	146	87	592
Se_2Br_2	Liquid	265	292	107	265	118	50	592

[a] Except for N_2F_2 and $[N_2O_2]^{2-}$, all the molecules listed take the $\mathbf{C}_2$ or $\mathbf{C}_{2v}$ structure.
[b] Solid.
[c] Gas.

Diazene (N_2H_2) exists in the *cis* and *trans* forms:

and the observed frequencies are 3116, 3025, 1347, and 1304 cm^{-1} for the *cis* form, and 3109 and 1333 cm^{-1} for the *trans* form.[596]

The Raman spectra of Se_2Cl_2 and Se_2Br_2 in CS_2 and CCl_4 solutions indicate that the former takes the $\mathbf{C}_2$ structure, whereas the latter takes the $\mathbf{C}_{2v}$ structure.[597] Although S_2Cl_2 does not dimerize at low temperatures, Se_2Cl_2 forms a dimer below −50°C. A new Raman band observed at 215 cm^{-1} was attributed to the dimer for which the following ring structure was proposed:[598]

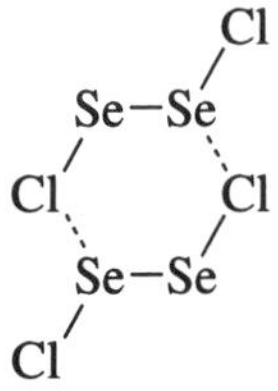

Ketelaar et al.[599] found that the liquid mixture of CS_2 and S_2Br_2, for example, exhibits IR spectra which show combination bands between the ν_3 of CS_2 (1515 cm^{-1}) and a series of S_2Br_2 vibrations. Such "simultaneous transitions" seem to suggest the formation of an intermolecular complex, at least on the vibrational time scale.

Normal coordinate analyses of H_2O_2[600] and S_2X_2 (X: a halogen)[601,592] have been carried out. Other $(XY)_2$-type compounds include dimeric metal halides such as $(LiF)_2$, which takes a planar ring structure (Sec. II-1).

(2) Planar WXYZ, XYZY, and XYYY Molecules (C_s)

Planar four-atom molecules of the WXYZ, XYZY, and XYYY types have six normal modes of vibrations, as shown in Fig. II-11. All these vibrations are both infrared- and Raman-active. In HXYZ and HYZY molecules, the XYZ and YZY skeletons may be linear (HNCO, HSCN) or nonlinear (HONO, HNSO). In the latter case, the molecule may take a *cis* or *trans* structure. Table II-5*b* lists the vibrational frequencies of molecules and ions belonging to these types. Normal coordinate analyses have been carried out for HN_3[619] and HONO.[610]

Teles et al.[602] measured the IR spectra of all four possible isomers of HNCO in Ar matrices at 13 K, and compared the observed frequencies with those

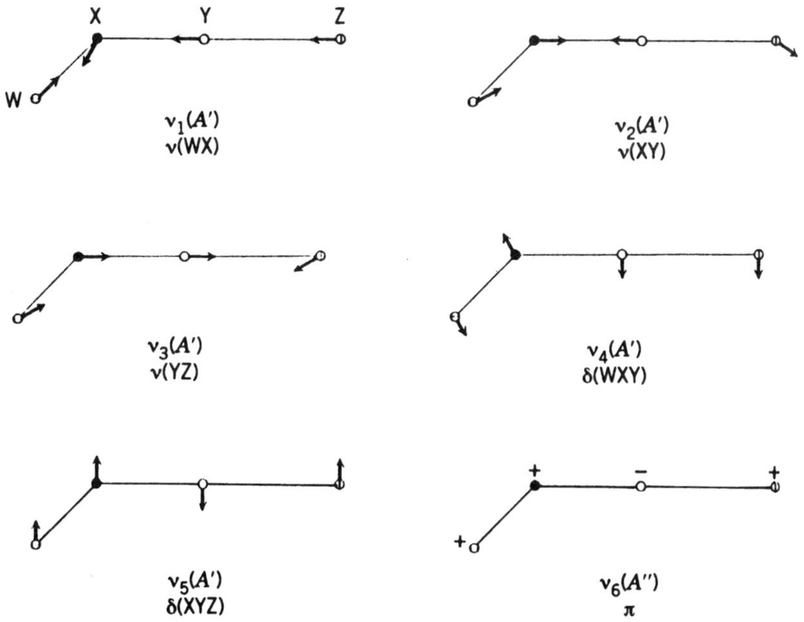

Fig. II-11. Normal modes of vibration of nonlinear WXYZ molecules.

obtained by *ab initio* calculations.

$$H \atop \diagdown N{=}C{=}O$$

Isocyanic acid

$$H \atop \diagdown O{-}C{\equiv}N$$

Cyanic acid

$$H{-}C{\equiv}N^{+}{-}O^{-}$$

Fulminic acid

$$H \atop \diagdown O{-}N^{+}{\equiv}C^{-}$$

Isofulminic acid

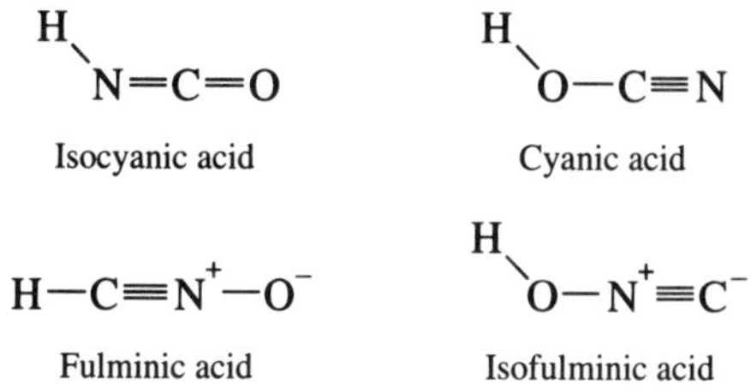

As shown above, fulminic acid is linear, while the remaining three acids take bent structures. For a detailed discussion on the structures and IR spectra of these acids, see the review by Teles et al.[602]

II-6. TETRAHEDRAL AND SQUARE-PLANAR FIVE-ATOM MOLECULES

(1) Tetrahedral XY$_4$ Molecules (T$_d$)

Figure II-12 illustrates the four normal modes of vibration of a tetrahedral XY$_4$ molecule. All four vibrations are Raman active, whereas only ν_3 and ν_4 are

Table II-5*b*. Vibrational Frequencies of Planar Four-Atom Molecules (cm^{-1})

Molecule WXYZ	State[a]	ν_1 ν(WX)	ν_2 ν(XY)	ν_3 ν(YZ)	ν_4 δ(WXY)	ν_5 δ(XYZ)	ν_6 π	Refs.
HNCO	Mat	3516.8 3505.7	2259.0	—	769.8	573.7	—	602
DNCO	Mat	2606.9	2231.0	—	578.6	475.4	—	602
HOCN	Mat	3569.6	1081.3	2286.3	1227.9	—	—	602, 603
DOCN	Mat	2635.0	949.4	2284.6	1077.8	—	—	602, 603
HCNO	Mat	3317.2	2192.7	1244.1	566.6	536.9 538.2	—	602
DCNO	Mat	2612.7	2063.2	1218.5	418.7	—	—	602
HONC	Mat	3443.7	628.4	2190.1	1232.4	361.2	379.3	602
DONC	Mat	2545.2	623.1	2190.3	902.6	357.3	362.1	602
FNCO	Mat	861	2172	(2160)	529	695	646	604
ClNCO	Gas	607.7	2212.2	1306.6	—	707.7	559.0	605
BrNCO	Mat	505	2196.0	1290.9	137.4	691.1	572.2	606
INCO	Gas	462.3	2201.1	1298.1	—	667.7	583.3	605
FNNN	Gas	873.5	1090.0	2037.0	241.0	658.0	504.0	607, 608
HNNN	Mat	3324	2150	1273	1168	527	588	609
DNNN	Mat	2466	—	1198	964	493	—	609
cis-HONO	Mat	3412	1633	1265	850	610	637	610
trans-HONO	Mat	3558	1684	1298	815	625	583	611
HNCS	Mat	3505	1979	988	577	461	—	612
DNCS	Mat	2623	1938	—	548	366	—	612
HCNS	Gas	3539	1989	857	615	469	539	613
DCNS	Gas	2645	1944	851	549	366	481	613
ClSCN	Sol'n	520	678	2162	—	353	—	605, 614
BrSCN	Sol'n	451	676	2157	—	369	—	605, 614
ISCN	Sol'n	372	700	2130	—	362	—	605, 614
cis-HNSO	Mat	3309	1083	1249	900	447	755	615
trans-HNSO	Mat	3308	982	1381	878	496	651	616
trans-HONS	Mat	3528.0	842.1	969.5	1363.3	476.5	531	617
FNSO	Liquid	825	995	1230	228	600	395	618
ClNSO	Liquid	526	989	1221	187	672	359	618
BrNSO	Liquid	451	1000	1214	161	624	342	618
INSO	Liquid	372	1028	1247	154	602	330	618

[a]Mat = inert gas matrix.

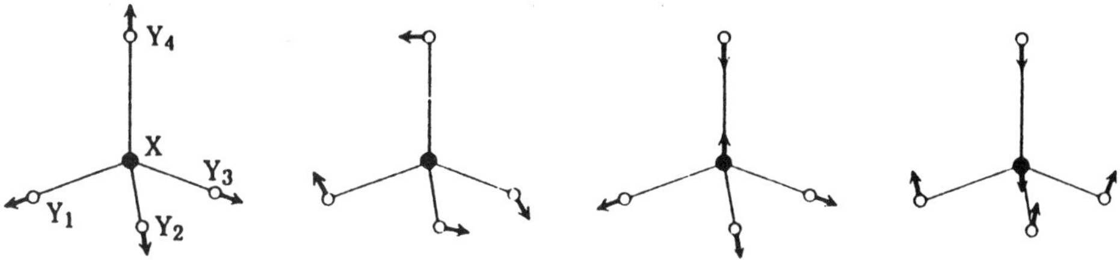

$\nu_1(A_1)$ $\nu_s(XY)$ | $\nu_2(E)$ $\delta_d(YXY)$ | $\nu_3(F_2)$ $\nu_d(XY)$ | $\nu_4(F_2)$ $\delta_d(YXY)$

Fig. II-12. Normal modes of vibration of tetrahedral XY_4 molecules.

infrared active. Appendix VII lists the **G** and **F** matrix elements for such a molecule.

Fundamental frequencies of XH_4-type molecules are listed in Table II-6*a*. The trends $\nu_3 > \nu_1$ and $\nu_2 > \nu_4$ hold for the majority of the compounds. The XH stretching frequencies are lowered whenever the XH_4 ions form hydrogen bonds with counterions. In the same family of the periodic table, the XH stretching frequency decreases as the mass of the X atom increases. Shirk and Shriver[622] noted, however, that the ν_1 frequency and the corresponding force

Table II-6*a*. Vibrational Frequencies of Tetrahedral XH_4 Molecules (cm^{-1})

Molecule	ν_1	ν_2	ν_3	ν_4	Refs.
$[^{10}BH_4]^-$	2270	1208	2250	1093	620, 621
$[^{10}BD_4]^-$	1604	856	1707	827	620, 621
$[AlH_4]^-$	1757	772	1678	760 or 766	622
$[AlD_4]^-$	1256	549	1220	560 or 556	622
$[GaH_4]^-$	1807	—	—	—	622
CH_4	2917	1534	3019	1306	623
CD_4	2085	1092	2259	996	624, 625
SiH_4	2180	970	2183	910	623, 626
SiH_4	—	—	2192	913	627
SiD_4	(1545)	(689)	1597	681	626, 628
GeH_4	2106	931	2114	819	623, 629
GeD_4	1504	665	1522	596	623
SnH_4	—	758	1901	677	630
SnD_4	—	539	1368	487	630
$[^{14}NH_4]^+$	3040	1680	3145	1400	623
$[^{15}NH_4]^+$	—	(1646)	3137	1399	631
$[ND_4]^+$	2214	1215	2346	1065	623
$[NT_4]^+$	—	976	2022	913	631
$[PH_4]^+$	2295	1086, 1026	2366, 2272	974, 919	632
$[PD_4]^+$	1654	772, 725	1732	677	632
$[AsH_4]^+$	2080	949	2142	818, 813	633

constant show an unusual trend in Group IIIB:

	$[BH_4]^-$	$[GaH_4]^-$	$[AlH_4]^-$
$\tilde{\nu}_1$ (cm^{-1})	2270	1807	1757
F_{11} (mdyn/Å)	3.07	1.94	1.84

In a series of ammonium halide crystals, the $\nu(NH_4^+)$ becomes lower and the $\delta(NH_4^+)$ becomes higher as the N—H···X hydrogen bond becomes weaker in the order of X = F > Cl > Br > I.

Figure II-13 shows the infrared spectrum of NH_4Cl, measured by Hornig et al.[634] They noted that the combination band between ν_4 (F_2) and ν_6 (rotatory lattice mode) is observed for NH_4F, NH_4Cl, and NH_4Br because the NH_4^+ ion does not rotate freely in these crystals. In NH_4I (phase I), however, this band is not observed because the NH_4^+ ion rotates freely.

Table II-6*b* lists the vibrational frequencies of a number of tetrahalogeno compounds. Except for $[TlBr_4]^-$ and UF_4, the trends $\nu_3 > \nu_1$ and $\nu_4 > \nu_2$ hold for all the compounds. The latter trend is opposite to that found for MH_4 compounds. The same table also shows the effect of changing the halogen. First, the MX stretching frequency decreases as the halogen is changed in the order F > Cl > Br > I. The average values of $\nu(MBr)/\nu(MCl)$ and $\nu(MI)/\nu(MCl)$ calculated from all the compounds listed in Table II-6*b* are 0.76 and 0.62, respectively, for ν_3, and 0.61 and 0.42, respectively, for ν_1. These values are very useful when we assign the MX stretching bands of halogeno complexes (Sec. III-23 of Part B). Second, the effect of changing the oxidation state on the MX stretching frequency is seen in pairs such as $[FeX_4]^-$ and $[FeX_4]^{2-}$ (X = Cl and Br);[670] the

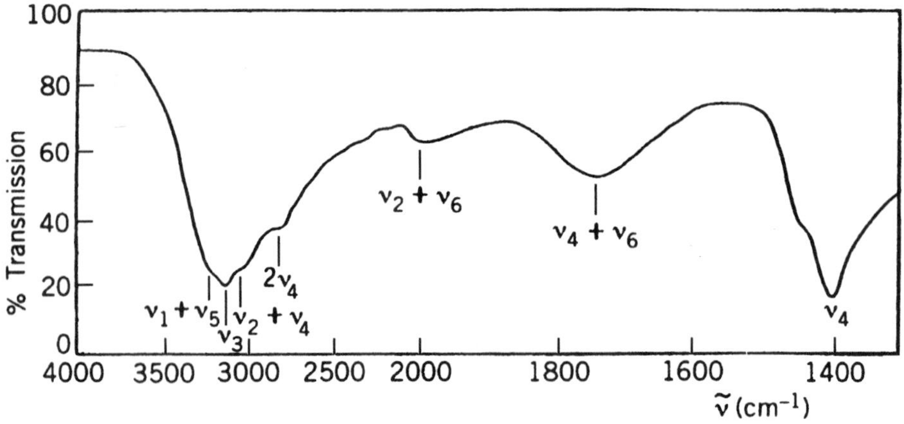

Fig. II-13. IR spectrum of crystalline NH_4Cl (ν_5, ν_6: lattice modes).

Table II-6*b*. Vibrational Frequencies of Tetrahedral Halogeno Compounds (cm^{-1})

Molecule	ν_1	ν_2	ν_3	ν_4	Refs.
$[BeF_4]^{2-}$	547	255	800	385	635
$[MgCl_4]^{2-}$	252	100	330	142	636
$[MgBr_4]^{2-}$	150	61	290	90	636
$[MgI_4]^{2-}$	107	42	259	60	636
$[BF_4]^-$	777	360	1070	533	637
$[BCl_4]^-$	406	192	722	278	638, 639
$[BBr_4]^-$	243	118	620	166	638, 640
$[BI_4]^-$	170	83	533	117	641
$[AlF_4]^-$	622	210	760	322	642
$[AlCl_4]^-$	348	119	498	182	643
$[AlBr_4]^-$	212	98	394	114	644
$[AlI_4]^-$	146	51	336	82	645
$[GaCl_4]^-$	343	120	370	153	646
$[GaBr_4]^-$	210	71	278	102	647
$[GaI_4]^-$	145	52	222	73	648
$[InCl_4]^-$	321	89	337	112	649
$[InBr_4]^-$	197	55	239	79	650
$[InI_4]^-$	139	42	185	58	648
$[TlCl_4]^-$	312	—	293	93, 110	651
$[TlBr_4]^-$	192	—	173, 185	78	651
$[TiI_4]^-$	130	—	146	60	651
CF_4	908.4	434.5	1283.0	631.2	652
CCl_4	460.0	214.2	792, 765	313.5	652
CBr_4	267	123	672	183	653
CI_4	178	90	555	123	654
SiF_4	800.8	264.2	1029.6	388.7	652
$SiCl_4$	423.1	145.2	616.5	220.3	652
$SiBr_4$	246.7	84.8	494.0	133.6	652
SiI_4	168.1	62	405.5	90.6	655
GeF_4	738	205	800	260	656
$GeCl_4$	396.9	125.0	459.1	171.0	652
$GeBr_4$	235.7	74.7	332.0	111.1	652
GeI_4	158.7	~60	264.1	79.0	655
$SnCl_4$	369.1	95.2	408.2	126.1	652
$SnBr_4$	222.1	59.4	284.0	85.9	652
SnI_4	147.7	42.4	210	63.0	652
$PbCl_4$	331	90	352	103	657
$[NF_4]^+$	848	443	1159	611	658, 641
$[NCl_4]^+$	430	233	635	283	659
$[PCl_4]^+$	458	178	662	255	660
$[PBr_4]^+$	254	116	503, 496	148	661, 662
$[PI_4]^+$	193	71	—	89	663
$[AsF_4]^+$	745	213	829	272	663a
$[AsCl_4]^+$	422	156	500	187	664, 665
$[AsBr_4]^+$	244	88	349	115	666
$[AsI_4]^+$	183	72	319	87	667
$[CuCl_4]^{2-}$	—	77	267, 248	136, 118	668, 669
$[CuBr_4]^{2-}$	—	—	216, 174	85	668
$[ZnCl_4]^{2-}$	276	80	277	126	670

Table II-6b. (Continued)

Molecule	ν_1	ν_2	ν_3	ν_4	Refs.
$[ZnBr_4]^{2-}$	171	—	204	91	670
$[ZnI_4]^{2-}$	118	—	164	—	670
$[CdCl_4]^{2-}$	261	84	249, 240	98	671
$[CdBr_4]^{2-}$	161	49	177	75, 61	671
$[CdI_4]^{2-}$	116	39	—	50	671
$[HgCl_4]^{2-}$	267	180	276	192	672
$[HgI_4]^{2-}$	126	35	140	41	673
$[ScI_4]^-$	129	37	—	54	674
TiF_4	712	185	793	209	675
$TiCl_4$	389	114	498	136	676
$TiBr_4$	231.5	68.5	393	88	676
TiI_4	162	51	323	67	676
ZrF_4	(725 ~ 600)	(200 ~ 150)	668	190	677
$ZrCl_4$	377	98	418	113	676
$ZrBr_4$	225.5	60	315	72	676
ZrI_4	158	43	254	55	676
$HfCl_4$	382	101.5	390	112	676
$HfBr_4$	235.5	63	273	71	676
HfI_4	158	55	224	63	676
VCl_4	383	128	475	128, 150	678
CrF_4	717	—	790	201	678a
$CrCl_4$	373	116	486	126	679
$CrBr_4$	224	60	368	71	679
$[MnCl_4]^{2-}$	256	—	278, 301	120	670, 680
$[MnBr_4]^{2-}$	195	65	209, 221	89	670, 680
$[MnI_4]^{2-}$	108	46	188, 193	56	670, 680
$[FeCl_4]^-$	330	114	~378	~136	670
$[FeBr_4]^-$	211	74	307/292	94	670, 6802
$[FeBr_4]^1$	211	74	307/292	94	670
$[FeCl_4]^{2-}$	266	82	286	119	670
$[FeBr_4]^{2-}$	162	—	219	84	670
$[FeI_4]^{2-}$	—	—	186	—	668
$[NiCl_4]^{2-}$	264	—	294, 280	119	668, 681
$[NiBr_4]^{2-}$	—	—	228	81	668, 681
$[NiI_4]^{2-}$	105	—	191	—	668
$[CoCl_4]^{2-}$	269	—	311, 291	135	681
$[CoBr_4]^{2-}$	166	—	231, 222	96	681
$[CoI_4]^{2-}$	118	—	202, 194	56	668
UF_4	614	340	420	180	682

MX stretching frequency increases as the oxidation state of the metal becomes higher. The ratio $\nu_3(FeX_4^-)/\nu_3(FeX_4^{2-})$ is 1.32 in this case.

In the solid state, ν_3 and ν_4 may split into two or three bands because of the site effect. In some cases, the MX_4 ions are distorted to a flattened tetrahedron ($\mathbf{D}_{2d}$) or a structure of lower symmetry ($\mathbf{C}_s$).[668,683] According to an X-ray analysis[684] the unit cell of $[(CH_3)_2CHNH_3]_2$ $[CuCl_4]$ contains two square-planar and four distorted tetrahedral $[CuCl_4]^{2-}$ ions.

Willett et al.[685] demonstrated by using IR spectroscopy that distorted tetrahedral ions can be pressed to square-planar ions under high pressure. In nitromethane, $(Et_4N)_2[CuCl_4]$ exhibits two bands at 278 and 237 cm^{-1}, indicating the distortion in solution.[686] A solution of $(Et_3NH)[GaCl_4]$ in 1,2-dichloroethylene exhibits three bands at 390, 383, and 359 cm^{-1} due to lowering of symmetry caused by NH···Cl hydrogen bonding.[687] The IR spectra of long-chain tertiary and quaternary ammonium salts of $[GaCl_4]^-$ and $[GaBr_4]^-$ ions in benzene solution show that the degree of distortion of these ions depends on the nature of the cation and the concentration.[688] Distortion of $[MCl_4]^{2-}$ ions (M = Fe, Co, Ni and Zn) is also reported for their cesium and rubidium salts.[689]

Matrix-isolation spectroscopy has provided structural information which is unique to XY_4-type molecules in inert gas environments. For example, the symmetry of the anions in $Cs^+BF_4^-$ and $Cs^+PF_4^-$ formed via salt/molecule reactions in Ar matrices are lowered to $\mathbf{C}_{3v}$ in the former,[690] and to a symmetry no higher than $\mathbf{C}_{2v}$ in the latter.[691] In inert gas matrices, the symmetry of UF_4 is distorted to $\mathbf{D}_{2d}$.[692] Irradiation of CCl_4 in Ar matrices at 12 K produced a new species, "isotetrachloromethane," for which Maier et al.[693] proposed the following structure:

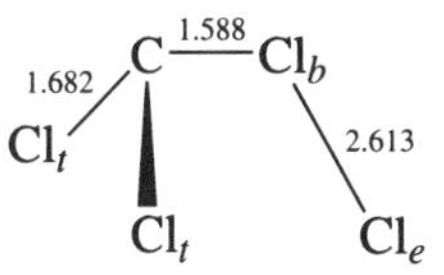

(Values shown are estimated bond distances (Å))

The $\nu(C—Cl_t)$ are observed at 1019.7 and 929.1 cm^{-1}, whereas the $\nu(C—Cl_b)$ and $\nu(Cl_b—Cl_e)$ are assigned at 501.9 and 246.4 cm^{-1}, respectively. Jacox and Thompson[694] presented IR evidence for the formation of the CO_4^- ion resulting from the reaction of CO_2, O_2, and excited Ne atoms in Ne matrices. The planar structure of $\mathbf{C}_s$ symmetry has been proposed:

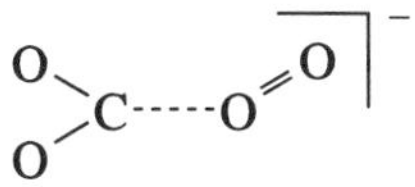

Table II-6*b* includes a number of data obtained by Clark et al. from their gas-phase Raman studies.[652,676] Some halides such as TiI_4[695], SnI_4[655,696], and $TiBr_4$[676], and VCl_4[697] have strong electronic absorptions in the visible region, and their resonance Raman spectra have been measured in solution. As discussed in Sec. I-23, these compounds exhibit a series of ν_1 overtones under rigorous resonance conditions.

A tetrahedral MCl_4 molecule in which M is isotopically pure and Cl is in natural abundance consists of five isotopic species because of the mixing of the ^{35}Cl (75.4%) and ^{37}Cl (24.6%) isotopes. Table II-6*c* lists their symmetries, percentages of natural abundance, and symmetry species of infrared-active modes corresponding to the ν_3 vibration of the $\mathbf{T}_d$ molecule. It has been established[698] that these nine bands overlap partially to give a "five-peak chlorine isotope pattern" whose relative intensity is indicated by the vertical lines shown in Fig. II-14*b*. If M is isotopically mixed, the spectrum is too complicated to assign by the conventional method. For example, tin is a mixture of 10 isotopes, none of which is predominant. Thus 50 bands are expected to appear in the ν_3 region of $SnCl_4$. It is almost impossible to resolve all these peaks, even in an inert gas matrix at 10 K. Königer and Müller,[699] therefore, prepared $^{116}SnCl_4$ and $^{116}Sn^{35}Cl_4$ on a milligram scale and measured their infrared spectra in Ar matrices. As expected, the former gave a "five-peak chlorine isotope pattern," whereas the latter showed a single peak at 409.8 cm^{-1}. This work was extended to $GeCl_4$, which consists of two Cl and five Ge isotopes. In this case, 25 peaks are expected to appear in the ν_3 region. However, the observed spectrum (Fig. II-14*a*) shows about 10 bands, Königer et al.,[700] therefore, prepared $^{74}GeCl_4$ and $Ge^{35}Cl_4$ and measured their spectra in Ar matrices. As expected, both compounds showed a "five-peak" spectrum. The *ism* (isotope shift per unit mass difference) values for Cl and Ge were found to be 3.8 and 1.2 cm^{-1}, respectively. Using these values, it was then possible to calculate the frequencies of all other isotopic molecules. Furthermore, the relative intensity of individual peaks is known from the relative concentration of each isotopic molecule. On the basis of this information, Tevault et al.[701] obtained a computer-simulation infrared spectrum of $GeCl_4$ in natural abundance (Fig. II-14).

Normal coordinate analyses of tetrahedral XY_4 molecules have been carried out by a number of investigators.[702] Thus far, Basile et al.[703] have made the most complete study. They calculated the force constants of 146 compounds by using GVF, UBF, and OVF fields (Sec. I-14), and discussed several factors that influence the values of the XY stretching force constants.

It has long been known that molecules such as SF_4, SeF_4, and TeF_4 assume a distorted tetrahedral structure ($\mathbf{C}_{2v}$) derived from a trigonal-bipyramidal geom-

Table II-6*c*. Infrared-Active Vibrations of $M^{35}Cl_n{}^{37}Cl_{4-n}$-Type Molecules

Species	Symmetry	Abundance (%)	IR-Active Modes
$M^{35}Cl_4$	$\mathbf{T}_d$	32.5	F_2
$M^{35}Cl_3^{37}Cl$	$\mathbf{C}_{3v}$	42.2	A_1, E
$M^{35}Cl_2^{37}Cl_2$	$\mathbf{C}_{2v}$	20.5	A_1, B_1, B_2
$M^{35}Cl^{37}Cl_3$	$\mathbf{C}_{3v}$	4.4	A_1, E
$M^{37}Cl_4$	$\mathbf{T}_d$	0.4	F_2

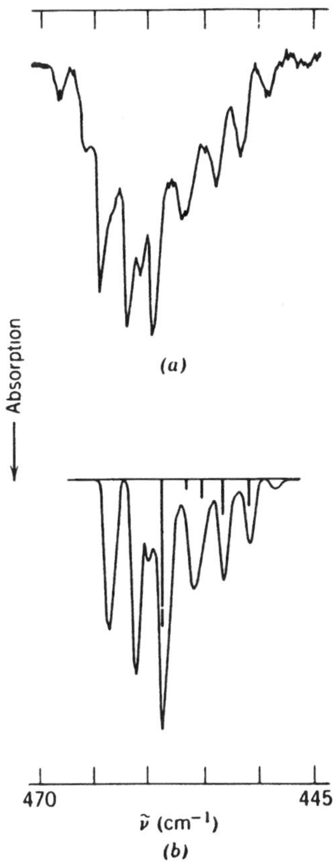

Fig. II-14. Matrix-isolation IR (*a*) and computer-simulation spectra (*b*) of $GeCl_4$. Vertical lines in (*b*) show the five-peak chlorine isotope pattern of $^{74}GeCl_4$.

etry with a lone pair of electrons occupying an equatorial position:

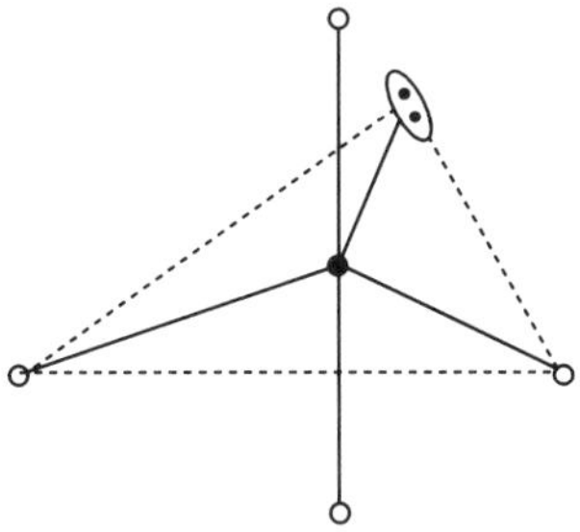

Table II-6*d* lists the vibrational frequencies of nine normal modes of such molecules. It should be noted that these compounds exhibit four stretching

Table II-6d. Vibrational Frequencies of Distorted Tetrahedral XY_4 Molecules[a] (cm^{-1})

C_{2v}	ν_1 A_1	ν_2 A_1	ν_3 A_1	ν_4 A_1	ν_5 A_2	ν_6 B_1	ν_7 B_1	ν_8 B_2	ν_9 B_2	Refs.
$[SbF_4]^-$	596	449	285	163	220	431	257	566	180	704
$[SbCl_4]^-$	339	296	147	—	—	321	199	246	—	704
	337	300	170	—	—	—	—	252	146	705
$[SbBr_4]^-$	228	190	—	—	—	201	140	169	—	704, 706
$[SbI_4]^-$	169	—	114	—	—	162	85	148	—	704, 706
SF_4	892	558	532	228	(437)	730	475	867	353	707
SeF_4	739	551	362	200	—	585	254	717	403	708
TeF_4	695	572	333	(152)	—	587	273	682	(185)	709
$[ClF_4]^+$	800	571	385	250	475	795	515	829	385	707
$[BrF_4]^+$	723	606	385	219	—	704	419	736	369	710
$[IF_4]^+$	728	614	345	263	—	—	388	719	311	710

[a]Whereas ν_1 and ν_8 are stretching modes of equatorial bonds, ν_2 and ν_6 are stretching modes of axial bonds. For the normal modes of bending vibrations, see Ref. 711.

modes, two of which are polarized in the Raman. Adams and Downs[709] carried out normal coordinate analyses on SeF_4 and TeF_4, and found that the axial bonds are weaker than the equatorial bonds. The vibrational spectra of tetraalkylammonium salts of $[AsX_4]^-$ (X = Cl and Br), $[BiX_4]^-$ (X = Cl, Br, and I), and $[SbX_4]^-$ (X = Cl, Br, and I) have been studied in the solid state and in solution.[706] Except for solid $(Et_4N)[AsCl_4]$ and $[(n\text{-}Bu)_4N][SbI_4]$, all these ions assume the distorted tetrahedral structure shown above.

The $[ClF_4]^+$, $[BrF_4]^+$, and $[IF_4]^+$ ions were found in the following adducts:[710]

$$ClF_5\cdot(AsF_5) = [ClF_4][AsF_6]$$

$$BrF_5\cdot(SbF_5)_2 = [BrF_4][Sb_2F_{11}]$$

$$IF_5\cdot(SbF_5) = [IF_4][SbF_6]$$

It should be noted that $SeCl_4$, $SeBr_4$, $TeCl_4$, and $TeBr_4$ consist of the pyramidal XY_3^+ cation and the Y^- anion in the solid state.[459,460]

Table II-6e lists the vibrational frequencies of tetrahedral MO_4-, MS_4-, and MSe_4-type compounds. The rules $\nu_3 > \nu_1$ and $\nu_4 > \nu_2$ hold for the majority of the compounds. Figure II-15 shows the Raman spectra of K_2SO_4 in the solid state as well as in aqueous solution. It should be noted that ν_2 and ν_4 are often too close to be observed as separate bands in Raman spectra. Weinstock et al.[727] showed that, in Raman spectra, ν_2 should be stronger than ν_4, and that ν_4 is hidden by ν_2 in $[MoO_4]^{2-}$ and $[ReO_4]^-$ ions.

Baran et al.[742,743] have found several relationships between the ν_1/ν_3 ratio and the negative charge of the anion or the mass of the central atom in a series

Table II-6*e*. Vibrational Frequencies of Tetrahedral MO_4-, MS_4-, and MSe_4-Type Compounds (cm^{-1})

Compound	ν_1	ν_2	ν_3	ν_4	Refs.
$[BO_4]^{5-}$	880	372	886	627	712, 713
$[CS_4]^{-}$	495	353	1000, 805	—	714
$[SiO_4]^{4-}$	819	340	956	527	715, 716
$[PO_4]^{3-}$	938	420	1017	567	717, 718
$[PS_4]^{3-}$	391	282	535, 512	317, 296	719
$[NO_4]^{3-}$	843	540, 500	1012, 988	669, 651	720
$[AsO_4]^{3-}$	837	349	878	463	721
$[AsS_4]^{3-}$	386	171	419	216	721
$[SbS_4]^{3-}$	366	156	380	178	721
$[SO_4]^{2-}$	983	450	1105	611	715
$[SeO_4]^{2-}$	833	335	875	432	715
$[ClO_4]^{-}$	928	459	1119	625	721
$[BrO_4]^{-}$	801	331	878	410	722, 723
$[IO_4]^{-}$	791	256	853	325	724
XeO_4	775.7	267	879.2	305.9	725
$[TiO_4]^{4-}$	761	306	770	371	726
$[ZrO_4]^{4-}$	792	332	846	387	726
$[HfO_4]^{4-}$	796	325	800	379	726
$[VO_4]^{3-}$	826	336	804	(336)	727
$[VO_4]^{4-}$	818	319	780	368	726
$[VS_4]^{3-}$	404.5	193.5	470	(193.5)	728
$[VSe_4]^{3-}$	(232)	121	365	(121)	728
$[NbS_4]^{3-}$	408	163	421	(163)	728
$[NbSe_4]^{3-}$	239	100	316	(100)	728
$[TaS_4]^{3-}$	424	170	399	(170)	728
$[TaSe_4]^{3-}$	249	103	277	(103)	728
$[CrO_4]^{2-}$	846	349	890	378	727
$[CrO_4]^{3-}$	830	330	765	330	729
$[MoO_4]^{2-}$	897	317	837	(317)	727
$[MoS_4]^{2-}$	458	184	472	(184)	730
$[MoSe_4]^{2-}$	255	120	340	120	731
$[WO_4]^{2-}$	931	325	838	(325)	727
$[WS_4]^{2-}$	479	182	455	(182)	730
$[WSe_4]^{2-}$	281	107	309	(107)	728
$[MnO_4]^{-}$	834	346	902	386	732, 733
$[MnO_4]^{2-}$	812	325	820	332	726
$[MnO_4]^{3-}$	789	308	778	332	734
$[TcO_4]^{-}$	912	325	912	336	727
$[ReO_4]^{-}$	971	331	920	(331)	727
$[ReS_4]^{-}$	501	200	486	(200)	735
$[FeO_4]^{2-}$	832	340	790	322	726
RuO_4	885.3	~319	921	336	736
$[RuO_4]^{-}$	830	339	845	312	726
$[RuO_4]^{2-}$	840	331	804	336	726

Table II-6e. (*Continued*)

Compound	ν_1	ν_2	ν_3	ν_4	Refs.
OsO_4	965.2	333.1	960.1	322.7	737
$[CoO_4]^{4-}$	670	320	633	320	738
$[B(OH)_4]^{-}$ [a]	754	379	945	533	739, 740
$[Al(OH)_4]^{-}$ [a]	615	310	(720)	(310)	741
$[Zn(OH)_4]^{2-}$ [a]	470	300	(570)	(300)	741

[a]Only MO_4 skeletal vibrations are listed for this ion.

of oxoanions listed in Table II-6*e*. These are:

1. For a given central atom, the ν_1/ν_3 ratio increases as the negative charge of the anion increases (e.g., $[MnO_4]^- < [MnO_4]^{2-} < [MnO_4]^{3-}$).

2. For anions of the same negative charge with the central atom belonging to the same group of the periodic table, the ν_1/ν_3 ratio increases with the mass of the central atom (e.g., $[PO_4]^{3-} < [AsO_4]^{3-}$).

3. For isoelectronic ions in which the mass of the central atom remains approximately constant, the ν_1/ν_3 ratio increases with the increasing negative charge of the anion (e.g., $[ReO_4]^- < [WO_4]^{2-}$).

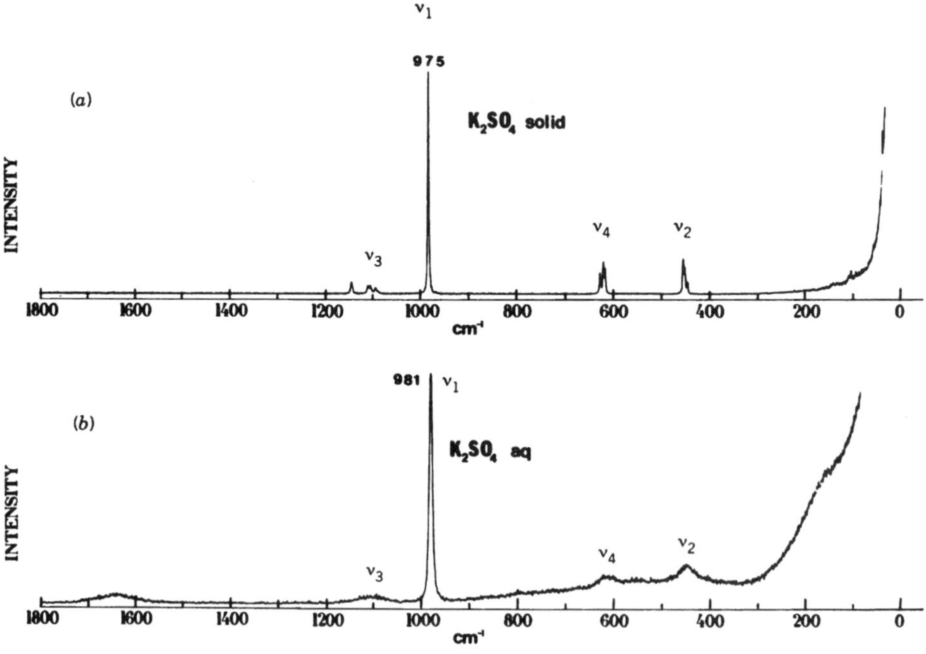

Fig. II-15. Raman spectra of K_2SO_4 in (*a*) the solid state and (*b*) aqueous solution.

These trends are very useful in making correct assignments of the ν_1 and ν_3 vibrations of tetraoxoanions.[743]

The IR spectra of matrix-isolated $M_2(SO_4)$ (M = K, Rb, and Cs) can be interpreted in terms of $\mathbf{D}_{2d}$ symmetry:

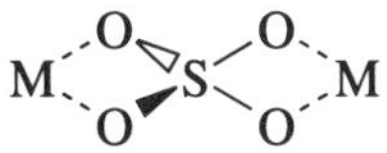

and their ν(M...O) vibrations are observed at 262–220, 215–190, and 195–148 cm^{-1}, respectively, for the K, Rb, and Cs salts.[744]

The IR spectrum and normal coordinate analysis of the SO_4 radical produced by the reaction of SO_3 with atomic oxygen in inert gas matrices are suggestive of the $\mathbf{C}_s$ structure[545] rather than the $\mathbf{C}_{2v}$ structure.[745]

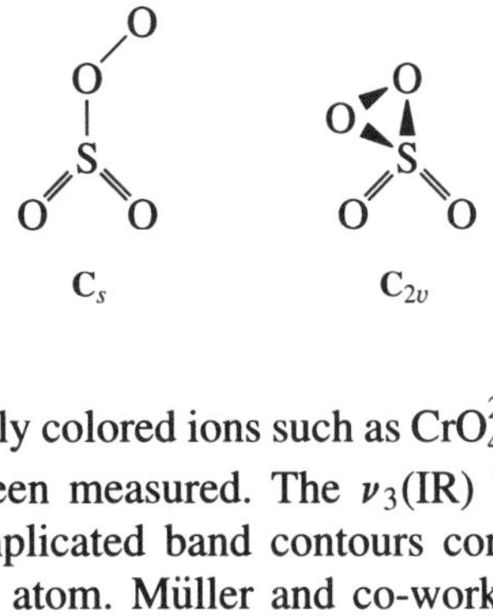

The RR spectra of highly colored ions such as CrO_4^{2-},[746] MoS_4^{2-},[747] VS_4^{3-},[748] and MnO_4^-,[732,746] have been measured. The ν_3(IR) bands of gaseous RuO_4[736] and XeO_4[749] exhibit complicated band contours consisting of individual isotope peaks of the central atom. Müller and co-workers[742,750,751] reviewed the vibrational spectra of transition-metal chalcogen compounds. Basile et al.[703] carried out normal coordinate analysis on more than 60 compounds of these types.

(2) Tetrahedral ZXY_3, Z_2XY_2, and $ZWXY_2$ Molecules

If one of the Y atoms of an XY_4 molecule is replaced by a Z atom, the symmetry of the molecule is lowered to $\mathbf{C}_{3v}$. If two Y atoms are replaced, the symmetry becomes $\mathbf{C}_{2v}$. In $ZWXY_2$ and ZWXYU types (X: central atom), the symmetry is further lowered to $\mathbf{C}_1$. As a result, the selection rules are changed as shown in Table II-6*f*. The number of infrared-active vibrations is six for ZXY_3 and eight for Z_2XY_2. Table II-6*g* lists the vibrational frequencies of ZXY_3-type molecules. The SO stretching frequency of the $[OSF_3]^+$ ion in $[OSF_3]SbF_6$ (1536 cm^{-1}) is the highest that has been observed, and corresponds to a force constant of 14.7 mdyn/Å.[770] It is also interesting to note that the structure of the $[OXeF_3]^+$ ion is $\mathbf{C}_s$[555], whereas that of the $[OXeF_3]^-$ ion is $\mathbf{C}_{2v}$[556] as

Table II-6f. Correlation Table[a] for T_d, C_{3v}, C_{2v}, and C_1

Point Group	ν_1	ν_2	ν_3	ν_4
T_d	A_1(R)	E(R)	F_2(IR, R)	F_2(IR, R)
C_{3v}	A_1(IR, R)	E(IR, R)	A_1(IR, R) +E(IR, R)	A_1(IR, R) +E(IR, R)
C_{2v}	A_1(IR, R)	A_1(IR, R) +A_2(R)	A_1(IR, R) +B_1(IR, R) +B_2(IR, R)	A_1(IR, R) +B_1(IR, R) +B_2(IR, R)
C_1	A(IR, R)	2A(IR, R)	3A(IR, R)	3A(IR, R)

[a]See Appendix IX.

shown below:

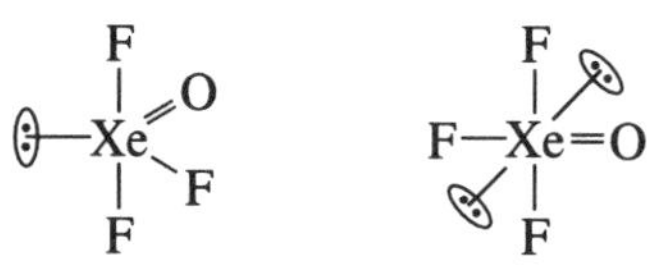

C_s C_{2v}

Vibrational spectra have been reported for a number of mixed halogeno complexes. Some references are as follows: $[AlCl_nBr_{4-n}]^-$ (800), SiF_nCl_{4-n} (801), $SiCl_nBr_{4-n}$ (802), and $[FeCl_nBr_{4-n}]^-$ (803). It is interesting to note that the SiFClBrI molecule exhibits the SiF, SiCl, SiBr, and SiI stretching bands at 910, 587, 486, and 333 cm^{-1}, respectively.[804] The vibrational spectrum of $OClF_3$ suggests a trigonal-bipyramidal structure which is similar to that of the $[OXeF_3]^+$ ion shown above.[805]

Table II-6g. Vibrational Spectra of ZXY_3 Molecules (cm^{-1})

C_{3v} ZXY_3	$\nu_1(A_1)$ $\nu(XY_3)$	$\nu_2(A_1)$ $\nu(XZ)$	$\nu_3(A_1)$ $\delta(XY_3)$	$\nu_4(E)$ $\nu(XY_3)$	$\nu_5(E)$ $\delta(XY_3)$	$\nu_6(E)$ $\rho_r(XY_3)$	Refs.
$[OCF_3]^-$	813	1560	595	960	422	576	752, 753
$FCCl_3$	537.6	1080.4	351.3	847.8	242.8	395.3	754
$BrCCl_3$	419.9	730.7	246.5	785.2	290.2	188.0	754
$ICCl_3$	390	684	224	755	284	188	755
$H^{28}SiF_3$	855.8	2315.6	425.3	997.8	843.6	306.2	756
$FSiCl_3$	465	948	239	640	282	167	757
$FSiBr_3$	318	912	163	520	226	110	757
$FSiI_3$	242	894	115	424	194	71	757
$BrSiF_3$	858	505	288	940	338	200	757
$^{79}BrSiH_3$	2198.8	929.8	430.8	2209.3	996.4	632.4	758
$[SSiCl_3]^-$	395	685	220	492	—	225	759
$[SGeCl_3]^-$	356	505	163	360	—	178	759

Table II-6g. (*Continued*)

C_{3v} ZXY$_3$	$\nu_1(A_1)$ $\nu(XY_3)$	$\nu_2(A_1)$ $\nu(XZ)$	$\nu_3(A_1)$ $\delta(XY_3)$	$\nu_4(E)$ $\nu(XY_3)$	$\nu_5(E)$ $\delta(XY_3)$	$\nu_6(E)$ $\rho_r(XY_3)$	Refs.
$HGeCl_3$	418.4	2155.7	181.8	708.6	454	145.0	760
ONF_3	743	1691	528	883	558	400	761
$[ClPBr_3]^+$	285	587	149	500	172	120	762
$[BrPCl_3]^+$	399	582	217	657	235	159	762
OPF_3	873	1415	473	990	485	345	763
$OPCl_3$	486	1290	267	581	337	193	764
$OPBr_3$	340	1261	173	488	267	118	764, 765
$[FPO_3]^-$	1001	794	534	1125	—	382	766
SPF_3	975.3	693.4	439.2	947.0	405.4	273.7	767
$SPCl_3$	435	753	250	542	250	167	768
$SPBr_3$	299	718	165	438	179	115	769
$[OSF_3]^+$	909	1540, 1532	535	1063	508	387	770
NSF_3	772.6	1522.9	524.6	814.6	432.3	346.2	771
$ISCl_3$	482	293	242	493	140	255	772
$[HSO_3]^-$	1038	2588	629	1200	509	1123	773
$[FSO_3]^-$	1142	862	571	1302	619	424	774
$[ClSO_3]^-$	1042	381	601	1300	553	312	775
$[SSO_3]^-$	995	446	669	1123	541	335	776
$[FSeO_3]^-$	896	580, 603	392	968, 974	409	301	777
$FClO_3$	1062.6	716.3	549.7	1317.9	589.8	403.9	778, 779
$FBrO_3$	875.2	605.0	(354)	974	(376)	(296)	780, 781
$[NClO_3]^{2-}$	815	1256	594	870	623	457	782
OVF_3	722	1058	258	806	308	204	783
$OVCl_3$	408	1035	165	504	249	129	784
$OVBr_3$	271	1025	120	400	83	212	785
$NVCl_3$	352	1033	—	430	—	—	786
$ONbCl_3$	395	997	106	448	225	110	784
$[FCrO_3]^-$	911	635	338	955	370	261	787
$[ClCrO_3]^-$	907	438	295	954	365	209	788
$[BrCrO_3]^-$	895	260	230	950	350	(175)	789
$[OMoS_3]^{2-}$	461	862	183	470	183	263	788
$[OMoSe_3]^{2-}$	293	858	120	355	120	188	790
$[SMoO_3]^{2-}$	900	472	318	846	318	239	791
$[SMoSe_3]^{2-}$	—	471	121	342	121	—	750
$[SeMoS_3]^{2-}$	349	458	—	473	150	183	792
$[OWS_3]^{2-}$	474	878	182	451	182	264	788
$[OWSe_3]^{2-}$	292	878	(120)	312	(120)	194	790
$[SWSe_3]^{2-}$	468	281	108	311	150	108	792
$FMnO_3$	903.6	715.5	339.0	950.6	380.0	264.0	793, 794
$ClMnO_3$	889.9	458.9	305	951.6	365	~200	791, 794
$FTcO_3$	962	696	317	951	347	231	795
$ClTcO_3$	948	451	300	932	342	197	796
$FReO_3$	1013.2	701	305.5	978.3	346.9	234.2	797
$ClReO_3$	994	436, 427	291	963	337	192	797
$BrReO_3$	997	350	195	963	332	168	787
$[NReO_3]^{2-}$	878	1022	315	830	273	380	798
$[SReO_3]^-$	948	528	322	906	322	(240)	787
$[NOsO_3]^-$	892.5	1026.2	310.0	872.0	299.7	371.5	799

Table II-6*h* lists the vibrational frequencies of tetrahedral Z_2XY_2 molecules. The vibrational spectrum of O_2XeF_2 can be interpreted on the basis of a trigonal-bipyramidal structure in which two F atoms are axial, and two O atoms and a pair of electrons are equatorial.[819] The structures of $[O_2ClF_2]^-$[817] and $[O_2BrF_2]^-$[818] are similar to that of O_2XeF_2, but that of $[O_2ClF_2]^+$[828] is pseudotetrahedral. The gas-phase Raman spectrum of Cl_2TeBr_2 at 310°C

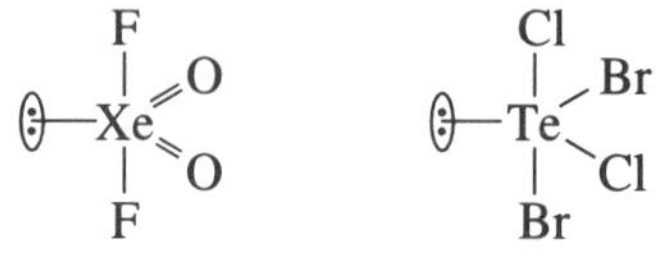

C_{2v} C_1

is indicative of the C_1 symmetry shown in the above diagram.[829] Table II-6*i* lists the vibrational frequencies of tetrahedral $ZWXY_2$ molecules. Other references are as follows: $SFPCl_2$ (836), $FOPCl_2$ (837), $FOPBr_2$ (838), $ClOPBr_2$ and $BrOPCl_2$ (764), $FBrSO_2$ (839), and $FClCrO_2$ (840).

(3) Square-Planar XY_4 Molecules (D_{4h})

Figure II-16 shows the seven normal modes of vibration of square-planar XY_4 molecules. Vibrations ν_3, ν_6, and ν_7 are infrared-active, whereas ν_1, ν_2, and ν_4 are Raman-active. Table II-6*j* lists the vibrational frequencies of some ions belonging to this group; XeF_4 (Sec. I-11) is an unusual example of a neutral molecule which takes a square-planar structure. This structure is predicted by the valence shell electron pair repulsion (VSEPR) theory[848] which states that two lone pairs in the valence shell occupy the axial positions because they exert greater mutual repulsion than single-bond pairs:

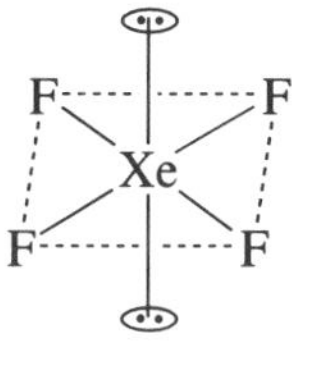

The structures of fluorides, oxides, and oxyfluorides of xenon discussed in other sections can also be rationalized based on the VSEPR theory. Figure II-17 shows the IR and Raman spectra of gaseous XeF_4 obtained by Claassen et al.[843,844]

Bosworth and Clark[849] measured the relative intensities of Raman-active fundamentals of some of these ions, and calculated their bond polarizability deriva-

TABLE II-6*h*. Vibrational Frequencies of Z_2XY_2 Molecules (cm^{-1})

Z_2XY_2	$\nu_1(A_1)$ $\nu(XY)$	$\nu_2(A_1)$ $\nu(XZ)$	$\nu_3(A_1)$ $\delta(XY_2)$	$\nu_4(A_1)$ $\delta(XZ_2)$	$\nu_5(A_2)$ $\rho_t(XY_2)$	$\nu_6(B_1)$ $\nu(XY)$	$\nu_7(B_1)$ $\rho_w(XY_2)$	$\nu_8(B_2)$ $\nu(XZ)$	$\nu_9(B_2)$ $\rho_r(XY_2)$	Refs.
$[F_2NH_2]^+$	2637	~1060	1543	528	—	2790	1176	1036	1487	806
$[O_2PH_2]^-$	2365	1046	1160	470	930	2308	820	1180	1093	807
F_2CI_2	605	1064	272	112	251	740	200	1110	278	808
H_2SiCl_2	942	2221	514	188	710	868	566	2221	602	809
F_2SiBr_2	414	891	270	115	187	540	241	974	257	810
H_2GeF_2	860	2155	720	(270)	(664)	814	720	2174	596	811
H_2GeCl_2	840	2132	404	163	648	772	420	2155	533	809
H_2GeBr_2	848	2122	298	104	105	757	324	2138	492	811
H_2GeI_2	821	2090	220	96	628	706	294	2110	451	811
$[Cl_2PBr_2]^+$	326	584	191	132	150	518	173	616	201	787
$[O_2PF_2]^-$	910	1179	269	567	—	962	492	1269	528	812
O_2SF_2	849	1270	385	553	—	886	544	1503	540	813
O_2SCl_2	405	1182	218	560	388	362	380	1414	282	814
F_2SCl_2	527	592	—	—	—	533	—	770	—	815
$[O_2SeF_2]^-$	396	859	241	445	—	—	—	833	304	816
$[O_2ClF_2]^-$	363	1070	198	559	480	510	~337	1221	~337	817
$[O_2BrF_2]^-$	374	885	201	429	405	448	339	912	314	818
O_2XeF_2	490	845	198	333	—	578	~313	902	~313	819
$[O_2VF_2]^-$	664	970	—	330	—	631	295	962	295	820
$[O_2VCl_2]^-$	438	970	—	332	—	431	230	959	199	821
$[O_2NbS_2]^{3-}$	464	897	246	356	—	514	(297)	872	(271)	822
O_2CrF_2	727	1006	208	364	(259)	789	304	1016	274	823
O_2CrCl_2	475	995	140	356	(224)	500	257	1002	215	824
O_2CrBr_2	399.0	982.6	—	(305)	—	403.5	—	995.4	—	825
O_2MoBr_2	262	995	147	373	—	338	161	970	184	826
$[O_2^{92}MoS_2]^{2-}$	473	819	200	307	267	506	267	801	246	827
$[O_2MoSe_2]^{2-}$	283	864	114	339	251	353	251	834	—	822
$[O_2WS_2]^{2-}$	454	886	196	310	280	442	280	848	235	822
$[O_2WSe_2]^{2-}$	282	888	116	319	235	329	235	845	156	822

TABLE II-6i. Vibrational Frequencies of $ZWXY_2$ Molecules (cm^{-1})

$ZWXY_2$[a]	$\nu_1(A_1)$ $\nu_s(XY_2)$	$\nu_2(A')$ $\nu(XW)$	$\nu_3(A')$ $\nu(XZ)$	$\nu_4(A')$ $\delta_s(XY_2)$	$\nu_5(A')$ $\delta_s(WXY)$	$\nu_6(A')$ $\delta_s(ZXY_2)$	$\nu_7(A'')$ $\nu_a(XY_2)$	$\nu_8(A'')$ $\delta(ZXW)$	$\nu_9(A'')$ $\delta_a(ZXY_2)$	Refs.
$OClPF_2$	900	623	1384	(419)	(274)	412	960	274	419	478, 830
$SClPF_2$	949	549	735	394	363	209	925	252	316	478, 830
$OBrPF_2$	884	561	1380	(413)	(240)	316	947	240	413	478, 830
$SBrPF_2$	938	477	719	389	288	175	911	231	297	478, 830
$OFPCl_2$	546	907	1358	205	330	382	626	253	374	478, 830
$SFPCl_2$	479	915	750	193	328	268	574	193	317	478, 830
$OFPBr_2$	472	888	1337	133	273	304	536	220	290	478, 830
$SFPBr_2$	377	887	713	129	274	218	470	162	254	478, 830
$OBrPCl_2$	545	432	1285	242	172	285	580	161	327	831
$SBrPCl_2$	493	372	743	(230)	150	206	536	150	230	831
$OClPBr_2$	391	552	1275	130	209	291	492	157	271	831
$SClPBr_2$	333	500	729	121	196	190	436	136	205	831
$[OSeMoS_2]^{2-}$	478^b	355	869	190	—	273	467^b	—	273	832, 833
$[OSeWS_2]^{2-}$	473	320	879	190	—	265^b	458	—	255^b	833, 834
$[OSMoSe_2]^{2-}$	360^b	461	865	—	—	—	320^b	—	—	835
$[OSWSe_2]^{2-}$	317^b	459	882	—	—	—	312^b	—	—	835

[a] X denotes the central atom.
[b] These assignments may be interchanged.

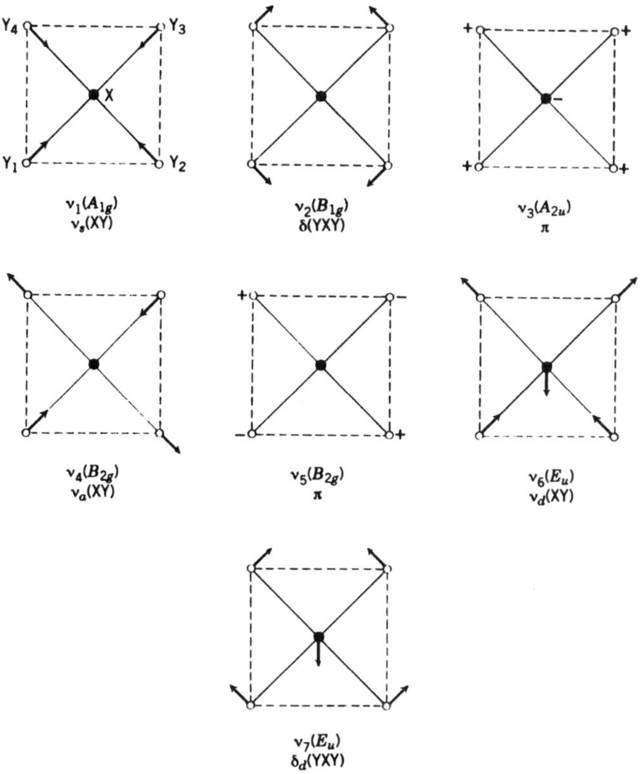

Fig. II-16. Normal modes of vibration of square-planar XY_4 molecules.

TABLE II-6j. Vibrational Frequencies of Square-Planar XY_4Molecules (cm^{-1})[a]

XY_4	$\nu_1(A_{1g})$ $\nu_s(XY)$	$\nu_2(B_{1g})$ $\delta(XY_2)$	$\nu_3(A_{2u})$ π	$\nu_4(B_{2g})$ $\nu_a(XY)$	$\nu_6(E_u)$ $\nu_d(XY)$	$\nu_7(E_u)$ $\delta_d(XY_2)$	Refs.
$[ClF_4]^-$	505	288	425	417	680–500	—	841
$[BrF_4]^-$	523	246	317	449	580–410	(194)	841
$[ICl_4]^-$	288	128	—	261	266	—	842
XeF_4	554.3	218	291	524	586	(161)	843, 844
$[AuCl_4]^-$	347	171	—	324	350	179	845, 845a
$[AuBr_4]^-$	212	102	—	196	252[b]	~ 110[b]	845, 845a
$[AuI_4]^-$	148	75	—	110	192	113	845
$[PdCl_4]^{2-}$	303	164	150	275	321	161	845a, 846
$[PdBr_4]^{2-}$	188	102	114	172	243	104	845, 846
$[PtCl_4]^{2-}$	330	171	147	312	313	165	845a, 846
$[PtBr_4]^{2-}$	208	106	105	194	227	112	845a, 846
$[PtI_4]^{2-}$	155	85	105	142	180	127	845, 846

[a]For these molecules ν_5 is inactive. The designations B_{1g} and B_{2g} may be interchanged, depending on the definition of symmetry axes involved.
[b]From Ref. 847.

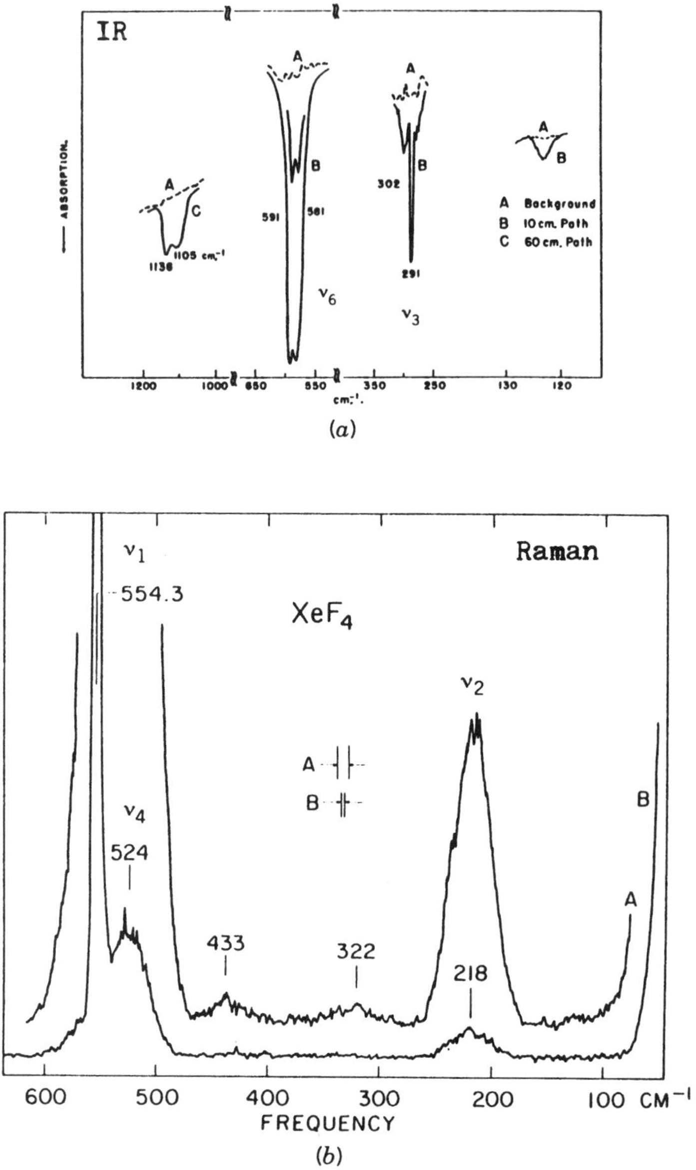

Fig. II-17. (*a*) IR[844] and (*b*) Raman[843] spectra of XeF_4 vapor (reproduced with permission).

tives and bond anisotropies. They[847] also measured the resonance Raman spectra of several $[AuBr_4]^-$ salts in the solid state, and observed progressions such as $n\nu_1$ (n = 1–9) and $\nu_2 + n\nu_1$ (n = 1–5). For normal coordinate analyses of square-planar XY_4 molecules, see Refs. 842, 850, and 851.

An empirical relationship between the antisymmetric M—Cl stretching frequency [ν_a(M—Cl), cm^{-1}] and the M—Cl distance (R_{M-Cl}, Å) has been derived using IR and structural data for a large number of planar Pt(II) and Pd(II) complexes of MCl_2L_2, MCl_2LL', and $MClL_3$ types (L: unidentate ligand).

$$[\nu_a(\text{M—Cl})]^2 = \frac{P}{(R_{M-Cl} - 1.6)^3}$$

where P is 38988 for Pt(II) and 41440 for Pd(II) complexes.[852]

II-7. TRIGONAL-BIPYRAMIDAL AND TETRAGONAL-PYRAMIDAL XY_5 AND RELATED MOLECULES

An XY_5 molecule may be a trigonal bipyramid ($\mathbf{D}_{3h}$) or a tetragonal pyramid ($\mathbf{C}_{4v}$). If it is trigonal-bipyramidal, only two stretching vibrations (A_2'' and E') are infrared-active. If it is tetragonal-pyramidal, three stretching vibrations (two A_1 and E) are infrared-active. As discussed in Sec. I-11, however, it is not always possible to make clear-cut distinctions of these structures based on selection rules since practical difficulties arise in counting the number of fundamental vibrations in infrared and Raman spectra.

(1) Trigonal-Bipyramidal XY_5 Molecules ($\mathbf{D}_{3h}$)

Figure II-18 shows the eight normal vibrations of a trigonal-bipyramidal XY_5 molecule. Six of these eight (A_1', E', and E'') are Raman-active and five (A_2'' and E') are infrared-active. Three stretching vibrations (ν_1, ν_2, and ν_5) are allowed in the Raman, whereas two (ν_3 and ν_5) are allowed in the infrared. Table II-7*a* lists the observed frequencies and band assignments of trigonal-bipyramidal XY_5 molecules. The IR spectrum of the ion-paired complex, $Cs^+[GeF_5]^+$, in inert gas matrices has been reported.[872] It takes a trigonal-bipyramidal structure perturbed by the presence of the Cs^+ ion in its vicinity.

The majority of trigonal-bipyramidal compounds show the following frequency trend:

ν_5	>	ν_3	>	ν_1	>	ν_2
equatorial stretch		axial stretch		equatorial stretch		axial stretch

Although the IR spectrum of $MoCl_5$ in the gaseous phase has been assigned based on a trigonal-bipyramidal structure, the corresponding spectrum in N_2 matrices suggests a tetragonal-pyramidal structure of $\mathbf{C}_{4v}$ symmetry. This is supported by the observation that it shows only two prominent IR bands at 473 (A_1) and 408 cm^{-1} (E) in the ν(Mo—Cl) region. Normal coordinate analy-

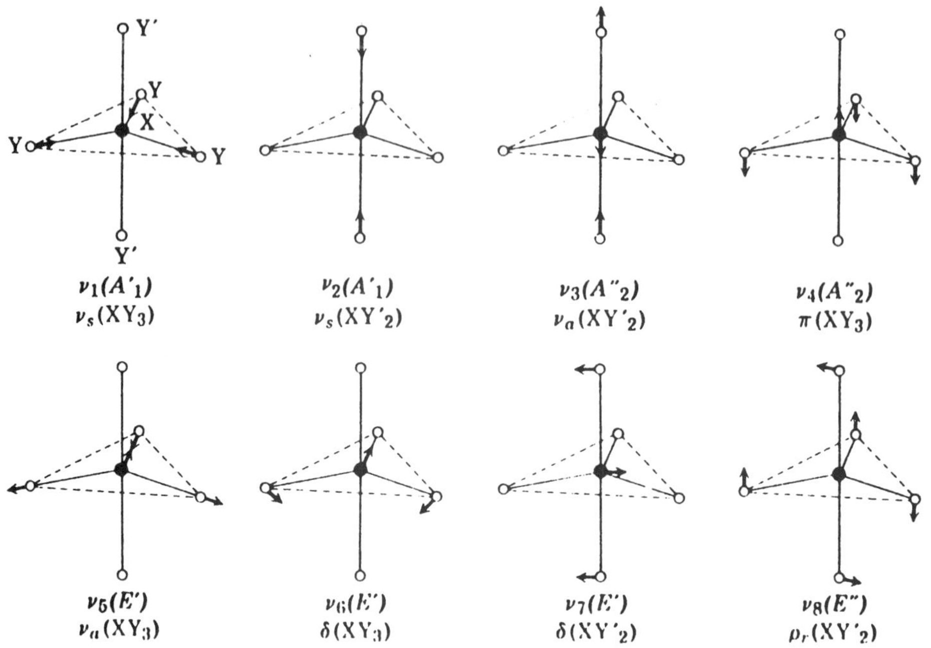

Fig. II-18. Normal modes of vibration of trigonal-bipyramidal XY_5 molecules.

TABLE II-7*a*. Vibrational Frequencies of Trigonal-Bipyramidal XY_5 Molecules (cm^{-1})

Molecule	Phase	ν_1	ν_2	ν_3	ν_4	ν_5	ν_6	ν_7	ν_8	Refs.
$[SiF_5]^-$	Sol'n[a]	708	519	785	481	874	449	—	—	853
$[SiCl_5]^-$	Sol'n[a]	372	—	395	271	550	250	—	—	854
$[GeCl_5]^-$	Solid	348	236	310	200	395	200	—	—	855
$[SnCl_5]^-$	Solid	340	—	314	160	350	150	66	169	856
$[SnBr_5]^-$	Solid	—	—	208	106	256	111	—	—	857
PF_5	Gas	817	640	944	575	1026	532	300	514	858
PCl_5	Sol'n[a]	392	281	443	300	580	272	102	261	859, 860
AsF_5	Gas	733	642	784	400	809	366	123	388	861
$AsCl_5$	Sol'n[a]	369	295	385	184	437	220	83	213	862
SbF_5	Gas	667	264	—	—	716 710	498	90	228	863
$SbCl_5$	Gas	355	309	—	—	400	173	58	120	864
$[CuCl_5]^{3-}$	Solid	260	—	268	—	170[b]	95[c]	—	—	865
$[CdCl_5]^{3-}$	Solid	251	—	236	—	157[b]	98[c]	—	—	865
$[TiCl_5]^-$	Sol'n[a]	348	302	355	178	411	190	66	166	866
$[TiCl_5]^-$	Solid	—	—	346	170	385	212	(83)	—	857
VF_5	Gas	719	608	784	331	810	282	(200)	350	867, 868
$NbCl_5$	Matrix	(349)	(293)	396	126	444	159	99	(139)	869
$NbBr_5$	Gas	234	178	288	(93)	315	119	67	101	864, 870
$TaCl_5$	Gas	406	324	—	—	—	181	54	127	864
$TaBr_5$	Gas	240	182	—	—	—	110	70	93	864
MoF_5	Matrix	—	—	683	—	713	261	112	—	871
$MoCl_5$	Gas	390	313	—	—	418	200	100	175	864

[a]Nonaqueous solution. [b]May be assigned to ν_7. [c]May be assigned to ν_8.

ses on trigonal-bipyramidal XY_5 molecules have been carried out by several investigators.[874–877] These calculations show that equatorial bonds are stronger than axial bonds.

Most neutral XY_5 molecules are dimerized or polymerized in the condensed phases. The molecules $MoCl_5$,[878] $NbCl_5$,[869, 879] $TaCl_5$,[880] and WCl_5[881] are dimeric in the liquid and solid states ($\mathbf{D}_{2h}$), whereas NbF_5 and TaF_5 are known to be tetrameric in the crystalline state.[882] Some of

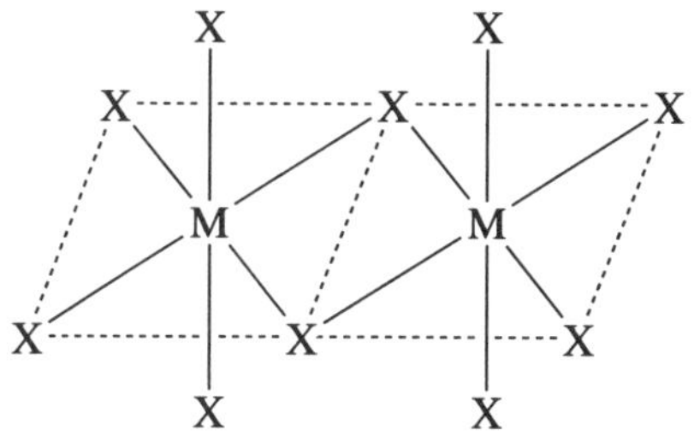

($\mathbf{D}_{2h}$)

these molecules are polymerized even in the gaseous phase. For example, SbF_5 is monomeric ($\mathbf{D}_{3h}$) at 350°C but polymeric at 140°C in the gaseous phase,[883] and NbF_5 and TaF_5 are polymeric in the gaseous phase if the temperature is below 350°C.[884] Although PCl_5 exists as a $\mathbf{D}_{3h}$ molecule in the gaseous and liquid states, it has an ionic structure consisting of $[PCl_4]^+[PCl_6]^-$ units in the crystalline state, as proved by Raman spectroscopy.[885] The importance of the ν_7 vibration in the intramolecular conversion of pentacoordinate molecules has been discussed by Holmes.[886]

Vibrational spectra of mixed halogeno compounds such as PF_nCl_{5-n},[887, 888] PF_3X_2(X = Cl, Br),[889] PH_nF_{5-n},[890, 891], PH_3F_2,[892, 893] $AsCl_3F_2$,[894] and $AsCl_4F$[895] have been assigned. The structure of OSF_4 is trigonal bipyramidal with two fluorines and one oxygen occupying the three equatorial positions ($\mathbf{C}_{2v}$).[896]

(2) Tetragonal-Pyramidal XY_5 and ZXY_4 Molecules ($\mathbf{C}_{4v}$)

Figure II-19 shows the nine normal modes of vibration of a tetragonal-pyramidal ZXY_4 molecule. Only A_1 and E vibrations are infrared-active, whereas all nine vibrations belonging to the A_1, B_1, B_2, and E species are Raman active. Table II-7*b* lists the vibrational frequencies of tetragonal-pyramidal XY_5 and ZXY_4 molecules. In the majority of XY_5 molecules, the axial stretching frequency (ν_1) is higher than the equatorial stretching frequencies (ν_2, ν_4, and ν_7). This is opposite to the trend found for trigonal-bipyramidal XY_5 molecules, discussed in the preceding section. For normal coordinate analyses on these compounds, see Refs. 900, 903, 920, and 921.

An adduct, $XeF_6 \cdot BiF_5$, is formulated as $[XeF_5]^+[BiF_6]^-$ in the solid state, its $[XeF_5]^+$ cation being tetragonal pyramidal.[907] This should be contrasted to the

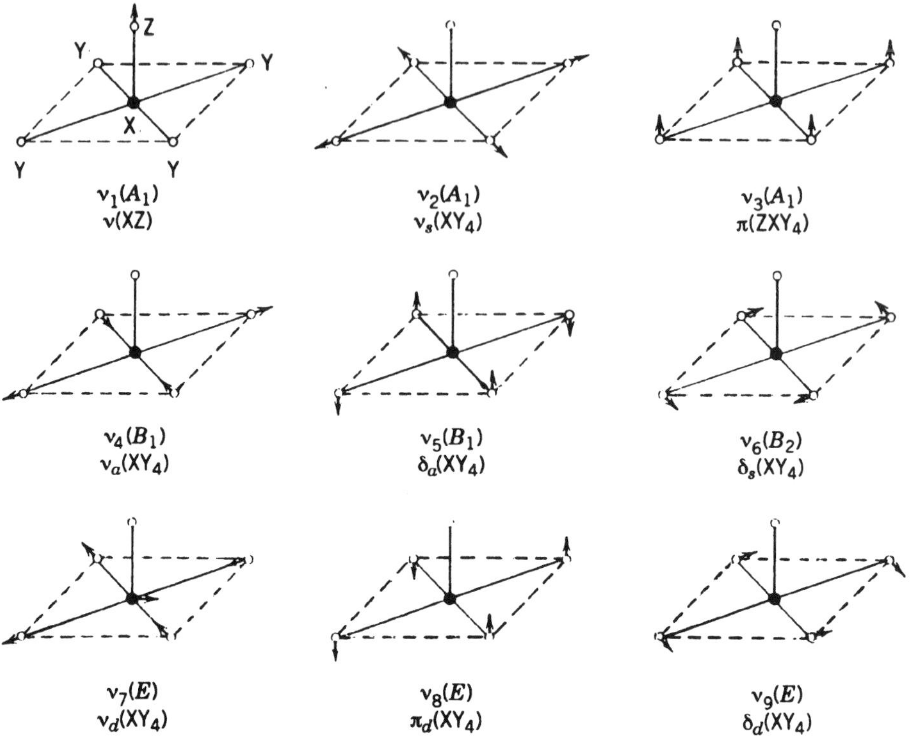

Fig. II-19. Normal modes of vibration of tetragonal-pyramidal ZXY_4 molecules.

$[XeF_5]^-$ anion which is pentagonal planar, as shown in the following section. In the $OCrF_4$—KrF_2 adduct, KrF_2 coordinates to the vacant axial position of $OCrF_4$ through a predominantly covalent bond:[922]

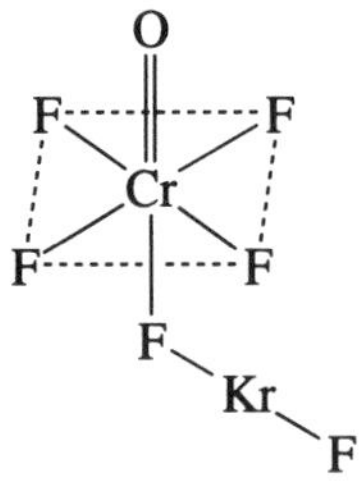

As a result, the KrF_2 vibrations are markedly shifted in frequency relative to those of free KrF_2[220]:

	ν_1	ν_2	ν_3	
Free KrF_2	449	233	596, 580	cm^{-1})
Bound KrF_2	487	176	550, 542	cm^{-1})

TABLE II-7*b*. Vibrational Frequencies of Tetragonal-Pyramidal XY_5 and ZXY_4 Molecules (cm^{-1})

XY_5 or ZXY_4	A_1 ν_1	A_1 ν_2	A_1 ν_3	B_1 ν_4	B_1 ν_5	B_2 ν_6	E ν_7	E ν_8	E ν_9	Refs.
$[InCl_5]^{2-}$	294	283	140	287	193	165	274	108	143	897
$[SbF_5]^{2-}$	557	427	278	388	—	220	375, 347	307	142	898
$[SbCl_5]^-$	445	285	180	420	—	117	300	255	90	899
$[SF_5]^-$	796	522	469	(435)	269	342	590	(435)	241	900
$[SeF_5]^-$	666	515	332	460	236	282	480	399	202	900
$[TeF_5]^-$	616	483	280	505	231	183	452	162	342	898, 901
$[TeCl_5]^-$	363	254	136	270	98	82	246	90	169	901
ClF_5	723	545	487	498	317	385	732	500	301	902–904
BrF_5	682	570	365	535	(281)	312	644	414	237	903, 905
IF_5	698	593	315	575	(257)	273	640	374	189	898, 903
NbF_5	740	686	513	—	—	—	729	261	103	906
$[XeF_5]^+$	660	598	355	635 628	242	291	650	415 395	218	907
$[OTeF_4]^{2-}$	837	461	—	390	—	190	335	—	129	908
$[OClF_4]^-$	1216	462	339	350	—	283	600, 550	415, 394	213	909
$[OBrF_4]^-$	932	525	312	459	236	248	506 483 }	421 407 398 }	194 164 }	910, 911
$[OIF_4]^-$	888	533	273	475	—	214	485	365	124	912
$OXeF_4$	920	567	285	527	(230)	233	608	365	161	903
$OCrF_4$	1028	686	277	—	—	—	744	320	271	913, 914
$OMoF_4$	1050	714	267	—	—	—	708	306	238	915–917
$OMoCl_4$	1015	450	143	400	148	220	396	256	172	918, 919
$[OMoCl_4]^-$	1008	354	184	327	158	167	364	240	114	920
OWF_4	1055	733	248	631	328	291	698	298	236	916, 917
$[OReI_4]^-$	1014	169	85	183	—	—	240	181	—	919a
$[NRuCl_4]^-$	1092	346	197	304	154	172	378	267	163	920
$[NRuBr_4]^-$	1088	224	156	187	103	128	304	211	98	920
$[NOsCl_4]^-$	1123	358	184	352	149	174	365	271	132	920
$[NOsBr_4]^-$	1119	162	122	156	110	120	220	273	98	920

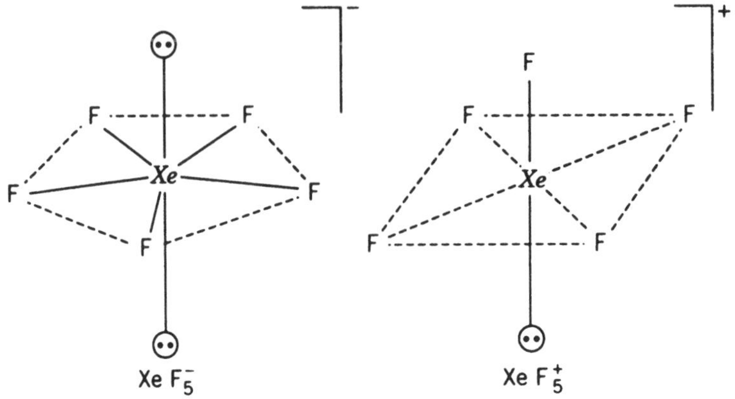

Fig. II-20. Structures of the XeF_5^- and XeF_5^+ ions.

(3) Pentagonal-Planar XY_5 Molecules ($\mathbf{D}_{5h}$)

The first example of a pentagonal-planar XY_5 type molecule was the XeF_5^- anion.[923] This anion was obtained as the 1:1 adduct of XeF_4 with $N(CH_3)_4F$, CsF or other alkali fluorides, and its structure was confirmed by X-ray analysis, IR/Raman, and NMR spectroscopy. Figure II-20 shows the structures of the XeF_5^- anion, which can be derived by replacing the two axial fluorine ligands of the pentagonal-bipyramidal IF_7 molecule with two sterically active lone-pair electrons. The difference in structure between this anion and the XeF_5^+ cation discussed in the preceding section may be understood based on the VSERP theory.[848]

The XeF_5^- anion has 12 normal vibrations which are classified into: A_1'(R) + A_2''(IR) + $2E_1'$(IR) + $2E_2'$(R) + E_2'' (inactive) under $\mathbf{D}_{5h}$ symmetry. Thus, only three vibrations (A_2'' and $2E_1'$) are IR-active and only three vibrations (A_1' + $2E_2'$) are Raman-active. In agreement with this prediction, the observed Raman spectrum of $(CH_3)N_4[XeF_5]$ exhibits three bands at 502 (symmetric stretch, A_1'), 423 (asymmetric stretch, E_2'), and 377 cm^{-1} (in-plane bending, E_2'). As discussed in Sec. I-11, the trigonal-bipyramidal and tetragonal-pyramidal structures can be ruled out since many more IR- and Raman-active bands are expected for these structures.

II-8. OCTAHEDRAL MOLECULES

(1) Octahedral XY_6 Molecules ($\mathbf{O}_h$)

Figure II-21 illustrates the six normal modes of vibration of an octahedral XY_6 molecule. Vibrations ν_1, ν_2, and ν_5 are Raman-active, whereas only ν_3 and ν_4 are infrared-active. Since ν_6 is inactive in both, its frequency is estimated from an analysis of combination and overtone bands. The **G** and **F** matrix elements of the octahedral XY_6 molecule are listed in Appendix VII.

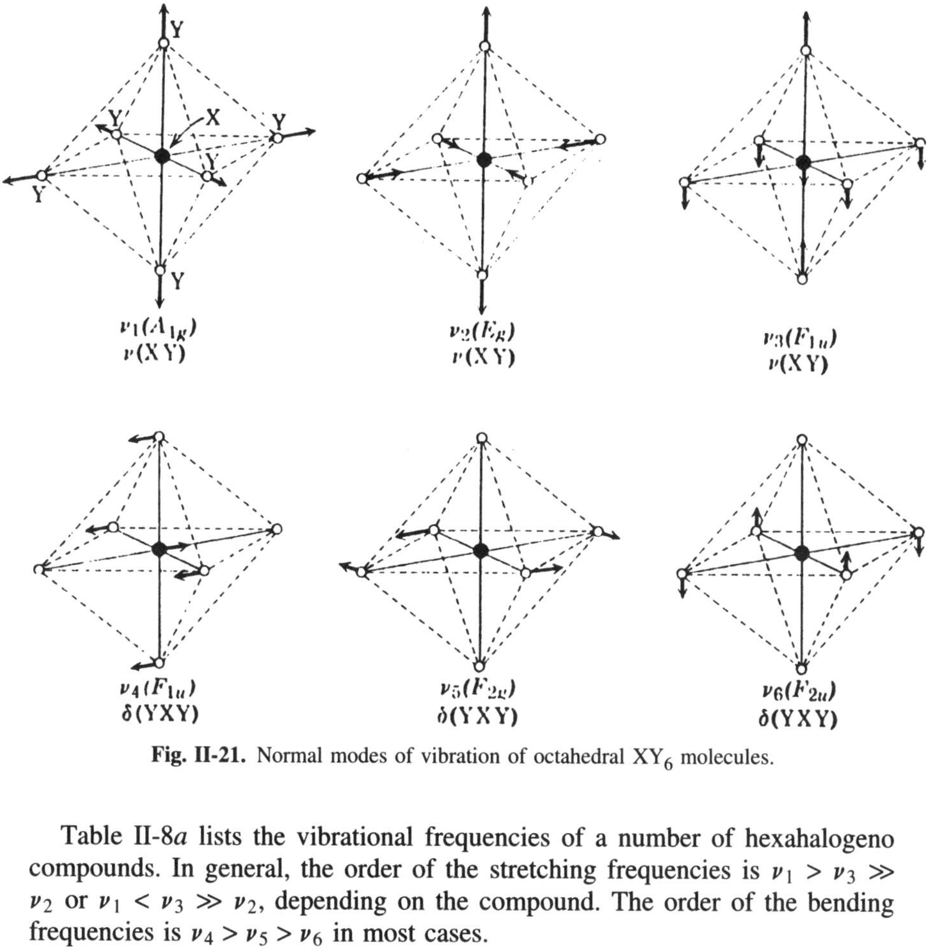

Fig. II-21. Normal modes of vibration of octahedral XY_6 molecules.

Table II-8*a* lists the vibrational frequencies of a number of hexahalogeno compounds. In general, the order of the stretching frequencies is $\nu_1 > \nu_3 \gg \nu_2$ or $\nu_1 < \nu_3 \gg \nu_2$, depending on the compound. The order of the bending frequencies is $\nu_4 > \nu_5 > \nu_6$ in most cases.

Several factors determine the trends in frequency:

(a) Mass of Central Atom Within the same family of the periodic table, the stretching frequencies decrease as the mass of the central atom increases, for example:

	$[AlF_6]^{3-}$		$[GaF_6]^{3-}$		$[InF_6]^{3-}$		$[TlF_6]^{3-}$
$\tilde{\nu}_1(cm^{-1})$	541	>	535	>	497	>	478
$\tilde{\nu}_3(cm^{-1})$	568	>	481	>	447	>	412

The trend in ν_1 directly reflects the trend in the stretching force constant (and bond strength) since the central atom is not moving in this mode. In ν_3, however, both X and Y atoms are moving, and the mass effect of the X atom cannot be ignored completely.

TABLE II-8a. Vibrational Frequencies of Octahedral XY_6 Molecules (cm^{-1})

Molecule	ν_1	ν_2	ν_3	ν_4	ν_5	ν_6[a]	Refs.
$[AlF_6]^{3-}$	541	(400)	568	387	322	(228)	924, 925
$[GaF_6]^{3-}$	535	(398)	481	298	281	(198)	924, 925
$[InF_6]^{3-}$	497	(395)	447	226	229	(162)	924, 925
$[InCl_6]^{3-}$	277	193	250	157	(149)	—	926
$[TlF_6]^{3-}$	478	387	412	202	209	(148)	924, 925
$[TlCl_6]^{3-}$	280	262	294	222, 246	155	(136)	927
$[TlBr_6]^{3-}$	161	153	190, 195	134, 156	95	(80)	927
$[SiF_6]^{2-}$	663	477	741	483	408	—	928
$[GeF_6]^{2-}$	624	471	603	339, 359	335	—	928
$[GeCl_6]^{2-}$	318	213	310	213	191	—	929
$[SnF_6]^{2-}$	592	477	559	300	252	—	928
$[SnCl_6]^{2-}$	311	229	303	166	158	—	930
$[SnBr_6]^{2-}$	190	144	224	118	109	—	931
$[SnI_6]^{2-}$	122	93	161	84	78	—	932
$[PbCl_6]^{2-}$	281	209	262	142	139	—	930
$[PF_6]^{-}$	756	585, 570	865, 835	559, 530	480 ~ 468	—	933
$[PCl_6]^{-}$	360	283	444	285	238	—	929, 933a
$[AsF_6]^{-}$	689	573	700	385	375	(252)	928
$[AsCl_6]^{-}$	337	289	333	220	202	—	929
$[SbF_6]^{-}$	668	558	669	350	294	—	928, 934
$[SbCl_6]^{-}$	330	282	353	180	175	—	935
$[SbCl_6]^{3-}$	327	274	—	—	137	—	936
$[SbBr_6]^{-}$	192	169	239, 224	119	103, 78	—	937
$[SbBr_6]^{3-}$	180	153	180	107	73	—	938, 939
$[SbI_6]^{3-}$	107	96	108	82	54	—	938, 939
$[BiF_6]^{-}$	590	547	585	—	231/247	—	934, 940
$[BiCl_6]^{3-}$	259	215	172	130	115	—	938, 939
$[BiBr_6]^{3-}$	156	130	128	75	62	—	938, 939
$[BiI_6]^{3-}$	114	103	96	(59)	54	—	938, 939
SF_6	775	643	939	614	524	(347)	941 ~ 943
SeF_6	708	658	780	437	403	(264)	941, 942
$[SeCl_6]^{2-}$	299	255	280	160–140	165	—	926
$[SeBr_6]^{2-}$	179	157	225	122	105	—	944
TeF_6	698	672	752	325	312	(197)	941, 942
$[TeCl_6]^{2-}$	298	250	250	136	140	—	945
$[TeCl_6]^{3-}$	264	192	230	146	(135)	—	920
$[TeBr_6]^{2-}$	174	153	198	—	75	—	945
$[ClF_6]^{-}$	525	384	—	—	289	—	946
$[ClF_6]^{+}$	679	630	890	582	513	—	947
$[BrF_6]^{+}$	658	660	—	—	405	—	948
$[BrF_6]^{-}$	568	454	400	204, 184	250	(138)	949
$[AuF_6]^{-}$	595	530	—	—	225	—	950
$[ScF_6]^{3-}$	495	375	458	257	235	—	951
$[ScI_6]^{3-}$	119	67	—	—	80	—	674
$[YF_6]^{3-}$	476	382	160	74	194	—	952
$[LaF_6]^{3-}$	443	334	130	63	171	—	952
$[GaF_6]^{3-}$	473	380	140	72	185	—	952
$[YbF_6]^{3-}$	491	370	156	70	196	—	952

TABLE II-8a. (Continued)

Molecule	ν_1	ν_2	ν_3	ν_4	ν_5	6[a]	Refs.
$[CeCl_6]^{2-}$	295	265	268	117	120	(86)	953
$[TiF_6]^{2-}$	618	(440)	615, 600	315, 281	308, 300	—	954
$[TiCl_6]^{2-}$	320	271	316	183	173	—	955
$[TiCl_6]^{3-}$	322	278	304, 290	—	175	—	956
$[TiBr_6]^{2-}$	192	—	244	119	115	—	955
$[ZrF_6]^{2-}$	589	(416)	537, 522	241, 192	258, 244	—	957
$[ZrCl_6]^{2-}$	327	237	290	150	153	—	958
$[ZrBr_6]^{2-}$	194	144	223	106	99	—	959
$[HfF_6]^{2-}$	572	(389)	448, 490	217, 184	259, 247	—	954
$[HfCl_6]^{2-}$	326	257	275	145	156	(80)	930
$[HfBr_6]^{2-}$	197	142	189	102	101	—	959
$[VF_6]^{-}$	676	538	646	300	330	—	960
$[VF_6]^{2-}$	584	—	578	273	—	—	960
$[VF_6]^{3-}$	533	—	511	292	—	—	960
$[VCl_6]^{2-}$	—	—	355, 305	—	—	—	961
$[NbF_6]^{-}$	683	562	602	244	280	—	962, 963
$[NbCl_6]^{-}$	368	288	333	162	183	—	930
$[NbCl_6]^{2-}$	—	—	314	165	—	—	963, 964
$[NbBr_6]^{-}$	—	—	240 ~ 216	—	—	—	965
$[NbBr_6]^{2-}$	—	—	236	112	—	—	963, 964
$[NbI_6]^{-}$	—	—	180	70, 66	—	—	963, 966
$[TaF_6]^{-}$	692	581	560	240	272	(192)	967
$[TaCl_6]^{-}$	378	298	330	158	180	—	930
$[TaCl_6]^{2-}$	—	—	297	160	—	—	930
$[TaBr_6]^{-}$	—	—	234 ~ 223	—	—	—	965
$[TaBr_6]^{2-}$	—	—	217	109	—	—	964
$[TaI_6]^{-}$	—	—	160	80	—	—	966
CrF_6	(720)	(650)	790	(266)	(309)	(110)	968, 969
$[CrCl_6]^{3-}$	286	237	315	199	162	182	970
$^{92}MoF_6$	741.8	652.0	749.5	265.7	317	117	971
$[MoF_6]^{2-}$	685	598	653	250	274	—	972
$[MoCl_6]^{-}$	356	—	327	162	—	—	973
$[MoCl_6]^{2-}$	329	—	308	168	154	—	973
$[MoCl_6]^{3-}$	305	—	268, 286, 302	167, 187	150	—	973
WF_6	770	676	711	258	321	(127)	941, 942
WCl_6	437	331	373	160	182	—	930
$[WCl_6]^{-}$	382	—	332, 312	157	168	—	973
$[WCl_6]^{2-}$	341	—	293	166, 150	—	—	973
$[MnF_6]^{2-}$	592	508	620	335	308	—	974
TcF_6	713	(639)	748	275	(297)	(145)	942
$[ReF_6]^{+}$	797	734	783	353	359	—	975
ReF_6	754	(671)	715	257	(295)	(147)	942
$[ReF_6]^{2-}$	611	530	535	249	221	(181)	976
$[ReCl_6]^{-}$	—	—	318	161	—	—	977
$[ReCl_6]^{2-}$	346	(275)	313	172	159	—	930, 978

TABLE II-8a. (*Continued*)

Molecule	ν_1	ν_2	ν_3	ν_4	ν_5	6^a	Refs.
$[ReBr_6]^{2-}$	213	(174)	217	118	104	—	979
$[NiF_6]^{2-}$	555	512	648	332	307, 298	—	980
$[PdCl_6]^{2-}$	318	289	346	200	178	—	981
$[PdBr_6]^{2-}$	198	176	253	130	100	—	982
PtF_6	656	(601)	705	273	(242)	(211)	942
$[PtF_6]^{2-}$	611	576	571	281	210	(143)	983
$[PtCl_6]^{2-}$	348	318	342	183	171	(88)	931
$[PtBr_6]^{2-}$	213	190	243	146	137	—	981
$[PtI_6]^{2-}$	—	—	186	46	—	—	984
$[FeF_6]^{3-}$	538	374	—	—	253	—	985
RuF_6	(675)	(624)	735	275	(283)	(186)	942, 986
$[RuCl_6]^{2-}$	—	—	346	188	—	—	986
OsF_6	731	(668)	720	268	(276)	(205)	942
$[OsCl_6]^{-}$	375	302	325	168	183	—	987
$[OsCl_6]^{2-}$	345.3	245.2	326	176	160	—	988, 989
$[OsBr_6]^{2-}$	210.6	169.2	227	122	100	—	988, 989, 989a
$[OsI_6]^{2-}$	152	121	170	91	80	—	990
RhF_6	(634)	(595)	724	283	(269)	(192)	942, 986
$[RhCl_6]^{2-}$	—	—	329	187	—	—	991
IrF_6	702	645	719	276	267	(206)	942
$[IrCl_6]^{2-}$	352	(225)	333	184	190	—	930
$[IrCl_6]^{3-}$	—	—	296	200	—	—	984
$[IrBr_6]^{2-}$	—	—	235	82	—	—	991
$[IrI_6]^{3-}$	149	133	175	87	88	—	992
$[ThCl_6]^{2-}$	294	255	259	—	114	—	993
UF_6	666	530	619	184	200	—	994
$[UF_6]^{-}$	—	—	525	173	—	—	995
$[UCl_6]^{-}$	343	273	310	122	136	—	995, 996
$[UCl_6]^{2-}$	299	237	262	114	121	(80)	993
$[UBr_6]^{-}$	—	—	214	87	—	—	995
NpF_6 [b]	646	525	618	198	208	169	997
NpF_6 [c]	643	574	604	191	—	138/149	997
$[NpCl_6]^{2-}$	310	—	265	117	128	—	998
PuF_6 [b]	625	519	612	200	209	177	999
PuF_6 [c]	615	—	—	202	198	172	1000

[a] The value of ν_6 can also be estimated by the relation $\nu_6 = \nu_5/\sqrt{2}$ (see Refs. 1001 and 1002).
[b] Ground state.
[c] Excited state.

(b) Oxidation State of Central Atom Across the periodic table, the stretching frequencies increase as the oxidation state of the central atom becomes higher. Thus we have:

	$[AlF_6]^{3-}$		$[SiF_6]^{2-}$		$[PF_6]^{-}$		SF_6
$\tilde{\nu}_1(cm^{-1})$	541	<	663	<	745	<	775
$\tilde{\nu}_3(cm^{-1})$	568	<	741	<	840	<	939

The effect of lowering the oxidation state of the same central atom is clearly seen in a series such as $[VF_6]^{n-}$ ($n = 1$, 2, and 3) and $[WCl_6]^{n-}$ ($n = 0$, 1, and 2):

	$[VF_6]^-$		$[VF_6]^{2-}$		$[VF_6]^{3-}$
$\tilde{\nu}_1(cm^{-1})$	676	>	584	>	533
$\tilde{\nu}_3(cm^{-1})$	646	>	.578	>	511

As in many other cases, the higher the oxidation state, the higher the frequency. The bending frequencies do not exhibit clear-cut trends.

(c) Halogen Series The effect of changing the halogen is seen in a number of series, for example:

	$[SnF_6]^{2-}$		$[SnCl_6]^{2-}$		$[SnBr_6]^{2-}$		$[SnI_6]^{2-}$
$\tilde{\nu}_1(cm^{-1})$	592	>	311	>	190	>	122
$\tilde{\nu}_3(cm^{-1})$	559	>	303	>	224	>	161

The stretching force constants also follow the same order. The ratios ν(MBr)/ν(MCl) and ν(MI)/ν(MCl) are about 0.61 and 0.42, respectively, for ν_1, and about 0.76 and 0.62, respectively, for ν_3.

(d) Electronic Structure In the $[MCl_6]^{3-}$ series, ν_3 and ν_4 change as follows:

	$Cr^{3+}(d^3)$	$Mn^{3+}(d^4)$	$Fe^{3+}(d^5)$	In^{3+}
$\tilde{\nu}_3(cm^{-1})$	315	342	248	248
$\tilde{\nu}_4(cm^{-1})$	200	183	184	161

All these metal ions are in the high-spin state. For $Fe^{3+}(t_{2g}^3 e_g^2)$, occupation of the antibonding orbitals lowers ν_3 drastically in comparison to the Cr^{3+} complex; its ν_3 is comparable to that of the In^{3+} complex, whose ν_3 is lowered because of the increased mass of the metal. On the other hand, the ν_3 of $[MnCl_6]^{3-}$ is higher than that of $[CrCl_6]^{3-}$ because the static Jahn–Teller effect of the Mn^{3+} ion causes a tetragonal distortion.[953]

(e) Electronic Excited States Table II-8*a* includes PuF_6 and NpF_6, for which vibrational frequencies have been determined both for the ground and excited states. The excited frequencies were obtained from the analysis of vibrational fine structure of electronic spectra.

The Raman intensity of an XY_6 molecule normally follows the order $I(\nu_1) > I(\nu_2) > I(\nu_5)$. Adams and Downs[1003] noted that $I(\nu_2)/I(\nu_1)$ is 0.5 to 1 for

$[TeCl_6]^{2-}$ and $[TeBr_6]^{2-}$, although it normally ranges from 0.05 to 0.1. Furthermore, they observed ν_3, which is not allowed in Raman spectra. From these and other items of evidence, they proposed that the $\mathbf{O}_h$ selection rule breaks down in $[TeX_6]^{2-}$ because less symmetrical electronic excited states perturb the $\mathbf{O}_h$ ground state. They also noted the distortion of $[SbX_6]^{3-}$ ions to $\mathbf{C}_{3v}$ symmetry from their Raman spectra in solution. Woodward and Creighton[1004] noted that $I(\nu_2) > I(\nu_1)$ holds in the aqueous Raman spectra of Na_2PtX_6 (X = Cl and Br) and Na_2PdCl_6, and attributed this unusual trend to the presence of six nonbonding d-electrons in the valence shell.

Table II-8a contains several XY_6-type ions which exhibit splitting of degenerate vibrations due to lowering of symmetry in the crystalline state. For example, the ν_3 vibration of the ReF_6^{2-} ion in K_2 $[ReF_6]$ splits into two bands (565 and 543 cm^{-1}) due to trigonal distortion in the crystalline state. Similar splitting is observed for the ν_4 vibration (222 and 257 cm^{-1}).[1005,1006]

Hunt et al.[1007] obtained two types of 1:1 adducts by co-depositing UF_6 with HF in excess Ar at 12 K; UF_6–HF exhibits a strong band at 3848 cm^{-1}, while UF_6–FH shows a weak, broad band at 3903 cm^{-1} in IR spectra. Similar experiments show that WF_6–HF exhibits a weak band at 3911 cm^{-1}, while WF_6–FH shows a strong band at 3884 cm^{-1}. These results indicate that, in hydrogen-bonded complexes, the order of proton affinity is $WF_6 < UF_6$, with HF serving as a proton donor and that, in anti-hydrogen-bonded complexes, the order of acidity is $UF_6 < WF_6$, with HF serving as a Lewis base. Relative yields of these two types are determined by the acidity/basicity of these fluorides.

Preresonance Raman spectra of 17 XY_6-type metal halides have been measured by Bosworth and Clark.[936] Hamaguchi et al.[1008] observed anomalous polarization (Sec. I-23) for all the Raman bands of the $IrCl_6^{2-}$ ion measured under resonance condition. Figure II-22 shows the polarized Raman spectra (488.0 nm excitation) of this ion in aqueous solution together with band assignments and depolarization ratios for each band. Since the Raman tensor for the ν_1, ν_2, and ν_5 vibrations cannot have an antisymmetric part, group theoretical arguments such as those used for heme proteins (Sec. I-23) are not applicable to octahedral XY_6 ions. A more detailed study by Hamaguchi[1009] shows that electronic degeneracy in the $IrCl_6^{2-}(5d^5)$ ion induces an antisymmetric part of vibrational Raman tensors.

Weinstock et al.[1010] noted that the combination bands $(\nu_1 + \nu_3)$ and $(\nu_2 + \nu_3)$ appear with similar frequencies, intensities, and shapes in the infrared spectra of $MoF_6(d^0)$ and $RhF_6(d^3)$. As is shown in Fig. II-23, however, $(\nu_2 + \nu_3)$ was very broad and weak in $TcF_6(d^1)$, $ReF_6(d^1)$, and $RuF_6(d^2)$, and $OsF_6(d^2)$. This anomaly was attributed to a dynamic Jahn–Teller effect. The static Jahn–Teller effect does not seem to operate in these compounds since no splittings of the triply degenerate fundamentals were observed.

Perhaps the most fascinating XY_6-type molecule is XeF_6. In their earlier work, Claassen et al.[1011] suggested the distortion of XeF_6 from $\mathbf{O}_h$ symmetry since they observed two stretching bands in infrared and three stretching bands in Raman spectra. It was not possible, however to determine the precise struc-

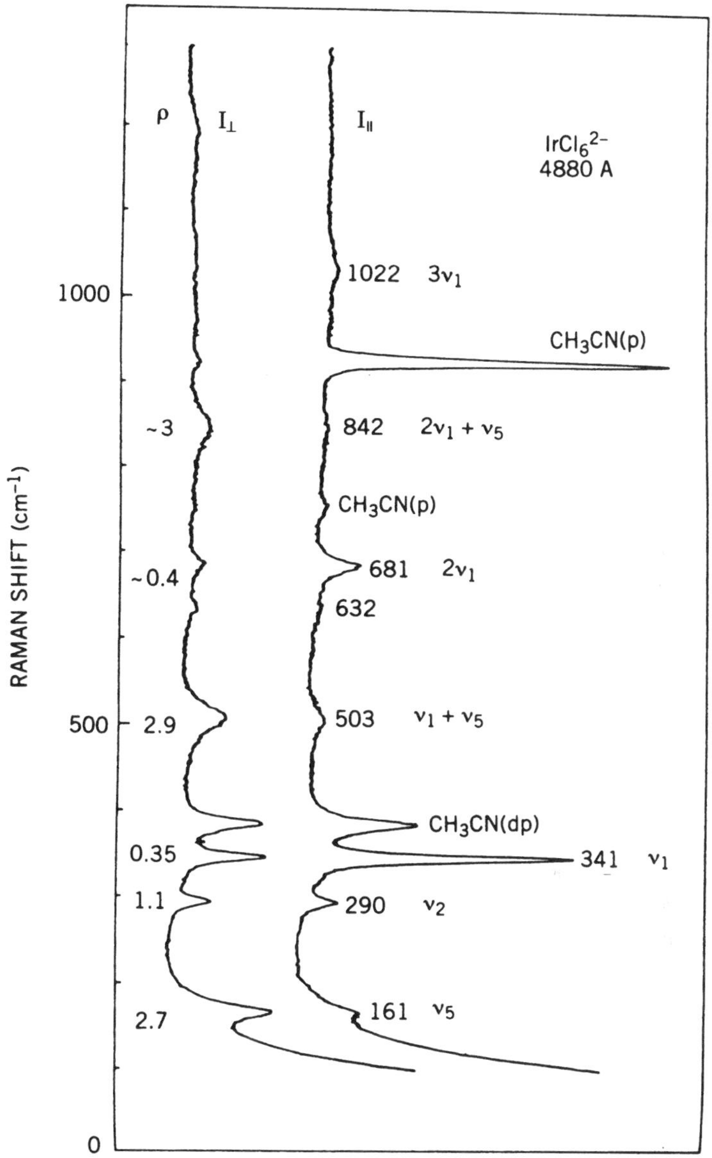

Fig. II-22. Polarized Raman spectra of $[(n\text{–}C_4H_9)_4N]_2\ IrCl_6$ in acetonitrile. (488.0 nm excitation) (reproduced with permission from Ref. 1008).

ture of XeF_6 until they[1012] carried out a detailed infrared, Raman, and electronic spectral study of XeF_6 vapor as a function of temperature. They were then able to show that XeF_6 consists of the three electronic isomers shown in Fig. II-24, and to explain subtle differences in spectra at different temperatures as a shift of equilibrium among these three isomers.

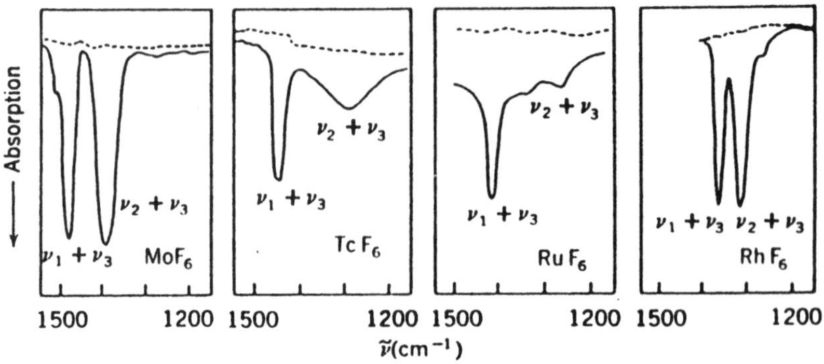

Fig. II-23. Band profiles for $(\nu_1 + \nu_3)$ and $(\nu_2 + \nu_3)$ for the $4d$ transition-series hexafluorides (reproduced with permission from Ref. 1010).

Metal isotope shifts for heavy metals such as uranium are expected to be extremely small. In fact, the ^{235}U-^{238}U isotope shift of the ν_3 band of gaseous UF_6 is only 0.65 ± 0.1 cm^{-1}.[1013]

Normal coordinate analyses on octahedral hexahalogeno compounds have been made by a number of investigators. Kim et al.[1014] calculated the force constants of 15 metal fluorides using the UBF and OVF fields (see Sec. I-14), and found that the latter is better than the former. LaBonville et al.[1015] calculated the force constants of 62 metal halides by using the UBF, OVF, GVF, modified UBF, and modified OVF fields, and found that the modified OVF field gives the best overall agreement with the observed frequencies. They also discussed the dependence of force constants on the mass of the halogen, the oxidation state of the metal, the number of nonbonding electrons in the valence shell, and the crystal-field stabilization energy.

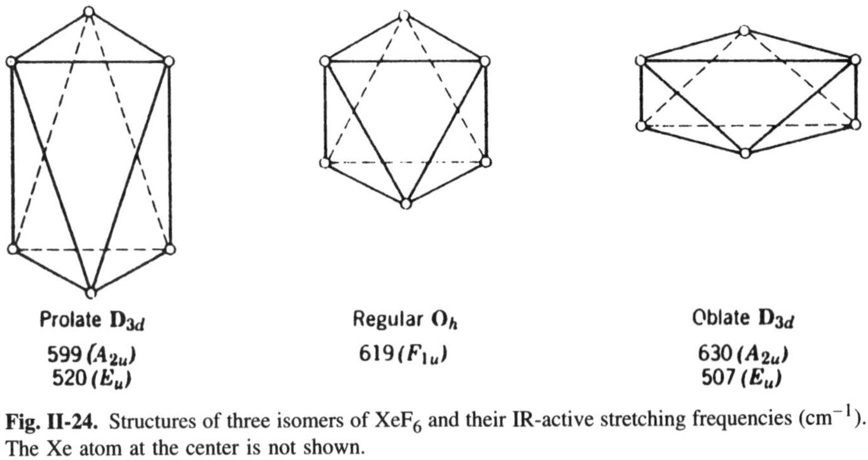

Fig. II-24. Structures of three isomers of XeF_6 and their IR-active stretching frequencies (cm^{-1}). The Xe atom at the center is not shown.

The ^{35}Cl NQR spectra provide information about the σ and π contributions to the covalent M—Cl bonding in $[MCl_6]^{n-}$-type ions, and these can be correlated with the force constants obtained from infrared and Raman studies.[930] Both infrared and NQR spectra suggest low-site symmetry of the $[MCl_6]^{3-}$ ion in $K_3[MCl_6]\cdot H_2O$ (M = Ir and Rh) crystals.[1016]

Coriolis coupling constants have been calculated from the vapor-phase Raman spectra of some XY_6-type molecules.[941]

Table II-8*b* lists the vibrational frequencies of MO_6-type ions. Hauck and Fadini calculated the force constants of these ions.[1017,1018]

(2) Octahedral XY_nZ_{6-n} Molecules

The XY_5Z molecule belongs to the $\mathbf{C}_{4v}$ point group, and its 11 normal vibrations are classified into $4A_1$, $2B_1$, B_2, and $4E$ modes, of which only A_1 and E are infrared-active; all are Raman-active. Table II-8*c* lists the observed frequencies of XY_5Z-type molecules. The symmetry of the $[XeF_5O]^-$ anion in its cesium salt is distorted to $\mathbf{C}_s$ owing to the presence of a sterically active lone pair on the octahedral faces adjacent to the axial fluorine.[1040]

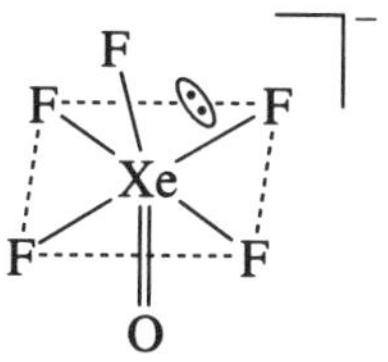

$\mathbf{C}_s$ symmetry

This results in splitting of the E modes, as shown in Table II-8*c*.

The XY_4Z_2 molecule may be *cis* ($\mathbf{C}_{2v}$) or *trans* ($\mathbf{D}_{4h}$). The *cis*-isomer is expected to give four XY stretching ($2A_1+B_1+B_2$) and two XZ stretching (A_1+B_1) modes, all of which are infrared- as well as Raman-active. The *trans*-isomer is expected to give three XY stretching ($A_{1g}+B_{1g}+E_u$) and two XZ stretching

TABLE II-8*b*. Vibrational Frequencies of Octahedral MO_6 Molecules (cm^{-1})

Compound	ν_1	ν_2	ν_3	ν_4	ν_5	Refs.
$Li_6[TeO_6]$	700	540	640	470	355	1017
$Li_6[WO_6]$	740	450	620	425	360	1017
α-$Li_6[ReO_6]$	680	505	620	425	360	1017
$Ca_4[PtO_6]$	—	530	575	425	345	1018
α-$Li[TeO_6]$	700	540	640	470	355	1018
$Ca_5[IO_6]_2$	771	490, 538	765, 695	451, 435	—	1018
$[NbO_6]^a$	620	542	600	350	269	1019

^{a}In $Sb[NbO_4]$ crystal.

TABLE II-8c. Vibrational Frequencies of Octahedral XY_5Z and XY_4WZ Molecules (cm^{-1})

XY_5Z or XY_4WZ	$\nu_1(A_1)$ $\nu(XZ)$	$\nu_2(A_1)$ $\nu(XW)^a$	$\nu_3(A_1)$ $\nu(XY_4)$	$\nu_4(A_1)$ $\pi(XY_4)$	$\nu_5(B_1)$ $\nu(XY_4)$	$\nu_6(B_1)$ $\pi(XY_4)$	$\nu_7(B_2)$ $\delta(XY_4)$	$\nu_8(E)$ $\nu(XY_4)$	$\nu_9(E)$ $\rho_w(XW)^a$	$\nu_{10}(E)$ $\rho_w(XZ)$	$\nu_{11}(E)$ $\delta(XY_4)$	Refs.
$[SbCl_5Br]^-$	219	308	334	151	287	—	—	344	—	—	—	1020, 1021
$[SbBr_5Cl]^-$	305	206	192	—	186	—	—	239	—	—	—	1020
SF_5Cl	402	855	707	602	625	271	505	909	597	399	441	1022
SF_5Br	272	848	691	586	620	—	500	898	575	222	419	1023
$[SF_5O]^-$	1153	722	697	506	541	472	452	780	607	530	325	1024, 1025
SeF_5Cl	729	654	440	384	636	—	380	745	421	334	213	1026
$[SeF_5O]^-$	919	559	649	—	556	—	—	639	—	—	—	1027
TeF_5Cl	708	662	312	410.5 404.0	651	—	302	726	324.6	259	167	1028
$[TeF_5O]^-$	867	576	643	345	—	260	—	635	331	320	279	1029
IF_5O	927	680	640	363	647	307	330	710	372	343	205	1030
$[TiF_5O]^{3-}$	920	379	520	290	—	—	—	520	138	335	235	1031
$[VF_5O]^{3-}$	943	383	525	317	—	—	—	525	139	342	237	1031
$[NbCl_5Br]^-$	210	310	365	181	285	120	134	352	161	153	75	1032
$[TaCl_5Br]^-$	204	318	368	183	300	(120)	168	325	151	143	73	1032

$[TaBr_5Cl]^-$	323	231	187	110	180	(73)	96	214	123	144	76	1032
$[CrF_5O]^-$	993	613	530	302	577	—	—	586	346	277	—	922
$[MoCl_5O]^{2-}$	998	318	331	168	336	159	164	321	233	137	147	1033
$[MoF_5O]^-$	973	662	492	300	580	—	—	580	324	252	—	1034
$[WF_5O]^-$	987	686	507	286	594	—	—	608	329	242	—	1034
WF_5Cl	407	744	703	257	644	182	377	661	290	227	307	1035
ReF_5O	990	739	643	309	652	234	334	713	260	365	125	1030
$[RuCl_5N]^-$	1048	284	318	192	307	168	184	337	233	154	174	1033
$[RuBr_5N]^-$	1046	201	207	156	181	136	147	204	257	110	144	1033
OsF_5O	963	716	644	281	644	210	332	701	263	367	164	1030, 1036
$[OsCl_5N]^{2-}$	1084	324	348	189	334	169	181	336	264	146	172	1033
$[OsBr_5N]^{2-}$	1085	192	198	156	172	136	149	234	217	115	144	1033
$[UF_5O]^-$	820	593	602	182	445	(161)	(283)	480	248	209	201	1037
$[XeF_5O]^-$	883	524	420	361	544	177	390	473 468 435	410 396	384 365	293 274	1038, 1039
$[MoCl_4OBr]^{2-}$	235	964	301	149	288	—	—	320	229	92	162	1040
$[WCl_4OBr]^{2-}$	233	960	326	149	—	—	187	298	230	92	162	1040

[a]For XY_5Z, W is regarded as Y *trans* to Z.
[b]Splitting due to lowering of symmetry (see text).

($A_{1g}+A_{2u}$) modes, of which E_u and A_{2u} are infrared-active and A_{1g} and B_{1g} are Raman-active. The selection rules for other XY_nZ_{6-n} molecules tabulated in Appendix V can be used to distinguish the structures of stereoisomers on the basis of their vibrational spectra.

Preetz and co-workers carried out an extensive study on the preparation, isolation, and vibrational spectra of a number of mixed-halogeno complexes: $[MCl_nBr_{6-n}]^{2-}$ ($n = 0 \sim 6$, M = Ru,[1041] Os[1042]), $[TeCl_nBr_{6-n}]^{2-}$ ($n = 0 \sim 6$),[1043] $[OsF_nCl_{6-n}]^{2-}$ ($n = 1 \sim 5$),[1044] $[MCl_nBr_{6-n}]^{3-}$ ($n = 1 \sim 5$, M = Rh,[1045] Ir[1046]), $[IrF_nCl_{6-n}]^{2-}$ ($n = 1 \sim 5$),[1047] $[PtF_nCl_{6-n}]^{2-}$ ($n = 1 \sim 5$),[1048] and $[PtCl_nBr_{6-n}]^{2-}$ ($n = 1 \sim 5$).[1049] These workers assigned the IR and Raman spectra and carried out normal coordinate analysis for these complexes.[1050] For example, the following stereoisomers were isolated for the $[OsF_nCl_{6-n}]^{2-}$ ions:

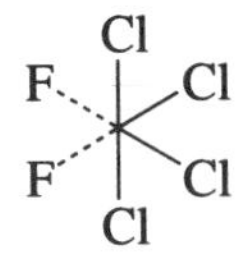

2a

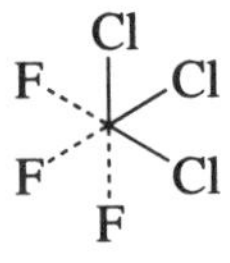

3a

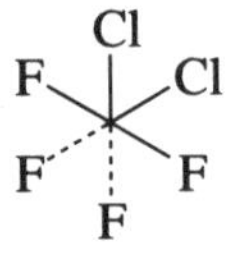

4a

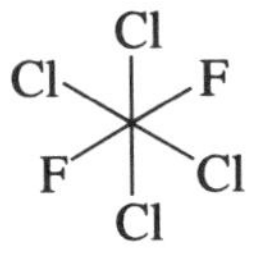

2b

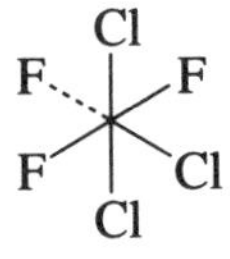

3b

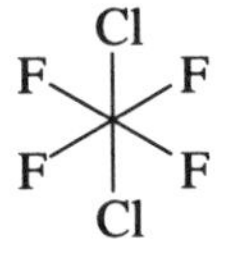

4b

Figure II-25 shows the IR and Raman spectra of these stereoisomers. In the Os—Cl stretching region (350–300 cm^{-1}), the *trans*-isomer (2*b*, $\mathbf{D}_{4h}$) exhibits one IR (E_u) and two Raman (A_{1g} and B_{1g}) bands, as expected from the selection rules. For the *cis*-isomer (2**a**, $\mathbf{C}_{2v}$), the selection rules predict four ($2A_1, B_1$, and B_2) bands both in IR and Raman spectra. However, only three IR and two Raman bands are apparent in Fig. II-24. Such discrepancies occur because bands are overlapped and IR/Raman active fundamentals are weak in some cases. Recently, Preetz and co-workers[1050a] reviewed the results of their extensive investigations on preparation and spectroscopic properties of mixed octahedral complexes and clusters. The Raman spectrum of crystalline $Rb_2[TeBr_{3.5}Cl_{2.5}]$ shows it to be a 1:1 mixture of $Rb_2[TeBr_3Cl_3]$ and $Rb_2[TeBr_4Cl_2]$.[1051]

II-9. XY_7 and XY_8 MOLECULES

The XY_7-type molecules are very rare. Both IF_7 and ReF_7 are known to be pentagonal-bipyramidal ($\mathbf{D}_{5h}$), and their vibrational spectra have been assigned completely, as shown in Table II-9. The A_1', E_1'', and E_2' vibrations are Raman-

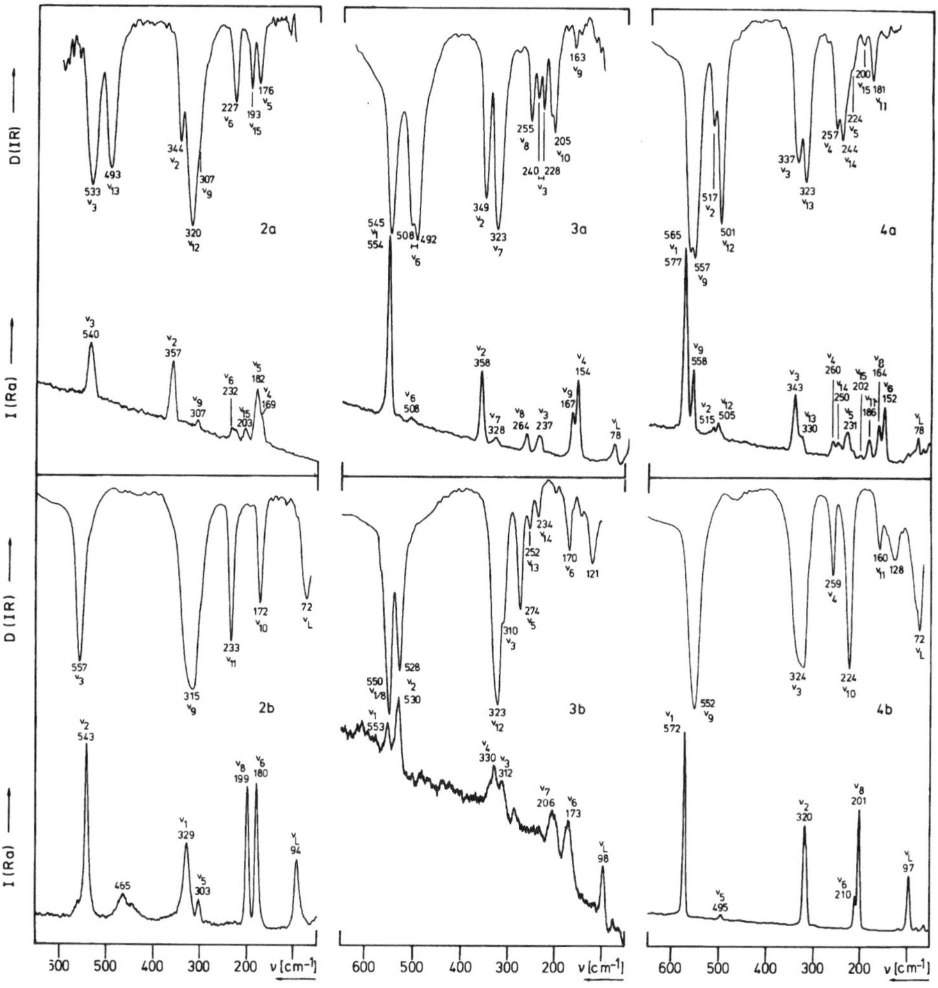

Fig. II-25. IR and Raman spectra of $Cs_2[OsF_nCl_{6-n}]$, $n = 2 \sim 4$, a: *cis*; b: *trans*. Exciting lines used are 488 nm (2a, 2b, and 3a), 647.1 nm (3b), and 514.5 nm (4a and 4b) (reproduced with permission from Ref. 1044).

active, and the A_2'' and E_1' vibrations are infrared-active. According to Eysel and Seppelt,[1055] IF_7 undergoes minor dynamic distortions from $\mathbf{D}_{5h}$ symmetry which cause violation of the $\mathbf{D}_{5h}$ selection rules for combination bands but not for the fundamentals. Normal coordinate analysis[1055] shows that the axial bonds are definitely stronger and shorter than the equatorial ones. Table II-9 also lists the frequencies and assignments of the $[UO_2F_5]^{2-}$ ion of $\mathbf{D}_{5h}$ symmetry.[1056] The vibrational spectra of the $[IF_6O]^-$,[1053,1056a] and $[TeF_6O]^{2-}$,[1056b] ions in which the axial positions of a pentagonal bipyramidal structure are occupied by a fluorine and an oxygen atom ($\mathbf{C}_{5v}$ symmetry) have been assigned.

TABLE II-9 Vibrational Frequencies[a] of XY_7 and XY_5Z_2 Molecules (cm^{-1})

D_{5h}	ReF_7 [1052]	TeF_7^- [1053, 1054]	IF_7 [1053, 1055]	$UF_5O_2^{3-}$ [1056]	Assignment[b]
$\nu_1(A_1')$	736	597	675	(668)	$\nu_s(MF_a)$
$\nu_2(A_1')$	645	640	629	816	$\nu_s(MF_e)$
$\nu_3(A_2'')$	703	695	746	873	$\nu_a(MF_a)$
$\nu_4(A_2'')$	299	332	363	380	$\delta(F_eMF_a)$
$\nu_5(E_1')$	703	625	672	740	$\nu_d(MF_e)$
$\nu_6(E_1')$	353	384	425	425	$\delta_d(F_eMF_e)$
$\nu_7(E_1')$	217	—	257	240	$\delta_d(F_aMF_a)$
$\nu_8(E_1'')$	597	299	308	—	$\delta_d(F_eMF_a)$
$\nu_9(E_2')$	489	458	509	—	$\nu_d(MF_e)$
$\nu_{10}(E_2')$	352	326	342	—	$\delta_d(F_eMF_e)$

[a]In these molecules $\nu_{11}(E_2'')$ is inactive.
[b]F_a and F_e denote the axial and equatorial F atoms, respectively.

The XY_8-type molecule may take the form of (I) a cube ($\mathbf{O}_h$), (II) an archimedean antiprism ($\mathbf{D}_{4d}$), (III) a dodecahedron ($\mathbf{D}_{2d}$), or (IV) a face-centered trigonal prism ($\mathbf{C}_{2v}$). Although XY_8 molecules are rare, X-ray analysis indicates that $[TaF_8]^{3-}$ and $[CrO_8]^{3-}$ ions take structures II and III, respectively.[1057,1058] The infrared and Raman spectra of crystalline $Na_3[TaF_8]$ are in accord with structure II, proposed by X-ray analysis.[1059] However, the vibrational spectra of the $[IF_8]^-$ and $[TeF_8]^{2-}$ ions support the archimedean antiprism structures[1053] shown below:

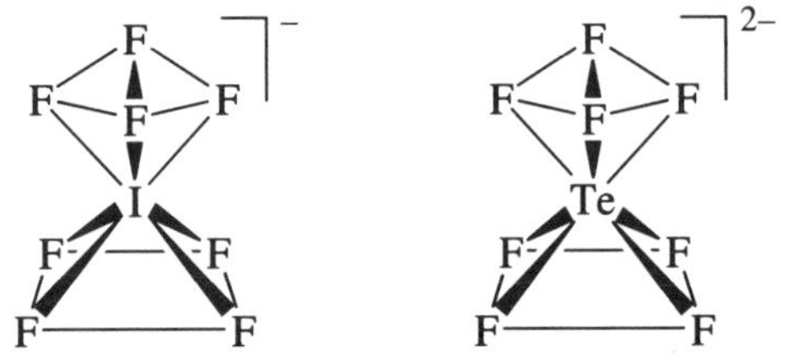

For normal coordinate analyses of a cubic and an archimedean antiprism XY_8 molecule, see Refs. 1060 and 1061, respectively.

II-10. X_2Y_4 AND X_2Y_6 MOLECULES

(1) X_2Y_4 Molecules

Depending upon the twisting angle (τ) between the two XY_2 planes, the symmetry of the $Y_2X—XY_2$ molecule may be $\mathbf{D}_{2h}$ ($\tau = 0°$, planar), $\mathbf{D}_{2d}$ ($\tau = 90°$, staggered), or $\mathbf{D}_2$ ($0° < \tau < 90°$, intermediate).

The $\mathbf{D}_{2h}$ structure may be confirmed if the infrared and Raman mutual exclusion rule holds. The $\mathbf{D}_{2d}$ and $\mathbf{D}_2$ structures can be distinguished by comparing the number of fundamentals with that predicted for each structure: 8 for $\mathbf{D}_2$ and 5 for $\mathbf{D}_{2d}$ in the infrared, and 12 for $\mathbf{D}_2$ and 9 for $\mathbf{D}_{2d}$ in the Raman.

B_2F_4[1002] and B_2Cl_4[1063] are staggered in the gaseous and liquid phases and planar in the solid state, whereas B_2Br_4[1064] is staggered in all phases. Apparently, steric hindrance plays a main role in determining the conformation. Both B_2F_4 and B_2Cl_4 are also staggered in Ar matrices.[1065] The vibrational spectra of gaseous and crystalline N_2O_4 have been assigned on the basis of $\mathbf{D}_{2h}$ symmetry.[1066,1066a] In a N_2 matrix, N_2O_4 is a mixture of the $\mathbf{D}_{2h}$, $\mathbf{D}_{2d}$, and ONO—NO_2 isomers.[67] The vibrational spectra of the oxalato ion ($C_2O_4^{2-}$) have been assigned based on $\mathbf{D}_{2d}$,[1067] $\mathbf{D}_{2h}$,[1068] and $\mathbf{D}_2$[1069] symmetry.

Molecules like N_2H_4 and N_2F_4 take the *trans* ($\mathbf{C}_{2h}$), *gauche* ($\mathbf{C}_2$), or *cis* ($\mathbf{C}_{2v}$) structure (Fig. II-26), depending on the angle of internal rotation. Evidently, the presence of lone-pair electrons on the X atom is responsible for the deviation of the X–XY_2 plane from the planarity. Most of these compounds exist as the *trans*- or *gauche*-isomer or as a mixture of both. The *trans*-isomer shows 6, whereas the *gauche*-isomer shows 12, fundamentals in the infrared.

In the gaseous, liquid, and solid states, N_2F_4 is a mixture of the *trans*- and *gauche*-isomers, and complete vibrational assignments have been made on each isomer.[1070] N_2H_4 is pure *gauche* in all physical states,[1071] and its IR spectrum in Ar matrices has been assigned.[1072] The IR spectrum of P_2H_4 in the gaseous state has been assigned on the basis of the *gauche* structure.[1073] However, it is *trans* in the solid state.[1074] The *trans* structure has been deduced from the vibrational spectra of P_2F_4 (all states),[1075] P_2Cl_4 (all states),[1076] and P_2I_4 (solid and solution).[1077] The dithionite ion, $[S_2O_4]^{2-}$, takes a structure of $\mathbf{C}_{2h}$ symmetry, both in solution and in the solid state, and its IR/Raman spectra have been assigned.[1078]

The infrared spectra of dimeric metal halides $(MX_2)_2$ isolated in inert gas matrices have been assigned on the basis of a planar-cyclic ring structure (Sec. II-2). Durig and co-workers reviewed the vibrational spectra of X_2Y_4-type molecules.[1079]

(2) Bridged X_2Y_6 Molecules ($\mathbf{D}_{2h}$)

Figure II-27 illustrates the 18 normal modes of vibration[1080] and band assignments for nonplanar bridging X_2Y_6-type molecules. The A_g, B_{1g}, B_{2g}, and

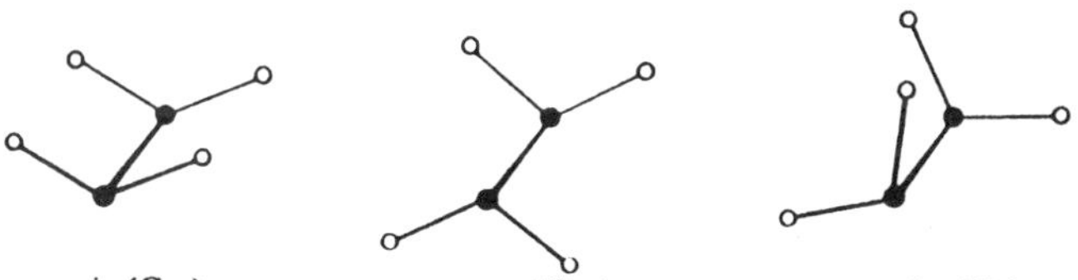

Figure II-26. Various conformations of hydrazine.

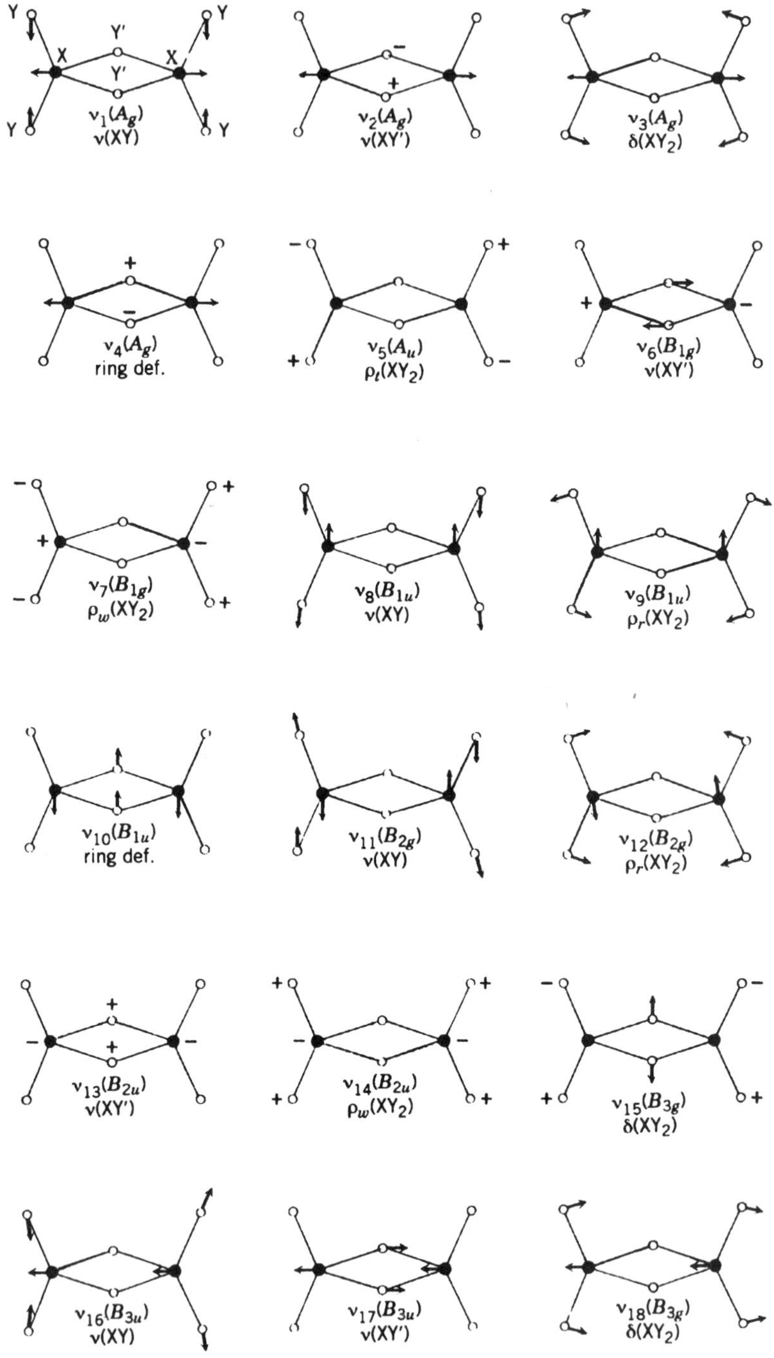

Fig. II-27. Normal modes of vibration of bridged X_2Y_6 molecules.[1080]

B_{3g} vibrations are Raman-active, whereas the B_{1u}, B_{2u}, and B_{3u} vibrations are infrared-active. Table II-10*a* lists the vibrational frequencies of molecules belonging to this type. In most compounds, the ν_1, ν_8, ν_{11}, and ν_{16} vibrations are largely due to the terminal XY_2 stretching motions, and their frequencies are higher than those of ν_2, ν_6, ν_{13}, and ν_{17}, which are mainly due to the vibrations of the bridging $X_2Y'_2$ group. As shown in Table II-10*a*, the ratios of the average terminal and bridging frequencies range from 0.74 to 0.61. Normal coordinate analyses of these compounds have been made by several investigators.[1085,1088,1090,1091]

It should be noted that In_2Cl_6 and In_2Br_6 are polymeric, and In_2I_6 is dimeric in the crystalline state, as shown by their spectra.[1092]

Table II-10*b* lists eight stretching frequencies of planar X_2Y_6 ions and molecules. In the $[M_2X_6]^{2-}$ (M = Pd and Pt; X = Cl, Br, and I) series, Goggin[1093] showed that the distinction between terminal and bridging vibrations is meaningless except for X = Cl, since these vibrations couple so strongly with each other. According to Forneris et al.[1094] the terminal and bridging stretching force constants of Au_2Cl_6 are 2.22 and 1.15 mdyn/Å, respectively, and those of I_2Cl_6 are 1.70 and 0.40 mdyn/Å, respectively (modified UBF). On the other hand, Adams and Churchill[1085] report values of 2.419 and 1.482 mdyn/Å, respectively, for the terminal and bridging force constants of Au_2Cl_6 (GVF).

(3) Singly Bridged X_2Y_6 Molecules

In the solid state, dichlorine hexaoxide, Cl_2O_6, exists as an ionic salt, $[ClO_2]^+$ $[ClO_4]^-$, and the vibrational frequencies of these ions have been listed in Tables II-2*b* and II-6*e*, respectively. In the gaseous state, Cl_2O_6 takes a singly bridged structure of $\mathbf{C}_s$ symmetry:

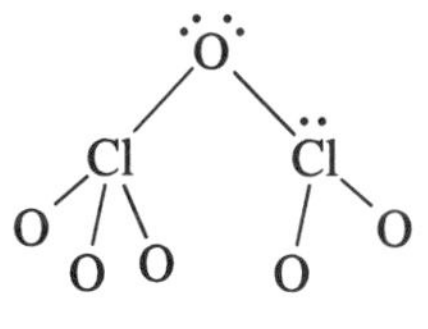

and its IR spectrum in inert gas matrices has been assigned.[1095]

(4) Ethane-Type X_2Y_6 Molecules ($\mathbf{D}_{3d}$)

The ethane-type X_2Y_6 molecule may be staggered ($\mathbf{D}_{3d}$), eclipsed ($\mathbf{D}_{3h}$), or *gauche* ($\mathbf{D}_3$). Figure II-28 shows the 12 normal modes of vibration of the staggered X_2Y_6 molecule. The A_{1g} and E_g vibrations are Raman-active, and the A_{2u} and E_u vibrations are infrared-active. Table II-10*c* lists the observed frequencies and band assignments based on $\mathbf{D}_{3d}$ symmetry. It should be noted that

TABLE II-10a. Vibrational Frequencies of Nonplanar Bridging X_2Y_6 Molecules (cm^{-1})

	B_2H_6	Al_2H_6	Al_2F_6	Al_2Cl_6	Al_2Br_6	Al_2I_6	Ga_2H_6	Ga_2Cl_6	Ga_2Br_6	Ga_2I_6	In_2I_6	Fe_2Cl_6
A_g	2526	2059	—	523	410	348	2082	413	291	229	187	422
	2096	1652	—	342	212	148	1571	318	204	143	134	305
	1187	816	—	219	142	95	792	167	119	85	69	150
	788	398	—	107	70	56	238	100	64	50	40	78
A_u	833	451	—	—	—	—	473	—	—	—	—	—
B_{1g}	1756	1492	—	284	354	—	1321	243	241	195	114	225
	860	532	—	166	82	82	374	125	85	64	55	112
B_{1u}	2613	2062	995	622	500	415	2075	464	347	273	228	467
	951	997	340	174	—	—	872	—	102	—	—	118
	367	249	—	—	—	—	239	—	—	—	—	24
B_{2g}	2597	2055	—	612	491	405	2068	462	339	265	232	450
	918	497	—	121	116	63	498	117	74	55	49	82
B_{2u}	1924	1350	600	422	341	291	1249	318	232	189	158	328
	974	694	—	135	90	64	693	114	82	61	49	99
B_{3g}	1023	844	—	—	—	54	821	215	158	68	44	80
B_{3u}	2518	2051	805	485	373	320	2074	390	269	213	178	406
	1615	1603	575	320	198	140	1400	282	188	134	125	280
	1175	766	300	144	110	81	729	156	90	77	59	116
ν_t/ν_b	0.72	0.74	—	0.61	0.62	—	0.67	0.67	0.69	0.67	0.65	0.65
Refs.	1081, 1082	1081	1083	1084	1085	1085	1081, 1086, 1087	1085, 1088	1085	1085	1085	462, 1089, 1090

TABLE II-10*b*. Vibrational Frequencies[a] of Planar Bridging X_2Y_6 Molecules (cm^{-1})

	$[Pd_2Cl_6]^{2-}$	$[Pd_2Br_6]^{2-}$	$[Pd_2I_6]^{2-}$	$[Pt_2Cl_6]^{2-}$	$[Pt_2Br_6]^{2-}$	$[Pt_2I_6]^{2-}$	I_2Cl_6	Au_2Cl_6
A_g								
$\nu_1, \nu(XY_t)$	346	262	219	349	241	196	344	379
$\nu_2, \nu(XY_b)$	302	194	143	316	211	160	198	328
B_{1g}								
$\nu_6, \nu(XY_t)$	328	253	—	333	238	196	314	366
$\nu_7, \nu(XY_b)$	265	173	130	294	193	145	142	289
B_{2u}								
$\nu_{12}, \nu(XY_t)$	335	257	218	330	236	196	327	364
$\nu_{13}, \nu(XY_b)$	262	178	—	300	192	147	170	309
B_{3u}								
$\nu_{16}, \nu(XY_t)$	343	264	218	341	239	196	340	374
$\nu_{17}, \nu(XY_b)$	297	192	140	312	210	157	205	309
Refs.	1093	1093	1093	1093	1093	1093	1094	1094

[a] XY_t and XY_b denote terminal and bridging XY stretching modes, respectively. When these two modes couple strongly, distinction between them is not clear (see Ref. 1093).

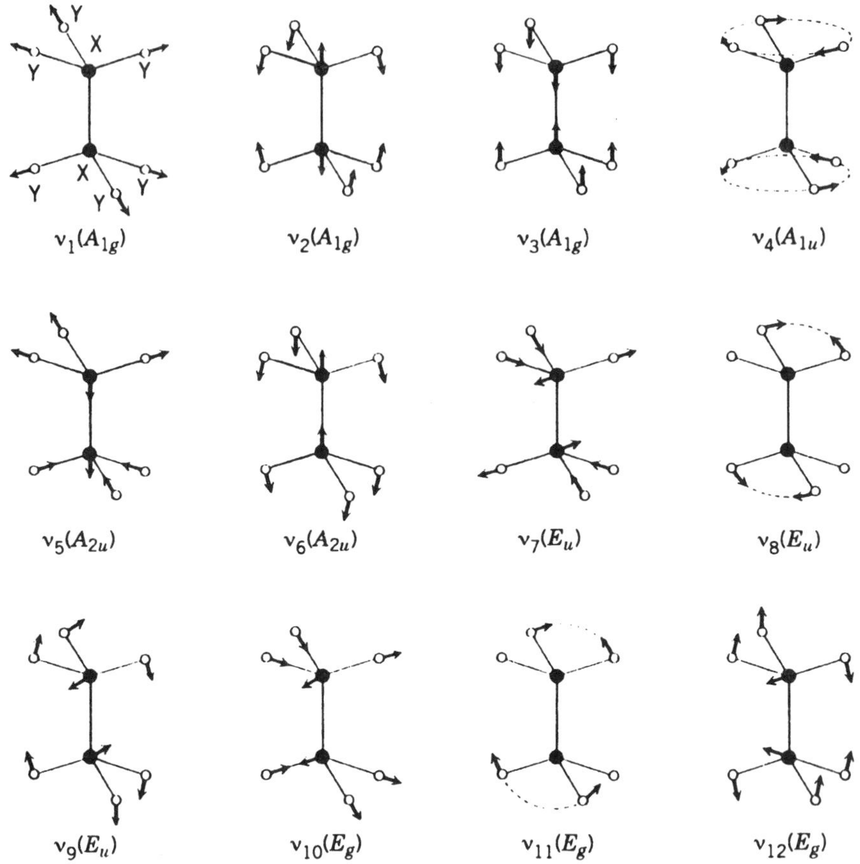

Fig. II-28. Normal modes of vibration of ethane-type X_2Y_6 molecules.

neutral Ga_2X_6 (X: a halogen) molecules take the bridging $\mathbf{D}_{2h}$ structure (Table II-10*a*), whereas $[Ga_2X_6]^{2-}$ ions take the ethane-like $\mathbf{D}_{3d}$ structure.

The structure of Si_2Cl_6 has been controversial; Griffiths[1107] prefers the $\mathbf{D}_{3h}$ or $\mathbf{D}'_{3h}$ structure* for liquid Si_2Cl_6, whereas Ozin[1108] favors the $\mathbf{D}_{3d}$ model for all phases. The $\mathbf{D}_{3h}$ and $\mathbf{D}_{3d}$ selection rules are similar except that the E_u modes of $\mathbf{D}_{3d}$ which are infrared-active become both infrared- and Raman-active (E') in $\mathbf{D}_{3h}$. The SiSi stretching mode was assigned at 354 cm^{-1} by Griffiths and at 627 cm^{-1} by Ozin. According to Höfler et al.,[1106] the SiSi stretching force constant increases in the order $Si_2H_6 < Si_2I_6 < Si_2Br_6 < Si_2Cl_6$. Normal coordinate analyses have also been made on Si_2Cl_6[1108] and $[Ga_2Cl_6]^{2-}$,[1104] The stretching frequencies of the $[In_2X_6]^{2-}$ ions (X = Cl, Br, and I) have been reported.[1109]

*If free rotation about the SiSi bond occurs, the symmetry plane (σ_h) is lost from $\mathbf{D}_{3h}$, and the molecule belongs to $\mathbf{D}'_{3h}$. Its selection rules are essentially the same as those of $\mathbf{D}_{3h}$.

TABLE II-10c. Vibrational Frequencies of Ethane-Type X_2Y_6 Molecules (cm^{-1})

	Molecule ($\mathbf{D}_{3d}$)														
	C_2H_6	Si_2H_6	Ge_2H_6	$N_2H_6^{2+}$	$P_2O_6^{4-}$	$P_2S_6^{4-}$	$P_2Se_6^{4-}$	$S_2O_6^{2-}$	$Ga_2Cl_6^{2-}$	$Ga_2Br_6^{2-}$	$Ga_2I_6^{2-}$	Si_2F_6	Si_2Cl_6	Si_2Br_6	Si_2I_6
A_{1g}															
$\nu_1, \nu(XY_3)$	2899	2152	(2070)	2650	1062	557	452	1102	375	316	285	910	351	223	154
$\nu_2, \delta(XY_3)$	(1375)	909	765	1524	670	197	116	710	106	70	42, 48	220	127	80	(51)
$\nu_3, \nu(XX)$	993	434	229	1027	275	374	217	293	233	164	118	541	624	562	510
A_{1u}															
$\nu_4, \rho_t(XY_3)$	275	—	144	455	—	—	—	—	—	—	—	—	—	—	—
A_{2u}															
$\nu_5, \nu(XY_3)$	2954	2154	2078	2600	942	444	440	1000	302	201	155	819	460	329	255
$\nu_6, \delta(XY_3)$	1379	844	755	1485	562	302	283	577	151	110	88	403	241	168	116
E_u															
$\nu_7, \nu(XY_3)$	2994	2179	2114	2739	1085	585/606	478	1240	327	237, 228	200	971	603	479	388
$\nu_8, \delta(XY_3)$	1486	940	898	1613	494	243	173	516	141	92	74	340	178	114	81
$\nu_9, \rho_r(XY_3)$	821	379	407	1096	200	87	132	204	89–66	64	(50)	203	74	50	(31)
E_g															
$\nu_{10}, \nu(XY_3)$	2963	2155	2150	2745	1168	578	474	1216	314	228	184	985	590	473	398
$\nu_{11}, \delta(XY_3)$	1460	929	875	1599	508	259	213	556	116	84	75	306	211	139	94
$\nu_{12}, \rho_r(XY_3)$	(1155)	625	417	1105	323	169	150	320	146	102	84	135	132	89	(53)
Refs.	1096	1097	1098	1099	1100	1101	1102	1103	1104	1104	1104	1105	1106	1106	1106

II-11. X_2Y_7, X_2Y_8, X_2Y_9, AND X_2Y_{10} MOLECULES

(1) X_2Y_7 Molecules

The X_2Y_7-(XY_3—Y—XY_3) type molecule belongs to the C_s, C_{2v}, or C_1 point group, depending on the relative orientation of the two XY_3 groups:

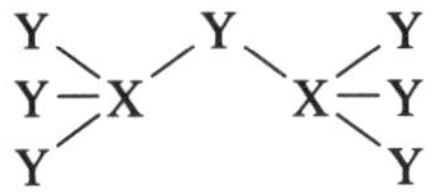

Seventeen vibrations are infrared-active in C_{2v}, while all 21 vibrations are infrared-active in C_s and C_1 symmetry. The 21 normal vibrations of the X_2Y_7 molecule may be classified into in-phase and out-of-phase coupling motions of terminal XY_3 group vibrations and the skeletal vibrations of the XYX bridge. Table II-11 lists the observed frequencies of these bridging vibrations.

On the basis of normal coordinate analyses, Brown and Ross[1111] have made complete assignments of the vibrational spectra of the $S_2O_7^{2-}$, $Se_2O_7^{2-}$, $V_2O_7^{4-}$, and $Cr_2O_7^{2-}$ ions (C_{2v} symmetry).

According to Wing and Callahan,[1117] the MOM bridge angle is always larger than 115°, and ν_a(MO) is at least 215 cm^{-1} higher than ν_s(MO). This separation increases as the MOM angle increases. In a series of $M_2^{2+}(V_2O_7)$ where M is Mg^{2+}, Ba^{2+}, Ni^{2+}, Co^{2+}, and so on, a linear relationship has been found between the V—O—V angle (α) and the Δ value defined below:[1118]

$$\Delta = -38.79 + 0.39\alpha$$

TABLE II-11 YXY Bridging Frequencies of X_2Y_7 Molecules (cm^{-1})

Compound	$\nu_a(YX_2)$	$\nu_s(YX_2)$	$\delta(YX_2)$	Refs
$Na_4[P_2O_7]$	915	730	—	1110
$Na_4[As_2O_7]$	735	550	245	1110
$Na_2[S_2O_7]$	825	725	182 or 116	1111
$Na_2[Se_2O_7]$	707	556	—	1111
$Na_4[V_2O_7]$	710	533	200	1111
$Na_2[Cr_2O_7]$	770	565, 554	220	1111
Re_2O_7	804	456	50	1112
Cl_2O_7	785	700	(165)	1113
$[Ga_2Cl_7]^-$	286	276	—	1114
$[Ga_2Br_7]^-$	222	195	—	1114
$[Si_2O_7]^{6-}$	1014	669	415	1115
$[Ge_2O_7]^{6-}$	896	520	323	1115
				1116

where

$$\Delta = [(\nu_a - \nu_s)(\nu_a + \nu_s)] \times 100$$

The molecule Re_2O_7 is monomeric in the gaseous and liquid states and polymeric in the solid state. Beattie and Ozin[1112] assigned the spectra of gaseous Re_2O_7 on the basis of $\mathbf{C}_{2v}$ symmetry. Vibrational analysis of Cl_2O_7 has been made by assuming $\mathbf{C}_2$[1119] or $\mathbf{C}_{2v}$[1113] symmetry. On the assumption of a linear OPO bridge, the vibrational assignments of divalent metal pyrophosphates ($M_2P_2O_7$) have been made in terms of $\mathbf{D}_{3h}$ or $\mathbf{D}_{3d}$ symmetry.[1120,1121] The vibrational spectra of the $Si_2O_7^{6-}$ and $Ge_2O_7^{6-}$ ions have also been assigned on the basis of $\mathbf{D}_{3d}$ symmetry.[1115] In contrast to the $[Ga_2Cl_7]^-$ ion, the Ga—I—Ga bond of the $[Ga_2I_7]^-$ ion in the molten and crystalline phases is linear.[1122]

(2) X_2Y_8 Molecules

The symmetry of the X_2Y_8 (Y_3X—Y—Y—XY_3) molecule may be low enough to activate all 24 normal vibrations in both infrared and Raman spectra. Thus far, the XYYX bridging frequencies have been assigned at 988 [ν_a(XYYX)], 784 [ν_s(XYYX)], 890 [ν(YY)], and 397 and 328 [δ(XYYX)] for the $[P_2O_8]^{4-}$ ion, and at 1062, 834, 854, and 328 and 236 cm^{-1}, respectively, for the $[S_2O_8]^{4-}$ ion.[1123]

In another type of X_2Y_8 ion, short, multiple M–M (metal–metal) bonds link two MX_4 units so that the overall symmetry becomes $\mathbf{D}_{4h}$ (eclipsed form).[1124] In Raman spectra, these metal–metal bonded compounds exhibit strong ν(M–M) vibrations in the low-frequency region because displacements of heavy metal ions which are linked by strong, covalent bonds produce large changes in polarizability. Furthermore, these compounds, under resonance conditions, exhibit a series of overtones of ν(M–M), ν(M–X) (X: a halogen) and combination bands from which one can calculate frequencies corrected for anharmonicity (ω_e) and anharmonicity constants (X_{11}) (Sec. I-23). Thus, RR spectra of the M_2X_8-type ions have been studied extensively: $[Mo_2Cl_8]^{4-}$,[1125,1125a] $[Re_2F_8]^{2-}$,[1126,1127] $[Re_2Cl_8]^{2-}$,[1128,1127] $[Re_2Br_8]^{2-}$,[1128,1127] $[Re_2I_8]^{2-}$,[1129] $[Te_2Cl_8]^{2-}$,[1130] $[Te_2Cl_8]^{3-}$,[1130] and $[Te_2Br_8]^{2-}$,[1130]. As an example, Fig. II-29 shows the RR spectra of the $[Re_2F_8]^{2-}$ ion obtained by Peters and Preetz.[1126] This ion exhibits an electronic absorption near 558 nm which is due to a transition from the $(\sigma)^2(\pi)^4(\delta)^2$ to the $(\sigma)^2(\pi)^4(\delta)(\delta*)$ state. Thus, the RR spectrum was measured using the 530.9-nm line of a Kr–ion laser. It is seen that the ν(Re–Re) at 320 cm^{-1} is markedly enhanced, and a series of overtones ($n\nu_1$, up to $n = 6$) and combination bands [ν_2 is the ν(Re–F) and the 182 cm^{-1} band is a skeletal bending] are observed. Using these data, the ω_e and X_{11} values were calculated to be 319.6 ± 0.6 cm^{-1} and 0.45 ± 0.05 cm^{-1},

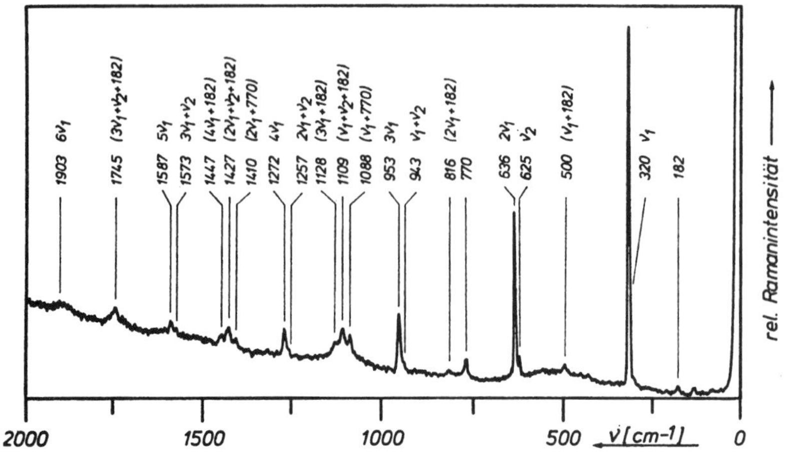

Fig. II-29. Resonance Raman spectrum of $[(n\text{-}C_4H_9)_4N]_2$ $(Re_2F_8)\cdot 4H_2O$ (530.9 nm excitation) (reproduced with permission from Ref. 1126).

respectively. In the $[Re_2X_8]^{2-}$ series, the ν(Re–Re) frequency shifts somewhat irregularly when the halogen (X) is changed; F (320 cm^{-1}), Cl (272 cm^{-1}), Br (277 cm^{-1}), and I (258 cm^{-1}). It is interesting to note that, under extremely high pressure (15 GPa), the ν(Re–Re) of the $[Re_2Cl_8]^{2-}$ ion is shifted to 357 cm^{-1} owing to shortening of the Re–Re bond.[1131]

Finally, it is possible to measure the ν(M–M) in the electronic excited state by using time-resolved resonance Raman (TR^3) spectroscopy (Ref. 54, Section I). For example, the absorption maximum of the $\delta \rightarrow \delta*$ transition of the $[Re_2Cl_8]^{2-}$ ion in CH_2Cl_2 is at 685 nm. Dallinger[1132] and co-workers[1133] saturated the $\delta\delta*$ ($^1A_{2u}$) state (lifetime, ~ 75 ns) by using a pump laser at 640 nm (5-ns pulses). The ν(Re–Re) of this electronic excited state was probed by using delayed 416-nm excitation. Three bands observed at 366, 204, and 138 cm^{-1} were assigned to ν(Re–Cl), ν(Re–Re), and δ(Cl–Re–Re), respectively. This result indicates that ν(Re–Re) becomes lower in going from the ground (δ^2, ^1A_g, 272 cm^{-1}) to the excited state ($\delta\delta*$, $^1A_{2u}$, 204 cm^{-1}) because an electron is promoted from a bonding (δ) to an antibonding ($\delta*$) orbital. In other cases, the ν(M–M) band is shifted to a higher frequency by electronic excitation because an electron is promoted from an antibonding to a bonding orbital. Examples of the latter are discussed in Part B (Coordination Compounds).

(3) X_2Y_9 Molecules

The X_2Y_9 molecule shown below belongs to the point group $\mathbf{D}_{3h}$, and its 27 normal vibrations are classified into 4 A_1'(R), A''(i.a.), A_2'(i.a.), 3 A_2''(IR), 5 E'(IR, R), and 4 E''(R). Two XY_t stretching (A_2'' and E') and two XY_b stretching (A_2'' and E') vibrations are infrared-active. Similarly, two XY_t stretching (A_1' and

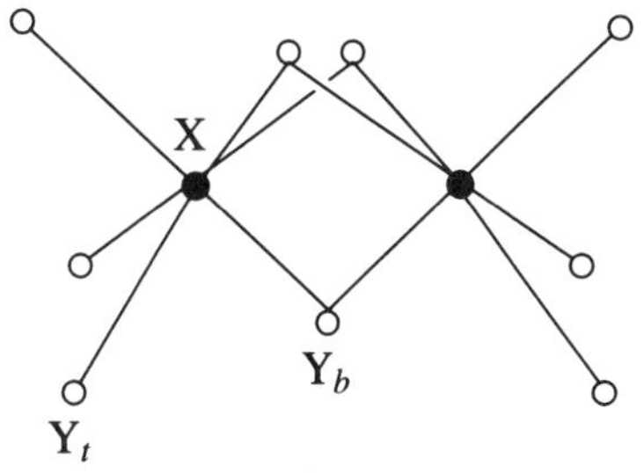

E'') and two XY_b stretching (A_1' and E'') are Raman-active. Table II-12 lists observed frequencies of these terminal and bridging stretching vibrations. Normal coordinate treatments have been made on this type of compound by several groups of investigators.[1135–1139] The results show that there is mixing of terminal and bridging, as well as stretching and bending, modes. Thus, the assignments given in the table are only approximate. The strength of the terminal and bridging bonds may be compared in terms of their stretching force constants. In the case of the $[Pt_2Br_9]^-$ ion, the Pt-Br_t and Pt-Br_b stretching force constants are 1.55 and 0.93 mdyn/Å, respectively.[1138] The IR and Raman spectra of the mixed-halogeno complex, $[Cr_2Cl_6Br_3]^{3-}$, can be assigned in terms of those of the $[Cr_2Cl_9]^{3-}$ and $[Cr_2Br_9]^{3-}$ ions.[1140]

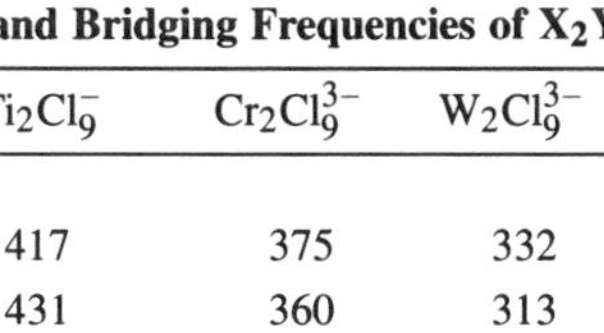

(C_{3v})

TABLE II-12. Terminal and Bridging Frequencies of X_2Y_9-Type Ions (cm^{-1})

Ion	$Ti_2Cl_9^-$	$Cr_2Cl_9^{3-}$	$W_2Cl_9^{3-}$	$Ir_2Cl_9^{3-}$	$Ir_2Br_9^{3-}$	$Pt_2Br_9^-$
Terminal stretch						
ν_1, A_1' (R)	417	375	332	340	232	253
ν_7, A_2'' (IR)	431	360	313	318	220	249
ν_{10}, E' (IR, R)	378	342	281	317	215	243
ν_{15}, E'' (R)	399, 390	320	294	298	—	(219)
Bridging stretch						
ν_2, A_1' (R)	321	280	257	280	211	211
ν_8, A_2'' (IR)	270	261	—	287	186	183
ν_{11}, E' (IR, R)	236	234	209	266	189	190
ν_{18}, E'' (R)	305	222	226	247	—	—
Refs	1134	1135, 1136	1135	1137	1137	1138

(4) X_2Y_{10} Molecules

The X_2Y_{10} molecule may take either one of the following structures:

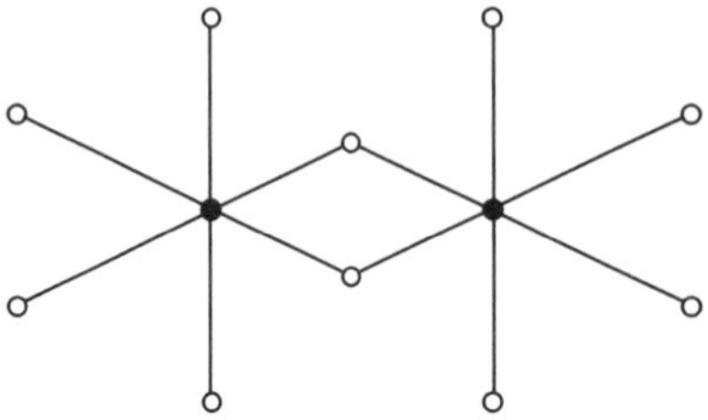

$\mathbf{D}_{2h}$

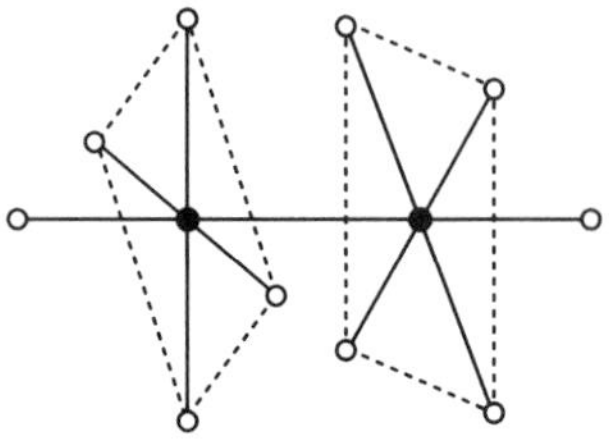

$\mathbf{D}_{4d}$

The 30 normal vibrations of the $\mathbf{D}_{2h}$ molecule are classified into $6A_g$, $2A_u$, $4B_{1g}$, $4B_{1u}$, $3B_{2g}$, $4B_{2u}$, $2B_{3g}$, and $5B_{3u}$, of which 13 vibrations (B_{1u}, B_{2u}, and B_{3u}) are infrared-active, and 15 vibrations ($6A_g + 4B_{1g} + 3B_{2g} + 2B_{3g}$) are Raman-active. These include four terminal ($B_{1u} + B_{2u} + 2B_{3u}$) and two bridging ($B_{2u} + B_{3u}$) vibrations which are IR-active, and five terminal ($2A_g + 2B_{1g} + B_{2g}$) and one bridging (A_g) vibrations which are Raman-active. Vibrational assignments have been reported for: Nb_2Cl_{10}[1141], Re_2Cl_{10}[1141], Os_2Cl_{10}[1142], $[Ti_2Cl_{10}]^{2-}$,[1134], $[Re_2Cl_{10}]^{2-}$,[1143], and $[Re_2Br_{10}]^{2-}$,[1143]. For example, the terminal and bridging vibrations of the $[Re_2Cl_{10}]^{2-}$ ion are found in the 367–321 and 278–250 cm^{-1} regions, respectively, although these modes may be mixed to some extent. Beattie et al.[1139] carried out normal coordinate analyses to assign the vibrational spectra of Nb_2X_{10} and Ta_2X_{10} (X = Cl and Br) based on $\mathbf{D}_{2h}$ symmetry.

The symmetry of the single-bridge XY_5—XY_5 type molecule may be $\mathbf{D}_{4d}$ (staggered) or $\mathbf{D}_{4h}$ (eclipsed), and these two structures cannot be distinguished by the selection rules. Under $\mathbf{D}_{4d}$ symmetry, the 30 normal vibrations are classified into $4A_1$, B_1, $3B_2$, $4E_1$, $3E_2$, and $4E_3$ species, of which B_2 and E_1 are infrared-active and A_1, E_2, and E_3 are Raman-active. Jones and Ekberg[1144] made complete assignments of infrared and Raman spectra of S_2F_{10} vapor based on $\mathbf{D}_{4d}$ symmetry.

II-12. METAL CLUSTER COMPOUNDS

Heavy metals such as Mo, W, Nb, and Ta form octahedral metal clusters. Figure II-30*a* shows the structure of the $[(M_6X_8)Y_6]^{2-}$ ion where M is Mo or W, X is a bridging halogen (Cl, Br), and Y is a terminal halogen (Cl, Br, and I). Under $\mathbf{O}_h$ symmetry, the 54 ($3 \times 20 - 6$) normal vibrations of this ion are classified into $3A_{1g} + 3E_g + 2F_{1g} + 4F_{2g} + A_{2u} + E_u + 5F_{1u} + 3F_{2u}$, of which 10 vibrations (A_{1g}, E_g, and F_{2g}) are Raman-active and 5 vibrations (F_{1u}) are IR-active. As to the M_6 skeleton, two ν(M–M) are Raman-active (A_{1g} and E_g) and one ν(M–M) is IR-active (F_{1u}). It is rather difficult to assign the $\nu(\text{M--M})$ empirically since strong coupling is expected among $\nu(\text{M--M})$, $\nu(\text{M--X})$, and $\nu(\text{M--Y})$ which have the same symmetry and similar frequencies.[1145] According to normal coordinate analysis by Mattes,[1146] the Mo—Mo stretching force constant is 0.3 mdyn/Å. In IR spectra, the $\nu(\text{Mo--Mo})$ vibrations are assigned at 99 and 85 cm^{-1}, respectively, for X = Cl and Br, whereas in Raman spectra, they are located in the 108–100 cm^{-1} region.[1147]

Figure II-30b shows the structure of the $[(\text{M}_6\text{X}_{12})\text{Y}_6]^{n-}$ ion where M is Nb and Ta and X and Y are bridging and terminal halogens, respectively. Under $\mathbf{O}_h$

(a) $(\text{M}_6\,\text{X}_8)\text{Y}_6^{2-}$

(b) $(\text{M}_6\,\text{X}_{12})\text{Y}_6^{n-}$

(c) $[\text{M}_6\text{O}_{19}]^{8-}$

(d) $[(\text{Pb}_6\text{O}(\text{OH})_6]^{4+}$

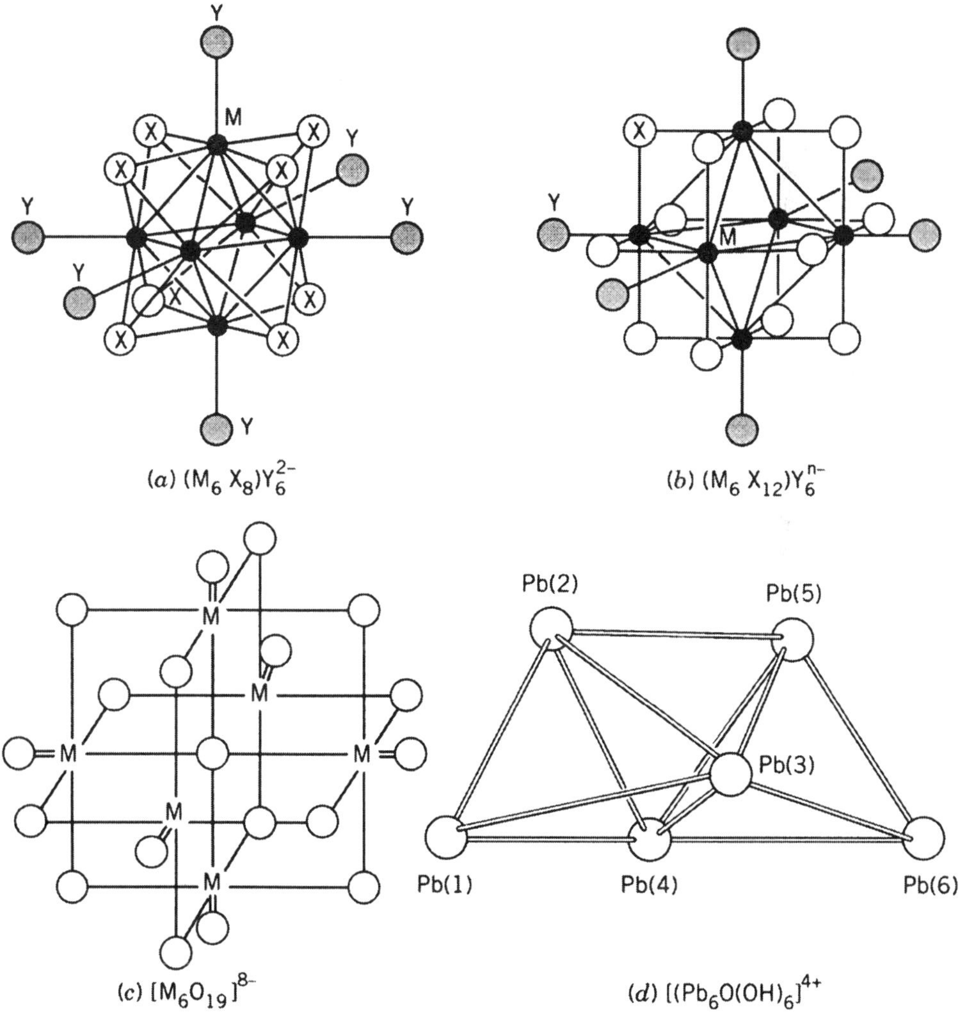

Fig. II-30. Structures of metal cluster compounds.

symmetry, the 66 (3 × 24 − 6) normal vibrations of such an ion are grouped into $3A_{1g} + A_{2g} + 4E_g + 3F_{1g} + 4F_{2g} + A_{2u} + E_u + 6F_{1u} + 4F_{2u}$. Again, two ν(M–M) are Raman-active (A_{1g} and E_g) and one ν(M–M) is IR-active (F_{1u}) for the M_6 skeleton. The IR spectra of these and similar metal clusters have been reported by several investigators.[1148–1152] The ν(Nb–Nb) of the Nb_6Cl_{12} cluster has been assigned empirically at 140 cm.$^{-1}$,[1149] However, it may be mixed with other modes.[1150] Mattes[1151] carried out normal coordinate analysis on the $(M_6X_{12})Y_n$ system (M = Nb, Ta; $n = 2 \sim 4$), and found their M–M stretching force constants to be less than 0.3 mdyn/Å. These ν(M–M) vibrations may be too low to be observed.

Figure II-30*c* shows the structure of the $[M_6O_{19}]^{8-}$ ion (M = Nb, Ta). Under $\mathbf{O}_h$ symmetry, this ion has 7 IR-active vibrations of F_{1u} symmetry and 11 Raman-active vibrations which are grouped into $3A_{1g}$, $4E_g$, and $4F_{2g}$. According to the results of normal coordinate analysis by Farrell et al.,[1153] the stretching force constants associated with three types of the Nb–O bonds are

K(Nb–O)			
5.66	2.92	0.91	*K*(Nb–Nb)
terminal	bridging	central	1.01 mdyn/Å

The ratio of the first three force constants is about 6:3:1. Although the NbNb stretching constant was estimated to be ca. 1 mdyn/Å, this value does not represent the strength of this bond, since such a value can be obtained without any M–M interaction. Rather, these workers suggest the absence of Nb–Nb interactions for d_0 [Nb(V)] ions because the M–M bond breathing mode which normally appears strongly in Raman spectra is weak. Mattes et al.[1154] also carried out normal coordinate analysis on the same system and obtained a ratio of 8:4:1 for the three M–O stretching force constants mentioned above.

The Raman spectrum of crystals having the composition $Pb_3(OH)_4(ClO_4)_2$ was originally interpreted in terms of an octahedral Pb_6 cluster.[1155] However, later X-ray analysis revealed the presence of the $[Pb_6O(OH)_6]^{4+}$ ion having the very unusual structure shown in Fig. II-30*d*.[1156] In this structure, three tetrahedra of Pb atoms share faces, the middle tetrahedron has an 0 atom (oxide) near the center while the OH groups lie on the faces of the two end tetrahedra. There are 12 Pb–Pb interactions, and their Pb–Pb distances range from 3.44 to 4.00 Å. Figure II-31 shows the low-frequency Raman spectrum in the region where the Pb–Pb vibrations appear.[1157] Under the idealized symmetry of $\mathbf{C}_2$, 12 cluster modes are expected. The strongest band at 150 cm^{-1} was assigned to the $\nu(Pb_3—Pb_4)$ since it is the shortest bond (3.44 Å). The remaining cluster modes were assigned based on crude normal coordinate calculations. This spectrum is unique because it shows an unusually large number of cluster vibrations.

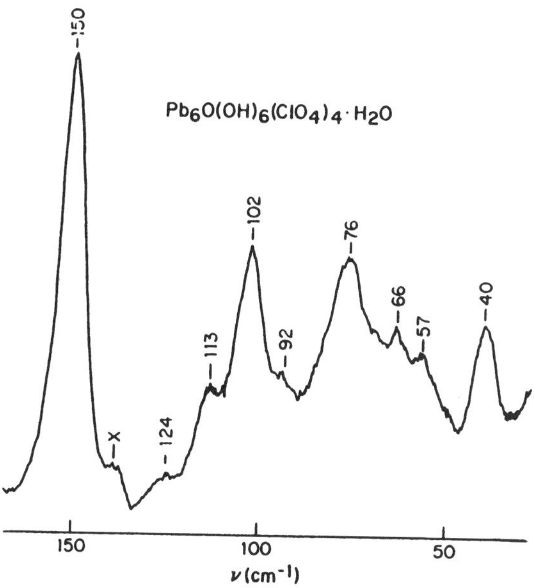

Fig. II-31. Low-frequency Raman spectrum of a single crystal of $Pb_6O(OH)_6(ClO_4)_4H_2O$ at −110°C with 632.8-nm excitation (reproduced with permission from Ref. 1157).

II-13. COMPOUNDS OF BORON

Boron compounds assume a wide variety of structures whose symmetries range from $\mathbf{C}_s$ to $\mathbf{I}_h$. Some examples are shown in Fig. II-32. Because of this, vibrational spectra of boron compounds have been studied extensively. The complications introduced by the natural abundances of ^{10}B (19.8%) and ^{11}B (80.2%) find use in making band assignments and in refining force constants for normal coordinate calculations. In the following, we list references for each group of compounds which have not been discussed in the preceding sections, and discuss band assignments for several selected compounds.

Boron oxides and sulfides B_2O_2 and B_2O_3[1158] B_2S_3,[1159] $B_3H_3O_3$ (boroxine),[1160] H_3BO_3 (boric acid),[1160a] $B_3O_3(OH)_3$ (metaboric acid),[1161] $H_2B_2O_3$,[1162] $CsBO_2$,[1163] $B_3O_6^{3-}$,[1164,1165] $B_2O_5^{4-}$, and $B_3O_7^{5-}$,[1166].

Boron halides B_4Cl_4[1167] $B_6X_6^{2-}$,[1168–1170] $B_6X_nY_{6-n}^{2-}$ (X, Y = Cl, Br, I),[1171] and $B_{12}X_{12}^{2-}$,[1172,1173]. Figure II-33 illustrates the IR and Raman spectra of the $B_6X_6^{2-}$ ions (X = Cl, Br, and I) obtained by Preetz and Fritze.[1169] Under $\mathbf{O}_h$ symmetry, the 30 (3 × 12 − 6) normal vibrations expected for these ions are classified into $2A_{1g} + 2E_g + F_{1g} + 2F_{2g} + 3F_{1u} + 2F_{2u}$. These vibrations can be subdivided into

B–B vibrations: $A_{1g}(\nu_1) + E_g(\nu_3) + F_{2g}(\nu_5) + F_{1u}(\nu_7) + F_{2u}$

B–X vibrations: $A_{1g}(\nu_2) + E_g(\nu_4) + F_{1g} + F_{2g}(\nu_6) + 2F_{1u}(\nu_8, \nu_9) + F_{2u}$

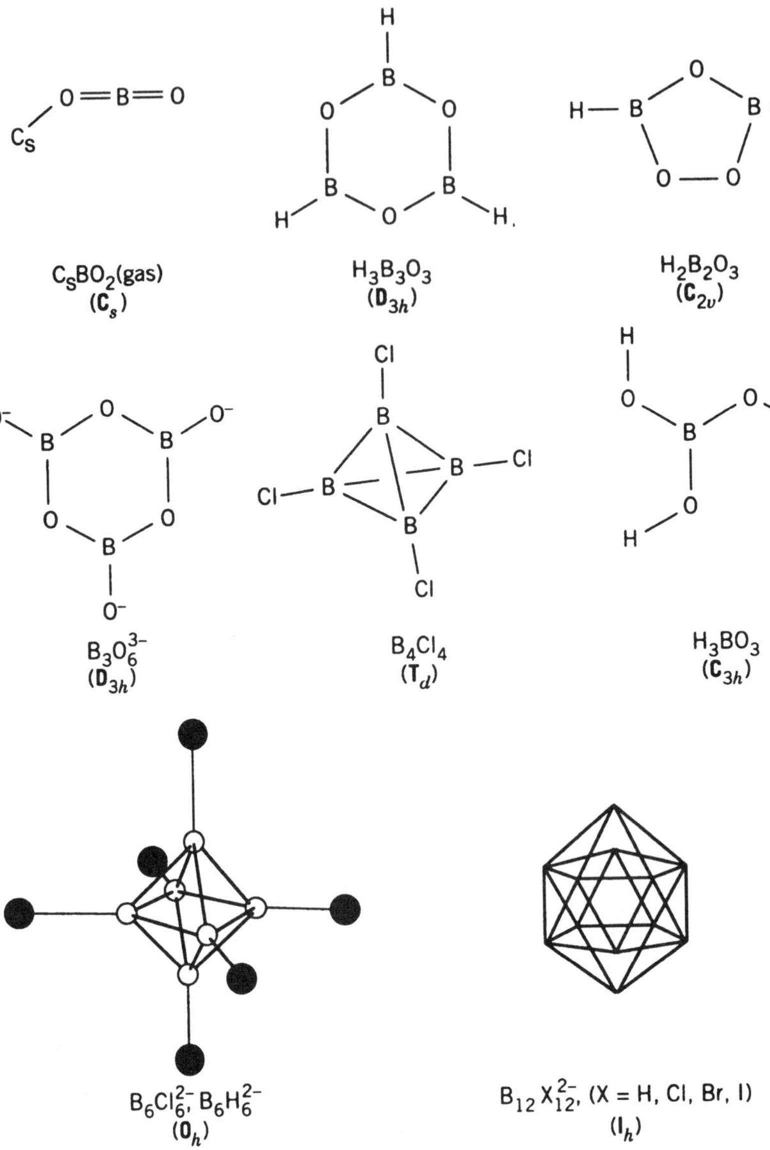

Fig. II-32. Structures of some boron oxides and halides.

The A_{1g}, E_g, and F_{2g} vibrations are Raman-active, whereas the F_{1u} vibrations are IR-active. As seen in Fig. II-33, the B–B vibrations are relatively insensitive, whereas the B–X vibrations are sensitive to halogen substitution.

The $B_{12}X_{12}^{2-}$ (X = Cl, Br, and I) ion (Fig. II-32) belongs to the highest symmetry ($\mathbf{I}_h$) point group, which consists of 120 symmetry operations. The 66 (3

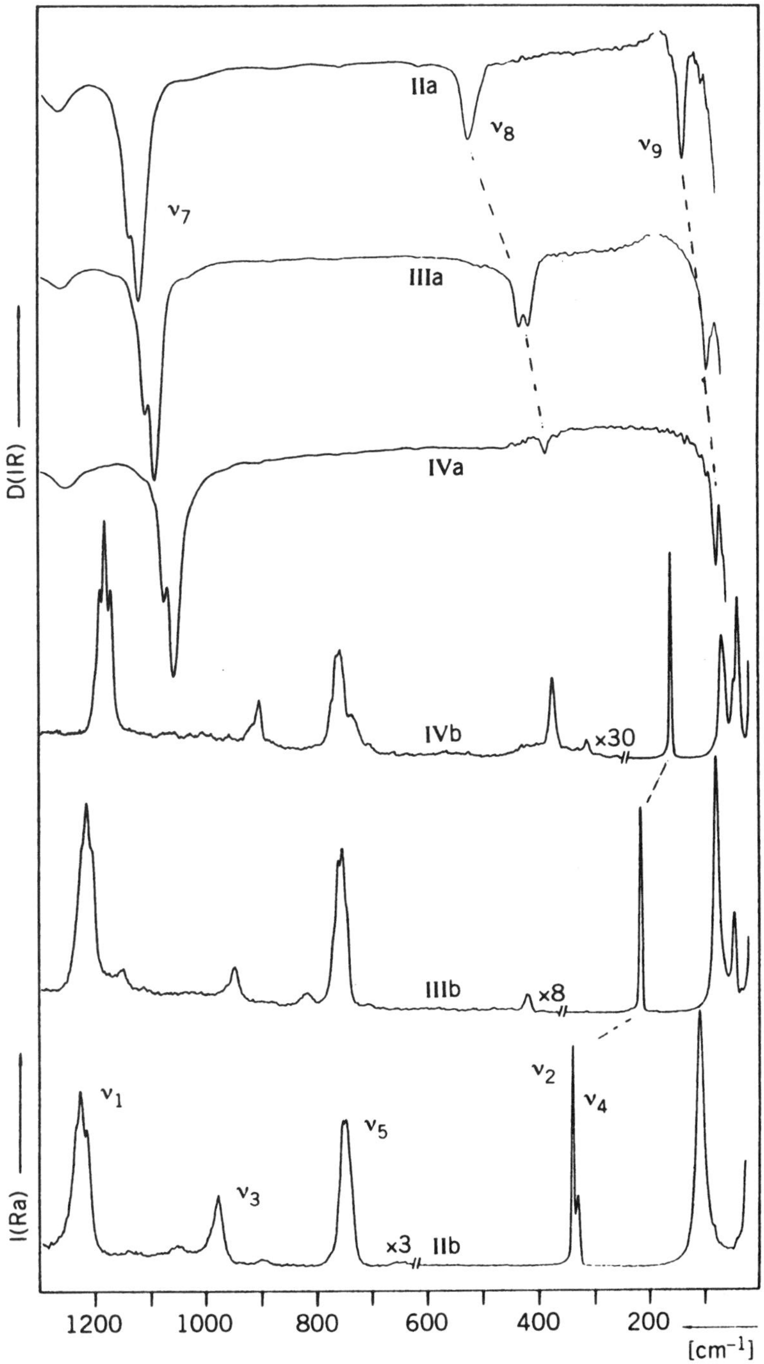

Fig. II-33. (*a*) IR and (*b*) Raman spectra of solid $Cs_2B_6Cl_6$ (II), $Cs_2B_6Br_6$ (III), and $Cs_2B_6I_6$ (IV) (reproduced with permission from Ref. 1169).

$\times$ 24 − 6) normal vibrations are classified into $2A_g + F_{1g} + 2G_g + 4H_g + 3F_{1u} + 2F_{2u} + 2G_u + 2H_u$. Here, G and H represent the four- and fivefold degenerate species, respectively. Owing to the high symmetry, its IR and Raman spectra are surprisingly simple; only three F_{1u} vibrations are IR-active, and only two A_g (polarized) and four H_g (depolarized) vibrations are Raman-active. Leites et al.[1172] reported the IR and Raman spectra together with empirical assignments, and Cyvin et al.[1173] carried out normal coordinate calculations on these ions.

Boron hydrides (Boranes) and their ions B_5H_9[1174], $B_{10}H_{14}$[1175], $B_2H_7^-$[1176], $B_6H_6^{2-}$[1177,1178], $B_6X_nH_{6-n}^{2-}$[1179], and $B_{12}H_{12}^{2-}$[1172,1173]. As is shown in Fig. II-34, the five boron atoms of B_5H_9 form a square-pyramidal skeleton, with four hydrogen atoms bridging four base boron atoms while five hydrogen atoms are bonded terminally to each boron atom, so that the symmetry of the whole molecule is C_{4v}. Complete band assignments have been made by Kalasinsky.[1174]

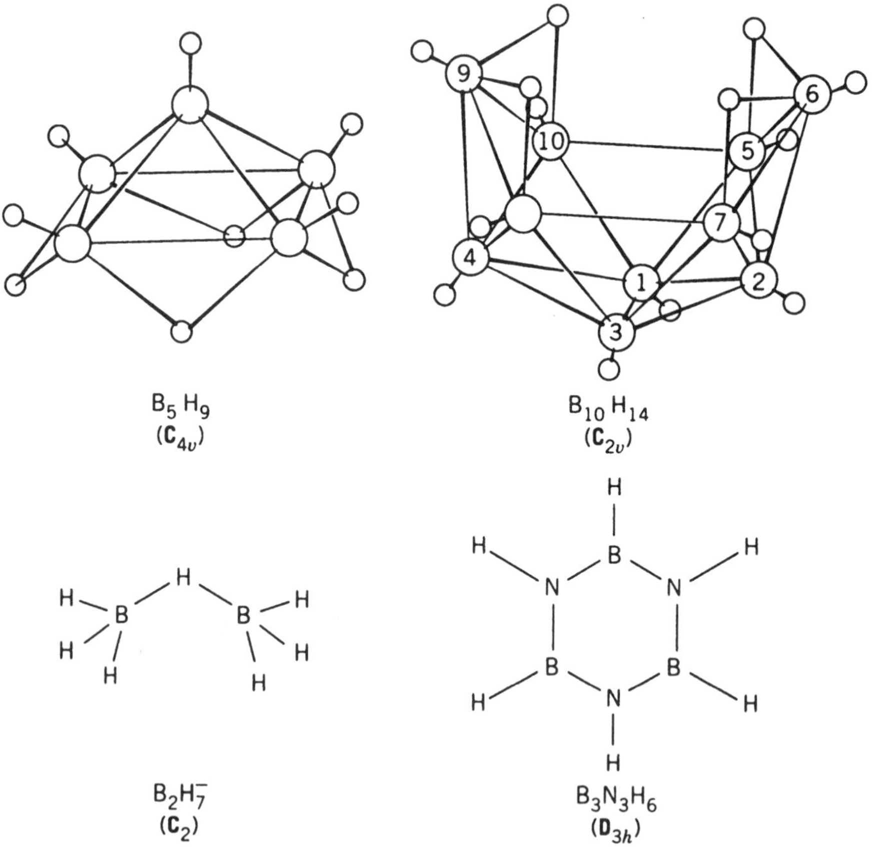

Fig. II-34. Structures of some boranes and borazine.

The structure of the $B_{12}H(D)_{12}^{2-}$ ion is the same as that of the $B_{12}X_{12}^{2-}$ ion discussed previously. For the spectra and band assignments, see Refs. 1172 and 1173.

Borazines $B_3N_3H_6$[1160,1180] and $B_3N_3H_3Cl_3$[1181]. Both molecules take planar ring structures of $\mathbf{D}_{3h}$ symmetry (Fig. II-34), and their 30 ($3 \times 12 - 6$) vibrations are classified into $4A_1'$(R) + $3A_2'$ (inactive) + $7E'$(IR, R) + $3E''$(R) + $3A_2''$(IR). Figure II-35 shows the matrix-isolated IR spectra of borazine obtained by Kaldor and Porter.[1160] Complete assignments based on normal coordinate calculations are found in the references given above.

Borohydride ion as a ligand As shown in Fig. II-36, the BH_4^- ion may coordinate to a metal, either as a bidentate or a tridentate via hydrogen bonding. For example, the Zr atom in $Zr(BH_4)_4$ is 12-coordinate because all of the BH_4^- ions are triply bridging,[1182] whereas the U atom in polymeric $U(BH_4)_4$ in the solid state is 14-coordinate because it is coordinated by two triply bridging (*cis*) and four doubly bridging BH_4^- ions.[1183] In these cases, the terminal and bridging ν(BH) bands are located in the 2600–2500 and 2200–2100 cm^{-1} regions, respectively.[1182–1185]

Several review articles are available on the vibrational spectra of boron compounds. Lehmann and Shapiro[1186] reviewed ν(BH), $\delta(BH_2)$, ν(BC), ν(BB),[1186] etc. of alkylboranes, and Bellamy et al.[1187] discussed ν(BN), ν(BH), ν(BX), ν(BC), etc. Meller[1188] lists ν(BN), ν(BH), ν(BX), and ν(BC) of a number of organoboron–nitrogen compounds. For vibrational spectra of boron compounds containing the B–P bond, see a review by Verkade.[1189] The group frequency charts shown in Appendix VIII include ν(BH), ν(BO), and ν(BX).

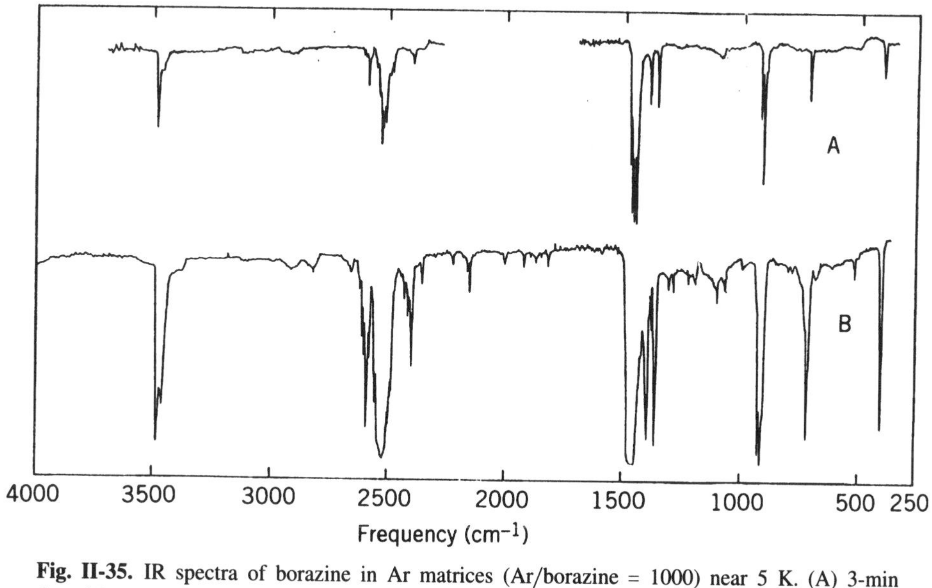

Fig. II-35. IR spectra of borazine in Ar matrices (Ar/borazine = 1000) near 5 K. (A) 3-min deposition; (B) 30-min deposition (reproduced with permission from Ref. 1160).

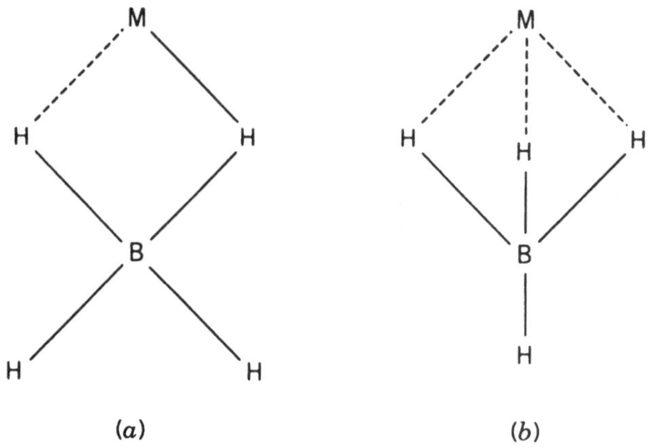

Fig. II-36. Modes of coordination of the BH_4^- ion. (*a*) Doubly bridging and (*b*) triply bridging.

II-14. COMPOUNDS OF CARBON

(1) Small Carbon Clusters

Vibrational studies on carbon clusters are highly important in interstellar and combustion research. An extensive review article on small carbon clusters (C_n, $n = 2 \sim 10$) is available.[1190] Table II-13 lists the structures and IR (antisymmetric stretching) frequencies observed in the gaseous phase and/or in inert gas matrices. It is seen that C_3, C_4, C_5, C_6, C_7, and C_9 take linear structures. Although not much is known about C_8 and C_{10}, theory predicts that cyclic structures are more favorable than linear structures for these even n clusters. Theory

TABLE II-13. Structures and IR Frequencies of Small Carbon Clusters

C_n	Structure	State	IR Frequency (cm^{-1})	Refs.
C_3	Linear	N_2 matrix	2042	1191, 1192
		Gas	63.42 (bending)	1193
C_4	Linear	Ar matrix	1543.4	1194, 1195
		Gas	1548.94	1196
C_5	Linear	Ar matrix	2164	1197, 1198
		Gas	2169.44	1199, 1200
		Ar matrix	1446.6 (ν_4)	1200a
C_5^+	Cyclicplanar	Ar matrix	2052	1197, 1201
C_6	Linear	Ar matrix	1952.0	1201a
			1197.3	
C_7	Linear	Gas	2138.3	1202, 1203
C_8	?	Ar matrix	1998	1198, 1204
C_9	Linear	Ar matrix	2128	1197
		Gas	2014.3	1205
C_{13}	Linear	Gas	1809	1206

also predicts that the $C_{10} \sim C_{20}$ clusters favor planar cyclic structures. However, recent work by Giesen et al.[1206] has shown that the C_{13} cluster is linear with the IR frequency near 1809 cm^{-1}.

Vala et al.[1197] confirmed the linear structure of the C_5 cluster by using isotope scrambling techniques. These workers trapped the C_5 cluster in Ar matrices (12 K) by evaporating graphite. As is seen in Fig. II-37 (lower part), a single IR peak is observed at 2164 cm^{-1} when a ^{12}C graphite is used (the 2128 cm^{-1} peak is due to $^{12}C_9$ contamination). However, this peak splits into 20 peaks (upper part) when a graphite sample consisting of a 1:1 mixture of ^{12}C and ^{13}C is used. This result provides definitive evidence for the linear structure since the number of possible isotopomers are 20, 12, and 8 for the linear, trigonal-bipyramidal, and cyclic structures, respectively. On the other hand, the C_5^+ ion takes a planar, cyclic structure ($\mathbf{D}_{5h}$) because its IR peak at 2052 cm^{-1} splits into 8 peaks when similar isotope scrambling experiments are carried out.[1201]

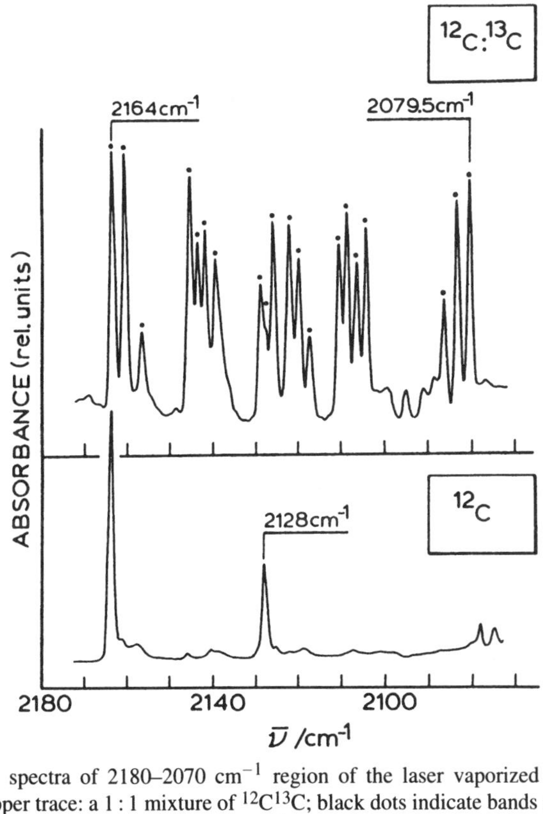

Fig. II-37. FTIR spectra of 2180–2070 cm^{-1} region of the laser vaporized graphite in an Ar matrix (12 K). Upper trace: a 1 : 1 mixture of $^{12}C^{13}C$; black dots indicate bands due to isotopomers of linear C_5; the bands at 2164 and 2079.5 cm^{-1} are due to $^{12}C_5$ and $^{13}C_5$, respectively. Lower trace: pure ^{12}C graphite (the band at 2128 cm^{-1} is due to $^{12}C_9$) (reproduced with permission from Ref. 1197).

(2) Large Carbon Clusters

Recently, C_{60} (Buckminsterfullerene) has attracted considerable attention because of its "soccer ball" structure (Fig. II-38).[1207,1208,1208a] Similar to the $B_{12}X_{12}^{2-}$ ion discussed in Sec. II-13, it belongs to the highest point group, $\mathbf{I}_h$. It has 20 hexagonal faces and 12 additional pentagonal faces, with the 60 carbon atoms being located at the vertices of a regular truncated icosahedron.

The 174 (3 × 60 − 6) normal vibrations are classified into $2A_g + A_u + 3F_{1g} + 4F_{1u} + 4F_{2g} + 5F_{2u} + 6G_g + 6G_u + 8H_g + 7H_u$. Because of its extremely high symmetry, only the four F_{1u} vibrations are IR-active and only the two A_g and eight H_g vibrations are Raman-active. Furthermore, the mutual exclusion rule holds because of the presence of a center of symmetry.

Figures II-39 (upper trace) and II-40 show the IR and Raman spectra, respectively, of C_{60} obtained by Bethune et al.[1209] As expected, the IR spectrum shows four bands at 1428, 1183, 577, and 527 cm^{-1}, and the Raman spectrum exhibits 10 bands at 1575, 1470, 1428, 1250, 1099, 774, 710, 496, 437, and 273 cm^{-1}. In general, the vibrations above ~1000 cm^{-1} are predominantly due to displacements tangential to the C_{60} surface, whereas those below 800 cm^{-1} are predominantly due to radial displacements.[1210] The two polarized (A_g) bands at 1470 and 496 cm^{-1} are assigned to the totally symmetric tangential stretching (or pentagonal pinching) and radial breathing modes, respectively. The lowest frequency band (depolarized) at 273 cm^{-1} is described as the cage-squashing mode (H_g) (Fig. II-41).[1208a] The IR[1211,1212] and Raman[1213] spectra of C_{60} have also been obtained by other workers.

Surface-enhanced Raman spectrum of C_{60} (on gold) exhibits 22 bands instead of 10 bands expected for $\mathbf{I}_h$ symmetry. Clearly, this is due to the lowering of symmetry and changes in electronic structure due to adsorption.[1214]

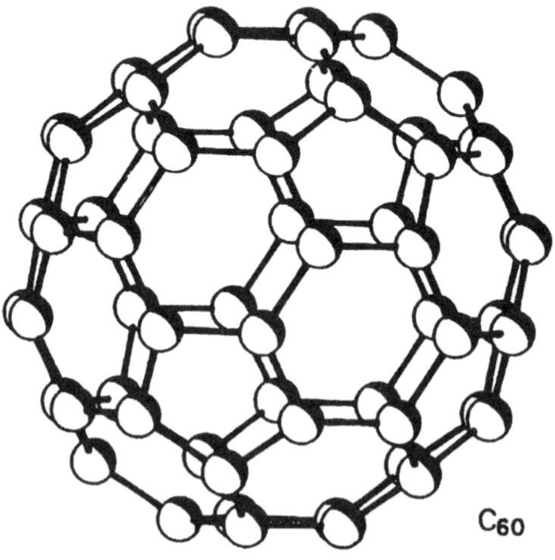

Fig. II-38. Structure of the C_{60} cluster (reproduced with permission from Ref. 1208a).

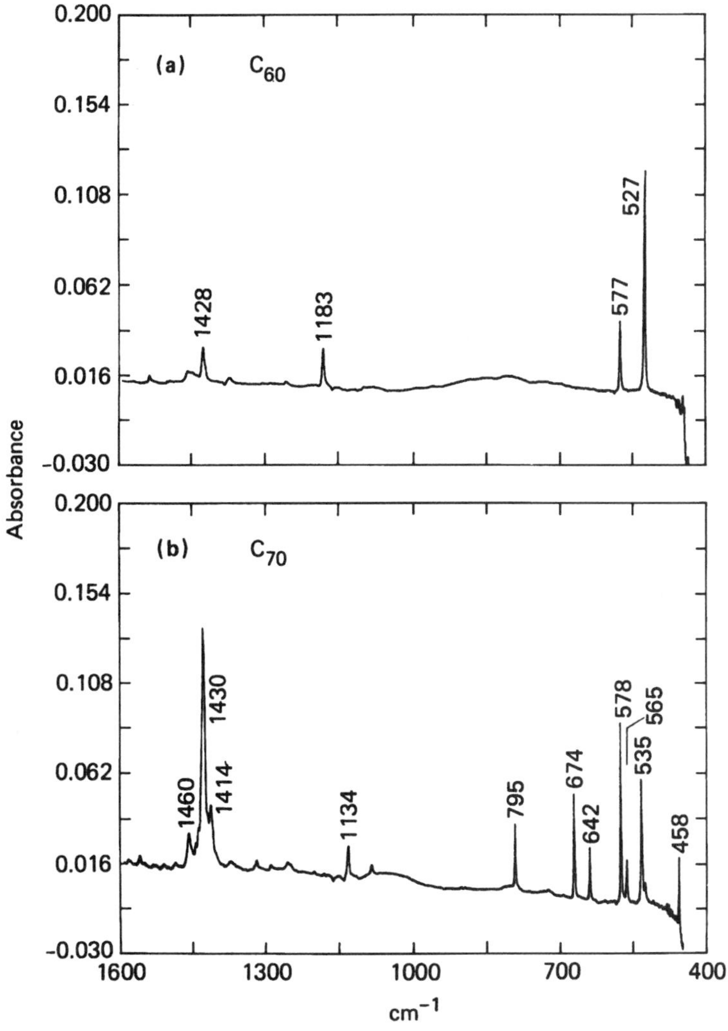

Fig. II-39. FTIR spectra of films of chromatographically separated (*a*) C_{60} and (*b*) C_{70} on KBr substrates. Absorbance values are for samples with undetermined thicknesses (reproduced with permission from Ref. 1209).

In $C_{60}Br_{24}$, the overall symmetry is lowered to $\mathbf{T}_h$ because the 24 Br atoms are bonded to the C_1 and C_4 atoms of the fused pairs of hexagons in the C_{60} framework. Under $\mathbf{T}_h$ symmetry, the ν(C=C) vibrations are grouped into $2A_g + 2E_g + 3F_u + F_g$, of which the three (F_u) vibrations are IR-active, and the five ($2A_g$, $2E_g$, and F_g) vibrations are Raman-active. In agreement with these predictions, the IR spectrum shows three bands at 1688, 1641, and 1610 cm^{-1}, while the Raman spectrum shows five bands at 1690, 1675, 1654, 1631, and 1607 cm^{-1}[1,1215].

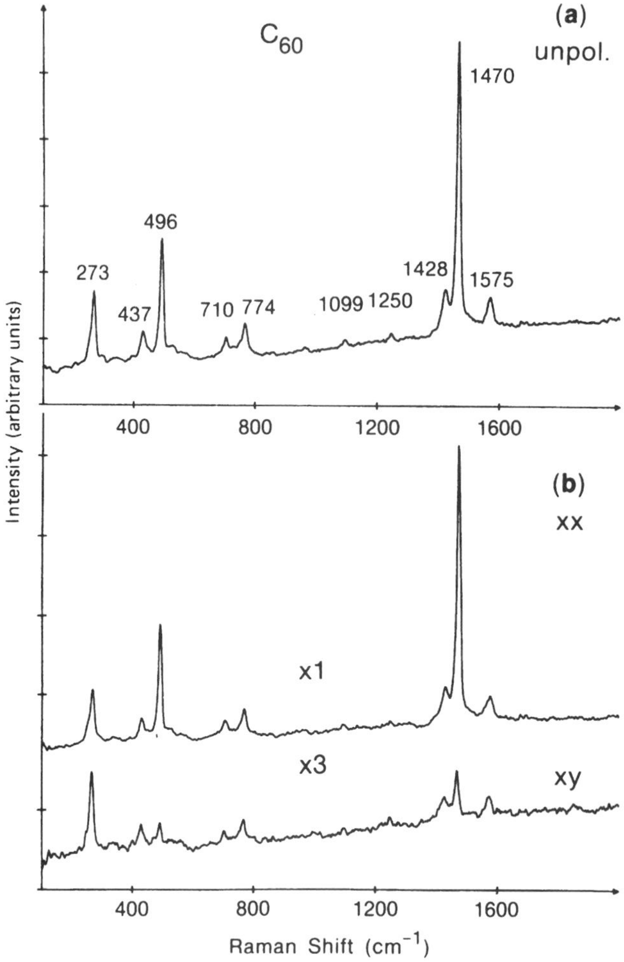

Fig. II-40. Raman spectra of a film of purified C_{60} on a suprasil substrate. (*a*) Unpolarized and (*b*) polarized spectra (reproduced with permission from Ref. 1209).

When C_{60} is doped with alkali metals, superconducting solids such as K_3C_{60} (T_c = 18.0 ~ 19.6 K) and Rb_3C_{60} (T_c = 28.0 ~ 29.8 K) are formed. Their Raman spectra[1216] show that the A_g mode at 1469 cm^{-1} is shifted to 1452 and 1448 cm^{-1}, respectively, in the K- and Rb-doped compounds, and that all the H_g bands are broadened considerably. The magnitude of the shift of the 1469 cm^{-1} band has been used to determine the stoichiometry of x in K_xC_{60}.[1217] These results indicate that, upon doping, electrons are transfered from the metal to the π-orbitals of the C_{60} surface, thus lengthening the CC bonds and that the C_{60} ion thus obtained undergoes symmetry lowering, which results in band splitting and/or broadening.

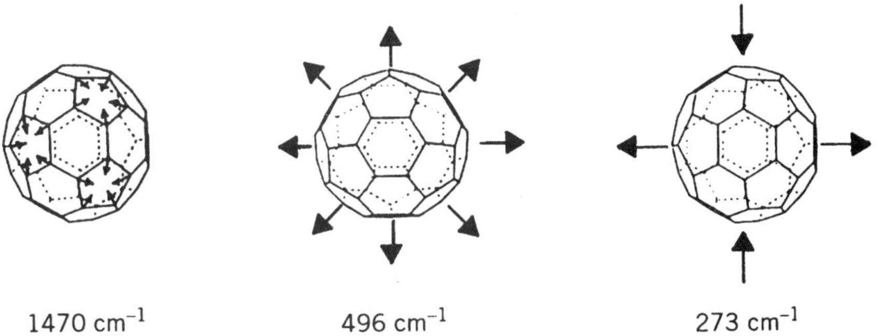

Fig. II-41. Vibrational modes of three prominent Raman bands (reproduced with permission from Ref. 1208a).

Another large carbon cluster, C_{70}, may take a structure of $\mathbf{D}_{5h}$ symmetry shown in Fig. II-42.[1208a] Then, the 204 (3 × 70 − 6) normal vibrations are classified into $12A_1' + 9A_2' + 9A_1'' + 10A_2'' + 21E_1' + 22E_2' + 19E_1'' + 20E_2''$, of which 31 ($A_2''$ and E_1') vibrations are IR-active and 53 (A_1', E_2', and E_1'') vibrations are Raman-active.[1218] Figures II-39 (lower trace) and II-43 show the IR and Raman spectra, respectively, of C_{70} obtained by Bethune et al.[1209,1219] It is seen that the numbers of the observed bands are much less than those predicted for the $\mathbf{D}_{5h}$ symmetry. This may suggest that the actual symmetry is higher than $\mathbf{D}_{5h}$ and/or that serious band overlaps occur in these spectra.

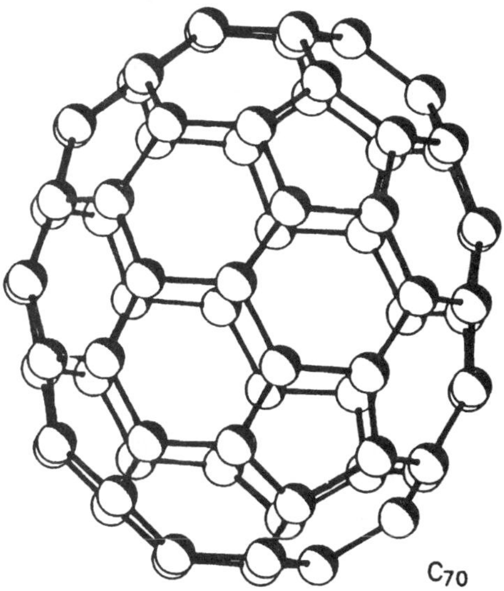

Fig. II-42. Structure of the C_{70} cluster (reproduced with permission from Ref. 1208a).

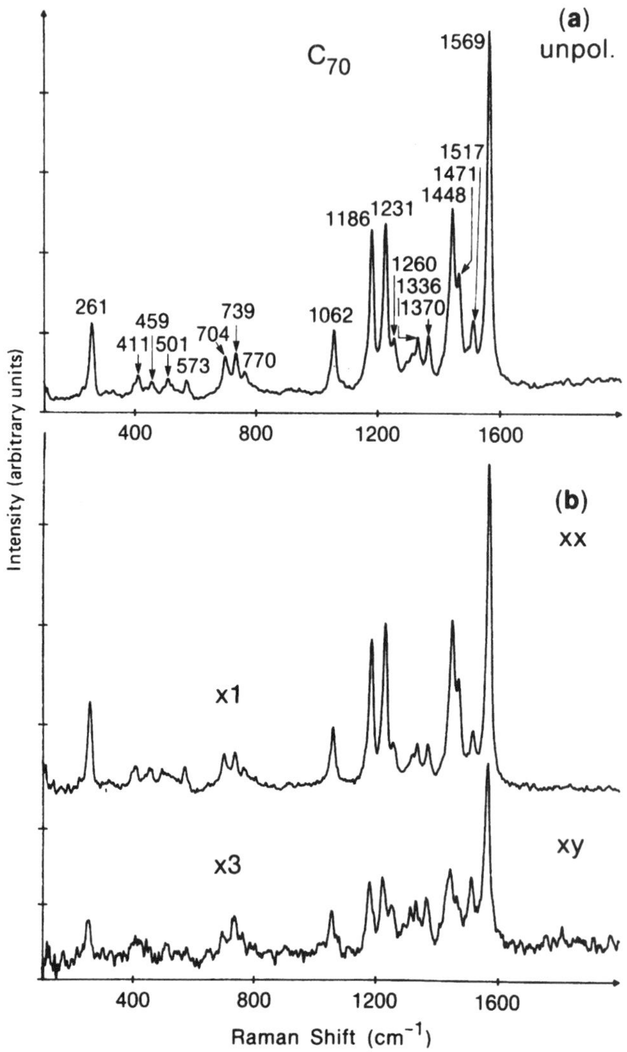

Fig. II-43. Raman spectra of a film of purified C_{70} on a suprasil substrate. (*a*) Unpolarized and (*b*) polarized spectra (reproduced with permission from Ref. 1209).

(3) Diamond and Graphite

Diamond and graphite are two crystalline forms of C_∞. In the former, carbon atoms are tetrahedrally bonded to four equivalent neighbors with a C–C bond distance of 1.54 Å throughout the crystal (Fig. II-44*a*). The primitive unit cell of diamond belongs to the factor group $\mathbf{O}_h^7$ and contains two carbon atoms.[1220] Then, at $k = 0$ (Sec. I-28), only three ($3 \times 2 - 3$) vibrations are expected. These vibrations belong to a triply degenerate F_{2g} species which is Raman-active. In agreement with this prediction, the Raman spectrum of diamond exhibits

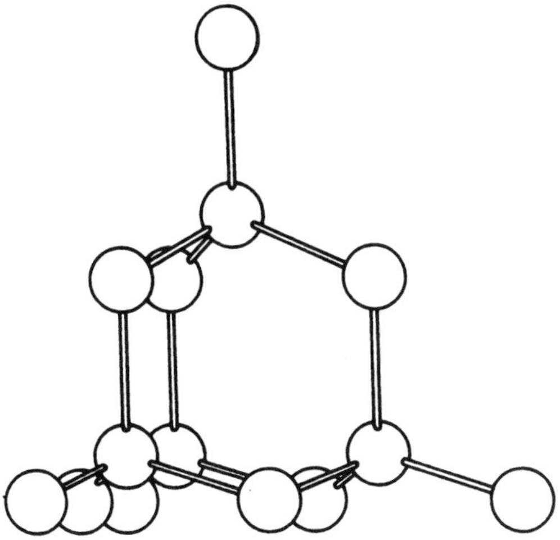

(a)

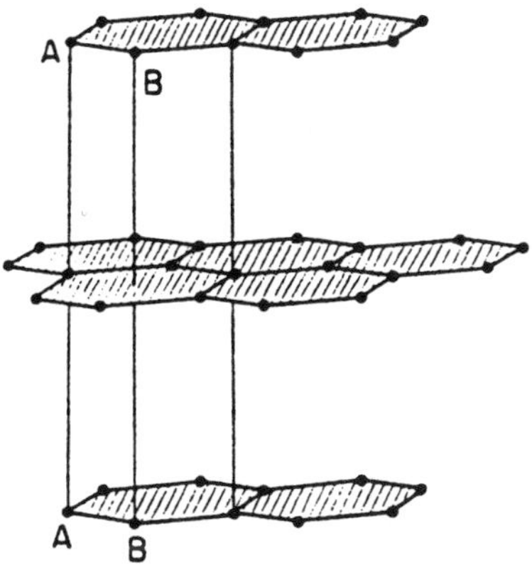

(b)

Fig. II-44. Structures of (a) diamond and (b) graphite.

only one sharp band at 1332 cm^{-1},[1221]. In high-pressure Raman spectroscopy, this band has practical utility as a pressure sensor for a diamond anvil cell;[1222] its frequency has been measured as a function of pressure and temperature up to 15 GPa and 400°C, respectively.[1223] Thick crystals used for diamond anvil cells[1222] show IR absorptions: type I diamond at ~2000 and ~1200 cm^{-1}; type II diamond only at ~2000 cm^{-1},[1224]. Thus, the latter is preferred for high-pressure IR studies. The absorptions at ~2000 and ~1200 cm^{-1} are attributed to N_2-induced defects and multiphonon effects, respectively.[1225]

In graphite, sheet structures are formed by linking each carbon atom to three equivalent neighbors in a trigonal planar fashion (C–C distance, 1.42 Å), and all sheets are parallel to each other with a distance of 3.35 Å, as shown in Fig. II-44*b*. If we consider the isolated sheet of $\mathbf{D}_{6h}$ symmetry, there are two carbon atoms in the repeat unit. Then, three (3 × 2 − 3) vibrations are grouped into the B_{2g} (inactive) and E_{2g} (Raman) species.[1226] In fact, the Raman spectrum of graphite exhibits only one band at 1575 cm^{-1}.[1227] This observation suggests that the intersheet interaction can be ignored for vibrational analysis. As expected, graphite shows no IR absorption bands.[1228]

(4) Inorganic Carbon Compounds

Carbon hydride radicals such as HCCC[1229] and HCCCC[1230] are of great interest in astronomy. The IR spectra of these radicals have been measured in Ar matrices. Their structures may be linear or cyclic. The IR spectra of carbon oxides are reported for the following compounds:

CCCO radical,[1231,1232] OCCCO,[1233~1235] OCCCCO,[1236] OCCCCCO[1237]

The structures of these oxides may be linear or pseudo-linear.[1235] The C_4O_2 molecule may be linear or bent, depending upon the electronic state.

The IR spectra and band assignments of several carbon sulfides have also been reported.

SCCCS,[1238] H_2CCS,[1239] H_2CCCS,[1240] OCCCS[1241,1242]

The structures of these compounds may be linear.

The three carbon oxide anions, $C_3O_3^{2-}$,[1243] $C_4O_4^{2-}$,[1244] and $C_5O_5^{2-}$,[1244] take the planar-ring structures shown in Fig. II-45. Because of aromaticity of these rings, their IR spectra exhibit the C=O stretching vibrations below 1600 cm^{-1}, which are much lower than normal ketones (~ 1700 cm^{-1}). The vibrational spectra of the $C_3X_3^+$ ion (X = Cl, Br, and I) have been assigned in terms of a $\mathbf{D}_{3h}$ structure similar to the $C_3O_3^{2-}$ ion.[1245] The $C(CN)_3^-$ ion takes a planar $\mathbf{D}_{3h}$ structure, and its vibrational spectra have been assigned completely.[1245a]

Vibrational spectra are reported for $F_3C{\equiv}SF$ [$\nu(C{\equiv}S)$, 1800 cm^{-1}],[1246] $(CF_3)_2(SF)^+$,[1247] and $(Cl_3C){-}SCl$.[1248]

II-15. COMPOUNDS OF SILICON AND GERMANIUM

(1) Small Silicon Compounds

Recently, small silicon clusters such as Si_4, Si_6, and Si_7 have been isolated and their structures determined by comparing Raman spectra predicted from ab

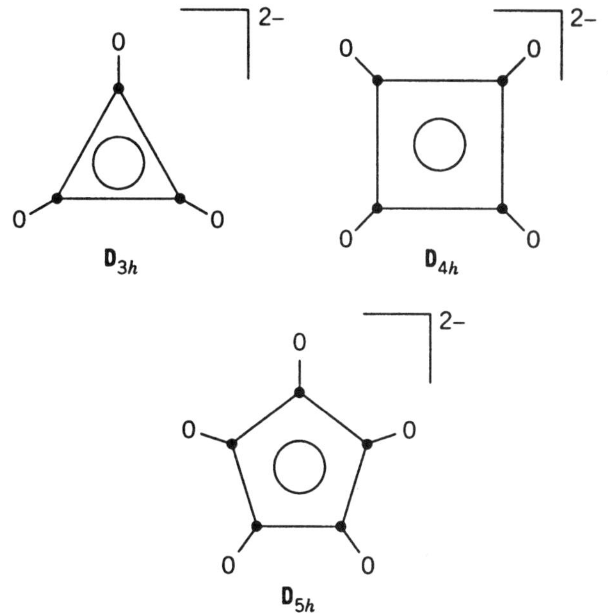

Fig. II-45. Structures of carbon oxide anions.

initio quantum-mechanical calculations with those observed in N_2 matrices.[1249] The numbers of Raman-active fundamentals should be three, five, and five for the predicted structures of a planar Si_4 rhombus ($\mathbf{D}_{2h}$), a distorted Si_6 octahedron ($\mathbf{D}_{4h}$), and a pentagonal bipyramidal Si_7 ($\mathbf{D}_{5h}$), respectively. Figure II-46 shows the Raman spectra of these clusters as obtained by Honea et al.[1249] It is seen that the observed spectra are in excellent agreement with the theoretical spectra, except for the B_{3g} mode of Si_4 (which may be hidden under the strong A_g band). In contrast to Si_4, the Si_4^{4-} ion is tetrahedral ($\mathbf{T}_d$), as its Raman spectrum shows the $\nu_1(A_1)$, $\nu_2(E)$, and $\nu_3(F_2)$ at 482, 285, and 356 cm^{-1}, respectively.[1250]

Silanes, such as Si_5H_{10} and Si_6H_{12}, take nonplanar ring structures consisting of the SiH_2 units. The vibrational spectra of these silanes have been assigned based on normal coordinate analysis assuming $\mathbf{D}_{5h}$ and $\mathbf{D}_{3d}$ symmetries, respectively.[1251,1252] Figure II-47 shows the IR and Raman spectra of Si_6H_{12} obtained by Hassler et al.[1252]

Vibrational spectra and band assignments are reported for: $(H_3Si)_3N$,[1253] $(H_3Si)_2O$,[1254] $(H_3Si)_2S$,[1255] $(H_3Si)PH_2$,[1256] $(H_3Si)_4N_2$,[1257] $(Cl_3Si)_2NH$,[1258] and $Si_2(NCO)_6$.[1259] It should be noted that trisilylamine is trigonal-planar while tetrasilylhydrazine takes a twisted $\mathbf{D}_{2d}$ structure (the dihedral angle being ~90°).

(2) Silicates and Silica

Vibrational spectra of orthosilicates (SiO_4^{4-}) and pyrosilicates ($Si_2O_7^{6-}$) have been discussed in preceding sections. Higher silicates take the variety of struc-

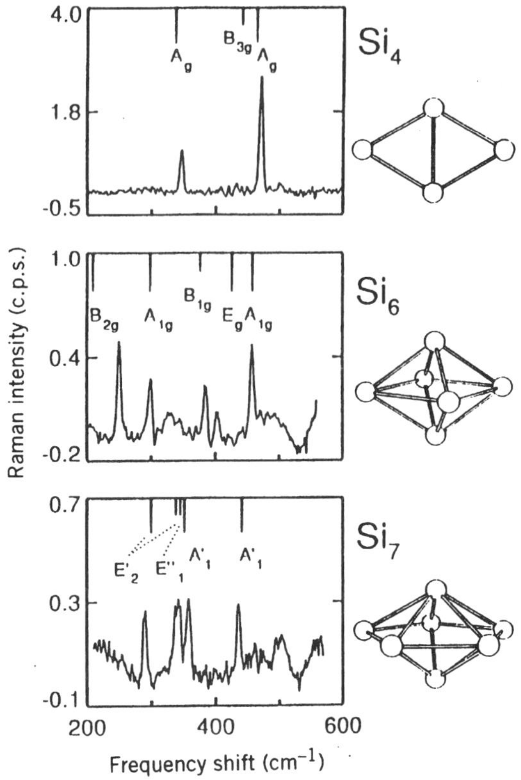

Fig. II-46. Raman spectra of Si_4, Si_6, and Si_7 deposited in a N_2 matrix, along with their predicted structures and Raman-allowed vibrational frequencies and intensities (reproduced with permission from Ref. 1249).

tures as shown in Fig. II-48. Etchepare[1260] has assigned the vibrational spectra of these silicates based on normal coordinate calculations. In general, the tetrahedral SiO_4 unit in these silicates exhibits the four fundamentals in the following regions: $\nu_1(A_1)$, 750–830 cm^{-1}; $\nu_2(E)$, 300–400 cm^{-1}; $\nu_3(F_2)$, 800–1000 cm^{-1}; $\nu_4(F_2)$, 450–600 cm^{-1}. According to McMillan,[1261] the structural units in silicate glasses can be distinguished based on Si–O stretching frequencies in Raman spectra:

Four terminal oxygens (SiO_4^{4-})	~850 cm^{-1}
Three terminal oxygens ($Si_2O_7^{6-}$)	~900 cm^{-1}
Two terminal oxygens ($Si_3O_9^{6-}$, chains)	950–1000 cm^{-1}
One terminal oxygen (sheets)	1050–1100 cm^{-1}

Aluminosilicates $[(Al, Si)O_2]_n$—such as zeolites, exhibit A1O vibrations (750–650 cm^{-1}) in addition to SiO vibrations.[1262] Their vibrational spectra and structure have been reviewed by Lazarev.[1262a] Condrate[1263,1264] reviewed the vibrational spectra of a variety of glasses.

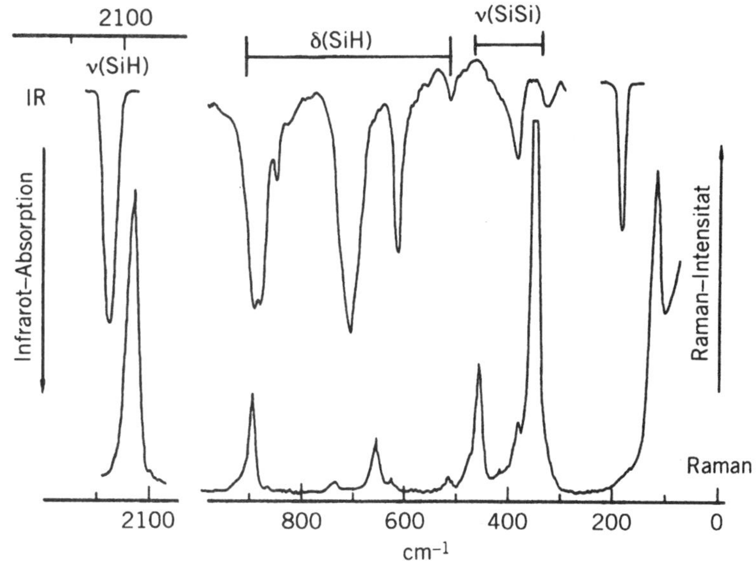

Fig. II-47. IR and Raman spectra of Si_6H_{12} (reproduced with permission from Ref. 1252).

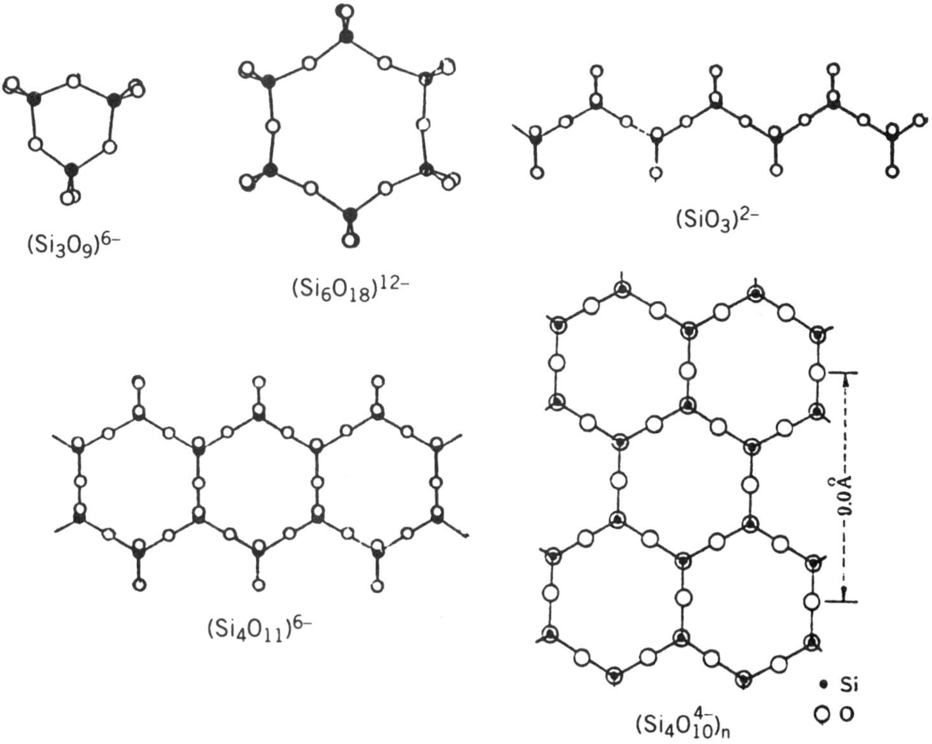

Fig. II-48. Structures of silicates.

Quartz—$(SiO_2)_\infty$—is a three-dimensional polymer in which the SiO_4 units are linked throughout the crystal. The vibrational spectra of α-quartz (including the ^{30}Si and ^{18}O derivatives)[1265] and β-quartz[1266] have been assigned based on their factor groups ($\mathbf{D}_3$ and $\mathbf{D}_6$, respectively). Figure II-49 shows the IR and Raman spectra of α-quartz obtained by Sato and McMillan.[1265]

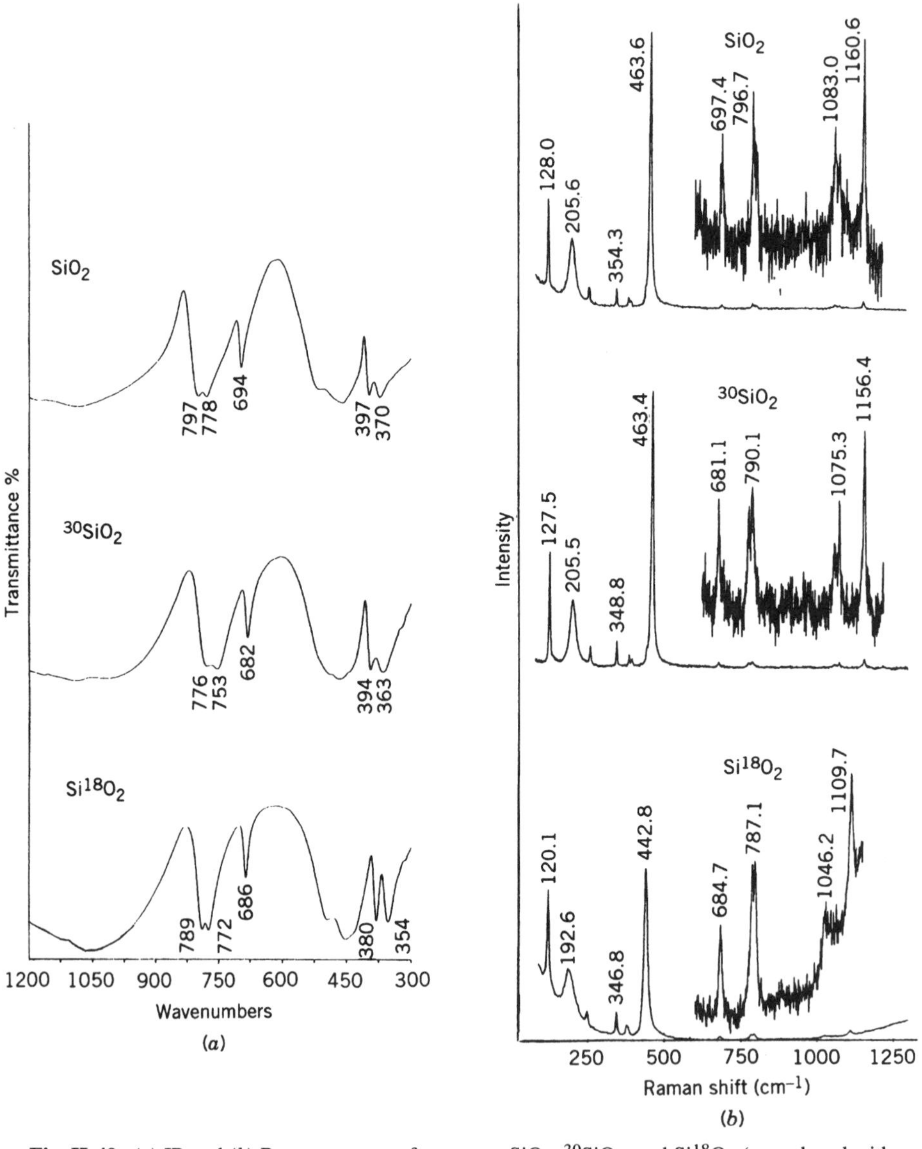

Fig. II-49. (*a*) IR and (*b*) Raman spectra of α-quartz, SiO_2, $^{30}SiO_2$, and $Si^{18}O_2$ (reproduced with permission from Ref. 1265).

Vibrational spectra of silicon compounds have been reviewed by Smith,[1267] Aylett (silicon hydrides),[1268] and Campbell-Ferguson and Ebsworth (halogeno-silane–amine adducts).[1269]

(3) Germanium Compounds

In addition to those mentioned in the earlier sections, vibrational spectra of germanium compounds such as $(H_3Ge)_3M$ (M = As, Sb)[1270] and $(Ge_4S_{10})^{4-}$ (cage structure of $\mathbf{T}_d$ symmetry)[1271] have been assigned. Schumann[1272] reviewed the ν(GeP) vibrations of a number of compounds.

II-16. COMPOUNDS OF NITROGEN

(1) Oxides

References on vibrational spectra of NO, NO_2, NO_3, N_2O, and N_2O_2 are given in preceding sections. The compound N_2O_3 takes the asymmetric form in the gaseous state, and its IR spectrum has been assigned based on $\mathbf{C}_s$ symmetry.[1273]

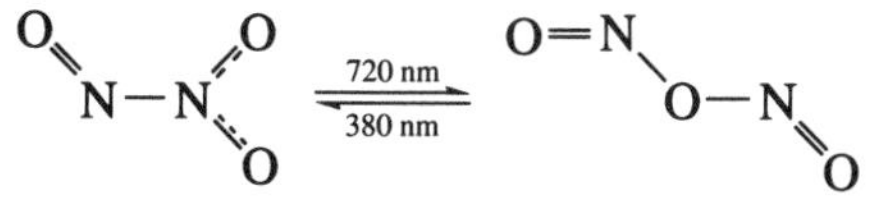

asymmetric form symmetric form

In N_2 matrices, however, the asymmetric form can be converted to the symmetric form by irradiation near 720 nm, and this form is reconverted to the asymmetric form by irradiation near 380 nm. The IR frequencies of both forms have been reported by Varetti and Pimentel.[1274]

As stated in Sec. II-10, N_2O_4 is a mixture of the three isomers: a planar ($\mathbf{D}_{2h}$) and its twisted (~90°, $\mathbf{D}_{2d}$) forms and the iso-form:

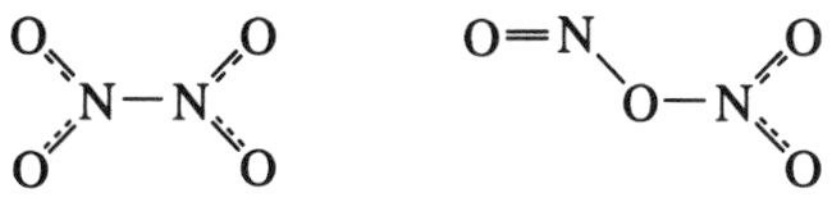

symmetric form iso-form

The IR spectrum of the latter has been assigned.[1275,1276]

Although N_2O_5 takes a planar structure in the gaseous state, it becomes an ionic crystal consisting of the NO_2^+ (nitronium) and NO_3^- ions:

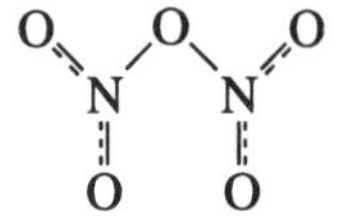

(C_{2v})

The IR spectra of N_2O_5 in both phases have been assigned by Hisatsune et al.[1277]

(2) Oxo-ions

References on NO^+, NO^-, NO_2^+, NO_2^-, and NO_3^- ions are found in preceding sections. The NO_4^{3-} ion is tetrahedral, and its IR and Raman spectra have been reported.[1278] As stated in Sec. II-5, the $N_2O_2^{2-}$ ion takes the *cis*- and *trans*-forms while the $N_2O_2^-$ ion obtained in Ar matrices may take a structure containing a terminal N=N bond:

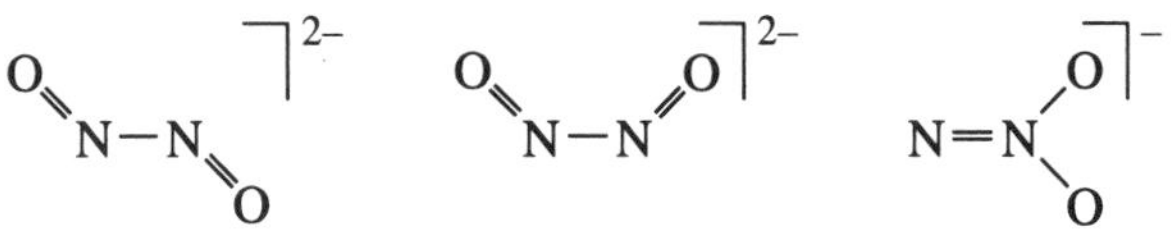

trans *cis*

The $N_2O_3^{2-}$ ion in $Na_2(N_2O_3)$ take two forms:

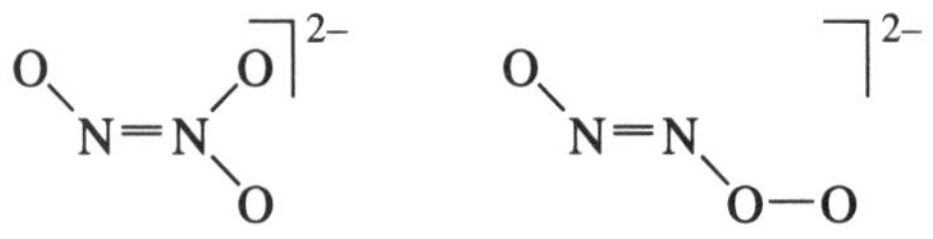

α-form β-form

The IR and Raman spectra of the former have been reported.[1278a] The vibrational spectra of nitrogen oxides and oxo-ions have been reviewed extensively by Laane and Ohlsen.[1279]

(3) Halogeno Compounds

Figure II-50 shows the structures of halogeno compounds for which complete vibrational assignments are found in the references cited.

(4) Amino Compounds

Figure II-51 shows the structures of nitrogen compounds containing amino groups. Complete band assignments are found in the references cited. Figure II-52 shows the IR spectrum of the heated vapor of cyanamide (H_2NCN)

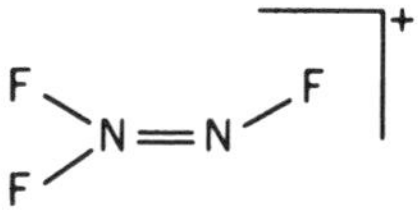

$N_2F_3^+$ (C_s)[1280]

$N(CF_3)_3$ (C_3)[1281]
(NC_3 pyramid is flat)

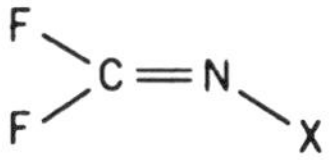

CF_2NX (C_s)[1282]
(X = Cl,Br)

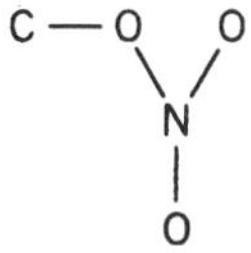

$XONO_2$(C_s)[1283,1284]
(X = Cl, Br)

Fig. II-50. Structures of halogeno compounds of nitrogen.

obtained by Birk and Winnewisser.[1286] It also shows some bands due to the tautomer, carbodiimide (HNCNH), which is more stable at lower temperatures.

(5) Sulfides and Selenides

Figure II-53 shows structures of five nitrogen compounds containing sulfur or selenium.

The NS_4^- ion may be a cyclic ring, a linear chain, or a branched chain:

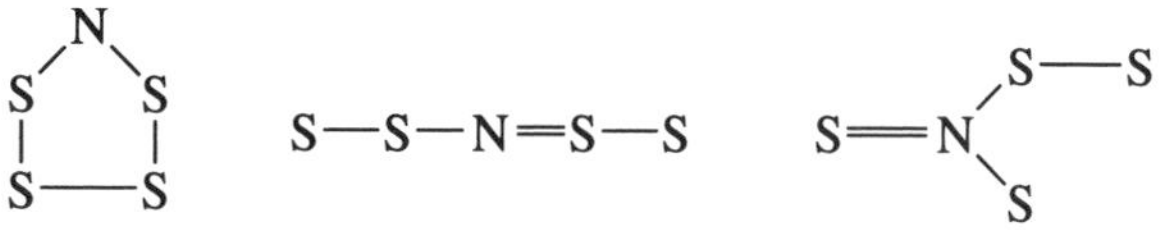

The IR/Raman spectra suggest the last structure.[1290] Although the $N_3S_3^-$ and

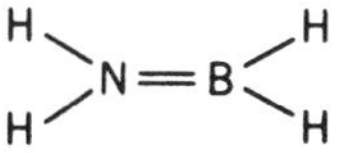

$(H_2N)BH_2$(C_s)[1285]

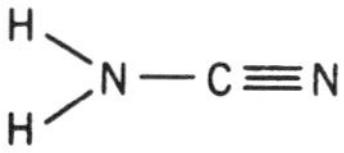

$(H_2N)CN$(C_{2v})[1286]

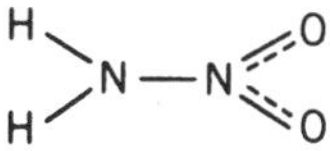

$(H_2N)NO_2$(C_s)[1287]

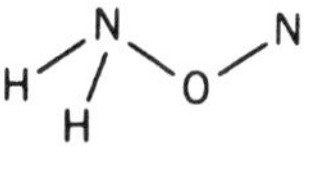

$(H_2N)OH$(C_s)[1288]

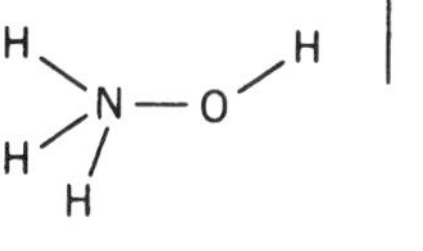

$[(H_3N)OH]$ Cl(C_s or C_1)[1289]

Fig. II-51. Structures of nitrogen compounds containing amino groups.

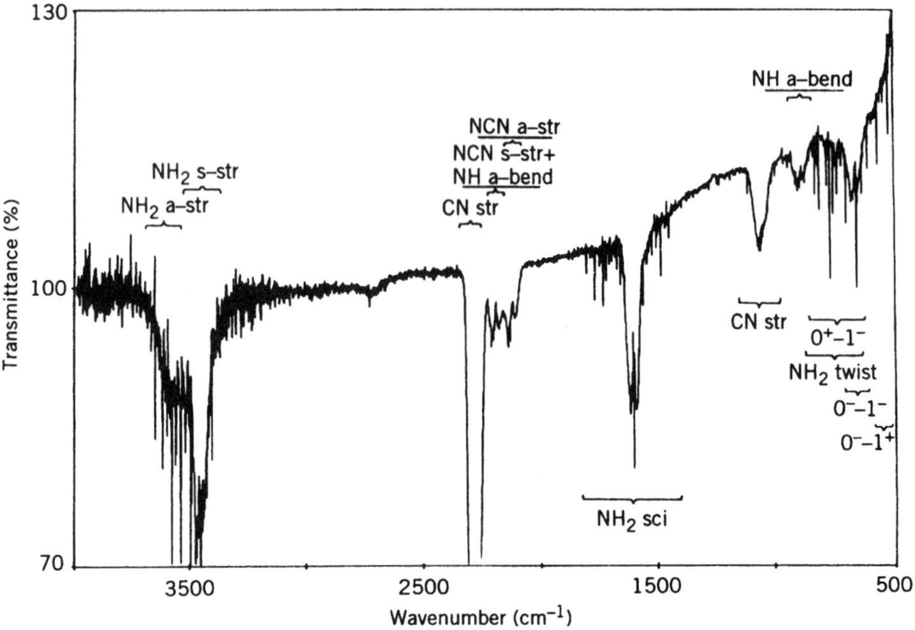

Fig. II-52. IR spectrum of heated cyanamide vapor. The underlined approximate description refer to absorptions due to the HNCNH tautomer (reproduced with permission from Ref. 1286).

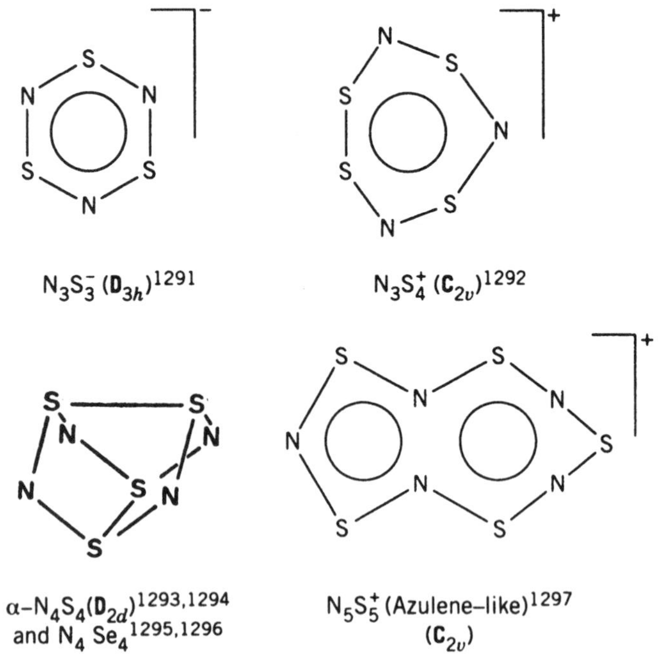

Fig. II-53. Structures of nitrogen sulfides and selenides.

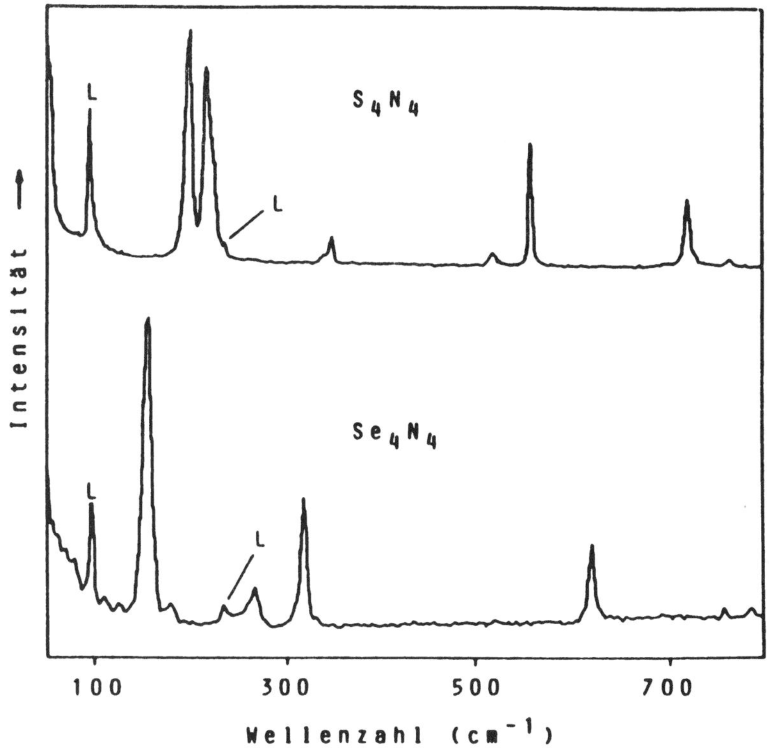

Fig. II-54. Raman spectra of N_4S_4 (crystal) and N_4Se_4 (powder) (L, laser line; 647-nm excitation) (reproduced with permission from Ref. 1296).

$N_3S_4^+$ ions take ring structures, the N_4S_4 and N_4Se_4 molecules assume a cage structure of $\mathbf{D}_{2d}$ symmetry, and their 18 ($3 \times 8 - 6$) normal vibrations are classified into $3A_1$(R) + $2A_2$(inactive) + $2B_1$(R) + $3B_2$(IR, R) + $4E$(IR, R). These vibrations have been assigned completely based on normal coordinate calculations.[1293] Figure II-54 shows the Raman spectra of N_4S_4 and N_4Se_4 obtained by Gowik and Klapötke.[1296] A highly unusual azulene-like structure of the $N_5S_5^+$ ion has been determined by X-ray analysis and characterized by IR spectroscopy.[1297]

II-17. COMPOUNDS OF PHOSPHORUS AND OTHER GROUP VB ELEMENTS

Vibrational spectra of phosphorus compounds have been reviewed extensively.[1298–1302] Figure II-55 shows a group frequency chart based on these review articles. Group frequency charts in Appendix VIII indicate the regions for ν(PH), ν(PO), and ν(PX)(X: a halogen).

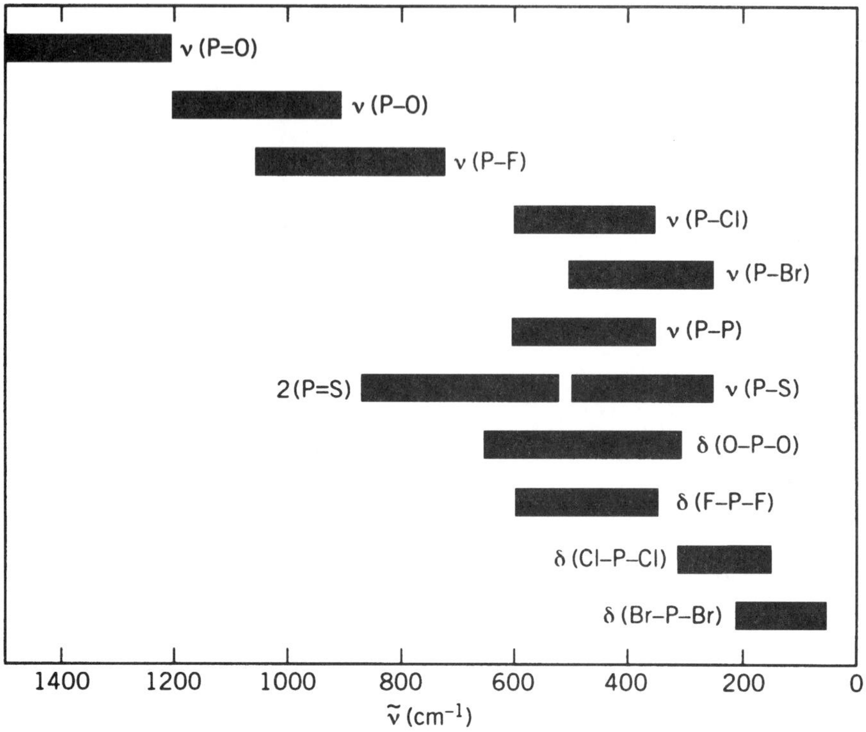

Fig. II-55. Group frequency chart of phosphorus compounds.

(1) Phosphorus Clusters

Figure II-56 shows the structures of phosphorus clusters for which vibrational assignments are available. The tetrahedral P_4 cluster exhibits the $\nu_1(A_1)$, $\nu_2(E)$, and $\nu_3(F_2)$ at 606, 363, and 464.5 cm^{-1}, respectively, in Raman spectra.[1309,1303a] The P_5^- ion takes a planar ring structure of $\mathbf{D}_{5h}$ symmetry, and the medium-intensity Raman (polarized) band at 463 cm^{-1} and the strong IR band at 815 cm^{-1} have been assigned to the A_1' and E_1' species, respectively.[1304] The P_6^{4-} ion takes a planar ring structure of $\mathbf{D}_{6h}$ symmetry, and exhibits three Raman bands at 356 (A_{1g}), 507 (E_{2g}), and 202 (E_{2g}) cm^{-1},[1305]. In the crystalline state, its symmetry is lowered to $\mathbf{D}_{2h}$. The P_{11}^{3-} ion takes a cage structure of $\mathbf{D}_3$ symmetry, and its 27 (3 × 11 − 6) vibrations are classified into $5A_1$(R) + $4A_2$(IR) + $9E$(IR, R). Complete band assignments have been made based on normal coordinate analysis of this ion.[1306]

(2) Oxides and Oxo-ions of Phosphorus

Reactions of P_2 and O_3 in inert gas matrices yield a variety of oxides, PO, PO_2, P_2O, P_2O_2, P_2O_3, P_2O_4, P_2O_5, and PO_2^- for which the key bands of each

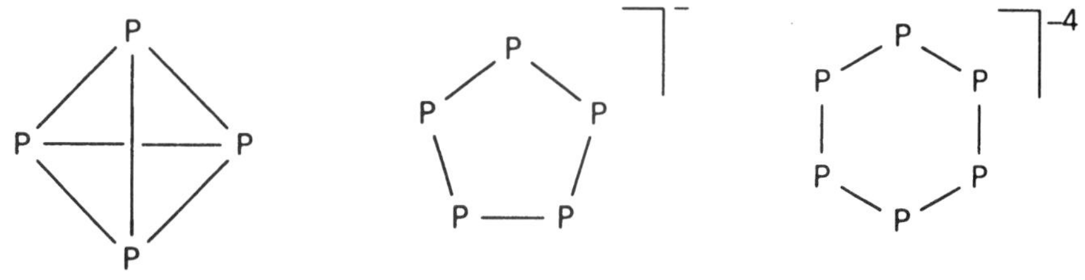

P_4 (T_d)[1303] P_5^- (D_{5h})[1304] P_6^{4-} (D_{6h})[1305]

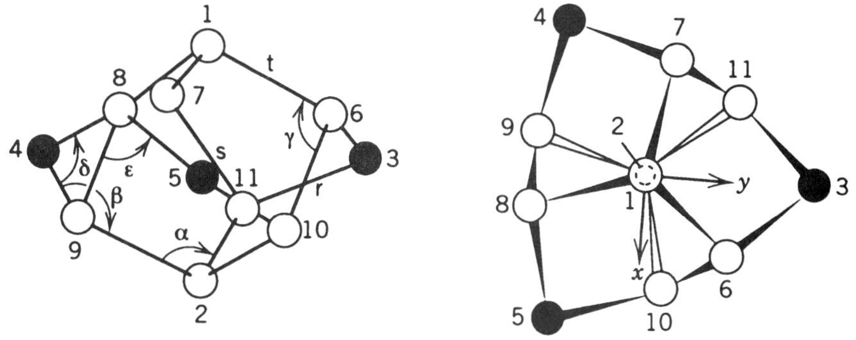

P_{11}^{4-} (D_3) — the vertical view along the C_3 axis is shown on the left[1306]

Fig. II-56. Structures of phosphorus clusters.

species have been identified in IR spectra.[1307] The IR spectrum of P_4O_{10} has been assigned based on the cage structure of $\mathbf{T}_d$ symmetry.[1307a] The higher oxo-ions take the ring structures shown in Fig. II-57. The vibrational spectra of the $P_3O_9^{3-}$ ion have been assigned based on $\mathbf{D}_{3h}$ symmetry.[1308] Although the highest symmetry expected for the $P_4O_{12}^{4-}$ ring is $\mathbf{D}_{4h}$, the spectra obtained in the crystalline state suggest lowering of symmetry to $\mathbf{S}_4$, $\mathbf{C}_{2h}$, $\mathbf{C}_{2v}$, and so on.[1309–1311] For example, it is $\mathbf{S}_4$ in the aluminium salt, and $\mathbf{C}_{2h}$ in the ammonium salt.[1310] The IR/Raman spectra are reported for the $P_{10}O_{30}^{10-}$ ion which takes a "cradle" structure of $\mathbf{C}_2$ symmetry.[1312,1313]

(3) Sulfides and Selenides of Phosphorus

Phosphorus sulfides and selenides take a variety of cage structures, a shown in Fig. II-58. The compound α-P_4S_3 belongs to the point group $\mathbf{C}_{3v}$, and its 15 ($3 \times 7 - 6$) vibrations are grouped into $4A_1$(IR, R) + A_2(inactive) + $5E$(IR, Raman). These vibrations have been assigned completely based on normal coordinate analysis.[1314] Raman spectra are reported and assigned for α-P_4S_4[1315] which takes the same structure as that of N_4S_4, discussed in the preceding section.[1293]

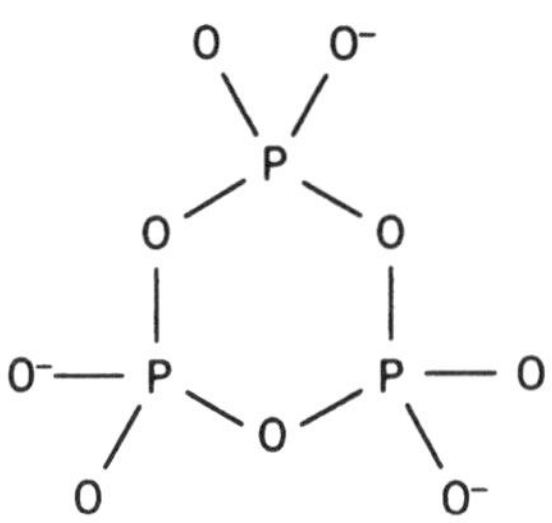

$P_3O_9^{3-}$ ($\mathbf{D}_{3h}$)[1308]

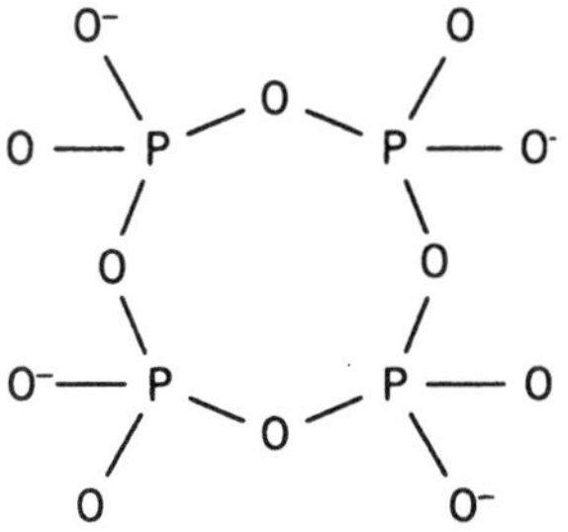

$P_4O_{12}^{4-}$ ($\mathbf{D}_{4h}$)[1309–1311]

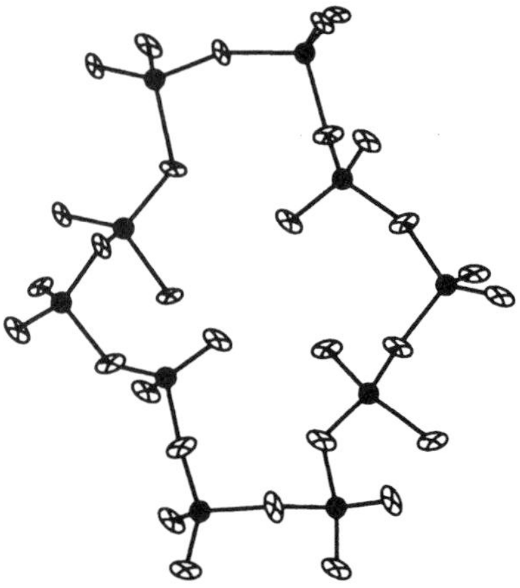

$P_{10}O_{30}^{10-}$ ($\mathbf{C}_2$)[1312, 1313]
(Puckered "cradle")

Fig. II-57. Structures of phosphorus oxide anions.

Gardner[1316] measured the IR and Raman spectra of P_4S_3, P_4S_5, P_4S_7, and P_4S_{10} in solids, melts, and vapors. Except for P_4S_3, these compounds decompose in melts and vapors. The compound P_4S_5 has no symmetry, and its 21 $(3 \times 9 - 6)$ vibrations are all IR/Raman-active; P_4S_7 also has no symmetry. In the free state, P_4S_{10} belongs to the point group $\mathbf{T}_d$, for which vibrational assignments have been made.[1316]

Raman spectra and vibrational assignments are available for α-P_4Se_3,[1317] which takes the same structure as α-P_4S_3 as well as for a series of $P_4S_xSe_{3-x}$ compounds ($x = 0$, 1, 2, and 3).[1318]

(4) Phosphorus Compounds Containing Nitrogen

Figure II-59 shows the structures of four nitrogen-containing phosphorus compounds for which complete vibrational assignments are available. The P_4N_4

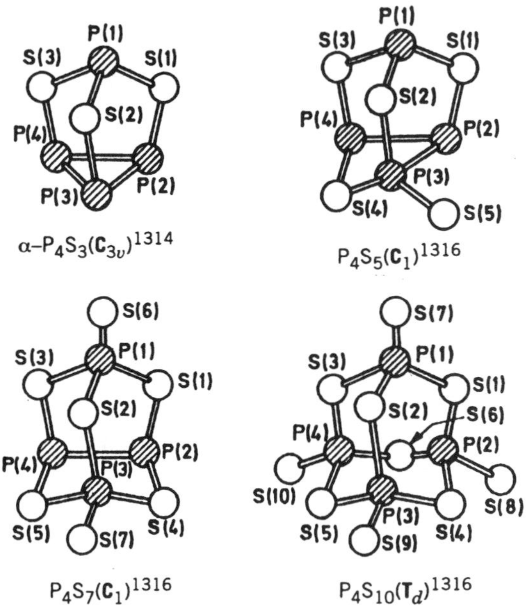

Fig. II-58. Structures of phosphorus sulfides (reproduced with permission from Ref. 1316).

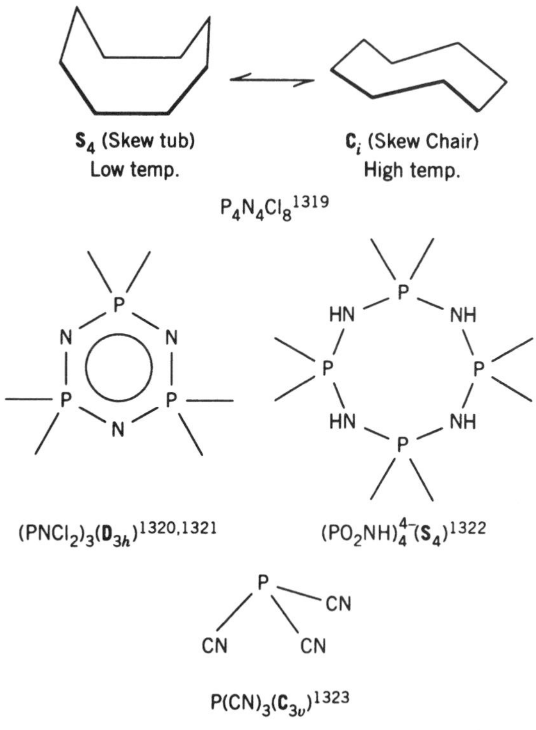

Fig. II-59. Structures of phosphorus compounds containing nitrogen.

skeleton of $P_4N_4Cl_8$ takes a skew-tub form at low temperatures and a skew-chair form at high temperatures. Varma et al.[1319] obtained the IR and Raman spectra of both forms (Figure II-60) and made complete assignments. Both $(PNCl_2)_3$[1320] and $(PNBr_2)_3$[1321] take a planar ring structure of $\mathbf{D}_{3h}$ symmetry. Although the highest symmetry expected for the $(PO_2NH)_4^{4-}$ ion is $\mathbf{D}_{4h}$, it is lowered to $\mathbf{S}_4$ in solution and to $\mathbf{C}_{2h}$ or $\mathbf{S}_4$ in the solid state.[1322] $P(CN)_3$ takes a pyramidal structure of $\mathbf{C}_{3v}$ symmetry.[1323]

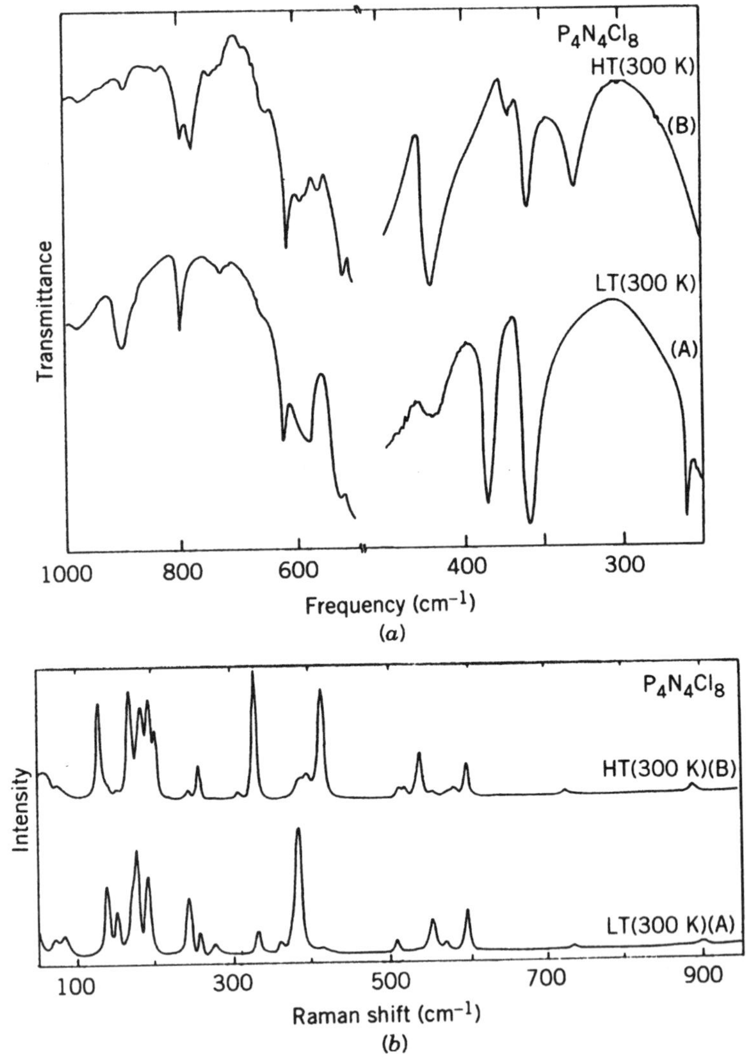

Fig. II-60. (*a*) IR and (*b*) Raman spectra of the low- and high-temperature forms of solid $P_4N_4Cl_8$ measured at 300K (reproduced with permission from Ref. 1319).

TABLE II-14. Vibrational Frequencies of Tetrahedral X_4-Type Molecules (cm^{-1})

Compound	$\nu_1(A_1)$	$\nu_2(E)$	$\nu_3(F_2)$	Ref.
P_4	606	353	464.5	1303
As_4	353	207.5	265.5	1325
Sb_4	241.5	137.1	178.5	1326
Bi_4	149.5	89.8	120.4	1327

(5) Hydrido Compound

A novel *trans*-$H_2PO_4^-$ ion was synthesized by Christe et al.[1324] and its IR/Raman spectra have been assigned completely via normal coordinate analysis based on a pseudo-octahedral model of $\mathbf{D}_{4h}$ symmetry.

(6) Compounds of Arsenic, Antimony, and Bismuth

The elements As_4[1325], Sb_4[1326] and Bi_4[1327] take on tetrahedral structures ($\mathbf{T}_d$) and their frequencies are tabulated in Table II-14. Vibrational spectra of the As_{11}^{3-} ion have been assigned based on the cage structure ($\mathbf{D}_3$) obtained for the P_{11}^{3-} ion.[1306] The gas-phase Raman spectrum of As_4O_6 can be interpreted in terms of $\mathbf{T}_d$ symmetry.[1328] Factor group analysis on crystalline $(As_2O_3)_n$ has been carried out to give complete assignments of its IR/Raman spectra.[1329] Vibrational assignments are available for $As_3O_9^{3-}$,[1330] and $As(CN)_3$[1323] which take the same structures as $P_3O_9^{3-}$ and $P(CN)_3$, respectively.

Gas-phase Raman spectra are reported for mixed As/P compounds such as As_3P ($\mathbf{C}_{3v}$), As_2P_2($\mathbf{C}_{2v}$), and AsP_3($\mathbf{C}_{3v}$).[1331] Far-IR spectra of the BiX_4^-, BiX_5^{2-}, and $Bi_2X_9^{3-}$ ions (X = Cl, Br) have been assigned empirically.[1332] Their structures in nonaqueous solutions may be $\mathbf{C}_{3v}$ or lower for BiX_4^-, a square pyramid of $\mathbf{C}_{4v}$ symmetry for BiX_5^{2-} and a structure lower than $\mathbf{D}_{3h}$ for $Bi_2X_9^{3-}$.

II-18. COMPOUNDS OF SULFUR AND SELENIUM

Compounds of sulfur and selenium are of great interest for spectroscopists because of their unusual structures and strong Raman bands involving S–S or Se–Se bonds. Steudel and co-workers have made an extensive study on vibrational spectra of these compounds. Figure II-61 shows a group frequency chart for sulfur compounds.

(1) Sulfur Clusters

Vibrational frequencies of diatomic S_2 and S_2^- and triatomic S_3, S_3^-, and S_3^{2-} are listed in Secs. II-1 and II-2, respectively. The IR spectra of the S_4 molecule produced in Ar matrices suggest the formation of two open-chain isomers.[1333]

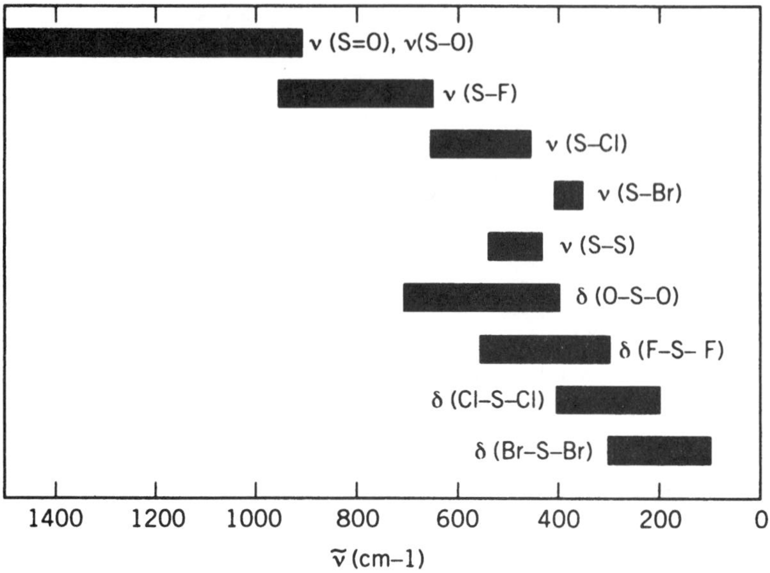

Fig. II-61. Group frequency chart of sulfur compounds.

Recent RR studies[1333a] show that the S_4 molecule is *trans*-planar in the gaseous phase. As is shown in Fig. II-62, the S_4^{2-} ion takes a nonplanar $\mathbf{C}_2$ structure, similar to that of H_2O_2 (dihedral angle, 98°), and its Raman spectrum has been assigned.[1334] In contrast, the S_4^{2+} ion is a square-planar ring of $\mathbf{D}_{4h}$ symmetry and its 6 (3 × 4 − 6) vibrations are classified into A_{1g}(R) + B_{1g}(R) + B_{2g}(R) + B_{2u}(inactive) + E_u(IR). These vibrations have been assigned completely based on normal coordinate analyses.[1335–1337] Table II-15 lists the vibrational frequencies of the S_4^{2+}, Se_4^{2+}, and Te_4^{2+} ions which have the same $\mathbf{D}_{4h}$ structures.

Figure II-62 shows possible conformations of the S_5^{2-} and S_6^{2-} ions. In $(NH_4)_2S_5$, the S_5^{2-} ion takes a helical chain of $\mathbf{C}_2$ symmetry, while in Na_2S_5, it takes a *cis*-conformation of $\mathbf{C}_s$ symmetry. The Raman spectrum of the former has been assigned based on normal coordinate analysis.[1338]

As is shown in Fig. II-62, larger cluster molecules such as S_6,[1339,1340] S_7,[1341] S_8[1342,1343] S_9,[1344] and S_{12}[1343] take puckered ring structures and their vibrational spectra have been reported. In general, the neutral molecules S_n (n = 6 ~ 12) take ring structures, while the anions S_n^{2-} (n = 4 ~ 6) take open-chain structures.

(2) Sulfur Oxides

As is shown in Fig. II-63, the S_8O molecule is an eight-membered crown-shaped S_8 ring to which an oxygen atom is bonded ($\mathbf{C}_s$ symmetry). Steudel and co-workers have measured the IR/Raman spectra of S_8O and made complete assignments based on normal coordinate analysis.[1345,1346] These work-

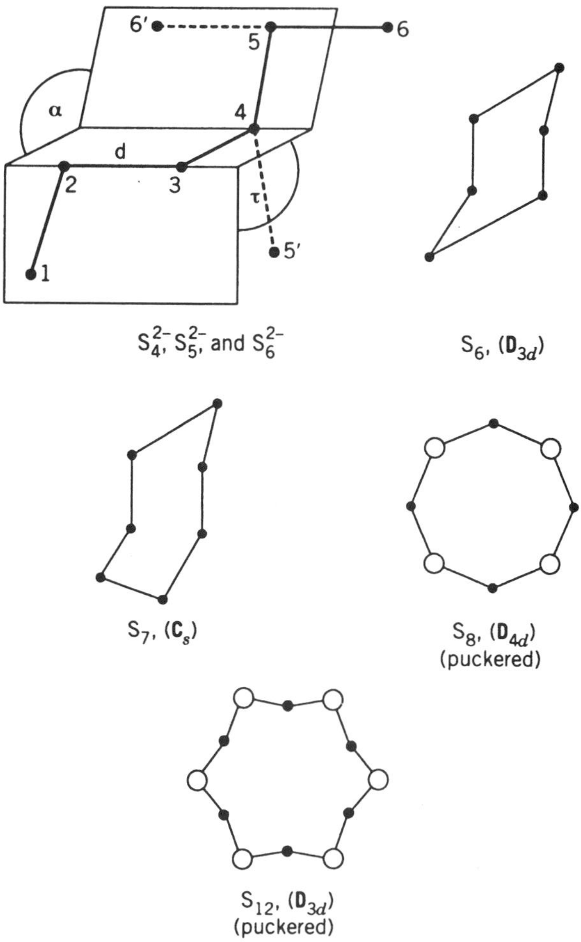

Fig. II-62. Structures of sulfur clusters.

TABLE II-15. Vibrational Frequencies of Square-Planar X_4-Type Ions (cm^{-1})

Ion		S_4^{2+}	Se_4^{2+}	Te_4^{2+}
$\nu_1(A_{1g})$		583.6	321.3	219
$\nu_2(B_{1g})$		371.2	182.3	109
$\nu_3(B_{2g})$		598.1	321.3	219
$\nu_5(E_u)$		542	302	187

[a]Taken from Ref. 1336.

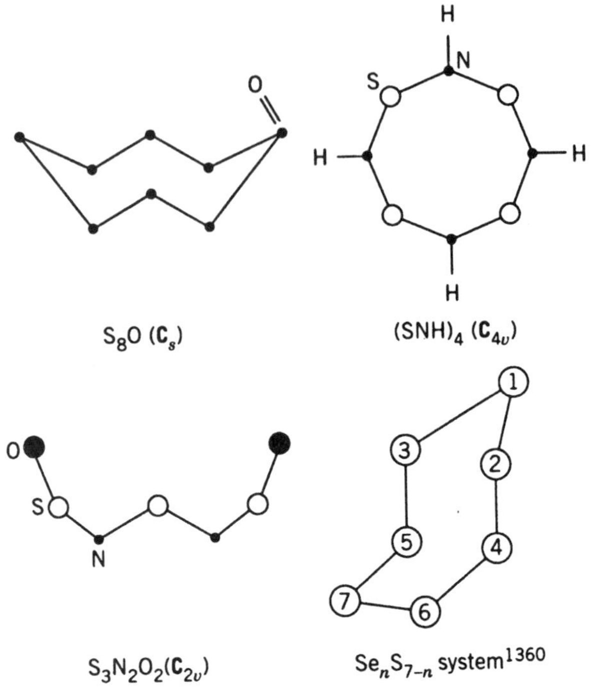

Fig. II-63. Structures of sulfur compounds containing oxygen, nitrogen, and selenium.

ers reported the Raman spectra of S_9O and $S_{10}O$ which may take similar ring structures.[1344] The Raman spectrum of the disulfite ion, $S_2O_5^{2-}$, viz. $(O_3S{-}SO_2)^{2-}$, has been assigned in terms of $\mathbf{C}_s$ symmetry.[1347]

(3) Sulfur Acids

The IR and Raman spectra of H_2S_3,[1348] H_2S_4[1348] H_2SO_4(gas),[1349] HSO_5^- (peroxymonosulfate ion),[1350] and $HOSO_2$[1351] have been assigned empirically or theoretically. The last compound was obtained by the gaseous reaction of OH and SO_2, and trapped in Ar matrices.

(4) Sulfur Compounds Containing Nitrogen and Halogen

Heptasulfur imide (S_7NH) takes an 8-membered crown-shaped ring structure with the S_2NH group almost planar. Steudel has made complete assignments of the IR/Raman spectra of S_7NH and its D and ^{15}N analogs using normal coordinate analysis.[1352] The IR/Raman spectra of the S_7X^+ ion (X: a halogen), which contain the S_7 ring, have been reported.[1353] The $(SNH)_4$ molecule forms the 8-membered crown-shaped ring of $\mathbf{C}_{4v}$ symmetry shown in Fig. II-63. Figure II-64 shows the Raman spectra of $(SNH)_4$ and $(SND)_4$ obtained by Steudel and

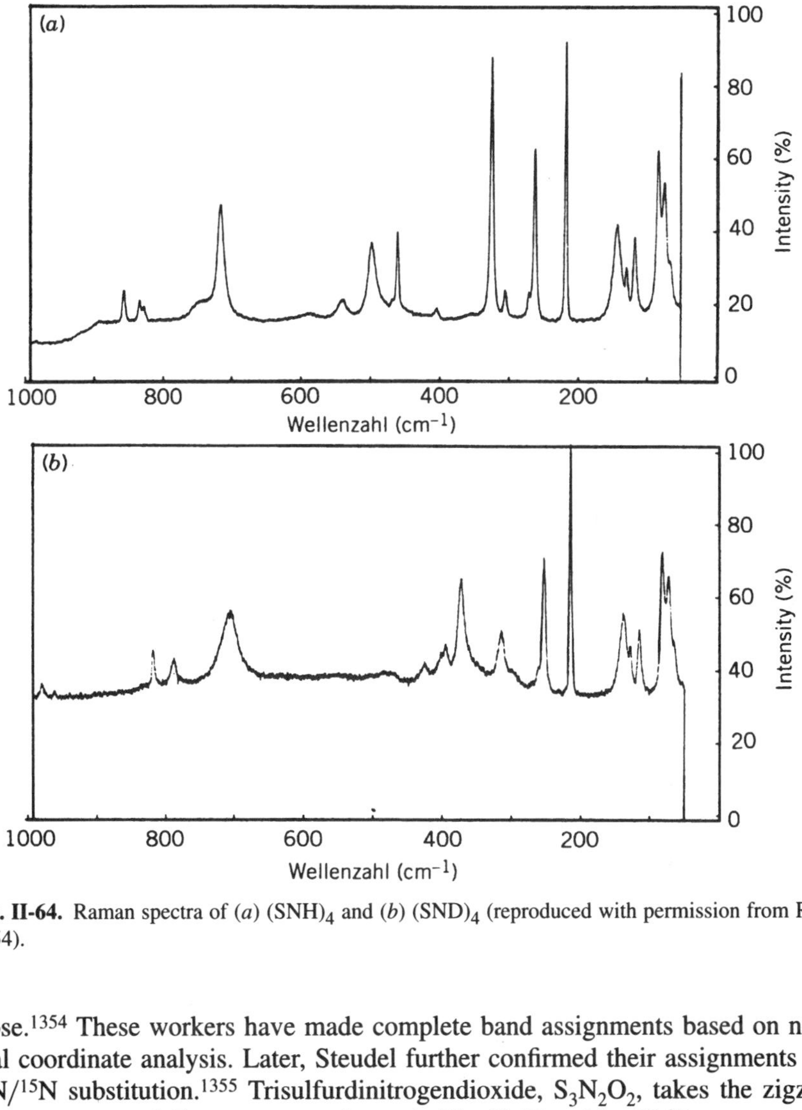

Fig. II-64. Raman spectra of (*a*) $(SNH)_4$ and (*b*) $(SND)_4$ (reproduced with permission from Ref. 1354).

Rose.[1354] These workers have made complete band assignments based on normal coordinate analysis. Later, Steudel further confirmed their assignments by $^{14}N/^{15}N$ substitution.[1355] Trisulfurdinitrogendioxide, $S_3N_2O_2$, takes the zigzag chain structure of $\mathbf{C}_{2v}$ symmetry shown in Fig. II-63 and its IR/Raman spectra have been assigned theoretically.[1356] Vibrational assignments are available for $ClO(SO_2F)$[1357] and $XN(SF_2)$ (X = F, Cl, Br),[1358] both of which belong to the $\mathbf{C}_s$ point group, and for the $[CF_3SO_3]^-$ ion, which may take the idealized $\mathbf{C}_{3v}$ symmetry.[1358a,1358b]

(5) Selenium Compounds

The vibrational frequencies of the Se_4^{2+} ion[1336,1359] are listed in Table II-15. Similar to S_8, the Se_8 molecule takes a crown-shaped ring structure of $\mathbf{D}_{4d}$

symmetry, and its IR/Raman spectra have been assigned via normal coordinate analysis.[1359] According to X-ray analysis, the Se_9^{2-} ion in [Sr(15-crown-5)$_2$] (Se_9) takes a zigzag chain structure. Four stretching (300–200 cm^{-1} region) and three bending (170–120 cm^{-1} region) vibrations have been reported for this ion.[1359a]

Using the UBF field, Laitinen et al.[1360] calculated vibrational frequencies of six-membered Se_nS_{6-n}(n = 1 ~ 5) rings for 11 possible isomers, and seven-membered 1, 2-Se_2S_5, and 1,2-Se_5S_2 rings for four possible isomers. All these compounds take chair conformations. Figure II-63 shows the conformation and numbering of atoms of the seven-membered ring system. The results of their calculations indicate that the stretching frequencies of various isomers are sufficiently different so that Raman spectroscopy can be used to distinguish these isomers. The Raman spectrum of 1, 2, 5, 6-Se_4S_4 has been assigned empirically.[1361] Simon and Paetzold[1362] made an extensive study on vibrational spectra of selenium compounds.

REFERENCES

1. The original references are found in G. Herzberg, *Molecular Spectra and Molecular Structure*. Vol. 1. *Spectra of Diatomic Molecules*, and K. P. Huber and G. Herzberg, *Molecular Spectra and Molecular Structure*. Vol. IV. *Constants of Diatomic Molecules*, Van Nostrand, Princeton, NJ, 1950 and 1979 resp.
2. A. G. Maki, W. B. Olson, and G. Thompson, *J. Mol. Spectrosc.*, **144,** 257 (1990).
3. C. Yamada and E. Hirota, *J. Chem. Phys.*, **88,** (1988).
4. A. G. Maki and W. B. Olson, *J. Chem. Phys.*, **90,** 6887 (1989).
5. U. Magg and H. Jones, *Chem. Phys. Lett.*, **146,** 415 (1988).
6. U. Magg, H. Birk, and H. Jones, *Chem. Phys. Lett.*, **151,** 503 (1988).
7. U. Magg and H. Jones, *Chem. Phys. Lett.*, **148,** 6 (1988).
8. B. Lemoine, C. Demuynck, J. L. Destombes, and P. B. Davies, *J. Chem. Phys.*, **89,** 673 (1988).
9. D. Petitprez, B. Lemoine, C. Demuynck, J. L. Destombes, and B. Macke, *J. Chem. Phys.*, **91,** 4462 (1989).
10. U. Magg, H. Birk, and H. Jones, *Chem. Phys. Lett.*, **151,** 263 (1988).
11. H. Birk, R. D. Urban, P. Polomsky, and H. Jones, *J. Chem. Phys.*, **94,** 5435 (1991).
12. U. Magg, H. Birk, and H. Jones, *Chem. Phys. Lett.*, **149,** 321 (1988).
13. R. D. Urban and H. Jones, *Chem. Phys. Lett.*, **163,** 34 (1989).
14. R. B. Wright, J. K. Bates, and D. M. Gruen, *Inorg. Chem.*, **17,** 2275 (1978).
15. H. Birk and H. Jones, *Chem. Phys. Lett.*, **161,** 27 (1989).
16. R. D. Urban, H. Birk, P. Polomsky, and H. Jones, *J. Chem. Phys.*, **94,** 2523 (1991).
17. R. D. Urban, U. Magg, H. Birk, and H. Jones, *J. Chem. Phys.*, **92,** 14 (1990).
18. J. L. Deutsch, W. S. Neil, and D. A. Ramsay, *J. Mol. Spectrosc.*, **125,** 115 (1987).

19. R. D. Urban, U. Magg, and H. Jones, *Chem. Phys. Lett.*, **154,** 135 (1989).
20. A. H. Bahnmaier, R. D. Urban, and H. Jones, *Chem. Phys. Lett.*, **155,** 269 (1989).
21. R. D. Urban, A. H. Bahnmaier, U. Magg, and H. Jones, *Chem. Phys. Lett.*, **158,** 443 (1989).
22. P. F. Barnath, *J. Chem. Phys.*, **86,** 4838 (1987).
23. A. E. Douglas, *Can. J. Phys.*, **35,** 71 (1957).
24. M. Petri, U. Simon, W. Zimmermann, W. Urban, J. P. Towle, and M. Brown, *Mol. Phys.*, **72,** 315 (1991).
25. U. Simon, M. Petri, W. Zimmermann, and W. Urban, *Mol. Phys.*, **71,** 1163 (1990).
26. W. Zimmermann, T. Neils, E. Bachem, R. Pahnke, and W. Urban, *Mol. Phys.*, **68,** 199 (1989).
27. U. Magg and H. Jones, *Chem. Phys. Lett.*, **166,** 253 (1990).
28. H. C. Miller and J. W. Farley, *J. Chem. Phys.*, **86,** 1167 (1987).
29. R. D. Urban, P. Polomsky, and H. Jones, *Chem. Phys. Lett.*, **181,** 485 (1991).
30. A. M. R. P. Bopegedra, C. R. Brazier, and P. F. Bernath, *Chem. Phys. Lett.*, **162,** 301 (1989).
31. B. D. Rehfuss, M. W. Crofton, and T. Oka, *J. Chem. Phys.*, **85,** 1785 (1986).
32. M. Gruebele, M. Polak, and R. J. Saykally, *J. Chem. Phys.*, **86,** 1698 (1987).
33. D. H. Rank, D. P. Eastman, B. S. Rao, and T. A. Wiggins, *J. Opt. Soc. Am.*, **52,** 1 (1962).
34. S. A. Rogers, C. R. Brazier, and P. F. Bernath, *J. Chem. Phys.*, **87,** 159 (1987).
35. W. Klemperer, W. G. Norris, A. Büchler, and A. G. Emslie, *J. Chem. Phys.*, **33,** 1534 (1960).
36. S. E. Veazey and W. Gordy, *Phys. Rev.*, **138A,** 1303 (1965).
37. S. A. Rice and W. Klemperer, *J. Chem. Phys.*, **27,** 573 (1957).
38. V. I. Baikov and K. P. Vasilevskii, *Opt. Spectrosc.*, **22,** 198 (1967).
39. A. G. Maki and W. B. Olson, *J. Mol. Spectrosc.*, **140,** 185 (1990).
40. H. G. Hedderich and C. E. Blom, *J. Chem. Phys.*, **90,** 4660 (1989).
41. H. G. Hedderich and C. E. Blom, *J. Mol. Spectrosc.*, **140,** 103 (1990).
42. A. D. Kirkwood, K. D. Bier, J. K. Thompson, T. L. Haslett, A. S. Huber, and M. Moskovits, *J. Phys. Chem.*, **95,** 2644 (1991).
43. C. Destoky, I. Dubois, and H. Bredohl, *J. Mol. Spectrosc.*, **136,** 216 (1989).
44. E. Mahieu, I. Dobois, and H. Bredohl, *J. Mol. Spectrosc.*, **134,** 317 (1989).
45. H. Uehara, K. Horiai, K. Nakagawa, and H. Suguro, *Chem. Phys. Lett.*, **178,** 553 (1991).
46. G. W. Lemire, G. A. Bishea, S. A. Heidecke, and M. D. Morse, *J. Chem. Phys.*, **92,** 121 (1990).
47. H. Uehara, K. Horiai, T. Mitani, and H. Suguro, *Chem. Phys. Lett.*, **162,** 137 (1989).
48. M. Gruebele, M. Polak, and R. J. Saykally, *Chem. Phys. Lett.*, **125,** 165 (1986).

49. M. Gruebele, M. Polak, G. A. Blake, and R. J. Saykally, *J. Chem. Phys.*, **85,** 6276 (1986).
50. P. B. Davies and P. A. Martin, *Mol. Phys.*, **70,** 89 (1990).
51. Y. Akiyama, K. Tanaka, and T. Tanaka, *Chem. Phys. Lett.*, **155,** 15 (1989).
52. Y. Akiyama, K. Tanaka, and T. Tanaka, *Chem. Phys. Lett.*, **165,** 335 (1990).
53. R. W. Martin and A. J. Merer, *Can. J. Phys.*, **51,** 125 (1973).
54. H. Birk and H. Jones, *Chem. Phys. Lett.*, **181,** 245 (1991).
55. C. Yamada, M. C. Chang, and E. Hirota, *J. Chem. Phys.*, **86,** 3804 (1987).
56. H. Kanamori, C. Yamada, J. E. Butler, K. Kawaguchi, and E. Hirota, *J. Chem. Phys.*, **83,** 4945 (1985).
57. K. Kawaguchi, E. Hirota, M. Ohishi, H. Suzuki, S. Takano, S. Yamamoto, and S. Saito, *J. Mol. Spectrosc.*, **130,** 81 (1988).
58. J. A. Coxon, S. Naxakis, and A. B. Yamashita, *Spectrochim. Acta*, **41A,** 1409 (1985).
59. E. H. Fink, K. D. Setzer, D. A. Ramsay, and M. Vervloet, *Chem. Phys. Lett.*, **179,** 103 (1991).
60. J. B. Burkholder, P. D. Hammer, C. J. Howard, and A. R. W. McKellar, *J. Mol. Spectrosc.*, **118,** 471 (1986).
61. H. Bürger, P. Schulz, E. Jacob, and M. Föhnle, *Z. Naturforsch.*, **41A,** 1015 (1986).
62. W. V. F. Brooks and B. Crawford, *J. Chem. Phys.*, **23,** 363 (1955).
63. J. A. Horsley and R. F. Barrow, *Trans. Faraday Soc.*, **63,** 32 (1967).
64. C. I. Frum, R. Engelman, and P. F. Bernath, *Chem. Phys. Lett.*, **167,** 356 (1990).
65. F. Ito, K. Sugawara, T. Nakanaga, H. Takeo, and C. Matsumura, *J. Mol. Spectrosc.*, **142,** 191 (1990).
66. M. E. Alikhani, L. Manceron, and J.-P. Perchard, *J. Chem. Phys.*, **92,** 22 (1990).
67. W. G. Fateley, H. A. Bent, and B. Crawford, Jr., *J. Chem. Phys.*, **31,** 204 (1959).
68. P. Brechignac, S. De Benedictis, N. Halberstadt, B. J. Whittaker, and S. Avrillier, *J. Chem. Phys.*, **83,** 2064 (1985).
69. P. C. Souers, D. Fearon, R. Garza, E. M. Kelly, P. E. Roberts, R. H. Sanborn, R. T. Tsugawa, J. L. Hunt, and J. D. Poll, *J. Chem. Phys.*, **70,** 1581 (1979).
70. R. A. Teichman, III, M. Epting, and E. R. Nixon, *J. Chem. Phys.*, **68,** 336 (1978).
71. G. A. Ozin, *Chem. Commun.*, 1325 (1969).
72. L. Andrews and R. R. Smardzewski, *J. Chem. Phys.*, **58,** 2258 (1973).
73. J. C. Evans, *Chem. Commun.*, **682,** (1969).
74. A. G. Hopkins and C. W. Brown, *J. Chem. Phys.*, **62,** 1598 (1975).
75. H. Fabian and F. Fisch, *J. Raman Spectrosc.*, **20,** 639 (1989).
76. W. F. Howard, Jr. and L. Andrews, *J. Am. Chem. Soc.*, **95,** 3045 (1973).
77. M. R. Clarke and G. Mamantov, *Inorg. Nucl. Chem. Lett.*, **7,** 993 (1971).
78. W. F. Howard, Jr. and L. Andrews, *J. Am. Chem. Soc.*, **95,** 2056 (1973).
79. R. J. Gillespie and M. J. Morton, *Chem. Commun.*, 1565 (1968).
80. W. Holzer, W. F. Murphy, and H. J. Bernstein, *J. Chem. Phys.*, **52,** 399 (1970).
81. R. J. Gillespie and M. J. Morton, *J. Mol. Spectrosc.*, **30,** 178 (1969).

82. W. F. Howard, Jr. and L. Andrews, *J. Am. Chem. Soc.*, **97,** 2956 (1975).
83. W. Schulze, H. U. Becker, R. Minkwitz, and K. Manzel, *Chem Phys. Lett.*, **55,** 59 (1978).
84. A. Givan and A. Loewenschuss, *Chem. Phys. Lett.*, **62,** 592 (1979).
85. R. D. Jones, D. A. Summerville, and F. Basolo, *Chem. Rev.*, **79,** 139 (1979).
86. A. J. Edwards, W. E. Falconer, J. E. Griffiths, W. A. Sunder, and M. J. Vasile, *J. Chem. Soc., Dalton Trans.*, 1129 (1974).
87. L. Andrews and R. R. Smardzewski, *J. Chem. Phys.*, **58,** 2258 (1973).
88. H. H. Eysel and S. Thym, *Z. Anorg. Allg. Chem.*, **411,** 97 (1975).
89. R. Steudel, *Z. Naturforsch.*, **30B,** 281 (1975).
90. L. Manceron, A. M. LeQuere, and J. P. Perchard, *J. Phys. Chem.*, **93,** 2960 (1989).
91. W. E. Thompson and M. E. Jacox, *J. Chem. Phys.*, **91,** 3826 (1989).
92. V. Adamantides, D. Neisius, and G. Verhaegen, *Chem. Phys.*, **48,** 215 (1980).
93. A. H. Jubert and E. L. Varetti, *An. Quim.*, **82,** 227 (1986).
94. W. E. Thompson and M. E. Jacox, *J. Chem. Phys.*, **93,** 3856 (1990).
95. R. B. Wright, J. K. Bates, and D. M. Gruen, *Inorg. Chem.*, **17,** 2275 (1978).
96. D. E. Milligan and M. E. Jacox, *J. Chem. Phys.*, **52,** 2594 (1970).
97. W. Seebass, J. Werner, W. Urban, E. R. Comben, and J. M. Brown, *Mol. Phys.*, **62,** 161 (1987).
98. K. Rosengren and G. C. Pimentel, *J. Chem. Phys.*, **43,** 507 (1965).
99. J. R. Anacona, P. B. Davies, and S. A. Johnson, *Mol. Phys.*, **56,** 989 (1985).
100. N. Acquista, L. J. Schoen, and D. R. Lide, Jr., *J. Chem. Phys.*, **48,** 1534 (1968).
101. B.-M. Cheng, Y.-P. Lee, and J. F. Ogilvie, *Chem. Phys. Lett.*, **151,** 109 (1988).
102. W. R. Busing, *J. Chem. Phys.*, **23,** 933 (1955).
103. J. C. Owrutsky, N. H. Rosenbaum, L. M. Tack, and R. J. Saykally, *J. Chem. Phys.*, **83,** 5338 (1985).
104. M. W. Crofton, R. S. Altman, M. F. Jagod, and T. Oka, *J. Phys. Chem.*, **89,** 3614 (1985).
105. Y. Aratono, M. Nakashima, M. Saeki, and E. Tachikawa, *J. Phys. Chem.*, **90,** 1528 (1986).
106. N. Acquista and L. J. Schoen, *J. Chem. Phys.*, **53,** 1290 (1970).
107. S. Civiš, C. E. Blom, and P. Jensen, *J. Mol. Spectrosc.*, **138,** 69 (1989).
108. D. Goddon, A. Groh, H. J. Hansen, M. Schneider, and W. Urban, *J. Mol. Spectrosc.*, **147,** 392 (1991).
109. R. J. Van Zee, T. C. DeVore, and W. Weltner, Jr., *J. Chem. Phys.*, **71,** 2051 (1979).
110. D. White, K. S. Seshadri, D. F. Dever, D. E. Mann, and M. J. Linevsky, *J. Chem. Phys.*, **39,** 2463 (1963).
111. K. S. Seshadri, D. White, and D. E. Mann, *J. Chem. Phys.*, **45,** 4697 (1966).
112. A. Snelson and K. S. Pitzer, *J. Phys. Chem.*, **67,** 882 (1963).
113. S. Schlick and O. Schnepp, *J. Chem. Phys.*, **41,** 463 (1964).

114. G. A. Thompson, A. G. Maki, W. B. Olson, and A. Weber, *J. Mol. Spectrosc.*, **124,** 130 (1987).
115. K. Horiai, T. Fujimoto, K. Nakagawa, and H. Uehara, *Chem. Phys. Lett.*, **147,** 133 (1988).
116. H. Uehara, K. Horiai, K. Nakagawa, and T. Fujimoto, *J. Mol. Spectrosc.*, **134,** 98 (1989).
117. D. E. Mann, G. V. Calder, K. S. Seshadri, D. White, and M. J. Linevsky, *J. Chem. Phys.*, **46,** 1138 (1967).
118. L. Andrews, E. S. Prochaska, and B. S. Ault, *J. Chem. Phys.*, **69,** 556 (1978).
119. S. Civiš, H. G. Hedderich, and C. E. Blom, *Chem. Phys. Lett.*, **176,** 489 (1991).
120. L. Andrews and B. S. Ault, *J. Mol. Spectrosc.*, **68,** 114 (1977).
121. W. Welter, Jr. and D. McLeod, Jr., *J. Phys. Chem.*, **69,** 3488 (1965).
122. F. W. Froben and F. Rogge, *Chem. Phys. Lett.*, **78,** 264 (1981).
123. D. W. Green, W. Korfmacher, and D. M. Gruen, *J. Chem. Phys.*, **58,** 404 (1973).
124. W. Weltner, Jr. and D. McLeod, Jr., *J. Chem. Phys.*, **42,** 882 (1965).
125. D. W. Green and K. M. Ervin, *J. Mol. Spectrosc.*, **89,** 145 (1981).
126. D. W. Green, G. T. Reedy, and J. G. Kay, *J. Mol. Spectrosc.*, **78,** 257 (1979).
127. J. G. Kay, D. W. Green, K. Duca, and G. L. Zimmerman, *J. Mol. Spectrosc.*, **138,** 49 (1989).
128. D. W. Green and G. T. Reedy, *J. Mol. Spectrosc.*, **74,** 423 (1979).
129. D. W. Green and G. T. Reedy, *J. Chem. Phys.*, **65,** 2921 (1976).
130. S. D. Gabelnick, G. T. Reedy, and M. G. Chasanov, *J. Chem. Phys.*, **58,** 4468 (1973).
131. D. M. Green and G. T. Reedy, *J. Chem. Phys.*, **69,** 552 (1978).
132. D. W. Green and G. T. Reedy, *J. Chem. Phys.*, **69,** 544 (1978).
133. E. S. Prochaska and L. Andrews, *J. Chem. Phys.*, **72,** 6782 (1980).
134. T. Nakanaga, H. Takeo, S. Kondo, and C. Matsumura, *Chem. Phys. Lett.*, **114,** 88 (1985).
135. S. J. Baros, M. Haak, and J. W. Nibler, *J. Chem. Phys.*, **82,** 670 (1985).
136. A. Snelson, *J. Phys. Chem.*, **71,** 3202 (1967).
137. J. M. Brom, Jr. and H. F. Franzen, *J. Chem. Phys.*, **54,** 2874 (1971).
138. A. G. Maki, *J. Mol. Spectrosc.*, **137,** 147 (1989).
139. M. E. Jacox and D. E. Milligan, *J. Chem. Phys.*, **50,** 3252 (1969).
140. H. Dubost and A. Abouaf-Marguin, *Chem. Phys. Lett.*, **17,** 269 (1972).
141. R. R. Steudel, *Z. Anorg. Allg. Chem.*, **361,** 180 (1968).
142. R. A. Penneman and L. H. Jones, *J. Chem. Phys.*, **24,** 293 (1956).
143. D. E. Milligan and M. E. Jacox, *J. Chem. Phys.*, **47,** 278 (1971).
144. C. Yamada, E. Hirota, S. Yamamoto, and S. Saito, *J. Chem. Phys.*, **88,** 46 (1988).
145. J. S. Anderson and J. S. Ogden, *J. Chem. Phys.*, **51,** 4189 (1969).
146. C. I. Frum, R. Engleman, and P. F. Bernath, *J. Chem. Phys.*, **93,** 5457 (1990).
147. J. S. Ogden and M. J. Ricks, *J. Chem. Phys.*, **52,** 352 (1970).

148. C. P. Marino, J. D. Guérin, and E. R. Nixon, *J. Mol. Spectrosc.*, **51,** 160 (1974).
149. J. S. Ogden and M. J. Ricks, *J. Chem. Phys.*, **53,** 896 (1970).
150. J. S. Ogden and M. J. Ricks, *J. Chem. Phys.*, **56,** 1658 (1972).
151. A. G. Maki and F. J. Lovas, *J. Mol. Spectrosc.*, **125,** 188 (1987).
152. V. P. Babaeva and V. Y. Rosolovskii, *Russ. J. Inorg. Chem. (Engl. Transl.)*, **16,** 471 (1971).
153. W. A. Guillory and C. E. Hunter, *J. Chem. Phys.*, **50,** 3516 (1969).
154. D. E. Tevault and L. Andrews, *J. Phys. Chem.*, **77,** 1646 (1973).
155. D. E. Tevault and L. Andrews, *J. Phys. Chem.*, **77,** 1640 (1973).
156. R. M. Atkins and P. L. Timms, *Spectrochim. Acta*, **33A,** 853 (1977).
157. H. Kanamori, E. Tiemann, and E. Hirota, *J. Chem. Phys.*, **89,** 621 (1988).
158. W. Holzer, W. F. Murphy, and H. J. Bernstein, *J. Mol. Spectrosc.*, **32,** 13 (1969).
159. M. E. Jacox and D. E. Milligan, *J. Chem. Phys.*, **46,** 184 (1967); **40,** 2461 (1964).
160. L. Andrews and J. I. Raymond, *J. Chem. Phys.*, **55,** 3078 (1971).
161. F. K. Chi and L. Andrews, *J. Phys. Chem.*, **77,** 3062 (1973).
162. D. E. Tevault, N. Walker, R. R. Smardzewski, and W. B. Fox, *J. Phys. Chem.*, **82,** 2733 (1978).
163. M. Feuerhahn, R. Minkwitz, and G. Vahl, *Spectrochim. Acta*, **36A,** 183 (1980).
164. G. A. Olah and M. B. Comisarow, *J. Am. Chem. Soc.*, **91,** 2172 (1969).
165. J. J. Orlando, J. B. Burkholder, A. M. R. P. Bopegedera, and C. J. Howard, *J. Mol. Spectrosc.*, **145,** 278 (1991).
166. W. Levason, J. S. Ogden, M. D. Spicer, and N. A. Young, *J. Chem. Soc., Dalton Trans.*, 349 (1990).
167. K. Nakagawa, K. Horiai, T. Konno, and H. Uehara, *J. Mol. Spectrosc.*, **131,** 233 (1988).
168. R. Minkwitz and F. W. Froben, *Chem. Phys. Lett.*, **39,** 473 (1976).
169. U. Magg, H. Birk, K. P. R. Nair, and H. Jones, *Z. Naturforsch.*, **44A,** 313 (1989).
170. W. Holzer, W. F. Murphy, and H. J. Bernstein, *J. Chem. Phys.*, **52,** 399 (1970).
171. M. J. Linevsky, *J. Chem. Phys.*, **34,** 587 (1961).
172. A. Snelson, *J. Chem. Phys.*, **46,** 3652 (1967).
173. M. Freiberg, A. Ron., and O. Schnepp, *J. Phys. Chem.*, **72,** 3526 (1968).
174. T. P. Martin and H. Schaber, *J. Chem. Phys.*, **68,** 4299 (1978).
175. J. R. Durig, K. K. Lau, G. Nagarajan, M. Walker, and J. Bragin, *J. Chem. Phys.*, **50,** 2130 (1969).
176. V. E. Bondybey and J. W. Nibler, *J. Chem. Phys.*, **58,** 2125 (1973).
177. D. Sellmann, A. Brandl, and R. Endell, *Angew. Chem., Int. Ed. Engl.*, **12,** 1019 (1973).
178. J. L. Hollenberg, *J. Chem. Phys.*, **46,** 3271 (1967); D. F. Smith, *ibid.*, **48,** 1429 (1968).
179. P. A. Giguère and N. Zengin, *Can. J. Chem.*, **36,** 1013 (1958).
180. D. F. Hornig and W. E. Osberg, *J. Chem. Phys.*, **23,** 662 (1955); G. L. Hiebert and D. F. Hornig, *ibid.*, **28,** 316 (1958).
181. R. D. Hunt and L. Andrews, *J. Chem. Phys.*, **82,** 4442 (1985).

181a. L. Andrews, S. R. Davis, and R. D. Hunt, *Mol. Phys.*, **77,** 993 (1992).
182. D. C. Clary, C. M. Lovejoy, S. V. O'Neil, and D. J. Nesbitt, *Phys. Rev. Lett.*, **61,** 1576 (1988).
183. C. M. Lovejoy and D. J. Nesbitt, *Chem. Phys. Lett.*, **147,** 490 (1988).
184. R. L. Robinson, D. Ray, D. H. Gwo, and R. J. Saykally, *J. Chem. Phys.*, **87,** 5149 (1987).
185. R. L. Robinson, D. H. Gwo, and R. J. Saykally, *J. Chem. Phys.*, **87,** 5156 (1987).
186. C. M. Lovejoy and D. J. Nesbitt, *Chem. Phys. Lett.*, **146,** 582 (1988).
187. R. L. Robinson, D. H. Gwo, and R. J. Saykally, *Mol. Phys.*, **63,** 1021 (1988).
188. C. M. Lovejoy, M. D. Schuder, and D. J. Nesbitt, *J. Chem. Phys.*, **85,** 4890 (1986).
189. V. Lorenzelli, *C. R. Hebd. Seances Acad. Sci.*, **258,** 5386 (1964).
190. W. B. Person, R. E. Humphrey, W. A. Deskin, and A. I. Popov, *J. Am. Chem. Soc.*, **80,** 2049 (1958); W. B. Person, R. E. Erickson, and R. E. Buckles, *ibid.*, **82,** 29 (1960); A. I. Popov, R. E. Humphrey, and W. B. Person, *ibid.*, 1850.
191. P. Klaboe, *J. Am. Chem. Soc.*, **89,** 3667 (1967).
192. L. H. Jones, *J. Chem. Phys.*, **22,** 217 (1954).
193. W. D. Stalleup and D. Williams, *J. Chem. Phys.*, **10,** 199 (1942).
194. R. A. Penneman and L. H. Jones, *J. Chem. Phys.*, **28,** 169 (1958).
195. L. H. Jones, *J. Mol. Spectrosc.*, **45,** 55 (1973); **49,** 82 (1974).
196. A. Snelson, *J. Phys. Chem.*, **72,** 250 (1968).
197. M. L. Lesiecki and J. W. Nibler, *J. Chem. Phys.*, **64,** 871 (1976).
198. G. V. Calder, D. E. Mann, K. S. Seshadri, M. Allavena, and D. White, *J. Chem. Phys.*, **51,** 2093 (1969).
199. D. White, G. V. Calder, S. Hemple, and D. E. Mann, *J. Chem. Phys.*, **59,** 6645 (1973).
199a. R. J. Konings and J. E. Fearon, *Chem. Phys. Lett.*, **206,** 57 (1993).
200. E. D. Samsonova, S. B. Osin, and V. F. Shevel'kov, *Russ. J. Inorg. Chem. (Engl. Transl.)*, **33,** 1598 (1988).
201. J. J. Habeeb and D. G. Tuck, *J. Chem. Soc., Dalton Trans.*, 866 (1976).
202. D. E. Milligan and M. E. Jacox, *J. Chem. Phys.*, **48,** 2265 (1968).
203. L. Andrews, *J. Chem. Phys.*, **48,** 979 (1968).
204. L. Andrews and T. G. Carver, *J. Chem. Phys.*, **49,** 896 (1968).
205. J. W. Hastie, R. H. Hauge, and J. L. Margrave, *J. Am. Chem. Soc.*, **91,** 2536 (1969).
206. D. E. Milligan and M. E. Jacox, *J. Chem. Phys.*, **49,** 1938 (1968).
207. G. Maass, R. H. Hauge, and J. L. Margrave, *Z. Anorg. Allg. Chem.*, **392,** 295 (1972).
208. J. W. Hastie, R. H. Hauge, and J. L. Margrave, *J. Phys. Chem.*, **72,** 4492 (1968).
209. L. Andrews and D. L. Frederick, *J. Am. Chem. Soc.*, **92,** 775 (1970).
210. R. H. Hauge, J. W. Hastie, and J. L. Margrave, *J. Mol. Spectrosc.*, **45,** 420 (1973).
211. G. A. Ozin and A. Vander Voet, *J. Chem. Phys.*, **56,** 4768 (1972).

212. I. R. Beattie and R. O. Perry, *J. Chem. Soc. A*, 2429 (1970).
213. M. D. Harmony and R. J. Myers, *J. Chem. Phys.*, **37,** 636 (1962).
214. J. K. Burdett, L. Hodges, V. Dunning, and J. H. Current, *J. Phys. Chem.*, **74,** 4053 (1970).
215. L. Andrews and D. L. Frederick, *J. Phys. Chem.*, **73,** 2774 (1969).
216. M. M. Rochkind and G. C. Pimentel, *J. Chem. Phys.*, **42,** 1361 (1965).
217. R. J. Glinski, C. D. Taylor, and F. W. Kutzler, *J. Phys. Chem.*, **94,** 6196 (1990).
217a. P. Hassanzadeh and L. Andrews, *J. Phys. Chem.*, **96,** 79 (1992).
218. M. Feuerhahn and G. Vahl, *Inorg. Nucl. Chem. Lett.*, **16,** 5 (1980).
219. G. A. Ozin and A. Vander Voet, *Chem. Commun.*, 896, (1970).
220. H. H. Claassen, G. L. Goodman, J. G. Malm, and F. Schreiner, *J. Chem. Phys.*, **42,** 1229 (1965).
221. P. A. Agron, G. M. Begun, H. A. Levy, A. A. Mason, C. G. Jones, and D. F. Smith, *Science*, **139,** 842 (1963).
222. L. Y. Nelson and G. C. Pimentel, *Inorg. Chem.*, **6,** 1758 (1967).
223. I. Person, M. Sandström, A. T. Steel, M. J. Zapetero, and R. Aakesson, *Inorg. Chem.*, **30,** 4075 (1991).
224. D. N. Waters and B. Basak, *J. Chem. Soc. A*, 2733 (1971).
225. P. Braunstein and R. J. H. Clark, *J. Chem. Soc., Dalton Trans.*, 1845 (1973).
226. A. Givan and A. Loewenschuss, *J. Chem. Phys.*, **72,** 3809 (1980).
227. A. Givan and A. Loewenschuss, *J. Chem. Phys.*, **68,** 2228 (1978).
228. A. Loewenschuss, A. Ron, and O. Schnepp, *J. Chem. Phys.*, **50,** 2502 (1969).
229. J. W. Hastie, R. Hauge, and J. L. Margrave, *J. Chem. Phys.*, **51,** 2648 (1969).
230. J. W. Hastie, R. H. Hauge, and J. L. Margrave, *Chem. Commun.*, 1452 (1969).
231. O. V. Blinova, V. G. Shklyarik, and L. D. Shcherbe, *Zh. Fiz. Khim.*, **62,** 1640 (1988).
232. M. E. Jacox and D. E. Milligan, *J. Chem. Phys.*, **51,** 4143 (1969).
233. I. Eliezer and A. Reger, *Coord. Chem. Rev.*, **9,** 189 (1972-1973).
234. J. W. Hastie, R. H. Hauge, and J. L. Margrave, *High Temp. Sci.*, **1,** 76 (1969).
235. I. R. Beattie, P. J. Jones, and N. A. Young, *Mol. Phys.*, **72,** 1309 (1991).
236. I. R. Beattie, P. J. Jones, and N. A. Young, *Inorg. Chem.*, **30,** 2250 (1991).
237. K. R. Thompson and K. D. Carlson, *J. Chem. Phys.*, **49,** 4379 (1968).
238. D. L. Cocke, C. A. Chang, and K. A. Gingerich, *Appl. Spectrosc.*, **27,** 260 (1973).
239. H. Huber, E. P. Kündig, G. A. Ozin, and A. Vander Voet, *Can. J. Chem.*, **52,** 95 (1974).
240. F. Ramondo, L. Bencivenni, and V. Rossi, *Chem. Phys.*, **124,** 291 (1988).
241. L. Andrews, *J. Phys. Chem.*, **73,** 3922 (1969).
242. A. Sommer, D. White, M. J. Linevsky, and D. E. Mann, *J. Chem. Phys.*, **38,** 87 (1963).
243. A. Snelson, *J. Phys. Chem.*, **74,** 2574 (1970).
244. A. J. Hinchcliffe and J. S. Ogden, *J. Phys. Chem.*, **77,** 2537 (1973); *Chem. Commun.*, 1053 (1969).

245. J. H. Taylor, W. S. Benedict, and J. Strong, *J. Chem. Phys.*, **20,** 1884 (1952).
246. A. H. Nielsen and R. J. Lagemann, *J. Chem. Phys.*, **22,** 36 (1954).
247. J. W. Johns, *Can. J. Phys.*, **42,** 1004 (1964).
248. M. E. Jacox and W. E. Thompson, *J. Chem. Phys.*, **91,** 1410 (1989).
249. T. Wentink, Jr., *J. Chem. Phys.*, **29,** 188 (1958).
250. H. Bürger and H. Willner, *J. Mol. Spectrosc.*, **128,** 221 (1988).
251. J. S. Anderson, A. Bos, and J. S. Ogden, *Chem. Commun.*, 1381 (1971).
252. A. R. Evans and D. B. Fitchen, *Phys. Rev.*, **B2,** 1074 (1970).
253. D. E. Tevault and L. Andrews, *Spectrochim. Acta*, **30A,** 969 (1974).
254. R. V. St. Louis and B. L. Crawford, Jr., *J. Chem. Phys.*, **42,** 857 (1965).
255. J. W. Nebgen, A. D. McElroy, and H. F. Klodowski, *Inorg. Chem.*, **4,** 1796 (1965).
256. D. F. Smith, Jr., J. Overend, R. C. Spiker, and L. Andrews, *Spectrochim. Acta*, **28A,** 87 (1972).
257. L. Andrews and R. C. Spiker, *J. Chem. Phys.*, **59,** 1863 (1973); D. M. Thomas and L. Andrews, *J. Mol. Spectrosc.*, **50,** 220 (1974).
258. A. Barbe, C. Secroun, and P. Jouve, *J. Mol. Spectrosc.*, **49,** 171 (1974).
259. P. Lenain, E. Piquenard, J. L. Lesne, and J. Corset, *J. Mol. Struct.*, **142,** 355 (1986).
260. R. J. H. Clark and D. G. Cobbold, *Inorg. Chem.*, **17,** 3169 (1978).
260a. G. J. Janz, J. W. Coutts, J. R. Downey, Jr., and E. Roduner, *Inorg. Chem.*, **15,** 1755 (1976).
261. D. Maillard, M. Allavena, and J. P. Perchard, *Spectrochim. Acta*, **31A,** 1523 (1975).
262. J. Lindenmayer, H. D. Rudolph, and H. Jones, *J. Mol. Spectrosc.*, **119,** 56 (1986).
263. L. Bencivenni, F. Ramondo, R. Teghil, and M. Pelino, *Inorg. Chim. Acta*, **121,** 207 (1986).
264. S. N. Cesaro, M. Spoliti, A. J. Hinchcliffe, and J. S. Ogden, *J. Chem. Phys.*, **55,** 5834 (1971).
265. M. Spoliti, S. N. Cesaro, and E. Ciffari, *J. Chem. Thermodyn.*, **4,** 507 (1972).
266. H. Uehara, K. Kawaguchi, and E. Hirota, *J. Chem. Phys.*, **83,** 5479 (1985).
267. R. D. Spratley, J. J. Turner, and G. C. Pimentel, *J. Chem. Phys.*, **44,** 2063 (1966).
268. J. S. Ogden and J. J. Turner, *J. Chem. Soc. A*, 1483 (1967).
269. M. M. Rochkind and G. C. Pimentel, *J. Chem. Phys.*, **42,** 1361 (1965).
270. H. S. P. Müller and H. Willner, *J. Phys. Chem.*, **97,** 10589 (1993).
270a. A. Arkell and I. Schwager, *J. Am. Chem. Soc.*, **89,** 5999 (1967).
271. W. Levason, J. S. Ogden, M. D. Spicer, and N. A. Young, *J. Chem. Soc., Dalton Trans.*, 349 (1990).
272. A. H. Nielsen and P. J. H. Woltz, *J. Chem. Phys.*, **20,** 1878 (1952).
273. K. M. Tobias and M. Jansen, *Z. Anorg. Allg. Chem.*, **550,** 16 (1987).
274. C. Campbell, J. P. M. Jones, and J. J. Turner, *Chem. Commun.*, 888 (1968).
275. W. Levason, J. S. Ogden, M. D. Spicer, and N. A. Young, *J. Am. Chem. Soc.*, **112,** 1019 (1990).

276. E. Jacob, *Angew. Chem., Int. Ed. Engl.*, **15,** 158 (1976).
277. S. D. Gabelnick, G. T. Reedy, and M. G. Chasanov, *J. Chem. Phys.*, **60,** 1167 (1974).
278. J. G. Kay, D. W. Green, K. Duca, and G. L. Zimmerman, *J. Mol. Spectrosc.*, **138,** 49 (1989).
279. N. S. McIntyre, K. R. Thompson, and W. Weltner, *J. Phys. Chem.*, **75,** 3243 (1971).
280. W. Weltner and D. McLeod, *J. Chem. Phys.*, **42,** 882 (1965).
281. W. Weltner and D. McLeod, *J. Mol. Spectrosc.*, **17,** 276 (1965).
282. S. D. Gabelnick, G. T. Reedy, and M. G. Chasanov, *J. Chem. Phys.*, **58,** 4468 (1973).
283. J. R. Ferraro and A. Walker, *J. Chem. Phys.*, **45,** 550 (1966).
284. D. W. Green and G. T. Reedy, *J. Chem. Phys.*, **69,** 544 (1978).
285. M. Somer, M. Hartweg, K. Peters, T. Popp, and H. G. von Schnering, *Z. Anorg. Allg. Chem.*, **595,** 217 (1991).
286. W. W. Wilson and K. O. Christe, *Inorg. Chem.*, **26,** 1631 (1987).
287. R. C. Spiker, Jr. and L. Andrews, *J. Chem. Phys.*, **59,** 1863 (1973).
288. D. M. Thomas and L. Andrews, *J. Mol. Spectrosc.*, **50,** 220 (1974).
289. R. A. Shepherd and W. R. M. Graham, *J. Chem. Phys.*, **88,** 3399 (1988).
290. D. W. Green and G. T. Reedy, in J. R. Ferraro and L. J. Basile, eds. *Fourier Transform Infrared Spectroscopy: Applications to Chemical Systems*, Vol. 1, Academic Press, New York, 1978.
291. D. W. Green and G. T. Reedy, *J. Chem. Phys.*, **65,** 2921 (1976).
292. D. W. Green and G. T. Reedy, *J. Chem. Phys.*, **69,** 552 (1978).
293. D. W. Green and G. T. Reedy, *J. Chem. Phys.*, **74,** 423 (1979).
294. L. H. Jones and R. A. Penneman, *J. Chem. Phys.*, **21,** 542 (1953).
295. L. H. Jones, *Spectrochim. Acta*, **11,** 409 (1959).
296. J. R. Bartlett and R. P. Cooney, *J. Mol. Struct.*, **193,** 295 (1989).
297. H. R. Hoekstra, *Inorg. Chem.*, **2,** 492 (1963).
298. J. I. Bullock, *J. Chem. Soc. A*, 781 (1969).
299. K. Ohwada, *Spectrochim. Acta*, **24A,** 595 (1968).
300. S. P. McGlynn, J. K. Smith, and W. C. Neely, *J. Chem. Phys.*, **35,** 105 (1961).
301. R. J. Gillespie and M. J. Morton, *Inorg. Chem.*, **9,** 616 (1970).
302. E. S. Prochaska and L. Andrews, *Inorg. Chem.*, **16,** 339 (1977).
303. K. O. Christe, W. Sawodny, and J. P. Guertin, *Inorg. Chem.*, **6,** 1159 (1967).
304. R. J. Gillespie and M. J. Morton, *Inorg. Chem.*, **9,** 811 (1970).
305. K. O. Christe and C. J. Schack, *Inorg. Chem.*, **9,** 2296 (1970).
306. T. Surles, L. A. Quaterman, and H. H. Hyman, *J. Inorg. Nucl. Chem.*, **35,** 668 (1973).
307. K. O. Christe and W. Sawodny, *Inorg. Chem.*, **8,** 212 (1969).
308. R. Minkwitz, J. Nowicki, H. Härtner, and W. Sawodny, *Spectrochim. Acta*, **47A,** 1673 (1991).
309. B. S. Ault and L. Andrews, *J. Chem. Phys.*, **64,** 4853 (1976).

310. W. Gabes and H. Gerding, *J. Mol. Struct.*, **14,** 267 (1972).
311. R. Forneris and Y. Tavares-Forneris, *J. Mol. Struct.*, **23,** 241 (1974).
312. W. W. Wilson and F. Aubke, *Inorg. Chem.*, **13,** 326 (1974).
313. K. O. Christe, R. Bau, and D. Zhao, *Z. Anorg. Allg. Chem.*, **593,** 46 (1991).
314. D. H. Boal and G. A. Ozin, *J. Chem. Phys.*, **55,** 3598 (1971).
315. A. G. Maki and R. Forneris, *Spectrochim. Acta*, **23A,** 867 (1967).
316. M. Couzi, J. C. Cornut, and P. V. Huong, *J. Chem. Phys.*, **56,** 426 (1972).
317. J. J. Rush, L. W. Schroeder, and A. J. Melveger, *J. Chem. Phys.*, **56,** 2793 (1972).
318. B. S. Ault, *J. Phys. Chem.*, **82,** 844 (1978).
319. K. Kawaguchi and E. Hirota, *J. Chem. Phys.*, **87,** 6838 (1987).
320. J. C. Evans and G. Y.-S. Lo, *J. Phys. Chem.*, **70,** 543 (1966).
321. G. C. Stirling, C. J. Ludman, and T. C. Waddington, *J. Chem. Phys.*, **52,** 2730 (1970).
322. J. W. Nibler and G. C. Pimentel, *J. Chem. Phys.*, **47,** 710 (1967).
323. J. C. Evans and G. Y.-S. Lo, *J. Phys. Chem.*, **71,** 3942 (1967).
324. B. S. Ault and L. Andrews, *J. Chem. Phys.*, **64,** 1986 (1976).
325. P. N. Noble, *J. Chem. Phys.*, **56,** 2088 (1972).
326. B. S. Ault and L. Andrews, *Inorg. Chem.*, **16,** 2024 (1977).
327. K. Kaya, N. Mikami, Y. Udagawa, and M. Ito, *Chem. Phys. Lett.*, **16,** 151 (1972).
328. W. Kiefer and H. J. Bernstein, *Chem. Phys. Lett.*, **16,** 5 (1972).
329. M. E. Heyde, L. Rimai, R. G. Kilponen, and D. Gill, *J. Am. Chem. Soc.*, **94,** 5222 (1972).
330. B. S. Ault, *Acc. Chem. Res.*, **15,** 103 (1982).
331. B. S. Ault, *J. Phys. Chem.*, **83,** 837 (1979).
331a. J. C. Evans and G. Y.-S. Lo, *J. Phys. Chem.*, **70,** 11 (1966).
332. Z. L. Xiao, R. H. Hauge, and J. L. Margrave, *J. Phys. Chem.*, **95,** 2696 (1991).
332a. Z. L. Xiao, R. H. Hauge, and J. L. Margrave, *J. Phys. Chem.*, **96,** 636 (1992).
333. J. M. Parnis and G. A. Ozin, *J. Phys. Chem.*, **93,** 1215 (1989).
334. L. Fredin, R. H. Hauge, Z. H. Kafafi, and J. L. Margrave, *J. Chem. Phys.*, **82,** 3542 (1985).
335. G. R. Smith and W. A. Guillory, *J. Chem. Phys.*, **56,** 1423 (1972).
336. D. E. Milligan and M. E. Jacox, *J. Chem. Phys.*, **43,** 4487 (1965).
337. T. Nakata and S. Matsushita, *J. Phys. Chem.*, **72,** 458 (1968).
338. A. Müller, R. Kebabcioglu, B. Krebs, P. Bouclier, J. Portier, and P. Hagenmuller, *Z. Anorg. Allg. Chem.*, **368,** 31 (1969).
339. H. Jacobs and K. M. Hassiepen, *Z. Anorg. Allg. Chem.*, **531,** 108 (1985).
340. D. Forney, M. E. Jacox, and W. E. Thompson, *J. Chem. Phys.*, **98,** 841 (1993).
340a. B. M. Dinelli, M. W. Crofton, and T. Oka, *J. Mol. Spectrosc.*, **127,** 1 (1988).
341. W. S. Benedict, N. Gailar, and E. K. Plyler, *J. Chem. Phys.*, **24,** 1139 (1956).
342. G. E. Walrafen, *J. Chem. Phys.*, **40,** 3249 (1964).
343. C. Haas and D. F. Hornig, *J. Chem. Phys.*, **32,** 1763 (1960); D. F. Hornig, H. F. White, and F. P. Reding, *Spectrochim. Acta*, **12,** 338 (1958).

344. S. Pinchas and M. Halmann, *J. Chem. Phys.*, **31,** 1692 (1959).
345. P. A. Staats, H. W. Morgan, and J. H. Goldstein, *J. Chem. Phys.*, **24,** 916 (1956).
346. I. Kanesaka, H. Hayashi, M. Kita, and K. Kawai, *J. Chem. Phys.*, **93,** 6113 (1990).
347. H. C. Allen and E. K. Plyler, *J. Chem. Phys.*, **25,** 1132 (1956).
348. A. J. Tursi and E. R. Nixon, *J. Chem. Phys.*, **53,** 518 (1970).
349. F. P. Reding and D. F. Hornig, *J. Chem. Phys.*, **27,** 1024 (1957).
350. A. H. Nielsen and H. H. Nielsen, *J. Chem. Phys.*, **5,** 277 (1937).
351. D. M. Cameron, W. C. Sears, and H. H. Nielsen, *J. Chem. Phys.*, **7,** 994 (1939).
352. E. Greinacher, W. Lüttke, and R. Mecke, *Z. Elektrochem.*, **59,** 23 (1955).
353. M. L. Josien and P. Saumagne, *Bull. Soc. Chim. Fr.*, 937 (1956).
354. G. E. Walrafen, M. S. Hokmabadi, and W.-H. Yang, *J. Chem. Phys.*, **85,** 6964 (1986).
355. G. E. Walrafen, *J. Phys. Chem.*, **94,** 2237 (1990).
356. G. E. Walrafen, *Encyclopedia of Earth System Science*, Vol. 4, Academic Press, San Diego, CA, 1992, p. 463.
357. J. E. Bertie, H. J. Labbé, and E. Whalley, *J. Chem. Phys.*, **49,** 775, 2141 (1968); J. E. Bertie and E. Whalley, *ibid.*, **40,** 1637 (1964).
358. G. P. Ayers and A. D. E. Pullin, *Spectrochim. Acta*, **32A,** 1629 (1976).
359. H. C. Allen, E. D. Tidwell, and E. K. Plyler, *J. Chem. Phys.*, **25,** 302 (1956).
360. J. Pacansky and G. V. Calder, *J. Phys. Chem.*, **76,** 454 (1972).
361. P. A. Staats, H. W. Morgan, and J. H. Goldstein, *J. Chem. Phys.*, **25,** 582 (1956).
362. R. E. Dodd and R. Little, *Spectrochim. Acta*, **16,** 1083 (1960).
363. W. O. Freitag and E. R. Nixon, *J. Chem. Phys.*, **24,** 109 (1956).
364. T. B. Freedman and E. R. Nixon, *J. Chem. Phys.*, **56,** 698 (1972).
365. S. Hemple and E. R. Nixon, *J. Chem. Phys.*, **47,** 4273 (1967).
366. Z. K. Ismail, R. H. Hauge, and J. L. Margrave, *J. Mol. Spectrosc.*, **45,** 304 (1973).
367. D. E. Milligan and M. E. Jacox, *J. Chem. Phys.*, **47,** 278 (1967).
368. J. B. Burkholder, A. Sinha, P. D. Hammer, and C. J. Howard, *J. Mol. Spectrosc.*, **126,** 72 (1987).
369. Z. K. Ismail, R. H. Hauge, and J. L. Margrave, *J. Chem. Phys.*, **57,** 5137 (1972).
370. E. R. Lory and R. F. Porter, *J. Am. Chem. Soc.*, **93,** 6301 (1971).
371. J. C. Owrutsky, C. S. Gudeman, C. C. Martner, L. M. Tack, N. H. Rosenbaum, and R. J. Saykally, *J. Chem. Phys.*, **84,** 605 (1986).
372. M. E. Jacox, D. E. Milligan, N. G. Mall, and W. E. Thompson, *J. Chem. Phys.*, **43,** 3734 (1965).
373. A. Maki, E. K. Plyler, and E. D. Tidwell, *J. Res. Natl. Bur. Stand.*, **66A,** 163 (1962).
374. N. H. Rosenbaum, J. C. Owrutsky, and R. J. Saykally, *J. Mol. Spectrosc.*, **133,** 365 (1989).
375. T. Wentink, Jr., *J. Chem. Phys.*, **29,** 188 (1958).
376. W. Beck, *Chem. Ber.*, **95,** 341 (1962); **98,** 298 (1965).

377. K. O. Christe, R. D. Wilson, and W. Sawodny, *J. Mol. Struct.*, **8,** 245 (1971).
378. D. E. Milligan, M. E. Jacox, and A. M. Bass, *J. Chem. Phys.*, **43,** 3149 (1965).
379. G. M. Begun and W. H. Fletcher, *J. Chem. Phys.*, **28,** 414 (1958).
380. V. E. Bondybey and J. H. English, *J. Chem. Phys.*, **67,** 664 (1979).
381. D. E. Milligan and M. E. Jacox, *J. Chem. Phys.*, **47,** 5157 (1967).
382. O. H. Ellestad, P. Klaeboe, E. E. Tucker, and J. Songstad, *Acta Chem. Scand.*, **26,** 1721 (1972).
383. P. Gray and T. C. Waddington, *Trans. Faraday Soc.*, **53,** 901 (1957).
384. R. Tian, J. C. Facelli, and J. Michl, *J. Phys. Chem.*, **92,** 4073 (1988).
385. P. O. Kinell and B. Strandberg, *Acta Chem. Scand.*, **13,** 1607 (1959).
386. H. W. Morgan, *J. Inorg. Nucl. Chem.*, **16,** 368 (1960).
387. O. H. Ellestad, P. Klaeboe, and J. Songstad, *Acta Chem. Scand.*, **26,** 1724 (1972).
388. R. Ahlrichs, S. Schuck, and H. Schnöckel, *Angew. Chem., Int. Ed. Engl.*, **27,** 421 (1988).
389. D. E. Milligan and M. E. Jacox, *J. Chem. Phys.*, **51,** 277 (1969).
390. J. F. Ogilvie, *Spectrochim. Acta*, **23A,** 737 (1967).
391. D. E. Milligan and M. E. Jacox, *J. Chem. Phys.*, **38,** 2627 (1963); *J. Mol. Spectrosc.*, **42,** 495 (1972).
392. P. N. Noble and G. C. Pimentel, *Spectrochim. Acta*, **24A,** 797 (1968).
393. A. Arkell and I. Schwager, *J. Am. Chem. Soc.*, **89,** 5999 (1967).
394. D. E. Tevault and R. R. Smardzewski, *J. Am. Chem. Soc.*, **100,** 3955 (1978).
395. J. A. Goleb, H. H. Claassen, M. H. Studier, and E. H. Appelman, *Spectrochim. Acta*, **28A,** 65 (1972).
396. I. Schwager and A. Arkell, *J. Am. Chem. Soc.*, **89,** 6006 (1967).
397. N. Walker, D. E. Tevault, and R. R. Smardzewski, *J. Chem. Phys.*, **69,** 564 (1978).
398. R. J. Isabel and W. A. Guillory, *J. Chem. Phys.*, **57,** 1116 (1972).
399. R. R. Smardzewski and W. B. Fox, *J. Am. Chem. Soc.*, **96,** 304 (1974); *J. Chem. Phys.*, **60,** 2104 (1974).
400. U. Magg, J. Lindenmeyer, and H. Jones, *J. Mol. Spectrosc.*, **126,** 270 (1987).
401. A. Müller, G. Nagarajan, O. Glemser, and J. Wegener, *Spectrochim. Acta*, **23A,** 2683 (1967); A. Müller, N. Mohan, S. J. Cyvin, N. Weinstock, and O. Glemser, *J. Mol. Spectrosc.*, **59,** 161 (1976).
402. R. R. Ryan and L. H. Jones, *J. Chem. Phys.*, **50,** 1492 (1969); L. H. Jones, L. B. Asprey, and R. R. Ryan, *ibid.*, **47,** 3371 (1967).
403. L. H. Jones, R. R. Ryan, and L. B. Asprey, *J. Chem. Phys.*, **49,** 581 (1968).
404. J. Laane, L. H. Jones, R. R. Ryan, and L. B. Asprey, *J. Mol. Spectrosc.*, **30,** 485 (1969).
405. M. Feuerhahn, W. Hilbig, R. Minkwitz, and U. Engelhardt, *Spectrochim. Acta*, **34A,** 1065 (1978).
406. R. Ahlrichs, R. Becherer, M. Binnewies, H. Borrmann, M. Lakenbrink, S. Schunck, and H. Schnöckel, *J. Am. Chem. Soc.*, **108,** 7905 (1986).
407. R. D. Verma and S. Nagaraj, *J. Mol. Spectrosc.*, **58,** 301 (1975).

408. D. E. Milligan, M. E. Jacox, A. M. Bass, J. J. Comeford, and D. E. Mann, *J. Chem. Phys.*, **42,** 3187 (1965).
409. M. E. Jacox and D. E. Milligan, *J. Chem. Phys.*, **43,** 866 (1965).
410. L. Andrews, F. K. Chi, and A. Arkell, *J. Am. Chem. Soc.*, **96,** 1997 (1974).
411. H. Schnöckel and S. Schunck, *Z. Anorg. Allg. Chem.*, **552,** 163 (1987).
412. H. Schnöckel and S. Schunck, *Z. Anorg. Allg. Chem.*, **552,** 155 (1987).
413. C. E. Smith, D. E. Milligan, and M. E. Jacox, *J. Chem. Phys.*, **54,** 2780 (1971).
414. S. C. Peake and A. J. Downs, *J. Chem. Soc., Dalton Trans.*, 859 (1974).
415. G. A. Ozin and A. Vander Voet, *J. Chem. Phys.*, **56,** 4768 (1972).
416. G. Herzberg, *Infrared and Raman Spectra of Polyatomic Molecules*, Van Nostrand, Princeton, NJ, 1945, p. 295.
417. F. P. Reding and D. F. Hornig, *J. Chem. Phys.*, **19,** 594 (1951); **22,** 1926 (1954).
418. H. W. Morgan, P. A. Staats, and J. H. Goldstein, *J. Chem. Phys.*, **27,** 1212 (1957).
419. S. Sundaram and F. F. Cleveland, *J. Mol. Spectrosc.*, **5,** 61 (1960).
420. E. Lee and C. K. Wu, *Trans. Faraday Soc.*, **35,** 1366 (1939).
421. W. H. Haynie and H. H. Nielsen, *J. Chem. Phys.*, **21,** 1839 (1953).
422. P. V. Huong and B. Desbat, *J. Raman Spectrosc.*, **2,** 373 (1974).
423. R. C. Taylor and G. L. Vidale, *J. Am. Chem. Soc.*, **78,** 5999 (1956).
424. R. Savoie and P. A. Giguère, *J. Chem. Phys.*, **41,** 2698 (1964).
425. P. A. Giguère and C. Madec, *Chem. Phys. Lett.*, **37,** 569 (1976).
425a. R. Minkwitz, A. Kornath and W. Sawodny, *Angew. Chem. Int. Ed. Engl.*, **31,** 643 (1992).
426. K. Burczyk and A. J. Downs, *J. Chem. Soc., Dalton Trans.*, 2351 (1990).
427. C. Yamada and E. Hirota, *Phys. Rev. Lett.*, **56,** 923 (1986).
428. T. Birchall and I. Drummond, *J. Chem. Soc. A*, 3162 (1971).
429. R. D. Gillard and G. Wilkinson, *J. Chem. Soc.*, 1640 (1964).
430. S. Sundaram, F. Suszek, and F. F. Cleveland, *J. Chem. Phys.*, **32,** 251 (1960).
431. G. DeAlti, G. Costa, and V. Galasso, *Spectrochim. Acta*, **20,** 965 (1964).
432. M. Pariseau, E. Wu, and J. Overend, *J. Chem. Phys.*, **39,** 217 (1963).
433. J. P. Perchard, R. B. Bohn, and L. Andrews, *J. Phys. Chem.*, **95,** 2707 (1991).
434. J. G. Contreras, J. S. Poland, and D. G. Tuck, *J. Chem. Soc., Dalton Trans.*, 922 (1973).
435. D. E. Milligan, M. E. Jacox, and J. J. Comeford, *J. Chem. Phys.*, **44,** 4058 (1966).
436. D. E. Milligan, M. E. Jacox, and W. A. Guillory, *J. Chem. Phys.*, **49,** 5330 (1968).
437. M. E. Jacox and D. E. Milligan, *J. Chem. Phys.*, **49,** 3130 (1968).
438. W. A. Guillory and C. E. Smith, *J. Chem. Phys.*, **53,** 1661 (1970).
439. P. S. Poskozim and A. L. Stone, *J. Inorg. Nucl. Chem.*, **32,** 1391 (1970).
440. I. Wharf and D. F. Shriver, *Inorg. Chem.*, **8,** 914 (1969).
441. C. J. Kallendorf and B. S. Ault, *J. Phys. Chem.*, **85,** 608 (1981).

442. M. J. Taylor, *J. Raman Spectrosc.*, **20,** 663 (1989).
443. B. S. Ault, *J. Phys. Chem.*, **85,** 3083 (1981).
444. B. Basak, *Inorg. Chim. Acta*, **45,** L47 (1980).
445. J. Shamir and H. H. Hyman, *Spectrochim. Acta*, **23A,** 1899 (1967).
446. P. J. Hendra and J. R. Mackenzie, *Chem. Commun.*, 760 (1968).
447. W. Sawondy, H. Härtner, R. Minkwitz, and D. Bernstein, *J. Mol. Struct.*, **213,** 145 (1989).
448. I. Tornieporth-Oetting and T. Klapötke, *Angew. Chem., Int. Ed. Engl.*, **29,** 677 (1990).
449. R. J. H. Clark and D. M. Rippon, *J. Mol. Spectrosc.*, **52,** 58 (1974).
450. H. Stammreich, R. Forneris, and Y. Tavares, *J. Chem. Phys.*, **25,** 580 (1956).
451. W. V. F. Brooks, J. Passmore, and E. K. Richardson, *Can. J. Chem.*, **57,** 3230 (1979).
452. C. J. Adams and A. J. Downs, *J. Chem. Soc. A*, 1534 (1971).
453. E. Denchik, S. C. Nyburg, G. A. Ozin, and J. T. Szymanski, *J. Chem. Soc. A*, 3157 (1971).
454. V. A. Maroni and P. T. Cunningham, *Appl. Spectrosc.*, **27,** 428 (1973).
455. T. R. Manley and D. A. Williams, *Spectrochim. Acta*, **21,** 1773 (1965).
456. J. A. Evans and D. A. Long, *J. Chem. Soc. A*, 1688 (1968).
457. H. E. Doorenbos, J. C. Evans, and R. O. Kagel, *J. Phys. Chem.*, **74,** 3385 (1970).
458. J. Passmore, E. K. Richardson, and P. Taylor, *Inorg. Chem.*, **17,** 1681 (1978).
459. E. A. Robinson and J. A. Ciruna, *Can. J. Chem.*, **46,** 3197 (1968).
460. D. M. Adams and P. J. Lock, *J. Chem. Soc. A*, 145 (1967).
461. J. Kouinis and A. G. Galinos, *Monatsh. Chem.*, **108,** 835 (1977).
462. A. Givan and A. Loewenschuss, *J. Raman Spectrosc.*, **6,** 84 (1977).
462a. R. J. M. Konings and A. S. Booij, *J. Mol. Struct.*, **271,** 183 (1992).
463. L. Peter and B. Meyer, *Inorg. Chem.*, **24,** 3071 (1985).
464. B. M. Stanbury, T. A. Holme, Z. H. Kafafi, and J. L. Margrave, *Chem. Phys. Lett.*, **129,** 181 (1986).
465. H. Siebert, *Z. Anorg. Allg. Chem.*, **275,** 225 (1955).
466. D. J. Gardiner, R. B. Girling, and R. E. Hester, *J. Mol. Struct.*, **13,** 105 (1972).
467. W. Sterzel and W. D. Schnee, *Z. Anorg. Allg. Chem.*, **383,** 231 (1971).
468. H. H. Claassen and G. Knapp, *J. Am. Chem. Soc.*, **86,** 2341 (1964).
469. C. Rocchiciolli, *Hebd. Seances Acad. Sci.*, **242,** 2922 (1956); **244,** 2704 (1957); **247,** 1108 (1958); **249,** 236 (1959).
470. W. E. Dasent and T. C. Waddington, *J. Chem. Soc.*, 2429, 3350 (1960).
471. J. J. Comeford, D. E. Mann, L. J. Schoen, and D. R. Lide, *J. Chem. Phys.*, **38,** 461 (1963).
472. G. E. Moore and R. M. Badger, *J. Am. Chem. Soc.*, **74,** 6076 (1952).
473. R. Minkwitz, V. Wölfel, R. Nass, H. Härtner, and W. Sawodny, *Z. Anorg. Allg. Chem.*, **570,** 127 (1989).
474. L. Andrews and R. Lascola, *J. Am. Chem. Soc.*, **109,** 6243 (1987).

475. J. J. Comeford, *J. Chem. Phys.*, **45,** 3463 (1966); J. J. Comeford, D. E. Mann, L. J. Schoen, and D. R. Lide, *ibid.*, **38,** 461 (1963).
476. L. Andrews and R. Withnall, *Inorg. Chem.*, **28,** 494 (1989).
477. L. Andrews and T. C. McInnis, *Inorg. Chem.*, **30,** 2990 (1991).
478. A. Müller, E. Niecke, B. Krebs, and O. Glemser, *Z. Naturforsch.*, **23B,** 588 (1968); A. Müller, K. Königer, S. J. Cyvin, and A. Fadini, *Spectrochim. Acta*, **29A,** 219 (1973).
479. C. R. S. Dean, A. Finch, and P. N. Crates, *J. Chem. Soc., Dalton Trans.*, 1384 (1972).
480. R. Minkwitz and G. Nowicki, *Angew. Chem., Int. Ed. Engl.*, **29,** 688 (1990).
481. M. Adelhelm and E. Jacob, *Angew. Chem., Engl. Int. Ed.*, **16,** 461 (1977).
482. J. K. O'Loane and M. K. Wilson, *J. Chem. Phys.*, **23,** 1313 (1955).
483. D. E. Martz and R. T. Lagemann, *J. Chem. Phys.*, **22,** 1193 (1954).
484. R. Steudel and D. Lautenbach, *Z. Naturforsch.*, **24,** 350 (1969).
485. H. Stammreich, R. Forneris, and Y. Tavares, *J. Chem. Phys.*, **25,** 1277 (1956).
486. K. H. Moock, D. Sülzle, and P. Klaeboe, *J. Fluorine Chem.*, **47,** 151 (1990).
486a. R. Minkwitz, G. Nowicki, B. Bäck, and W. Sawodny, *Inorg. Chem.*, **32,** 787 (1993).
487. D. F. Burow, *Inorg. Chem.*, **11,** 573 (1972).
488. R. J. Gillespie, P. H. Spekkens, J. B. Milne, and D. M. Moffett, *J. Fluorine Chem.*, **7,** 43 (1976).
489. L. E. Alexander and I. R. Beattie, *J. Chem. Soc., Dalton Trans.*, 1745 (1972).
490. J. A. Rolfe and L. A. Woodward, *Trans. Faraday Soc.*, **51,** 779 (1955).
491. A. Simon and R. Paetzold, *Z. Anorg. Allg. Chem.*, **301,** 246 (1959); *Naturwissenschaften*, **44,** 108 (1957).
492. M. Falk and P. A. Giguère, *Can. J. Chem.*, **34,** 1680 (1958).
493. D. F. Smith, G. M. Begun, and W. H. Fletcher, *Spectrochim. Acta*, **20,** 1763 (1964).
494. R. J. Gillespie and P. H. Spekkens, *J. Chem. Soc., Dalton Trans.*, 1539 (1977).
495. K. O. Christe, E. C. Curtis, and C. J. Schack, *Inorg. Chem.*, **11,** 2212 (1972).
496. R. Bougon, J. Isabey, and P. Plurien, *C. R. Hebd. Seances Acad. Sci.*, **273C,** 415 (1971).
497. K. O. Christe and W. W. Wilson, *Inorg. Chem.*, **27,** 2714 (1988).
498. R. Minkwitz and G. Nowicki, *Z. Naturforsch.*, **44B,** 1343 (1989).
499. J. Breidung and W. Thiel, *J. Phys. Chem.*, **92,** 5597 (1988).
500. B. S. Ault, *J. Am. Chem. Soc.*, **100,** 2426 (1978).
501. R. Steudel and D. Lautenbach, *Z. Naturforsch.*, **24B,** 350 (1969).
502. M. Goldstein and G. C. Tok, *J. Chem. Soc. A*, 2303 (1971).
503. A. Kaldor and R. F. Porter, *J. Am. Chem. Soc.*, **93,** 2140 (1971).
504. J. Vanderryn, *J. Chem. Phys.*, **30,** 331 (1959).
505. D. A. Dows, *J. Chem. Phys.*, **31,** 1637 (1959).
506. R. J. H. Clark and P. D. Mitchell, *J. Chem. Phys.*, **56,** 2225 (1972).
507. D. A. Dows and G. Bottger, *J. Chem. Phys.*, **34,** 689 (1961).

508. T. Wentink, Jr. and V. H. Tiensuu, *J. Chem. Phys.*, **28,** 826 (1958).
508a. G. V. Chertihin and L. Andrews, *J. Phys. Chem.*, **97,** 10295 (1993).
509. Y. S. Yang and J. S. Shirk, *J. Mol. Spectrosc.*, **54,** 39 (1975).
510. I. R. Beattie, H. E. Blayden, S. M. Hall, S. N. Jenny, and J. S. Ogden, *J. Chem. Soc., Dalton Trans.*, 666 (1976).
511. I. R. Beattie and J. R. Horder, *J. Chem. Soc. A*, 2655 (1969).
512. M. Somer, K. Peters, T. Popp, and H. G. von Schnering, *Z. Anorg. Allg. Chem.*, **597,** 201 (1991).
513. G. K. Selivanov and A. A. Mal'tsev, *Zh. Strukt. Khim.*, **14,** 943 (1973).
514. R. G. Pong, A. E. Shirk, and J. S. Shirk, *J. Mol. Spectrosc.*, **66,** 35 (1977).
515. W. Blase, G. Cordier, K. Peters, M. Somer, and H. G. von Schnering, *Angew. Chem., Int. Ed. Engl.*, **30,** 326 (1991).
516. J. E. D. Davies and D. A. Long, *J. Chem. Soc. A*, 2050 (1968).
517. L. Andrews and G. C. Pimentel, *J. Chem. Phys.*, **47,** 3637 (1967).
518. J. E. D. Davies and D. A. Long, *J. Chem. Soc. A*, 2054 (1968).
519. P. Biscarini, L. Fusina, G. Nivellini, and G. Pelizzi, *J. Chem. Soc., Dalton Trans.*, 664 (1977).
520. R. Becker and W. Brockner, *Z. Naturforsch.*, **39A,** 1120 (1984).
521. R. D. Wesley and C. W. DeKock, *J. Chem. Phys.*, **55,** 3866 (1971).
522. M. Lesiecki, J. W. Nibler, and C. W. DeKock, *J. Chem. Phys.*, **57,** 1352 (1972).
523. C. E. Sjögren and E. Rytter, *Spectrochim. Acta*, **41A,** 1277 (1985).
524. K. Shimizu and H. Shingu, *Spectrochim. Acta*, **22,** 1999 (1966).
525. S. Konaka, Y. Murata, K. Kuchitsu, and Y. Morino, *Bull. Chem. Soc. Jpn.*, **39,** 1134 (1966).
526. L. Beckmann, L. Gutjahr, and R. Mecke, *Spectrochim. Acta*, **21,** 141 (1965).
527. I. W. Levin and S. Abramowitz, *J. Chem. Phys.*, **43,** 4213 (1965).
528. J. L. Duncan, *J. Mol. Spectrosc.*, **13,** 338 (1964).
529. C. D. Bass, L. Lynds, T. Wolfram, and R. E. DeWames, *J. Chem. Phys.*, **40,** 3611 (1964).
530. W. C. Steele and J. C. Decius, *J. Chem. Phys.*, **25,** 1184 (1956).
531. J. P. Laperches and P. Tarte, *Spectrochim. Acta*, **22,** 1201 (1966).
532. P. E. Bethell and N. Sheppard, *Trans. Faraday Soc.*, **51,** 9 (1959).
533. S. Bhagavantum and T. Venkatarayudu, *Proc. Indian Acad. Sci.*, **9A,** 224 (1939).
534. A. Müller and M. Stockburger, *Z. Naturforsch.*, **20A,** 1242 (1965).
535. A. Müller, N. Mohan, P. Cristophliemk, I. Tossidis, and M. Dräger, *Spectrochim. Acta*, **29A,** 1345 (1973).
536. A. Müller, G. Gattow, and H. Seidel, *Z. Anorg. Allg. Chem.*, **347,** 24 (1966).
537. I. Nakagawa and J. L. Walter, *J. Chem. Phys.*, **51,** 1389 (1969).
538. T. Ishiwata, I. Fujiwara, Y. Naruge, K. Obi, and I. Tanaka, *J. Phys. Chem.*, **87,** 1349 (1983).
539. R. R. Friedl and S. P. Sander, *J. Phys. Chem.*, **91,** 2721 (1987).
540. K. Kawaguchi, T. Ishiwata, I. Tanaka, and E. Hirota, *Chem. Phys. Lett.*, **180,** 436 (1991).

541. V. E. Bondybey and J. H. English, *J. Mol. Spectrosc.*, **109,** 221 (1985).
542. K. Stopperka, *Z. Anorg. Allg. Chem.*, **345,** 277 (1966).
543. A. Kalder, A. G. Maki, A. J. Dorney, and I. M. Mills, *J. Mol. Spectrosc.*, **45,** 247 (1973).
544. N. J. Brassington, H. G. M. Edwards, and V. Fawcett, *Spectrochim. Acta*, **43A,** 451 (1987).
545. P. LaBonville, R. Kugel, and J. R. Ferraro, *J. Chem. Phys.*, **67,** 1477 (1977).
546. M. E. Jacox and D. E. Milligan, *J. Chem. Phys.*, **54,** 919 (1971).
547. C. C. Addison and B. M. Gatehouse, *J. Chem. Soc.*, 613 (1960).
548. J. R. Ferraro and A. Walker, *J. Chem. Phys.*, **45,** 550 (1966).
549. G. E. Walrafen and D. E. Irish, *J. Chem. Phys.*, **40,** 911 (1964).
550. G. J. Janz and T. R. Kozlowski, *J. Chem. Phys.*, **40,** 1699 (1964).
551. D. Smith, D. W. James, and J. P. Devlin, *J. Chem. Phys.*, **54,** 4437 (1971).
552. C. J. Peacock, A. Müller, and R. Kebabcioglu, *J. Mol. Struct.*, **2,** 163 (1968).
553. P. Thirugnanasambandam and G. J. Srinivasan, *J. Chem. Phys.*, **50,** 2467 (1969).
554. R. A. Frey, R. L. Redington, and A. L. K. Aljibury, *J. Chem. Phys.*, **54,** 344 (1971).
555. R. J. Gillespie, B. Landa, and G. J. Schrobilgen, *Inorg. Chem.*, **15,** 1256 (1976).
556. R. J. Gillespie and G. J. Schrobilgen, *Chem. Commun.*, 595 (1977).
557. D. W. Green, G. T. Reedy, and S. D. Gabelnick, *J. Chem. Phys.*, **73,** 4207 (1980).
558. K. H. Lee, H. Takeo, S. Kondo, and C. Matsumura, *Bull. Chem. Soc. Jpn.*, **58,** 1772 (1985).
558a. R. Köppe and H. Schnöckel, *J. Chem. Soc., Dalton Trans.*, 3393 (1992).
558b. R. Köppe, M. Tacke, and H. Schnöckel, *Z. Anorg. Allg. Chem.*, **605,** 35 (1991).
559. B. S. Ault, *Inorg. Chem.*, **21,** 756 (1982).
560. D. L. Bernitt, K. O. Hartman, and I. C. Hisatsune, *J. Chem. Phys.*, **42,** 3553 (1965).
561. K. Itoh and H. J. Bernstein, *Can. J. Chem.*, **34,** 170 (1956).
562. P. D. Mallinson, D. C. McKean, J. H. Holloway, and I. A. Oxton, *Spectrochim. Acta*, **31A,** 143 (1975).
563. N. C. Craig, *Spectrochim. Acta*, **44A,** 1225 (1988).
564. J. Overend and J. C. Evans, *Trans. Faraday Soc.*, **55,** 1817 (1959).
565. R. F. Stratton and A. H. Nielsen, *J. Mol. Spectrosc.*, **4,** 373 (1960).
566. M. E. Jacox and D. E. Milligan, *J. Mol. Spectrosc.*, **58,** 142 (1975).
567. A. J. Downs, *Spectrochim. Acta*, **19,** 1165 (1963).
568. I. S. Butler and A. M. English, *Spectrochim. Acta*, **33A,** 545 (1977).
569. A. Haas, B. Koch, N. Welcman, and H. Willner, *Spectrochim. Acta*, **32A,** 497 (1976).
570. A. Müller, N. Mohan, P. Cristophliemk, I. Tossidis, and M. Dräger, *Spectrochim. Acta*, **29A,** 1345 (1973).
571. H. Schnöckel, H. J. Göcke, and R. Köppe, *Z. Anorg. Allg. Chem.*, **578,** 159 (1989).
572. R. Köppe and H. Schnöckel, *Z. Anorg. Allg. Chem.*, **592,** 179 (1991).

573. W. A. Guillory and M. L. Bernstein, *J. Chem. Phys.*, **62,** 1058 (1975).
574. K. O. Christe, C. J. Schack, and R. D. Wilson, *Inorg. Chem.*, **13,** 2811 (1974).
575. C. A. Wamser, W. B. Fox, B. Sukornick, J. R. Holmes, B. B. Stewart, R. Juurik, N. Vanderkooi, and D. Gould, *Inorg. Chem.*, **8,** 1249 (1969).
576. A. Allan, J. L. Duncan, J. H. Holloway, and D. C. McKean, *J. Mol. Spectrosc.*, **31,** 368 (1969).
577. R. Minkwitz, D. Bernstein, W. Sawodny, and H. Härtner, *Z. Anorg. Allg. Chem.*, **580,** 109 (1990).
578. S. Schunck, H. J. Göcke, R. Köppe, and H. Schnöckel, *Z. Anorg. Allg. Chem.*, **579,** 66 (1989).
579. R. Withnall and L. Andrews, *J. Am. Chem. Soc.*, **107,** 2567 (1985); **108,** 8118 (1986).
580. D. F. Wolfe and G. L. Humphrey, *J. Mol. Struct.*, **3,** 293 (1969).
581. R. G. S. Pong, A. E. Shirk, and J. S. Shirk, *Ber. Bunsenges. Phys. Chem.*, **82,** 79 (1978).
582. J. Goubeau and K. Laitenberger, *Z. Anorg. Allg. Chem.*, **320,** 78 (1963).
583. J. E. Rauch and J. C. Decius, *Spectrochim. Acta*, **22,** 1963 (1966).
584. G. E. McGraw, D. L. Bernitt, and I. C. Hisatsune, *Spectrochim. Acta*, **23A,** 25 (1967).
585. S. T. King and J. Overend, *Spectrochim. Acta*, **23A,** 61 (1967).
586. S. T. King and J. Overend, *Spectrochim. Acta*, **22,** 689 (1966).
587. P. A. Giguère and T. K. K. Srinivasan, *J. Raman Spectrosc.*, **2,** 125 (1974).
588. D. J. Gardiner, N. J. Lawrence, and J. J. Turner, *J. Chem. Soc. A*, 400 (1971).
589. N. Zengin and P. A. Giguère, *Can. J. Chem.*, **37,** 632 (1959).
590. R. D. Brown and G. P. Pez, *Spectrochim. Acta*, **26A,** 1375 (1970).
591. S. G. Frankiss and D. S. Harrison, *Spectrochim. Acta*, **31A,** 161 (1975).
592. R. Forneris and C. E. Hennies, *J. Mol. Struct.*, **5,** 449 (1970).
593. W. A. Guillory and C. E. Hunter, *J. Chem. Phys.*, **50,** 3516 (1969).
594. J. R. Ohlsen and J. Laane, *J. Am. Chem. Soc.*, **100,** 6948 (1978).
595. D. E. Milligan and M. E. Jacox, *J. Chem. Phys.*, **55,** 3404 (1971).
596. N. Wiberg, G. Fischer, and H. Bachhuber, *Angew. Chem., Int. Ed. Engl.*, **16,** 780 (1977).
597. W. Kiefer, *Spectrochim. Acta*, **27A,** 1285 (1971).
598. R. Steudel, D. Jensen, and B. Plinke, *Z. Naturforsch.*, **42B,** 163 (1987).
599. J. A. A. Ketelaar, F. N. Hooge, and G. Blasse, *Recl. Trav. Chim. Pays-Bas*, **75,** 220 (1956).
600. P. A. Giguère and O. Bain, *J. Phys. Chem.*, **56,** 340 (1952).
601. C. A. Frenzel and K. E. Blick, *J. Chem. Phys.*, **55,** 2715 (1971).
602. J. H. Teles, G. Maier, B. A. Hess, Jr., L. J. Schaad, M. Winnewisser, and B. P. Winnewisser, *Chem. Ber.*, **123,** 753 (1989).
603. M. E. Jacox and D. E. Milligan, *J. Chem. Phys.*, **40,** 2457 (1964).
604. K. Gholivand, H. Willner, D. Bielefeldt, and A. Haas, *Z. Naturforsch.*, **39B,** 1211 (1984).

605. T. C. Devore, *J. Mol. Struct.*, **162,** 287 (1987).
606. M. Gerke, G. Schatte, and H. Willner, *J. Mol. Spectrosc.*, **135,** 539 (1989).
607. D. Christen, H. G. Mack, G. Schatte, and H. Willner, *J. Am. Chem. Soc.*, **110,** 707 (1988).
608. K. Gholivand, G. Schatte, and H. Willner, *Inorg. Chem.*, **26,** 2137 (1987).
609. C. B. Moore and K. Rosengren, *J. Chem. Phys.*, **44,** 4108 (1966).
610. W. A. Guillory and C. E. Hunter, *J. Chem. Phys.*, **54,** 598 (1971).
611. R. T. Hall and G. C. Pimentel, *J. Chem. Phys.*, **38,** 1889 (1963).
612. J. R. Durig and D. W. Wertz, *J. Chem. Phys.*, **46,** 3069 (1967).
613. G. R. Draper and R. L. Werner, *J. Mol. Spectrosc.*, **50,** 369 (1974).
614. M. J. Nielsen and A. D. E. Pullin, *J. Chem. Soc.*, 604 (1960).
615. P. O. Tchir and R. D. Spratley, *Can. J. Chem.*, **53,** 2311 (1975).
616. P. O. Tchir and R. D. Spratley, *Can. J. Chem.*, **53,** 2331 (1975).
617. M. Nonella, J. R. Huber, and T. K. Ha, *J. Phys. Chem.*, **91,** 5203 (1987).
618. H. H. Eysel, *J. Mol. Struct.*, **5,** 275 (1970).
619. W. T. Thompson and W. H. Fletcher, *Spectrochim. Acta*, **22,** 1907 (1966).
620. A. R. Emery and R. C. Taylor, *J. Chem. Phys.*, **28,** 1029 (1958).
621. C. J. H. Schutte, *Spectrochim. Acta*, **16,** 1054 (1960); J. A. A. Ketelaar and C. J. H. Schutte, *ibid.*, **17,** 1240 (1961).
622. A. E. Shirk and D. F. Shriver, *J. Am. Chem. Soc.*, **95,** 5904 (1973).
623. Landolt-Börnstein, *Physikalisch-chemische Tabellen*, Vol. 2, Springer, Berlin, (1951).
624. G. E. MacWood and H. C. Urey, *J. Chem. Phys.*, **4,** 402 (1936).
625. H. M. Kaylor and A. H. Nielsen, *J. Chem. Phys.*, **23,** 2139 (1955).
626. I. F. Kovalev, *Opt. Spektrosk.*, **2,** 310 (1957).
627. R. E. Wilde, T. K. K. Srinivasan, R. W. Harral, and S. G. Sankar, *J. Chem. Phys.*, **55,** 5681 (1971).
628. J. H. Meal and M. K. Wilson, *J. Chem. Phys.*, **24,** 385 (1956).
629. L. P. Lindemann and M. K. Wilson, *J. Chem. Phys.*, **22,** 1723 (1954).
630. I. W. Levin and H. Ziffer, *J. Chem. Phys.*, **43,** 4023 (1965).
631. H. W. Morgan, P. A. Staats, and J. H. Goldstein, *J. Chem. Phys.*, **27,** 1212 (1957).
632. J. R. Durig, D. J. Antion, and F. G. Baglin, *J. Chem. Phys.*, **49,** 666 (1968).
633. J. R. Durig, C. B. Pate, and Y. S. Li, *J. Chem. Phys.*, **54,** 1033 (1971).
634. E. L. Wagner and D. F. Hornig, *J. Chem. Phys.*, **18,** 296, 305 (1950); R. C. Plumb and D. F. Hornig, *ibid.*, **21,** 366 (1953); **23,** 947 (1955); W. Vedder and D. F. Hornig, *ibid.*, **35,** 1560 (1961).
635. A. S. Quist, J. B. Bates, and G. E. Boyd, *J. Phys. Chem.*, **76,** 78 (1972).
636. V. A. Maroni, *J. Chem. Phys.*, **55,** 4789 (1971).
637. A. S. Quist, J. B. Bates, and G. E. Boyd, *J. Chem. Phys.*, **54,** 4896 (1971).
638. R. J. H. Clark, S. Joss, and M. J. Taylor, *Spectrochim. Acta*, **42A,** 927 (1986).

639. M. C. Dhamelincourt and M. Migeon, *C. R. Hebd. Seances Acad. Sci.*, **281,** C79 (1975).
640. J. A. Creighton, *J. Chem. Soc.*, 6589 (1965).
641. K. O. Christe, M. D. Lind, N. Thorup, D. R. Russell, J. Fawcett, and R. Bau, *Inorg. Chem.*, **27,** 2450 (1988).
642. B. Gilbert, G. Mamantov, and G. M. Begun, *Inorg. Nucl. Chem. Lett.*, **10,** 1123 (1974).
643. E. Rytter and H. A. Øye, *J. Inorg. Nucl. Chem.*, **35,** 4311 (1973).
644. D. H. Brown and D. T. Stewart, *Spectrochim. Acta*, **26A,** 1344 (1970).
645. G. M. Begun, C. R. Boston, G. Torsi, and G. Mamantov, *Inorg. Chem.*, **10,** 886 (1971).
646. H. A. Øye and W. Bues, *Inorg. Nucl. Chem. Lett.*, **8,** 31 (1972).
647. L. A. Woodward and A. A. Nord, *J. Chem. Soc.*, 2655 (1955).
648. L. A. Woodward and G. H. Singer, *J. Chem. Soc.*, 716 (1958).
649. L. A. Woodward and M. J. Taylor, *J. Chem. Soc.*, 4473 (1960).
650. L. A. Woodward and P. T. Bill, *J. Chem. Soc.*, 1699 (1955).
651. D. M. Adams and D. M. Morris, *J. Chem. Soc. A*, 694 (1968).
652. R. J. H. Clark and D. M. Rippon, *Chem. Commun.*, 1295 (1971); R. J. H. Clark and P. D. Mitchell, *J. Chem. Soc., Faraday Trans. 2*, **71,** 515 (1975).
653. R. R. Haun and W. D. Harkins, *J. Am. Chem. Soc.*, **54,** 3917 (1932).
654. H. Stammreich, Y. Tavares, and D. Bassi, *Spectrochim. Acta*, **17,** 661 (1961).
655. R. J. H. Clark and T. J. Dines, *Inorg. Chem.*, **19,** 1681 (1980).
656. A. D. Caunt, L. N. Short, and L. A. Woodward, *Trans. Faraday Soc.*, **48,** 873 (1952); *Nature (London)*, **168,** 557 (1951).
657. R. J. H. Clark and B. K. Hunter, *J. Mol. Struct.*, **9,** 354 (1971).
658. K. O. Christe, *Spectrochim. Acta*, **36A,** 921 (1980).
659. R. Minkwitz, D. Bernstein, and W. Sawodny, *Angew. Chem., Int. Ed. Engl.*, **29,** 181 (1990).
660. P. Van Huong and B. Desbat, *Bull. Soc. Chim. Fr.*, 2631 (1972).
661. M. Delahaye, P. Dhamelincourt, and J. C. Merlin, *C. R. Hebd. Seances Acad. Sci.*, **272B,** 370 (1971).
662. W. Gabes and H. Gerding, *Recl. Trav. Chim. Pays-Bas*, **90,** 157 (1971); W. Gabes, K. Olie, and H. Gerding, *ibid.*, **91,** 1367 (1972).
663. I. Tornieporth-Oetting and T. M. Klapötke, *J. Chem. Soc., Chem. Commun.*, 132 (1990).
663a. A. Schulz and T. M. Klapötke, *Spectrochim. Acta*, **51A,** 905 (1995).
664. A. Müller and A. Fadini, *Z. Anorg. Allg. Chem.*, **349,** 164 (1967).
665. J. Weidlein and K. Dehnicke, *Z. Anorg. Allg. Chem.*, **337,** 113 (1965).
666. T. Klapötke, J. Passmore, and E. G. Awere, *J. Chem. Soc., Chem. Commun.*, 1426 (1988).
667. I. Tornieporth-Oetting and T. Klapötke, *Angew. Chem., Int. Ed. Engl.*, **28,** 1671 (1989).
668. A. Sabatini and L. Sacconi, *J. Am. Chem. Soc.*, **86,** 17 (1964).

669. I. R. Beattie, T. R. Gilson, and G. A. Ozin, *J. Chem. Soc. A*, 534 (1969).

670. J. S. Avery, C. D. Burbridge, and D. M. L. Goodgame, *Spectrochim. Acta*, **24A,** 1721 (1968).

671. P. L. Goggin, R. J. Goodfellow, and K. Kessler, *J. Chem. Soc., Dalton Trans.*, 1914 (1977).

672. G. J. Janz and D. W. James, *J. Chem. Phys.*, **38,** 905 (1963).

673. D. A. Long and J. Y. H. Chau, *Trans. Faraday Soc.*, **58,** 2325 (1962).

674. M. M. Metallinou, L. Nalbandian, G. N. Papatheodorou, W. Voigt, and H. H. Emons, *Inorg. Chem.*, **30,** 4260 (1991).

675. L. E. Alexander and I. R. Beattie, *J. Chem. Soc., Dalton Trans.*, 1745 (1972).

676. R. J. H. Clark, B. K. Hunter, and D. M. Rippon, *Inorg. Chem.*, **11,** 56 (1972); R. J. H. Clark and D. M. Rippon, *J. Mol. Spectrosc.*, **44,** 479 (1972).

677. A. Büchler, J. B. Berkowitz-Mattuck, and D. H. Dugre, *J. Chem. Phys.*, **34,** 2202 (1961).

678. M. F. A. Dove, J. A. Creighton, and L. A. Woodward, *Spectrochim. Acta*, **18,** 267 (1962).

678a. J. Jacobs, H. S. Müller, H. Willner, E. Jacob, and H. Bürger, *Inorg. Chem.*, **31,** 5357 (1992).

679. B. Cuoni, F. P. Emmenegger, C. Rohrbasser, C. W. Schläpfer, and P. Studer, *Spectrochim. Acta*, **34A,** 247 (1978).

680. H. G. M. Edwards, M. J. Ware, and L. A. Woodward, *Chem. Commun.*, 540 (1968).

680a. G. Thiele, D. Honert, and H. Rotter, *Z. Anorg. Allg. Chem.*, **616,** 195 (1992).

681. H. G. M. Edwards, L. A. Woodward, M. J. Gall, and M. J. Ware, *Spectrochim. Acta*, **26A,** 287 (1970).

682. W. Krasser and H. W. Nürnberg, *Spectrochim. Acta*, **26A,** 1059 (1970).

683. D. M. Adams and P. J. Lock, *J. Chem. Soc. A*, 620 (1967).

684. D. N. Anderson and R. D. Willett, *Inorg. Chim. Acta*, **8,** 167 (1974).

685. R. D. Willett, J. R. Ferraro, and M. Choca, *Inorg. Chem.*, **13,** 2919 (1974).

686. D. Forster, *Chem. Commun.*, 113 (1967).

687. P. L. Goggin and T. G. Buick, *Chem. Commun.*, 290 (1967).

688. R. A. Work, III and M. L. Good, *Spectrochim. Acta*, **28A,** 1537 (1972).

689. J. T. R. Dunsmuir and A. P. Lane, *J. Chem. Soc. A*, 404, 2781 (1971).

690. R. L. Hunt and B. S. Ault, *Spectrochim. Acta*, **37A,** 63 (1981).

691. P. Wermer and B. S. Ault, *Inorg. Chem.*, **20,** 970 (1981).

692. V. N. Bukhmarina, Yu. B. Predtchenskii, and L. D. Shcherba, *J. Mol. Struct.*, **218,** 33 (1990).

693. G. Maier, H. P. Reisenauer, J. Hu, B. A. Hess, and L. J. Schaad, *Tetrahedron Lett.*, **30,** 4105 (1989).

694. M. E. Jacox and W. E. Thompson, *J. Phys. Chem.*, **95,** 2781 (1991).

695. R. J. H. Clark and P. D. Mitchell, *J. Am. Chem. Soc.*, **95,** 8300 (1973); *J. Raman Spectrosc.*, **2,** 399 (1974).

696. R. J. H. Clark and P. D. Mitchell, *Chem. Commun.*, 762 (1973).

697. T. Kamisuki and S. Maeda, *Chem. Phys. Lett.*, **21,** 330 (1973).
698. S. T. King, *J. Chem. Phys.*, **49,** 1321 (1968).
699. F. Königer and A. Müller, *J. Mol. Spectrosc.*, **56,** 200 (1975).
700. F. Königer, A. Müller, and K. Nakamoto, *Z. Naturforsch.*, **30B,** 456 (1975).
701. D. Tevault, J. D. Brown, and K. Nakamoto, *Appl. Spectrosc.*, **30,** 461 (1976).
702. A. Müller and B. Krebs, *J. Mol. Spectrosc.*, **24,** 180 (1967); A. Müller and A. Fadini, *Z. Anorg. Allg. Chem.*, **349,** 164 (1967).
703. L. J. Basile, J. R. Ferraro, P. LaBonville, and M. C. Wall, *Coord. Chem. Rev.*, **11,** 21 (1973).
704. C. J. Adams and A. J. Downs, *J. Chem. Soc. A*, 1534 (1971).
705. J. Milne, *Can. J. Chem.*, **53,** 888 (1975).
706. G. Y. Ahlijah and M. Goldstein, *J. Chem. Soc. A*, 326 (1970); *Chem. Commun.*, 1356 (1968).
707. K. O. Christe, H. Willner, and W. Sawodny, *Spectrochim. Acta*, **35A,** 1347 (1979).
708. K. Seppelt, *Z. Anorg. Allg. Chem.*, **416,** 12 (1975).
709. C. J. Adams and A. J. Downs, *Spectrochim. Acta*, **28A,** 1841 (1972).
710. K. O. Christe and W. Sawodny, *Inorg. Chem.*, **12,** 2879 (1973).
711. L. E. Alexander and I. R. Beattie, *J. Chem. Soc., Dalton Trans.*, 1745 (1972).
712. A. M. Heyns, K. J. Range, and M. Widenauer, *Spectrochim. Acta*, **46A,** 1621 (1990).
713. S. D. Ross, *Spectrochim. Acta*, **28A,** 1555 (1972).
714. M. Robineau and D. Zins, *C. R. Hebd. Seances Acad. Sci.*, **280,** C759 (1975).
715. Landolt-Börnstein, *Physikalisch-chemische Tabellen*, Vol . 2, Springer, Berlin, 1951.
716. D. Fortnum and J. O. Edwards, *J. Inorg. Nucl. Chem.*, **2,** 264 (1956).
717. E. Steger and K. Herzog, *Z. Anorg. Allg. Chem.*, **331,** 169 (1964).
718. E. Steger and W. Schmidt, *Ber. Bunsenges. Phys. Chem.*, **68,** 102 (1964).
719. O. Sala and M. L. A. Temperini, *Chem. Phys. Lett.*, **36,** 652 (1975).
720. M. Jansen, *Angew. Chem., Int. Ed. Engl.*, **16,** 534 (1977).
721. H. Siebert, *Z. Anorg. Allg. Chem.*, **275,** 225 (1954).
722. L. C. Brown, G. M. Begun, and G. E. Boyd, *J. Am. Chem. Soc.*, **91,** 2250 (1969).
723. E. H. Appelman, *Inorg. Chem.*, **8,** 223 (1969).
724. H. Siebert, *Z. Anorg. Allg. Chem.*, **273,** 21 (1953).
725. R. S. McDowell and L. B. Asprey, *J. Chem. Phys.*, **57,** 3062 (1972).
726. F. Gonzalez-Vilchez and W. P. Griffith, *J. Chem. Soc., Dalton Trans.*, 1416 (1972).
727. N. Weinstock, H. Schulze, and A. Müller, *J. Chem. Phys.*, **59,** 5063 (1973).
728. A. Müller, K. H. Schmidt, K. H. Tytko, J. Bouwma, and F. Jellinek, *Spectrochim. Acta*, **28A,** 381 (1972).
729. A. Müller, R. Kebabcioglu, M. J. F. Leroy, and G. Kaufmann, *Z. Naturforsch.*, **23B,** 740 (1968).

730. A. Müller, N. Weinstock, and H. Schulze, *Spectrochim. Acta*, **28A,** 1075 (1972); K. H. Schmidt and A. Müller, *ibid.*, 1829.
731. A. Müller, B. Krebs, R. Kebabcioglu, M. Stockburger, and O. Glemser, *Spectrochim. Acta*, **24A,** 1831 (1968).
732. H. Homborg and W. Preetz, *Spectrochim. Acta*, **32A,** 709 (1976).
733. P. J. Hendra, P. Le Barazer, and A. Crookell, *J. Raman Spectrosc.*, **20,** 35 (1989).
734. E. J. Baran and S. G. Manca, *Spectrosc. Lett.*, **15,** 455 (1982).
735. A. Müller, E. Diemann, and U. V. K. Rao, *Chem. Ber.*, **103,** 2961 (1970).
736. R. S. McDowell, L. B. Asprey, and L. C. Hoskins, *J. Chem. Phys.*, **56,** 5712 (1972).
737. R. S. McDowell, *Inorg. Chem.*, **6,** 1759 (1967); R. S. McDowell and M. Goldblatt, *ibid.*, **10,** 625 (1971).
738. E. J. Baran, *Z. Anorg. Allg. Chem.*, **399,** 57 (1973).
739. J. O. Edwards, G. C. Morrison, V. F. Ross, and J. W. Schultz, *J. Am. Chem. Soc.*, **77,** 266 (1955).
740. N. J. Campbell, J. Flaganan, and W. P. Griffith, *J. Chem. Phys.*, **83,** 3712 (1985).
741. E. R. Lippincott, J. A. Psellos, and M. C. Tobin, *J. Chem. Phys.*, **20,** 536 (1952).
742. A. Müller, E. J. Baran, and R. O. Carter, *Struct. Bonding (Berlin)*, **26,** 81 (1976).
743. E. J. Baran, *Inorg. Chem.*, **20,** 4453 (1981).
744. F. Ramondo, L. Bencivenni, V. Rossi, and H. M. Nagarathna-Naik, *Mol. Phys.*, **64,** 1145 (1988).
745. R. Kugel and H. Taube, *J. Phys. Chem.*, **79,** 2130 (1975).
746. W. Kiefer and H. J. Bernstein, *Mol. Phys.*, **23,** 835 (1972).
747. A. Ranade and M. Stockburger, *Chem. Phys. Lett.*, **22,** 257 (1973).
748. A. Ranade, W. Krasser, A. Müller, and E. Ahlborn, *Spectrochim. Acta*, **30A,** 1341 (1974).
749. R. S. McDowell and L. B. Asprey, *J. Chem. Phys.*, **57,** 3062 (1972).
750. K. H. Schmidt and A. Müller, *Coord. Chem. Rev.*, **14,** 115 (1974).
751. A. Müller, E. Diemann, R. Jostes, and H. Bögge, *Angew. Chem., Int. Ed. Engl.*, **20,** 934 (1981).
752. K. O. Christe, E. C. Curtis, and C. J. Schack, *Spectrochim. Acta*, **31A,** 1035 (1975).
753. B. S. Ault, *J. Phys. Chem.*, **84,** 3448 (1980).
754. R. J. H. Clark and O. H. Ellestad, *J. Mol. Spectrosc.*, **56,** 386 (1975).
755. R. H. Mann and P. M. Harris, *J. Mol. Spectrosc.*, **45,** 65 (1973).
756. H. Bürger, S. Biedermann, and A. Ruoff, *Spectrochim. Acta*, **30A,** 1655 (1974).
757. J. Goubeau, F. Haenschke, and A. Ruoff, *Z. Anorg. Allg. Chem.*, **366,** 113 (1969).
758. F. Lattanzi, C. DiLauro, and H. Bürger, *Mol. Phys.*, **72,** 575 (1991).
759. U. Müller and V. Krug, *Z. Naturforsch.*, **40B,** 1015 (1985).
760. A. Ruoff, H. Bürger, S. Biedermann, and J. Cichon, *Spectrochim. Acta*, **30A,** 1647 (1974).
761. E. C. Curtis, D. Philipovich, and W. H. Maberly, *J. Chem. Phys.*, **46,** 2904 (1967).

762. A. Finch, P. N. Gates, F. J. Ryan, and F. F. Bentley, *J. Chem. Soc., Dalton Trans.*, 1863 (1973).

763. H. S. Gutowsky and A. D. Liehr, *J. Chem. Phys.*, **26,** 329 (1957).

764. M. L. Delwaulle and F. François, *C. R. Hebd. Seances Acad. Sci.*, **220,** 817 (1945).

765. H. Gerding and M. van Driel, *Recl. Trav. Chim. Pays-Bas*, **61,** 419 (1942).

766. J. Durand, L. Beys, P. Hillaire, S. Aleonard, and L. Cot, *Spectrochim. Acta*, **34A,** 123 (1978).

767. F. Königer and A. Müller, *Spectrochim. Acta*, **33A,** 971 (1977).

768. H. Gerding and R. Westrik, *Recl. Trav. Chim. Pays-Bas*, **61,** 842 (1942).

769. M. L. Delwaulle and F. François, *C. R. Hebd. Seances Acad. Sci.*, **226,** 896 (1948).

770. M. Brownstein, P. A. W. Dean, and R. J. Gillespie, *Chem. Commun.*, 9 (1970).

771. F. Königer, A. Müller, and O. Glemser, *J. Mol. Struct.*, **46,** 29 (1978).

772. Y. Tavares-Forneris and R. Forneris, *J. Mol. Struct.*, **24,** 205 (1965).

773. I. C. Hisatsune and J. Heicklen, *Can. J. Chem.*, **53,** 2646 (1975).

774. C. S. Alleyne, K. O. Mailer, and R. C. Thompson, *Can. J. Chem.*, **52,** 336 (1974).

775. D. J. Stufkens and H. Gerding, *Recl. Trav. Chim. Pays-Bas*, **89,** 417 (1970).

776. E. Steger, I. C. Ciurea, and A. Fadini, *Z. Anorg. Allg. Chem.*, **350,** 225 (1967).

777. M. Černík and K. Dostál, *Z. Anorg. Allg. Chem.*, **425,** 37 (1976).

778. W. F. Murphy and H. Katz, *J. Raman Spectrosc.*, **7,** 76 (1978).

779. K. Burczyk, H. Bürger, M. LeGuennec, G. Wlodarczak, and J. Demaison, *J. Mol. Spectrosc.*, **148,** 65 (1991).

780. H. H. Claassen and E. H. Appelman, *Inorg. Chem.*, **9,** 622 (1970).

781. H. Bürger, G. Pawelke, and E. H. Appelman, *J. Mol. Spectrosc.*, **144,** 201 (1990).

782. J. Goubeau, E. Kilcioglu, and E. Jacob, *Z. Anorg. Allg. Chem.*, **357,** 190 (1968).

783. H. Selig and H. H. Claassen, *J. Chem. Phys.*, **44,** 1404 (1966).

784. G. A. Ozin and D. J. Reynolds, *Chem. Commun.*, 884 (1969).

785. F. A. Miller and W. K. Baer, *Spectrochim. Acta*, **17,** 114 (1961).

786. K. D. Scherfise and K. Dehnicke, *Z. Anorg. Allg. Chem.*, **538,** 119 (1986).

787. H. Stammreich, O. Sala, and D. Bassi, *Spectrochim. Acta*, **19,** 593 (1963).

788. H. Stammriech, O. Sala, and K. Kawai, *Spectrochim. Acta*, **17,** 226 (1961).

789. A. Müller, K. H. Schmidt, E. Ahlborn, and C. J. L. Lock, *Spectrochim. Acta*, **29A,** 1773 (1973).

790. K. H. Schmidt and A. Müller, *Spectrochim. Acta*, **28A,** 1829 (1972).

791. A. Müller, N. Mohan, H. Dornfeld, and C. Tellez, *Spectrochim. Acta*, **34A,** 561 (1978).

792. A. Müller, K. H. Schmidt, and U. Zint, *Spectrochim. Acta*, **32A,** 901 (1976).

793. E. L. Varetti, R. R. Filgueira, and A. Müller, *Spectrochim. Acta*, **37A,** 369 (1981).

794. E. L. Varetti, *J. Raman Spectrosc.*, **22,** 307 (1991).

795. J. Binenboym, U. El-Gad, and H. Selig, *Inorg. Chem.*, **13,** 319 (1974).
796. A. Guest, H. E. Howard-Lock, and C. J. L. Lock, *J. Mol. Spectrosc.*, **43,** 273 (1972).
797. I. R. Beattie, R. A. Crocombe, and J. S. Ogden, *J. Chem. Soc., Dalton Trans.*, 1481 (1977).
798. A. Müller, B. Krebs, and W. Höltje, *Spectrochim. Acta*, **23A,** 2753 (1967).
799. K. H. Schmidt, V. Flemming, and A. Müller, *Spectrochim. Acta*, **31A,** 1913 (1975).
800. R. H. Bradley, P. N. Brier, and D. E. H. Jones, *J. Chem. Soc. A*, 1397 (1971).
801. K. Hamada, G. A. Ozin, and E. A. Robinson, *Bull. Chem. Soc. Jpn.*, **44,** 2555 (1971).
802. F. Höfler, *Z. Naturforsch.*, **26A,** 547 (1971).
803. C. A. Clausen and M. L Good, *Inorg. Chem.*, **9,** 220 (1970).
804. F. Höfler and W. Veigl, *Angew. Chem., Int. Ed. Engl.*, **10,** 919 (1971).
805. K. O. Christe and E. C. Curtis, *Inorg. Chem.*, **11,** 2196 (1972).
806. K. O. Christe, *Inorg. Chem.*, **14,** 2821 (1975).
807. M. Abenoza and V. Tabacik, *J. Mol. Struct.*, **26,** 95 (1975).
808. I. McAlpine and H. Sutcliffe, *Spectrochim. Acta*, **25A,** 1723 (1969).
809. J. E. Drake, C. Riddle, and D. E. Rogers, *J. Chem. Soc. A*, 910 (1969).
810. M.-L. Dubois, M.-B. Delhaye, and F. Wallart, *C. R. Hebd. Seances Acad. Sci.*, **269B,** 260 (1969).
811. J. E. Drake and C. Riddle, *J. Chem. Soc. A*, 2114 (1969).
812. J. Weidlein, *Z. Anorg. Allg. Chem.*, **358,** 13 (1968).
813. S. Sportouch, R. J. H. Clark, and R. Gaufres, *J. Raman Spectrosc.*, **2,** 153 (1974).
814. D. E. Martz and R. T. Lagemann, *J. Chem. Phys.*, **22,** 1193 (1954).
815. R. Minkwitz, U. Nass, and J. Sawatzki, *J. Fluorine Chem.*, **31,** 175 (1986).
816. R. J. Gillespie, J. B. Milne, D. Moffett, and P. Spekkens, *J. Fluorine Chem.*, **7,** 43 (1976).
817. K. O. Christe and E. C. Curtis, *Inorg. Chem.*, **11,** 35 (1972).
818. R. Bougon, P. Joubert, and G. Tantot, *J. Chem. Phys.*, **66,** 1562 (1977).
819. H. H. Claassen, E. L. Gasner, H. Kim, and J. L. Huston, *J. Chem. Phys.*, **49,** 253 (1968).
820. E. Ahlborn, E. Diemann, and A. Müller, *Chem. Commun.*, 378 (1972).
821. V. D. Fenske, A.-F. Shihada, H. Schwab, and K. Dehnicke, *Z. Anorg. Allg. Chem.*, **471,** 140 (1980).
822. M. Muller, M. J. F. Leroy, and R. Rohmer, *C. R. Hebd. Seances Acad. Sci.*, **270C,** 1458 (1970).
823. S. D. Brown, G. L. Gard, and T. M. Loehr, *J. Chem. Phys.*, **64,** 1219 (1976).
824. M. Spoliti, J. H. Thirtle, and T. M. Dunn, *J. Mol. Spectrosc.*, **52,** 146 (1974).
825. E. G. Hope, W. Levason, J. S. Ogden, and M. Tajik, *J. Chem. Soc., Dalton Trans.*, 1587 (1986).
826. V. V. Kovba and A. A. Mal'tsev, *Russ. J. Inorg. Chem. (Engl. Transl.)*, **20,** 11 (1975).

827. A. Müller, N. Weinstock, K. H. Schmidt, K. Nakamoto, and C. W. Schläpfer, *Spectrochim. Acta*, **28A,** 2289 (1972).
828. K. O. Christe, R. D. Wilson, and E. C. Curtis, *Inorg. Chem.*, **12,** 1358 (1973).
829. G. A. Ozin and A. Vander Voet, *Chem. Commun.*, 1489 (1970).
830. A. Müller, B. Krebs, E. Niecke, and A. Ruoff, *Ber. Bunsenges, Phys. Chem.*, **71,** 571 (1967).
831. M. L. Delwaulle and F. François, *J. Chim. Phys.*, **46,** 87 (1949); *Hebd. Seances Acad. Sci.*, **226,** 894 (1948).
832. A. Müller and E. Diemann, *Z. Naturforsch.*, **24B,** 353 (1969).
833. A. Müller and E. Diemann, *Chem. Ber.*, **102,** 2603 (1969).
834. E. Diemann and A. Müller, *Inorg. Nucl. Chem. Lett.*, **5,** 339 (1969).
835. A. Müller and E. Diemann, *Z. Anorg. Allg. Chem.*, **373,** 57 (1970).
836. J. R. Durig and J. W. Clark, *J. Chem. Phys.*, **46,** 3057 (1967).
837. M. L. Delwaulle and F. François, *C. R. Hebd. Seances Acad. Sci.*, **222,** 1193 (1946).
838. A. Müller, E. Niecke, and O. Glemser, *Z. Anorg. Allg. Chem.*, **350,** 246 (1967).
839. T. T. Crow and R. T. Lagemann, *Spectrochim. Acta*, **12,** 143 (1958).
840. G. D. Flesch and H. J. Svec, *J. Am. Chem. Soc.*, **80,** 3189 (1958).
841. K. O. Christe and C. J. Schack, *Inorg. Chem.*, **9,** 1852 (1970).
842. H. Stammreich and R. Forneris, *Spectrochim. Acta*, **16,** 363 (1960).
843. P. Tsao, C. C. Cobb, and H. H. Claassen, *J. Chem. Phys.*, **54,** 5247 (1971).
844. H. H. Claassen, C. L. Chernick, and J. G. Malm, *J. Am. Chem. Soc.*, **85,** 1927 (1963).
845. P. J. Hendra, *J. Chem. Soc. A*, 1298 (1967); *Spectrochim. Acta*, **23A,** 2871 (1967).
845a. L. A. Degen and A. J. Rowlands, *Spectrochim. Acta*, **47A,** 1263 (1991).
846. P. L. Goggin and J. Mink, *J. Chem. Soc., Dalton Trans.*, 1479 (1974).
847. Y. M. Bosworth and R. J. H. Clark, *J. Chem. Soc., Dalton Trans.*, 381 (1975).
848. R. J. Gillespie, *J. Chem. Educ.*, **40,** 295 (1963); **47,** 18 (1970).
849. Y. M. Bosworth and R. J. H. Clark, *Inorg. Chem.*, **14,** 170 (1975).
850. J. Hiraishi and T. Shimanouchi, *Spectrochim. Acta*, **22,** 1483 (1966).
851. A. N. Pandey and U. P. Verma, *J. Mol. Struct.*, **42,** 171 (1977).
852. L. V. Konovalov, V. Yu. Kukushkin, V. K. Bel'skii, and V. E. Konovalov, *Russ. J. Inorg. Chem. (Engl. Transl.)*, **35,** 863 (1990).
853. H. C. Clark, K. R. Dixon, and J. G. Nicolson, *Inorg. Chem.*, **8,** 450 (1969).
854. I. R. Beattie and K. M. Livingston, *J. Chem. Soc. A*, 859 (1969).
855. I. R. Beattie, T. Gilson, K. Livingston, V. Fawcett, and G. A. Ozin, *J. Chem. Soc. A*, 712 (1967).
856. J. I. Bullock, N. J. Taylor, and F. W. Parrett, *J. Chem. Soc., Dalton Trans.*, 1843 (1972).
857. J. A. Creighton and J. H. S. Green, *J. Chem. Soc. A*, 808 (1968).
858. I. R. Beattie, K. M. S. Livingston, and D. J. Reynolds, *J. Chem. Phys.*, **51,** 4269 (1969).

859. N. A. Chumaevskii, *Russ. J. Inorg. Chem. (Engl. Transl.)*, **29,** 1415 (1984).
860. P. van Huong and B. Desbat, *Bull. Soc. Chim. Fr.*, 2631 (1972).
861. L. C. Hoskins and R. C. Lord, *J. Chem. Phys.*, **46,** 2402 (1967).
862. K. Seppelt, *Z. Anorg. Allg. Chem.*, **434,** 5 (1977).
863. J. Gaunt and J. B. Ainscough, *Spectrochim. Acta*, **10,** 57 (1957).
864. I. R. Beattie and G. A. Ozin, *J. Chem. Soc. A*, 1691 (1969).
865. T. V. Long, A. W. Herlinger, E. F. Epstein, and I. Bernal, *Inorg. Chem.*, **9,** 459 (1970).
866. C. S. Creaser and J. A. Creighton, *J. Chem. Soc., Dalton Trans.*, 1402 (1975).
867. H. H. Claassen and H. Selig, *J. Chem. Phys.*, **44,** 4039 (1965).
868. E. G. Hope, *J. Chem. Soc., Dalton Trans.*, 723 (1990).
869. R. D. Werder, R. A. Frey, and H. Günthard, *J. Chem. Phys.*, **47,** 4159 (1967).
870. E. M. Nour, *Spectrochim. Acta*, **42A,** 1411 (1986).
871. N. Acquista and S. Abramowitz, *J. Chem. Phys.*, **58,** 5484 (1973).
872. A. M. McNair and B. S. Ault, *Inorg. Chem.*, **21,** 2603 (1982).
873. A. K. Brisdon, J. T. Graham, E. G. Hope, D. M. Jenkins, W. Levason, and J. S. Ogden, *J. Chem. Soc., Dalton Trans.*, 1529 (1990).
874. R. A. Condrate and K. Nakamoto, *Bull. Chem. Soc. Jpn.*, **39,** 1108 (1966).
875. R. R. Holmes, R. M. Deiters, and J. A. Golen, *Inorg. Chem.*, **8,** 2612 (1969).
876. I. W. Levin, *J. Mol. Spectrosc.*, **33,** 61 (1970).
877. H. Selig, J. H. Holloway, J. Tyson, and H. H. Claassen, *J. Chem. Phys.*, **53,** 2559 (1970).
878. D. E. Sands and A. Zalkin, *Acta Crystallogr.*, **12,** 723 (1959).
879. A. Zalkin and D. E. Sands, *Acta Crystallogr.*, **11,** 615 (1958).
880. R. A. Walton and B. J. Brisdon, *Spectrochim. Acta*, **23A,** 2489 (1967).
881. P. M. Boorman, N. N. Greenwood, M. A. Hildon, and H. J. Whitfield, *J. Chem. Soc. A*, 2017 (1967).
882. A. J. Edwards, *J. Chem. Soc.*, 3714 (1964).
883. L. E. Alexander and I. R. Beattie, *J. Chem. Phys.*, **56,** 5829 (1972).
884. L. E. Alexander, I. R. Beattie, and P. J. Jones, *J. Chem. Soc., Dalton Trans.*, 210 (1972).
885. H. Gerding and H. Houtgraaf, *Recl. Trav. Chim.*, **74,** 5 (1955).
886. R. R. Holmes, *Acc. Chem. Res.*, **5,** 296 (1972).
887. R. R. Holmes, *J. Chem. Phys.*, **46,** 3718, 3724, 3730 (1967).
888. C. Macho, R. Minkwitz, J. Rohmann, B. Steger, V. Wölfel, and H. Oberhammer, *Inorg. Chem.*, **25,** 2828 (1986).
889. J. A. Salthouse and T. C. Waddington, *Spectrochim. Acta*, **23A,** 1069 (1967).
890. J. Breidung, W. Thiel, and A. Komornicki, *J. Phys. Chem.*, **92,** 5603 (1988).
891. R. R. Holmes and C. J. Hora, Jr., *Inorg. Chem.*, **11,** 2506 (1972).
892. R. Winkwitz and A. Liedtke, *Inorg. Chem.*, **28,** 4238 (1989).
893. H. Beckers, J. Breidung, H. Bürger, R. Kuna, A. Rahner, W. Schneider, and W. Thiel, *J. Chem. Phys.*, **93,** 4603 (1990).

894. R. Minkwitz and H. Prenzel, *Z. Anorg. Allg. Chem.*, **548,** 103 (1987).
895. R. Minkwitz and H. Prenzel, *Z. Anorg. Allg. Chem.*, **534,** 150 (1986).
896. K. O. Christe, C. J. Schack, and E. C. Curtis, *Spectrochim. Acta*, **33A,** 323 (1977).
897. D. M. Adams and R. R. Smardzewski, *J. Chem. Soc. A*, 714 (1971).
898. L. E. Alexander and I. R. Beattie, *J. Chem. Soc. A*, 3091 (1971).
899. H. A. Szymanski, R. Yelin, and L. Marabella, *J. Chem. Phys.*, **47,** 1877 (1967).
900. K. O. Christe, E. C. Curtis, C. J. Schack, and D. Pilipovich, *Inorg. Chem.*, **11,** 1679 (1972).
901. T. Schönherr, *Z. Naturforsch.*, **43B,** 159 (1988).
902. O. V. Blinova, S. L. Dobycin, and L. D. Shcherba, *Opt. Spektrosk.*, **61,** 756 (1986).
903. G. M. Begun, W. H. Fletcher, and D. F. Smith, *J. Chem. Phys.*, **42,** 2236 (1965).
904. K. O. Christe, *Spectrochim. Acta*, **27A,** 631 (1971).
905. R. A. Frey, R. L. Redington, and A. L. Khidir Aljibury, *J. Chem. Phys.*, **54,** 344 (1971).
906. N. Acquista and S. Abramowitz, *J. Chem. Phys.*, **56,** 5221 (1972).
907. B. Družina and B. Žemva, *J. Fluorine Chem.*, **39,** 309 (1988).
908. J. B. Milne and D. Moffett, *Inorg. Chem.*, **12,** 2240 (1973).
909. K. O. Christe and E. C. Curtis, *Inorg. Chem.*, **11,** 2209 (1972).
910. R. Bougon, T. B. Huy, P. Charpin, and G. Tantot, *C. R. Hebd. Seances Acad. Sci.*, **283,** C71 (1976).
911. W. W. Wilson and K. O. Christe, *Inorg. Chem.*, **26,** 916 (1987).
912. J. B. Milne and D. M. Moffett, *Inorg. Chem.*, **15,** 2165 (1976).
913. E. G. Hope, P. J. Jones, W. Levason, J. S. Ogden, M. Tajik, and J. W. Turff, *J. Chem. Soc., Dalton Trans.*, 529 (1985).
914. K. O. Christe, W. W. Wilson, and R. A. Bougon, *Inorg. Chem.*, **25,** 2163 (1986).
915. W. Levason, R. Narayanaswamy, J. S. Ogden, A. J. Rest, and J. W. Turff, *J. Chem. Soc., Dalton Trans.*, 2501 (1981).
916. L. E. Alexander, I. R. Beattie, A. Bukovszky, P. J. Jones, C. J. Marsden, and G. J. Van Schalkwyk, *J. Chem. Soc., Dalton Trans.*, 81 (1974).
917. R. T. Paine and R. S. McDowell, *Inorg. Chem.*, **13,** 2366 (1974).
918. I. R. Beattie, K. M. S. Livingston, D. J. Reynolds, and G. A. Ozin, *J. Chem. Soc. A*, 1210 (1970).
919. K. Iijima and S. Shibata, *Bull. Chem. Soc., Jpn.*, **48,** 666 (1975).
920. R. J. Collin, W. P. Griffith, and D. Pawson, *J. Mol. Struct.*, **19,** 531 (1973).
920a. W. Preetz and W. Lierka, *Z. Naturfotsch.*, **48B,** 44 (1993).
921. M. G. Krishna Pillai and P. Parameswaran Pillai, *Can. J. Chem.*, **46,** 2393 (1968).
922. K. O. Christe, W. W. Wilson, and R. A. Bougon, *Inorg. Chem.*, **25,** 2163 (1986).
923. K. O. Christe, E. C. Curtis, D. A. Dixon, H. P. Mercier, J. C. P. Sanders, and G. J. Schrobilgen, *J. Am. Chem. Soc.*, **113,** 3351 (1991).
924. M. J. Reisfeld, *Spectrochim. Acta*, **29A,** 1923 (1973).

925. E. J. Baran and A. E. Lavat, *Z. Naturforsch.*, **36A,** 677 (1981).
926. T. Barrowcliffe, I. R. Beattie, P. Day, and K. Livingston, *J. Chem. Soc. A*, 1810 (1967).
927. T. G. Spiro, *Inorg. Chem.*, **6,** 569 (1967).
928. C. Naulin and R. Bougon, *J. Chem. Phys.*, **64,** 4155 (1976).
929. I. R. Beattie, T. Gilson, K. Livingston, V. Fawcett, and G. A. Ozin, *J. Chem. Soc. A*, 712 (1967).
930. T. L. Brown, W. G. McDugle, Jr., and L. G. Kent, *J. Am. Chem. Soc.*, **92,** 3645 (1970).
931. M. Debeau and M. Krauzman, *C. R. Hebd. Seances Acad. Sci.*, **264B,** 1724 (1967).
932. I. Wharf and D. F. Shriver, *Inorg. Chem.*, **8,** 914 (1969).
933. A. M. Heyns, *Spectrochim. Acta*, **33A,** 315 (1977).
933a. A. S. Muir, *Polyhedron*, **10,** 2217 (1991).
934. A. I. Popov, A. V. Shcharabarin, V. F. Sukhoverkov, and N. A. Chumaevsky, *Z. Anorg. Allg. Chem.*, **576,** 242 (1989).
935. M. Burgard and J. MacCordick, *Inorg. Nucl. Chem. Lett.*, **6,** 599 (1970).
936. Y. M. Bosworth and R. J. H. Clark, *J. Chem. Soc., Dalton Trans.*, 1749 (1974).
937. R. J. H. Clark and M. L. Duarte, *J. Chem. Soc., Dalton Trans.*, 790 (1977).
938. M. A. Hooper and D. W. James, *Aust. J. Chem.*, **26,** 1401 (1973).
939. M. A. Hooper and D. W. James, *J. Inorg. Nucl. Chem.*, **35,** 2335 (1973).
940. T. Surles, L. A. Quaterman, and H. H. Hyman, *J. Inorg. Nucl. Chem.*, **35,** 670 (1973).
941. Y. M. Bosworth, R. J. H. Clark, and D. M. Rippon, *J. Mol. Spectrosc.*, **46,** 240 (1973).
942. H. H. Claassen, G. L. Goodman, J. H. Holloway, and H. Selig, *J. Chem. Phys.*, **53,** 341 (1970).
943. A. Aboumajd, H. Berger, and R. Saint-Loup, *J. Mol. Spectrosc.*, **78,** 486 (1979).
944. P. J. Hendra and Z. Jović, *J. Chem. Soc. A*, 600 (1968).
945. R. W. Berg, F. W. Poulsen, and N. J. Bjerrum, *J. Chem. Phys.*, **67,** 1829 (1977).
946. K. O. Christe, W. W. Wilson, R. V. Chirak, J. C. P. Sanders, and G. J. Schrobilgen, *Inorg. Chem.*, **29,** 3306 (1990).
947. K. O. Christe, *Inorg. Chem.*, **12,** 1580 (1973).
948. R. J. Gillespie and G. J. Schrobilgen, *Inorg. Chem.*, **13,** 1230 (1974).
949. R. Bougon, P. Charpin, and J. Soriano, *C. R. Hebd. Seances Acad. Sci.*, **272C,** 565 (1971).
950. N. Bartlett and K. Leary, *Rev. Chim. Minér.*, **13,** 82 (1976).
951. S. Turrell, S. Hafsi, P. Conflant, P. Barbier, M. Drache, and J. C. Champarnaud-Mesjard, *J. Mol. Struct.*, **174,** 449 (1988).
952. R. Becker, A. Lentz, and W. Sawodny, *Z. Anorg. Allg. Chem.*, **420,** 210 (1976).
953. D. M. Adams and D. M. Morris, *J. Chem. Soc. A*, 694 (1968).
954. I. W. Forrest and A. P. Lane, *Inorg. Chem.*, **15,** 265 (1976).
955. R. J. H. Clark, L. Maresca, and R. J. Puddephatt, *Inorg. Chem.*, **7,** 1603 (1968).

956. P. C. Crouch, G. W. A. Fowles, and R. A. Walton, *J. Chem. Soc. A*, 972 (1969).
957. P. A. W. Dean and D. F. Evans, *J. Chem. Soc. A*, 698 (1967).
958. D. M. Adams and D. C. Newton, *J. Chem. Soc. A*, 2262 (1968).
959. W. von Bronswyk, R. J. H. Clark, and L. Maresca, *Inorg. Chem.*, **8,** 1395 (1969).
960. R. Becker and W. Sawodny, *Z. Naturforsch.*, **28B,** 360 (1973).
961. R. A. Walton and B. J. Brisdon, *Spectrochim. Acta*, **23A,** 2222 (1967).
962. O. L. Keller, *Inorg. Chem.*, **2,** 783 (1963).
963. E. Hahn and R. Hebisch, *Spectrochim. Acta*, **47A,** 1097 (1991).
964. S. M. Horner, R. J. H. Clark, B. Crociani, D. B. Copley, W. W. Horner, F. N. Collier, and S. Y. Tyree, *Inorg. Chem.*, **7,** 1859 (1968).
965. D. Brown and P. J. Jones, *J. Chem. Soc. A*, 247 (1967).
966. G. A. Ozin, G. W. A. Fowles, D. J. Tidmarsh, and R. A. Walton, *J. Chem. Soc. A*, 642 (1969).
967. O. L. Keller and A. Chetham-Strode, *Inorg. Chem.*, **5,** 367 (1966).
968. B. Weinstock and G. L. Goodman, *Adv. Chem. Phys.*, **11,** 169 (1965).
969. F. G. Hope, P. J. Jones, W. Levason, J. S. Ogden, M. Tajik, and J. W. Turff, *J. Chem. Soc., Dalton Trans.*, 1443 (1985).
970. H. H. Eysel, *Z. Anorg. Allg. Chem.*, **390,** 210 (1972).
971. R. S. McDowell, R. J. Sherman, L. B. Asprey, and R. C. Kennedy, *J. Chem. Phys.*, **62,** 3974 (1975).
972. R. R. Smardzewski, R. E. Noftle, and W. B. Fox, *J. Mol. Spectrosc.*, **62,** 449 (1976).
973. J. A. Creighton and T. J. Sinclair, *Spectrochim. Acta*, **35A,** 507 (1979).
974. C. D. Flint, *J. Mol. Spectrosc.*, **37,** 414 (1971).
975. E. Jacob and M. Fähnle, *Angew. Chem., Int. Ed. Engl.*, **15,** 159 (1976).
976. J. A. LoMenzo, S. Strobridge, H. H. Patterson, and E. Engstrom, *J. Mol. Spectrosc.*, **66,** 150 (1977).
977. P. W. Frais, C. J. L. Lock, and A. Guest, *Chem. Commun.*, 1612 (1970).
978. G. L. Bottger and C. V. Damsgard, *Spectrochim. Acta*, **28A,** 1631 (1972).
979. L. A. Woodward and M. J. Ware, *Spectrochim. Acta*, **20,** 711 (1964).
980. K. O. Christe, *Inorg. Chem.*, **16,** 2238 (1977).
981. M. Debeau and H. Poulet, *Spectrochim. Acta*, **25A,** 1553 (1969).
982. P. J. Hendra and P. J. D. Park, *Spectrochim. Acta*, **23A,** 1635 (1967).
983. L. A. Woodward and M. J. Ware, *Spectrochim. Acta*, **19,** 775 (1963).
984. D. M. Adams and H. A. Gebbie, *Spectrochim. Acta*, **19,** 925 (1963).
985. K. Wieghardt and H. H. Eysel, *Z. Naturforsch.*, **25B,** 105 (1970).
986. J. M. Fletcher, W. E. Gardner, A. C. Fox, and G. Topping, *J. Chem. Soc. A*, 1038 (1967).
987. W. Preetz and M. Bruns, *Z. Naturforsch.*, **38B,** 680 (1983).
988. Y. M. Bosworth and R. J. H. Clark, *J. Chem. Soc., Dalton Trans.*, 1749 (1974).
989. D. A. Kelly and M. L. Good, *Spectrochim. Acta*, **28A,** 1529 (1972).
989a. H. Homborg, *Z. Anorg. Allg. Chem.*, **460,** 17 (1980).

990. N. J. Campbell, V. A. Davis, W. P. Griffith, and T. J. Townend, *J. Chem. Soc., Dalton Trans.*, 1673 (1985).
991. M. Debeau, *Spectrochim. Acta*, **25A,** 1311 (1969).
992. W. Preetz and H. J. Steinebach, *Z. Naturforsch.*, **41B,** 260 (1986).
993. L. A. Woodward and M. J. Ware, *Spectrochim. Acta*, **24A,** 921 (1968).
994. R. T. Paine, R. S. McDowell, L. B. Asprey, and L. H. Jones, *J. Chem. Phys.*, **64,** 3081 (1976).
995. J. L. Ryan, *J. Inorg. Nucl. Chem.*, **33,** 153 (1971).
996. E. Stumpp and G. Piltz, *Z. Anorg. Allg. Chem.*, **409,** 53 (1974).
997. R. N. Mulford, H. J. Dewey, and J. E. Barefield, *J. Chem. Phys.*, **94,** 4790 (1991).
998. B. W. Berringer, J. B. Gruber, T. M. Loehr, and G. P. O'Leary, *J. Chem. Phys.*, **55,** 4608 (1971).
999. H. J. Dewey, J. E. Barefield, and W. W. Rice, *J. Chem. Phys.*, **84,** 684 (1986).
1000. S. J. David and K. C. Kim, *J. Chem. Phys.*, **89,** 1780 (1988).
1001. D. M. Yost, C. S. Steffens, and S. T. Gross, *J. Chem. Phys.*, **2,** 311 (1934).
1002. A. Fadini and S. Kemmler-Sack, *Spectrochim. Acta*, **34A,** 853 (1978).
1003. C. J. Adams and A. J. Downs, *Chem. Commun.*, 1699 (1970).
1004. L. A. Woodward and J. A. Creighton, *Spectrochim. Acta*, **17,** 594 (1961).
1005. M. Bettinelli, L. Disipio, A. Pasquetto, G. Ingletto, and A. Montenero, *Inorg. Chim. Acta*, **99,** 37 (1985).
1006. M. Bettinelli, L. Disipio, G. Ingletto, and C. Razzetti, *Inorg. Chim. Acta*, **133,** 7 (1987).
1007. R. D. Hunt, L. Andrews, and L. MacToth, *J. Phys. Chem.*, **95,** 1183 (1991).
1008. H. Hamaguchi, I. Harada, and T. Shimanouchi, *Chem. Phys. Lett.*, **32,** 103 (1975).
1009. H. Hamaguchi, *J. Chem. Phys.*, **69,** 569 (1978).
1010. B. Weinstock, H. H. Claassen, and C. L. Chernick, *J. Chem. Phys.*, **38,** 1470 (1963).
1011. H. Kim, H. H. Claassen, and E. Pearson, *Inorg. Chem.*, **7,** 616 (1968).
1012. H. H. Claassen, G. L. Goodman, and H. Kim, *J. Chem. Phys.*, **56,** 5042 (1972).
1013. D. M. Cox and J. Elliott, *Spectrosc. Lett.*, **12,** 275 (1979).
1014. H. Kim, P. A. Souder, and H. H. Claassen, *J. Mol. Spectrosc.*, **26,** 46 (1968).
1015. P. LaBonville, J. R. Ferraro, M. C. Wall, and L. J. Basile, *Coord. Chem. Rev.*, **7,** 257 (1972).
1016. P. J. Cresswell, J. E. Fergusson, B. R. Penfold, and D. E. Scaife, *J. Chem. Soc., Dalton Trans.*, 254 (1972).
1017. J. Hauck and A. Fadini, *Z. Naturforsch.*, **25B,** 422 (1970).
1018. J. Hauck, *Z. Naturforsch.*, **25B,** 224, 468, 647 (1970).
1019. P. Ayyub, M. S. Multani, V. R. Palkar, and R. Vijayaraghavan, *Phys. Rev.*, **B34,** 8137 (1986).
1020. C. J. Adams and A. J. Downs, *J. Inorg. Nucl. Chem.*, **34,** 1829 (1972).
1021. G. Goetz, M. Deneux, and M. J. F. Leroy, *Bull. Soc. Chim. Fr.*, **29,** (1971).

1022. J. E. Griffith, *Spectrochim. Acta*, **23A,** 2145 (1967).

1023. K. O. Christe, E. C. Curtis, and C. J. Schack, *Spectrochim. Acta*, **33A,** 69 (1977).

1024. K. O. Christe, C. J. Schack, D. Pilipovich, E. C. Curtis, and W. Sawodny, *Inorg. Chem.*, **12,** 620 (1973).

1025. W. Heilemann, R. Mews, S. Pohl, and W. Saak, *Chem. Ber.*, **122,** 427 (1989).

1026. K. O. Christe, C. J. Schack, and E. C. Curtis, *Inorg. Chem.*, **11,** 583 (1972).

1027. K. Seppelt, *Z. Anorg. Allg. Chem.*, **399,** 87 (1973).

1028. W. V. F. Brooks, M. E. Eshaque, C. Lau, and J. Passmore, *Can. J. Chem.*, **54,** 817 (1976).

1029. P. K. Miller, K. D. Abney, A. K. Rappe, O. P. Anderson, and S. H. Strauss, *Inorg. Chem.*, **27,** 2255 (1988).

1030. J. H. Holloway, H. Selig, and H. H. Claassen, *J. Chem. Phys.*, **54,** 4305 (1971).

1031. K. Dehnicke, G. Pausewang,and W. Rüdorff, *Z. Anorg. Allg. Chem.*, **366,** 64 (1969).

1032. G. A. Ozin, G. W. A. Fowles, D. J. Tidmarsh, and R. A. Walton, *J. Chem. Soc. A*, 642 (1969).

1033. R. J. Collin, W. P. Griffith, and D. Pawson, *J. Mol. Struct.*, **19,** 531 (1973).

1034. A. Beuter and W. Sawodny, *Z. Anorg. Allg. Chem.*, **427,** 37 (1976).

1035. D. M. Adams, G. W. Fraser, D. M. Morris, and R. D. Peacock, *J. Chem. Soc. A*, 1131 (1968).

1036. E. G. Hope, W. Levason, and J. S. Ogden, *J. Chem. Soc., Dalton Trans.*, 61 (1988).

1037. P. Joubert, R. Bougon, and B. Gaudreau, *Can. J. Chem.*, **56,** 1874 (1978).

1038. G. J. Schrobilgen, *Chem. Commun.*, 894 (1980).

1039. J. H. Holloway, V. Kaucic, D. Martin-Rovet, D. R. Russell, G. J. Schrobilgen, and H. Selig, *Inorg. Chem.*, **24,** 678 (1985).

1040. J. P. Brunnette and M. J. F. Leroy, *J. Inorg. Nucl. Chem.*, **36,** 289 (1974).

1041. W. Preetz and H. N. von Allwörden, *Z. Naturforsch.*, **42B,** 381 (1987); H. N. von Allwörden and W. Preetz, *ibid.*, **42A,** 597 (1987).

1042. W. Preetz and K. Irmer, *Z. Naturforsch.*, **45B,** 283 (1990).

1043. W. Preetz and A. Wendt, *Z. Naturforsch.*, **46B,** 1496 (1991).

1044. W. Preetz, D. Ruf, and D. Tensfeldt, *Z. Naturforsch.*, **39B,** 1100 (1984).

1045. W. Preetz and W. Kuhr, *Z. Naturforsch.*, **44B,** 1221 (1989).

1046. W. Preetz and H.-J. Steinebach, *Z. Naturforsch.*, **40B,** 745 (1985).

1047. D. Tensfeldt and W. Preetz, *Z. Naturforsch.*, **39B,** 1185 (1984).

1048. W. Preetz and P. Erlhöfer, *Z. Naturforsch.*, **44B,** 412 (1989); P. Erlhöfer and W. Preetz, *ibid.*, 619, 1214.

1049. W. Preetz and G. Rimkus, *Z. Naturforsch.*, **37B,** 579 (1982).

1050. K. Irmer and W. Preetz, *Z. Naturforsch.*, **46B,** 1200 (1991).

1050a. W. Preetz, G. Peters and D. Bublitz, *Chem. Rev.*, **96,** 977 (1996).

1051. W. Abriel and H. Ehrhardt, *Z. Naturforsch.*, **43B,** 557 (1988).

1052. H. H. Claassen, E. L. Gasner, and H. Selig, *J. Chem. Phys.*, **49,** 1803 (1968).

1053. K. O. Christe, J. C. P. Sanders, G. J. Schrobiligen, and W. W. Wilson, *J. Chem. Soc., Chem. Commun.*, 837 (1991).

1054. A. R. Mahjoub and K. Seppelt, *J. Chem. Soc., Chem. Commun.*, 840 (1991).

1055. H. H. Eysel and K. Seppelt, *J. Chem. Phys.*, **56,** 5081 (1972).

1056. Nguyen-Quy-Dao, *Bull. Soc. Chim. Fr.*, 3976 (1968).

1056a. K. O. Christe, D. A. Dixon, A. R. Mamjoub, H. P. A. Mercier, J. C. P. Sanders, K. Seppelt, G. J. Schrobiligen, and W. W. Wilson, *J. Am. Chem. Soc.*, **115,** 2696 (1993).

1056b. K. O. Christe, D. A. Dixon, J. C. P. Sanders, G. J. Schrobiligen, and W. W. Wilson, *Inorg. Chem.*, **32,** 4089 (1993).

1057. J. L. Hoard, W. G. Martin, M. E. Smith, and J. E. Whitney, *J. Am. Chem. Soc.*, **76,** 3820 (1954).

1058. R. Stomberg and C. Brosset, *Acta Chem. Scand.*, **14,** 441 (1960).

1059. K. O. Hartman and F. A. Miller, *Spectrochim. Acta*, **24A,** 669 (1968).

1060. C. W. F. T. Pistorius, *Bull. Soc. Chim. Belg.*, **68,** 630 (1959).

1061. H. L. Schlaefer and H. F. Wasgestian, *Theor. Chim. Acta*, **1,** 369 (1963).

1062. J. R. Durig, J. W. Thompson, J. D. Witt, and J. D. Odom, *J. Chem. Phys.*, **58,** 5339 (1973).

1063. D. E. Mann and L. Fano, *J. Chem. Phys.*, **26,** 1665 (1957).

1064. J. D. Odom, J. E. Saunders, and J. R. Durig, *J. Chem. Phys.*, **56,** 1643 (1972).

1065. L. A. Minon, K. S. Seshadri, R. C. Taylor, and D. White, *J. Chem. Phys.*, **53,** 2416 (1970).

1066. F. Mélen, R. Pokorni, and M. Herman, *Chem. Phys. Lett.*, **194,** 181 (1992).

1066a. B. Andrews and A. Anderson, *J. Chem. Phys.*, **74,** 1534 (1981).

1067. G. M. Begun and W. H. Fletcher, *Spectrochim. Acta*, **19,** 1343 (1963).

1068. H. Murata and K. Kawai, *J. Chem. Phys.*, **25,** 589, 796 (1956).

1069. R. E. Hester and R. A. Plane, *Inorg. Chem.*, **3,** 513 (1964).

1070. D. N. Shchepkin, L. A. Zhygula, and L. P. Belozerskaya, *J. Mol. Struct.*, **49,** 265 (1978).

1071. J. R. Durig, S. F. Bush, and E. E. Mercer, *J. Chem. Phys.*, **44,** 4238 (1966).

1072. R. Lascola, R. Withnall, and L. Andrews, *Inorg. Chem.*, **27,** 642 (1988).

1073. E. R. Nixon, *J. Phys. Chem.*, **60,** 1054 (1956).

1074. S. G. Frankiss, *Inorg. Chem.*, **7,** 1931 (1968).

1075. K. H. Rhee, A. M. Snider, Jr., and F. A. Miller, *Spectrochim. Acta*, **29A,** 1029 (1973).

1076. S. G. Frankiss and F. A. Miller, *Spectrochim. Acta*, **21,** 1235 (1965).

1077. S. G. Frankiss, F. A. Miller, H. Stammreich, and Th. T. Sans, *Spectrochim. Acta*, **23A,** 543 (1967).

1078. W. C. Hodgeman, J. B. Weinrach, and D. W. Bennett, *Inorg. Chem.*, **30,** 1611 (1991).

1079. J. R. Durig, B. M. Gimarc, and J. D. Odom, in J. R. Durig, ed., *Vibrational Spectra and Structure*, Vol. 2, Marcel Dekker, New York, 1975, p. 1.

1080. R. P. Bell and H. C. Longuet-Higgins, *Proc. R. Soc. London, Ser. A*, **183,** 357 (1945).
1081. C. Liang, R. D. Davy, and H. F. Schaefer, *Chem. Phys. Lett.*, **159,** 393 (1989).
1082. J. L. Duncan, D. C. McKean, and I. Torto, *J. Mol. Spectrosc.*, **85,** 16 (1981).
1083. A. Snelson, *J. Phys. Chem.*, **71,** 3202 (1967).
1084. M. Tranquille and M. Fouassier, *J. Chem. Soc., Faraday Trans. 2*, **76,** 26 (1980).
1085. D. M. Adams and R. G. Churchill, *J. Chem. Soc. A*, 697 (1970).
1086. A. J. Downs, M. J. Goode, and C. R. Pulham, *J. Am. Chem. Soc.*, **111,** 1936 (1989).
1087. C. R. Pulham, A. J. Downs, M. J. Goode, D. W. H. Rankin, and H. E. Robertson, *J. Am. Chem. Soc.*, **113,** 5149 (1991).
1088. I. R. Beattie, T. Gilson, and P. Cocking, *J. Chem. Soc. A*, 702 (1967).
1089. L. Nalbandian and G. N. Papatheodorou, *High Temp. Sci.*, **28,** 49 (1988).
1090. R. A. Frey, R. D. Werder, and H. H. Günthard, *J. Mol. Spectrosc.*, **35,** 260 (1970).
1091. T. Onishi and T. Shimanouchi, *Spectrochim. Acta*, **20,** 721 (1964).
1092. N. N. Greenwood, D. J. Prince, and B. P. Straughan, *J. Chem. Soc. A*, 1694 (1968).
1093. P. L. Goggin, *J. Chem. Soc., Dalton Trans.*, 1483 (1974).
1094. R. Forneris, J. Hiraishi, F. A. Miller, and M. Uehara, *Spectrochim. Acta*, **26A,** 581 (1970).
1095. M. Jansen, G. Schatte, K. M. Tobias, and H. Willner, *Inorg. Chem.*, **27,** 1703 (1988).
1096. B. L. Crawford, W. H. Avery, and J. W. Linnett, *J. Chem. Phys.*, **6,** 682 (1938).
1097. G. W. Bethke and M. K. Wilson, *J. Chem. Phys.*, **26,** 1107 (1957).
1098. D. A. Dows and R. M. Hexter, *J. Chem. Phys.*, **24,** 1029 (1956).
1099. R. G. Snyder and J. C. Decius, *Spectrochim. Acta*, **13,** 280 (1959).
1100. W. G. Palmer, *J. Chem. Soc.*, 1552 (1961).
1101. S. J. Cyvin, B. N. Cyvin, C. Wibbelmann, R. Becker, W. Brockner, and M. Parensen, *Z. Naturforsch.*, **40A,** 709 (1985).
1102. W. Brockner, L. Ohse, U. Pätzmann, B. Eisenmann, and H. Schäfer, *Z. Naturforsch.*, **40A,** 1248 (1985).
1103. K. Buijs, *J. Chem. Phys.*, **36,** 861 (1962).
1104. C. A. Evans, K. H. Tan, S. P. Tapper, and M. J. Taylor, *J. Chem. Soc., Dalton Trans.*, 988 (1973).
1105. F. Höfler, S. Waldhör, and E. Hengge, *Spectrochim. Acta*, **28A,** 29 (1972).
1106. F. Höfler, W. Sawodny, and E. Hengge, *Spectrochim. Acta*, **26A,** 819 (1970).
1107. J. E. Griffiths, *Spectrochim. Acta*, **25A,** 965 (1969).
1108. G. A. Ozin, *J. Chem. Soc. A*, 2952 (1969).
1109. B. H. Freeland, J. H. Hencher, D. G. Tuck, and J. G. Contreras, *Inorg. Chem.*, **15,** 2144 (1976).
1110. W. Bues, K. Buchler, and P. Kuhnle, *Z. Anorg. Allg. Chem.*, **325,** 8 (1963).
1111. R. G. Brown and S. D. Ross, *Spectrochim. Acta*, **28,** 1263 (1972).

1112. I. R. Beattie and G. A. Ozin, *J. Chem. Soc. A*, 2615 (1969).
1113. J. Roziere, J.-L. Pascal, and A. Potier, *Spectrochim. Acta*, **29A,** 169 (1973).
1114. A. Grodzicki and A. Potier, *J. Inorg. Nucl. Chem.*, **35,** 61 (1973).
1115. P. Tarte, M. J. Pottier, and A. M. Proces, *Spectrochim. Acta*, **29A,** 1017 (1973).
1116. R. Saez-Puche, M. Bijkerk, F. Fernandez, E. J. Baran, and I. L. Botto, *J. Alloys Compounds*, **184,** 25 (1992).
1117. R. M. Wing and K. P. Callahan, *Inorg. Chem.*, **8,** 871 (1969).
1118. E. J. Baran, *J. Mol. Struct.*, **48,** 441 (1978).
1119. J. D. Witt and R. M. Hammaker, *J. Chem. Phys.*, **58,** 303 (1973).
1120. A. Hezel and S. D. Ross, *Spectrochim. Acta*, **23A,** 1583 (1967).
1121. R. W. Mooney and R. L. Goldsmith, *J. Inorg. Nucl. Chem.*, **31,** 933 (1969).
1122. D. Mascherpa-Corral and A. Potier, *J. Inorg. Nucl. Chem.*, **38,** 211 (1976).
1123. A. Simon and H. Richter, *Z. Anorg. Allg. Chem.*, **304,** 1 (1960); **315,** 196 (1962).
1124. F. A. Cotton and G. Wilkinson, *Advanced Inorganic Chemistry*, 3rd ed., Wiley, New York, 1972, p. 552.
1125. R. J. H. Clark and M. L. Franks, *J. Am. Chem. Soc.*, **97,** 2691 (1975).
1125a. P. S. Santos, M. L. A. Temperini, and O. Sala, *Chem. Phys. Lett.*, **56,** 148 (1978).
1126. G. Peters and W. Preetz, *Z. Naturforsch.*, **34B,** 1767 (1970).
1127. R. J. H. Clark and M. J. Stead, *Inorg. Chem.*, **22,** 1214 (1983).
1128. S. D. Conradson, A. P. Sattelberger, and W. H. Woodruff, *J. Am. Chem. Soc.*, **110,** 1309 (1988).
1129. W. Preetz, G. Peters, and L. Rudzik, *Z. Naturforsch.*, **34B,** 1240 (1979).
1130. W. Preetz and G. Peters, *Z. Naturforsch.*, **35B,** 797 (1980).
1131. D. E. Morris, A. P. Sattelberger, and W. H. Woodruff, *J. Am. Chem. Soc.*, **108,** 8270 (1986).
1132. R. F. Dallinger, *J. Am. Chem. Soc.*, **107,** 7202 (1985).
1133. J. R. Schoonover, R. F. Dallinger, P. M. Killough, A. P. Sattelberger, and W. H. Woodruff, *Inorg. Chem.*, **30,** 1093 (1991).
1134. J. Shamir, S. Schneider, A. Bino, and S. Cohen, *Inorg. Chim. Acta*, **111,** 141 (1986).
1135. R. J. Ziegler and W. M. Risen, Jr., *Inorg. Chem.*, **11,** 2796 (1972).
1136. J. D. Black, J. T. R. Dunsmuir, I. W. Forrest, and A. P. Lane, *Inorg. Chem.*, **14,** 1257 (1975).
1137. H. J. Steinebach and W. Preetz, *Z. Anorg. Allg. Chem.*, **530,** 155 (1985).
1138. P. Hollmann, W. Preetz, H. Hillebrecht, and G. Thiele, *Z. Anorg. Allg. Chem.*, **611,** 28 (1992).
1139. I. R. Beattie, T. R. Gilson, and G. A. Ozin, *J. Chem. Soc.*, 2765 (1968).
1140. S. E. Butler, P. W. Smith, R. Stranger, and I. E. Grey, *Inorg. Chem.*, **25,** 4375 (1986).
1141. D. A. Edwards and R. T. Ward, *J. Chem. Soc. A*, 1617 (1970).
1142. R. C. Burns and T. A. O'Donnell, *Inorg. Chem.*, **18,** 3081 (1979).
1143. W. Kelm and W. Preetz, *Z. Anorg. Allg. Chem.*, **117,** 568 (1989).
1144. L. H. Jones and S. A. Ekberg, *Spectrochim. Acta*, **36A,** 761 (1980).

1145. D. Hartley and M. J. Ware, *J. Chem. Soc., Chem. Commun.*, 912 (1967).
1146. R. Mattes, *Z. Anorg. Allg. Chem.*, **357,** 30 (1968).
1147. A. Zelverte, S. Mancour, and P. Caillet, *Spectrochim. Acta*, **42A,** 837 (1986).
1148. P. M. Boorman and B. P. Straughan, *J. Chem. Soc.*, 1514 (1966).
1149. R. A. MacKay and R. F. Schneider, *Inorg. Chem.*, **7,** 455 (1968).
1150. P. B. Fleming, J. L. Meyer, W. K. Grindstaff, and R. E. McCarley, *Inorg. Chem.*, **9,** 1769 (1970).
1151. R. Mattes, *Z. Anorg. Allg. Chem.*, **364,** 279 (1969).
1152. P. Caillet, S. Ihmaine, and C. Perrin, *J. Mol. Struct.*, **216,** 27 (1990).
1153. F. J. Farrell, V. A. Maroni, and T. G. Spiro, *Inorg. Chem.*, **8,** 2638 (1969).
1154. R. Mattes, H. Bierbüsse, and J. Fuchs, *Z. Anorg. Allg. Chem.*, **385,** 230 (1971).
1155. V. A. Maroni and T. G. Spiro, *J. Am. Chem. Soc.*, **89,** 45 (1967); *Inorg. Chem.*, **7,** 188 (1968).
1156. T. G. Spiro, D. H. Templeton, and A. Zalkin, *Inorg. Chem.*, **8,** 856 (1969).
1157. T. G. Spiro, V. A. Maroni, and C. O. Quicksall, *Inorg. Chem.*, **8,** 2524 (1969).
1158. W. Weltner, Jr. and J. R. W. Warn, *J. Chem. Phys.*, **37,** 292 (1962).
1159. I. R. Beattie, P. J. Jones, D. J. Wild, and T. R. Gilson, *J. Chem. Soc., Dalton Trans.*, 267 (1987).
1160. A. Kaldor and R. F. Porter, *Inorg. Chem.*, **10,** 775 (1971).
1160a. J. R. Durig, W. H. Green, and A. L. Marston, *J. Mol. Struct.*, **2,** 19 (1968).
1161. J. L. Parsons, *J. Chem. Phys.*, **33,** 1860 (1960).
1162. F. A. Grimm and R. F. Porter, *Inorg. Chem.*, **8,** 731 (1969).
1163. R. J. M. Konings, A. S. Booij, and E. H. P. Cordfunke, *J. Mol. Spectrosc.*, **145,** 451 (1991).
1164. I. Hisatsune and N. A. Suarez, *Inorg. Chem.*, **3,** 168 (1964).
1165. B. Tian, G. Wu, and R. Xu, *Spectrochim. Acta*, **43A,** 65 (1987).
1166. W. Bues, G. Foerster, and R. Schmitt, *Z. Anorg. Allg. Chem.*, **344,** 148 (1966).
1167. F. R. Brown, F. A. Miller, and C. Sourisseau, *Spectrochim. Acta*, **32A,** 125 (1976).
1168. J. Thesing, J. Baurmeister, W. Preetz, D. Thiery, and H. G. von Schnering, *Z. Naturforsch.*, **46B,** 800 (1991).
1169. W. Preetz and J. Fritze, *Z. Naturforsch.*, **39B,** 1472 (1984).
1170. J. Thesing, W. Preetz, and J. Baurmeister, *Z. Naturforsch.*, **46B,** 19 (1991).
1171. J. Thesing, M. Stallbaum, and W. Preetz, *Z. Naturforsch.*, **46B,** 602 (1991).
1172. L. A. Leites, S. S. Bukalov, A. P. Kurbakova, M. M. Kaganski, Yu. L. Gaft, N. T. Kuznetzov, and I. A. Zakharova, *Spectrochim. Acta*, **38A,** 1047 (1982).
1173. S. J. Cyvin, B. N. Cyvin, and T. Mogstad, *Spectrochim. Acta*, **42A,** 985 (1986).
1174. V. F. Kalasinsky, *J. Phys. Chem.*, **83,** 3239 (1979).
1175. W. E. Keller and H. L. Johnson, *J. Chem. Phys.*, **20,** 1749 (1952).
1176. Y. Matsui and R. C. Taylor, *Spectrochim. Acta*, **45A,** 299 (1989).
1177. W. Preetz and J. Thesing, *Z. Naturforsch.*, **44B,** 121 (1989).
1178. J. Thesing and W. Preetz, *Z. Naturforsch.*, **45B,** 641 (1990).

1179. J. Fritze and W. Preetz, *Z. Naturforsch.*, **42B,** 293 (1987).

1180. B. Roussel, A. Chapput, and G. Fleury, *J. Mol. Struct.*, **31,** 371 (1976).

1181. K. E. Blick, K. Niedenzu, W. Sawodny, T. Takasuka, T. Totani, and H. Watanabe, *Inorg. Chem.*, **10,** 1133 (1971).

1182. B. E. Smith, H. F. Shurvell, and B. D. James, *J. Chem. Soc., Dalton Trans.*, 711 (1978).

1183. R. T. Paine, R. W. Light, and M. Nelson, *Spectrochim. Acta*, **35A,** 213 (1979).

1184. A. G. Csaszar, L. Hedberg, K. Hedberg, R. C. Burns, A. T. Wen, and M. J. McGlinchey, *Inorg. Chem.*, **30,** 1371 (1991).

1185. C. J. Dain, A. J. Downs, M. J. Goode, D. G. Evans, K. T. Nicholls, D. W. H. Rankin, and H. E. Robertson, *J. Chem. Soc., Dalton Trans.*, 967 (1991).

1186. W. J. Lehmann and I. Shapiro, *Spectrochim. Acta*, **17,** 396 (1961).

1187. L. J. Bellamy, W. Gerrard, M. F. Lappert, and R. L. Williams, *J. Chem. Soc.*, 2412 (1958).

1188. A. Meller, *Organomet. Chem. Rev.*, **2,** 1 (1967).

1189. J. G. Verkade, *Coord. Chem. Rev.*, **9,** 1 (1972).

1190. W. Weltner, Jr. and R. J. Van Zee, *Chem. Rev.*, **89,** 1713 (1989).

1191. W. Weltner, Jr., P. N. Walsh, and C. L. Angell, *J. Chem. Phys.*, **40,** 1299 (1964); W. Weltner, Jr. and D. McLeod, Jr., *ibid.*, 1305.

1192. H. Sasada, T. Amano, C. Jarman, and P. F. Bernath, *J. Chem. Phys.*, **94,** 2401 (1991).

1193. C. A. Schuttenmaier, R. C. Cohen, N. Pugliano, J. R. Heath, A. L. Cooksy, K. L. Busarow, and R. J. Saykally, *Science*, **249,** 897 (1990).

1194. P. A. Withey, L. N. Shen, and W. R. M. Graham, *J. Chem. Phys.*, **95,** 820 (1991).

1195. L. N. Shen, P. A. Withey, and W. R. M. Graham, *J. Chem. Phys.*, **94,** 2395 (1991).

1196. J. R. Heath and R. J. Saykally, *J. Chem. Phys.*, **94,** 3271 (1991).

1197. M. Vala, T. M. Chandrasekhar, J. Szczepanski, R. J. van Zee, and W. Weltner, Jr., *J. Chem. Phys.*, **90,** 595 (1989).

1198. J. Szczepanski and M. Vala, *J. Phys. Chem.*, **95,** 2792 (1991).

1199. N. Moazzen-Ahmedi, A. R. W. McKellar, and T. Amano, *J. Chem. Phys.*, **91,** 2140 (1989).

1200. J. R. Heath, A. L. Cooksy, M. H. W. Gruebele, C. A. Schuttenmaier, and R. J. Saykally, *Science*, **244,** 564 (1989).

1200a. R. H. Kranze and W. R. M. Graham, *J. Chem. Phys.*, **96,** 2517 (1992).

1201. M. Vala, Y. M. Chandraskhar, J. Szczepanski, and R. Pellow, *J. Mol. Struct.*, **222,** 209 (1990).

1201a. R. H. Kranze and W. R. M. Graham, *J. Chem. Phys.*, **98,** 71 (1993).

1202. J. R. Heath and R. J. Saykally, *J. Chem. Phys.*, **94,** 1724 (1991).

1203. J. R. Heath, R. A. Sheeks, A. L. Cooksy, and R. J. Saykally, *Science*, **249,** 895 (1990).

1204. J. Kurtz and D. R. Huffman, *J. Chem. Phys.*, **92,** 30 (1990).

1205. J. R. Heath and R. J. Saykally, *J. Chem. Phys.*, **93,** 8392 (1990).

1206. T. F. Giesen, A. Van Orden, H. J. Hwang, R. S. Fellers, R. A. Provençal, and R. J. Saykally, *Science*, **265,** 756 (1994).

1207. H. W. Kroto, J. E. Fischer, and D. E. Cox, *The Fullerenes*, Pergamon Press, Oxford, England, 1993.

1208. D. Koruga, S. Hameroff, J. Withers, R. Loutfy, and M. Sundareshan, *Fullerene C_{60}: History, Physics, Nanobiology, and Nanotechnology*, North-Holland, Amsterdam, 1993.

1208a. E. J. Maggio, "Bouncing Balls of Carbon: The Discovery and Promise of Fullerenes," in *A Positron Named Priscilla*, National Academy Press, Washington, DC, 1994.

1209. D. S. Bethune, G. Meijer, W. C. Tang, H. J. Rosen, W. G. Golden, H. Seki, C. A. Brown, and M. S. de Vries, *Chem. Phys. Lett.*, **179,** 181 (1991).

1210. R. E. Stanton and M. D. Newton, *J. Phys. Chem.*, **92,** 2141 (1988).

1211. J. P. Hare, T. J. Dennis, H. W. Kroto, R. Taylor, A. W. Allaf, S. Balm, and D. R. Walton, *J. Chem. Soc., Chem. Commun.*, 412 (1991).

1212. C. I. Frum, R. Engleman, Jr., H. G. Hedderich, P. F. Bernath, L. D. Lamb, and D. R. Huffman, *Chem. Phys. Lett.*, **176,** 504 (1991).

1213. T. J. Dennis, J. P. Hare, H. W. Kroto, R. Taylor, D. R. Walton, and P. J. Hendra, *Spectrochim. Acta*, **47A,** 1289 (1991).

1214. R. L. Garrell, T. M. Herne, C. A. Szafrabski, F. Diederich, F. Ettl, and R. L. Whetten, *J. Am. Chem. Soc.*, **113,** 6302 (1991).

1215. F. N. Tebbe, R. L. Harlow, D. B. Chase, D. L. Thorn, G. C. Campbell, Jr., J. C. Calabrese, N. Herron, R. J. Young, and E. Wasserman, *Science*, **256,** 822 (1992).

1216. P. Zhou, K. A. Wang, A. M. Rao, P. C. Eklund, G. Dresselhaus, and M. S. Dresselhaus, *Phys. Rev.*, **B45,** 10838 (1992).

1217. P. C. Eklund, P. Zou, K. A. Wang, G. Dresselhaus, and M. S. Dresselhaus, *J. Phys. Chem. Solids*, **53,** 1391 (1992).

1218. Z. Slanina, J. M. Rudziński, M. Togashi, and E. Ōsawa, *J. Mol. Struct. (Theochem.)*, **202,** 169 (1989).

1219. D. S. Bethune, G. Meijer, W. C. Tang, and H. J. Rosen, *Chem. Phys. Lett.*, **174,** 219 (1990).

1220. J. C. Decius and R. M. Hester, *Molecular Vibrations in Crystals*, McGraw-Hill, New York, 1977, p. 114.

1221. D. S. Knight and W. B. White, *J. Mater. Res.*, **4,** 385 (1989).

1222. J. R. Ferraro, *Vibrational Spectroscopy at High External Pressures*, Academic Press, New York, 1984.

1223. A. Tardieu, F. Cansell, and J. P. Petitet, *J. Appl. Phys.*, **68,** 3243 (1990).

1224. E. R. Lippincott, F. E. Welsh, and C. E. Weir, *Anal. Chem.*, **33,** 137 (1961).

1225. X. X. Bi, P. C. Eklund, J. G. Zhang, A. M. Rao, T. A. Perry, and C. P. Beetz, Jr., *J. Mater. Res.*, **5,** 811 (1990).

1226. W. G. Fateley, F. R. Dollish, N. T. McDevitt, and F. F. Bentley, *Infrared and Raman Selection Rules for Molecular and Lattice Vibrations*, Wiley-Interscience, New York, 1972, p. 163.

1227. F. Tuinstra and J. L. Koenig, *J. Chem. Phys.*, **53,** 1126 (1970).

1228. K. Omori, *Sci. Rep. Tohoku Univ., Ser. 3*, **1,** 102 (1961).
1229. J. W. Huang and W. R. M. Graham, *J. Chem. Phys.*, **93,** 1583 (1990).
1230. L. N. Shen, T. J. Doyle, and W. R. M. Graham, *J. Chem. Phys.*, **93,** 1597 (1990).
1231. P. Botschwina and H. P. Reisenauer, *Chem. Phys. Lett.*, **183,** 217 (1991).
1232. R. D. Brown, D. E. Pullin, E. H. N. Rice, and M. Rodler, *J. Am. Chem. Soc.*, **107,** 7877 (1985).
1233. W. H. Smith and J. J. Barrett, *J. Chem. Phys.*, **51,** 1475 (1969).
1234. W. H. Weber, P. D. Maker, and C. W. Peters, *J. Chem. Phys.*, **64,** 2149 (1976).
1235. W. R. Thorson and I. Nakagawa, *J. Chem. Phys.*, **33,** 994 (1960).
1236. G. Maier, H. P. Reisenauer, H. Balli, W. Brandt, and R. Janoschek, *Angew. Chem., Int. Ed. Engl.*, **29,** 905 (1990).
1237. G. Maier, H. P. Reisenauer, U. Schafer, and H. Balli, *Angew. Chem., Int. Ed. Engl.*, **27,** 566 (1988).
1238. J. B. Bates and W. H. Smith, *Chem. Phys. Lett.*, **14,** 362 (1972).
1239. H. W. Kroto and D. McNaughton, *J. Mol. Spectrosc.*, **114,** 473 (1985).
1240. E. Suzuki and F. Watari, *Chem. Phys. Lett.*, **168,** 1 (1990).
1241. F. Holland and M. Winnewisser, *J. Mol. Spectrosc.*, **149,** 45 (1991).
1242. F. Holland, M. Winnewisser, and J. W. Johns, *Can. J. Phys.*, **68,** 435 (1990).
1243. R. West, D. Eggerding, J. Perkins, D. Handy, and E. C. Tuazon, *J. Am. Chem. Soc.*, **101,** 1710 (1979).
1244. M. Ito and R. West, *J. Am. Chem. Soc.*, **85,** 2580 (1963).
1245. M. J. Taylor, P. N. Gates, and P. M. Smith, *Spectrochim. Acta*, **48A,** 205 (1992).
1245a. K. W. Hipps and A. T. Aplin, *J. Phys. Chem.*, **89,** 5459 (1985).
1246. D. Christen, H. G. Mack, C. J. Marsden, H. Oberhammer, G. Schatte, K. Seppelt, and R. Willner, *J. Am. Chem. Soc.*, **109,** 4009 (1987).
1247. R. Minkwitz and A. Werner, *J. Fluorine Chem.*, **39,** 141 (1988).
1248. C. O. Della Vedova and P. J. Aymonino, *J. Raman Spectrosc.*, **17,** 485 (1986).
1249. E. C. Honea, A. Ogura, C. A. Murray, K. Raghavachari, W. O. Sprenger, M. F. Jarrold, and W. L. Brown, *Nature (London)*, **366,** 42 (1993).
1250. G. Kliche, M. Schwarz, and H. G. von Schnering, *Angew. Chem., Int. Ed. Engl.*, **26,** 349 (1987).
1251. F. Höfler, G. Bauer, and E. Hengge, *Spectrochim. Acta*, **32A,** 1435 (1976).
1252. K. Hassler, E. Hengge, and D. Kovar, *Spectrochim. Acta*, **34A,** 1193 (1978).
1253. F. A. Miller, J. Perkins, G. A. Gibbon, and B. A. Swisshelm, *J. Raman Spectrosc.*, **2,** 93 (1974).
1254. J. R. Durig, M. J. Flanagan, and V. F. Kalasinksy, *J. Chem. Phys.*, **66,** 2775 (1977).
1255. J. R. Durig, M. J. Flanagan, and V. F. Kalasinsky, *Spectrochim. Acta*, **34A,** 63 (1978).
1256. J. R. Durig, Y. S. Li, M. M. Chen, and J. D. Odom, *J. Mol. Spectrosc.*, **59,** 74 (1976).
1257. J. R. Durig, K. S. Kalasinsky, and V. F. Kalasinsky, *J. Mol. Struct.*, **35,** 201 (1976).

1258. K. Hamada, *J. Mol. Struct.*, **48,** 191 (1978).
1259. F. Höfler and W. Peter, *Z. Naturforsch.*, **30B,** 282 (1975).
1260. J. Etchepare, *Spectrochim. Acta*, **26A,** 2147 (1970).
1261. P. McMillan, *Am. Mineral.*, **69,** 622 (1984).
1262. P. K. Dutta and B. Del Barco, *J. Chem. Soc., Chem. Commun.*, 1297 (1985).
1262a. A. N. Lazarev, *Vibrational Spectra and Structure of Silicates*, Consultants Bureau, New York, 1972.
1263. R. A. Condrate, *J. Non-Cryst. Solids*, **84,** 26 (1986).
1264. R. A. Condrate, "The Infrared and Raman Spectra of Glasses," in L. D. Pye, H. J. Stevens, and W. C. LaCourse, eds., *Introduction to Glass Science*, Plenum Press, New York, 1972.
1265. R. K. Sato and P. F. McMillan, *J. Phys. Chem.*, **91,** 3494 (1987).
1266. J. B. Bates and A. S. Quist, *J. Chem. Phys.*, **56,** 1528 (1972).
1267. A. L. Smith, *Spectrochim. Acta*, **16,** 87 (1960).
1268. B. J. Aylett, *Adv. Inorg. Chem. Radiochem.*, **11,** 262 (1968).
1269. H. J. Campbell-Ferguson and E. A. V. Ebsworth, *J. Chem. Soc. A*, 705 (1967).
1270. E. A. V. Ebsworth, D. W. H. Rankin, and G. M. Sheldrick, *J. Chem. Soc. A*, 2828 (1968).
1271. S. Pohl and B. Krebs, *Z. Anorg. Allg. Chem.*, **424,** 265 (1876).
1272. H. Schumann, *Angew. Chem., Int. Ed. Engl.*, **8,** 937 (1969).
1273. C. H. Bibart and G. E. Ewing, *J. Chem. Phys.*, **61,** 1293 (1974).
1274. E. L. Varetti and G. C. Pimentel, *J. Chem. Phys.*, **55,** 3813 (1971).
1275. I. C. Hisatsune, J. P. Devlin, and Y. Wada, *J. Chem. Phys.*, **33,** 714 (1960).
1276. G. R. Smith and W. A. Guillory, *J. Mol. Spectrosc.*, **68,** 223 (1977).
1277. I. C. Hisatsune, J. P. Devlin, and Y. Wada, *Spectrochim. Acta*, **18,** 1641 (1962).
1278. T. Bremm and M. Jansen, *Z. Naturforsch.*, **46B,** 1031 (1991); *Z. Anorg. Allg. Chem.*, **608,** 49 (1992).
1278a. H. R. Hunt, J. R. Cox, and J. D. Ray, *Inorg. Chem.*, **1,** 938 (1962).
1279. J. Laane and J. R. Ohlsen, *Prog. Inorg. Chem.*, **27,** 465 (1980).
1280. K. O. Christe and C. J. Schack, *Inorg. Chem.*, **17,** 2749 (1978).
1281. H. Bürger, H. Niepel, G. Pawelke, and H. Oberhammer, *J. Mol. Struct.*, **54,** 159 (1979).
1282. C. O. Della Vedova, S. E. Ulic, A. Ben Altabef, and P. J. Aymonino, *Z. Phys. Chem.*, **268,** 445 (1987).
1283. D. Tevault and R. R. Smardzewski, *J. Phys. Chem.*, **82,** 375 (1978).
1284. W. W. Wilson and K. O. Christe, *Inorg. Chem.*, **26,** 1573 (1987).
1285. M. C. L. Gerry, W. Lewis-Bevan, A. J. Merer, and N. P. C. Westwood, *J. Mol. Spectrosc.*, **110,** 153 (1985).
1286. M. Birk and W. Winnewisser, *Chem. Phys. Lett.*, **123,** 382 (1986).
1287. M. Nonella, R. P. Müller, and J. R. Huber, *J. Mol. Spectrosc.*, **112,** 142 (1985).
1288. R. Withnall and L. Andrews, *J. Phys. Chem.*, **92,** 2155 (1988).
1289. D. L. Frasco and E. L. Wagner, *J. Chem. Phys.*, **30,** 1124 (1959).

1290. T. Chivers and I. Drummond, *Inorg. Chem.*, **13,** 1222 (1974).

1291. J. Bojes, T. Chivers, W. G. Laidlaw, and M. Trsic, *J. Am. Chem. Soc.*, **101,** 4517 (1979).

1292. I. Nevitt, H. S. Rzepa, and J. D. Woollins, *Spectrochim. Acta*, **45A,** 367 (1989).

1293. R. Steudel, *Z. Naturforsch.*, **36A,** 850 (1981).

1294. A. Turowski, R. Appel, W. Sawodny, and K. Molt, *J. Mol. Struct.*, **48,** 313 (1978).

1295. J. Adel, C. Ergezinger, R. Figge, and K. Dehnicke, *Z. Naturforsch.*, **43B,** 639 (1988).

1296. P. K. Gowik and T. M. Klapötke, *Spectrochim. Acta*, **46A,** 1371 (1990).

1297. U. Patt-Siebel, S. Ruangsuttinarupap, U. Müller, J. Pebler, and K. Dehnicke, *Z. Naturforsch.*, **41B,** 1191 (1986).

1298. D. E. C. Corbridge, *Top. Phosphorus Chem.*, **6,** 235 (1969).

1299. D. E. C. Corbridge, *The Structural Chemistry of Phosphors*, Elsevier, Amsterdam, 1974.

1300. L. C. Thomas, *Interpretation of the Infrared Spectra of Organophosphorus Compounds*, Heyden, London, 1974.

1301. L. C. Thomas and R. A. Chittenden, *Spectrochim. Acta*, **26A,** 781 (1970).

1302. J. Goubeau, *Pure Appl. Chem.*, **44,** 393 (1975).

1303. Y. M. Bosworth, R. J. H. Clark, and D. M. Rippon, *J. Mol. Spectrosc.*, **46,** 240 (1973).

1303a. H. G. M. Edwards, *J. Mol. Struct.*, **295,** 95 (1993).

1304. M. Baudler, S. Akapoglu, D. Ouzounis, F. Wasgestian, B. Meinigke, H. Budzikiewicz, and H. Münster, *Angew. Chem., Int. Ed. Engl.*, **27,** 280 (1988).

1305. H. G. von Schnering, T. Meyer, W. Hoenle, W. Schmettow, U. Hinze, W. Bauhofer, and G. Kliche, *Z. Anorg. Allg. Chem.*, **553,** 261 (1987).

1306. H. G. von Schnering, M. Somer, G. Kliche, W. Hönle, T. Meyer, J. Wolf, L. Ohse, and P. B. Kempa, *Z. Anorg. Allg. Chem.*, **601,** 13 (1991).

1307. Z. Mielke, M. McCluskey, and L. Andrews, *Chem. Phys. Lett.*, **165,** 146 (1990).

1307a. R. J. M. Konings, E. H. P. Cordfunke, and A. S. Booij, *J. Mol. Spectrosc.*, **152,** 29 (1992).

1308. W. P. Griffith and K. J. Rutt, *J. Chem. Soc. A*, 2331 (1968).

1309. N. Santha and V. U. Nayar, *J. Raman Spectrosc.*, **21,** 517 (1990).

1310. G. Foumakoye, R. Cahay, and P. Tarte, *Spectrochim. Acta*, **46A,** 1245 (1990).

1311. X. Mathew and V. U. Nayar, *J. Raman Spectrosc.*, **20,** 633 (1989).

1312. C. I. Cabello and E. J. Baran, *Spectrochim. Acta*, **41A,** 1359 (1985).

1313. M. Bagieu-Beucher, A. Durif, and I. C. Guitel, *J. Solid State Chem.*, **45,** 159 (1982).

1314. G. R. Burns, J. R. Rollo, and R. W. G. Syme, *J. Raman Spectrosc.*, **19,** 345 (1988).

1315. W. Bues, M. Somer, and W. Brockner, *Z. Naturforsch.*, **36A,** 842 (1981).

1316. M. Gardner, *J. Chem. Soc., Dalton Trans.*, 691 (1973).

1317. G. R. Burns, J. R. Rollo, J. D. Sarfrati, and K. R. Morgan, *Spectrochim. Acta*, **47A,** 811 (1991).
1318. G. R. Burns, J. R. Rollo, and J. D. Sarfrati, *Inorg. Chim. Acta*, **161,** 35 (1989).
1319. V. Varma, J. P. Fernandes, and C. N. R. Rao, *J. Mol. Struct.*, **198,** 403 (1989).
1320. I. C. Hisatsune, *Spectrochim. Acta*, **21,** 18 (1965).
1321. T. R. Manley and D. A. Williams, *Spectrochim. Acta*, **23A,** 149 (1967).
1322. E. Steger and K. Lunkwitz, *J. Mol. Struct.*, **3,** 67 (1969).
1323. H. G. M. Edwards, J. S. Ingman, and D. A. Long, *Spectrochim. Acta*, **32A,** 731 (1976).
1324. K. O. Christe, C. J. Schack, and E. C. Curtis, *Inorg. Chem.*, **15,** 843 (1976).
1325. K. Manzel, W. Schulze, V. Wolfel, and R. Minkwitz, *Z. Naturforsch.*, **37B,** 1127 (1982).
1326. H. Sontag and R. Weber, *Chem. Phys.*, **70,** 23 (1982).
1327. V. E. Bondybey and J. H. English, *J. Chem. Phys.*, **73,** 42 (1980).
1328. S. H. Brumbach and G. M. Rosenblatt, *J. Chem. Phys.*, **56,** 3110 (1972).
1329. R. Mercier and C. Sourisseau, *Spectrochim. Acta*, **34A,** 337 (1978).
1330. W. P. Griffith, *J. Chem. Soc. A*, 905 (1967).
1331. G. A. Ozin, *J. Chem. Soc. A*, 2307 (1970).
1332. R. A. Work and M. L. Good, *Spectrochim. Acta*, **29A,** 1547 (1973).
1333. G. D. Brabson, Z. Mielke, and L. Andrews, *J. Phys. Chem.*, **95,** 79 (1991).
1333a. E. Picquenard, M. S. Boumedien, and J. Corset, *J. Mol. Struct.*, **293,** 63 (1993).
1334. R. Steudel, *J. Phys. Chem.*, **80,** 1516 (1976).
1335. R. C. Burns and R. J. Gillespie, *Inorg. Chem.*, **21,** 3877 (1982).
1336. R. J. H. Clark, T. J. Dines, and L. T. H. Ferris, *J. Chem. Soc., Dalton Trans.*, 2237 (1982).
1337. R. Minkwitz, J. Nowicki, W. Sawodny, and K. Härtner, *Spectrochim. Acta*, **47A,** 151 (1991).
1338. R. Steudel and F. Schuster, *Z. Naturforsch.*, **32A,** 1313 (1977).
1339. P. Lenain, E. Picquenard, J. Corset, D. Jensen, and R. Steudel, *Ber. Bunsenges. Phys. Chem.*, **92,** 859 (1988).
1340. L. A. Nimon, V. D. Neff, R. E. Cantley, and R. O. Buttlar, *J. Mol. Spectrosc.*, **22,** 105 (1967).
1341. R. Steudel and F. Schuster, *J. Mol. Struct.*, **44,** 143 (1978).
1342. G. A. Ozin, *J. Chem. Soc. A*, 116 (1969).
1343. R. Steudel and H.-J. Mäusle, *Z. Naturforsch.*, **33A,** 951 (1978).
1344. R. Steudel, T. Sandow, and J. Steidel, *Z. Naturforsch.*, **40B,** 594 (1985).
1345. R. Steudel and D. F. Eggers, Jr., *Spectrochim. Acta*, **31A,** 871 (1975).
1346. R. Steudel and M. Rebsch, *J. Mol. Spectrosc.*, **51,** 334 (1974).
1347. A. W. Herlonger and T. V. Long, *Inorg. Chem.*, **8,** 2661 (1969).
1348. H. Wieser, P. J. Krueger, E. Muller, and J. B. Hyne, *Can. J. Chem.*, **47,** 1633 (1969).
1349. K. Stopperka and F. Kilz, *Z. Anorg. Allg. Chem.*, **370,** 49 (1969).

1350. E. M. Appelman, L. J. Basile, H. Kim, and J. R. Ferraro, *Spectrochim. Acta*, **41A,** 1295 (1985).

1351. Y. P. Kuo, B. M. Cheng, and Y. P. Lee, *Chem. Phys. Lett.*, **177,** 195 (1991).

1352. R. Steudel, *J. Phys. Chem.*, **81,** 343 (1977).

1353. R. Minkwitz and J. Nowicki, *Inorg. Chem.*, **29,** 2361 (1990).

1354. R. Steudel and F. Rose, *Spectrochim. Acta*, **33A,** 979 (1977).

1355. R. Steudel, *J. Mol. Struct.*, **87,** 97 (1982).

1356. R. Steudel, J. Steidel, and N. Rautenberg, *Z. Naturforsch.*, **35B,** 792 (1980).

1357. K. O. Christe, C. J. Schack, and E. C. Curtis, *Spectrochim. Acta*, **26A,** 2367 (1970).

1358. R. Kebabcioglu, R. Mews, and O. Glemser, *Spectrochim. Acta*, **28A,** 1593 (1972).

1358a. D. H. Johnson and D. F. Shriver, *Inorg. Chem.*, **32,** 1045 (1993).

1358b. S. P. Gejji, K. Hermansson, and J. Lindgren, *J. Phys. Chem.*, **97,** 3712 (1993).

1359. R. Steudel, *Z. Naturforsch.*, **30A,** 1481 (1975).

1359a. V. Müller, C. Crebe, U. Müller, and K. Dehnicke, *Z. Anorg. Allg. Chem.*, **619,** 416 (1993).

1360. R. Laitinen, R. Steudel, and E.-M. Strauss, *J. Chem. Soc., Dalton Trans.*, 1869 (1985).

1361. D. M. Giolando, M. Papavassiliou, J. Pickardt, T. B. Rauchfuss, and R. Steudel, *Inorg. Chem.*, **27,** 2596 (1988).

1362. A. Simon and R. Paetzold, *Z. Anorg. Allg. Chem.*, **303,** 39, 46, 53, 72, 79 (1960); *Z. Elektrochem.*, **64,** 209 (1960).

APPENDICES

APPENDIX I. POINT GROUPS AND THEIR CHARACTER TABLES

The following are the character tables of the point groups that appear frequently in this book. The species (or the irreducible representations) of the point group are labeled according to the following rules: A and B denote nondegenerate species (one-dimensional representation). A represents the symmetric species (character = +1) with respect to rotation about the principal axis (chosen as z axis), whereas B represents the antisymmetric species (character = −1) with respect to rotation about the principal axis; E and F denote doubly degenerate (two-dimensional representation) and triply degenerate species (three-dimensional representation), respectively. If two species in the same point group differ in the character of C (other than the principal axis), they are distinguished by subscripts 1, 2, 3, ... If two species differ in the character of σ (other than σ_v), they are distinguished by $'$ and $''$. If two species differ in the character of i, they are distinguished by subscripts g and u. If these rules allow several different labels, g and u take precedence over 1, 2, 3, ... , which in turn take precedence over $'$ and $''$. The labels of species of point groups $\mathbf{C}_{\infty v}$ and $\mathbf{D}_{\infty h}$ (linear molecules) are exceptional and are taken from the notation for the component of the electronic orbital angular momentum along the molecular axis.

$\mathbf{C}_1$	I
A	1

$\mathbf{C}_2$	I	$C_2(z)$		
A	+1	+1	T_z, R_z	$\alpha_{xx}, \alpha_{yy}, \alpha_{zz}, \alpha_{xy}$
B	+1	−1	T_x, T_y, R_x, R_y	α_{yz}, α_{xz}

$\mathbf{C}_s$	I	$\sigma(xy)$		
A'	+1	+1	T_x, T_y, R_z	$\alpha_{xx}, \alpha_{yy}, \alpha_{zz}, \alpha_{xy}$
A''	+1	−1	T_z, R_x, R_y	α_{yz}, α_{xz}

$\mathbf{C}_i \equiv \mathbf{S}_2$	I	i		
A_g	+1	+1	R_x, R_y, R_z	All components of α
A_u	+1	−1	T_x, T_y, T_x	

$\mathbf{C}_{2v}$	I	$C_2(z)$	$\sigma_v(xz)$	$\sigma_v(yz)$		
A_1	+1	+1	+1	+1	T_z	$\alpha_{xx}, \alpha_{yy}, \alpha_{zz}$
A_2	+1	+1	−1	−1	R_z	α_{xy}
B_1	+1	−1	+1	−1	T_x, R_y	α_{xz}
B_2	+1	−1	−1	+1	T_y, R_x	α_{yz}

$\mathbf{C}_{2h}$	I	$C_2(z)$	$\sigma_h(xy)$	i		
A_g	+1	+1	+1	+1	R_z	$\alpha_{xx}, \alpha_{yy}, \alpha_{zz}, \alpha_{xy}$
A_u	+1	+1	−1	−1	T_z	
B_g	+1	−1	−1	+1	R_x, R_y	α_{yz}, α_{xz}
B_u	+1	−1	+1	−1	T_x, T_y	

$\mathbf{D}_2 \equiv \mathbf{V}$	I	$C_2(z)$	$C_2(y)$	$C_2(x)$		
A	+1	+1	+1	+1		$\alpha_{xx}, \alpha_{yy}, \alpha_{zz}$
B_1	+1	+1	−1	−1	T_z, R_z	α_{xy}
B_2	+1	−1	+1	−1	T_y, R_y	α_{xz}
B_3	+1	−1	−1	+1	T_x, R_x	α_{yz}

$\mathbf{D}_{2h} \equiv \mathbf{V}_h$	I	$\sigma(xy)$	$\sigma(xz)$	$\sigma(yz)$	i	$C_2(z)$	$C_2(y)$	$C_2(x)$		
A_g	+1	+1	+1	+1	+1	+1	+1	+1		$\alpha_{xx}, \alpha_{yy}, \alpha_{zz}$
A_u	+1	−1	−1	−1	−1	+1	+1	+1		
B_{1g}	+1	+1	−1	−1	+1	+1	−1	−1	R_z	α_{xy}
B_{1u}	+1	−1	+1	+1	−1	+1	−1	−1	T_z	
B_{2g}	+1	−1	+1	−1	+1	−1	+1	−1	R_y	α_{xz}
B_{2u}	+1	+1	−1	+1	−1	−1	+1	−1	T_y	
B_{3g}	+1	−1	−1	+1	+1	−1	−1	+1	R_x	α_{yz}
B_{3u}	+1	+1	+1	−1	−1	−1	−1	+1	T_x	

$\mathbf{C}_3$	I	C_3	C_3^2		
A	1	1	1	T_z, R_z	$\alpha_{xx}+\alpha_{yy}, \alpha_{zz}$
E	$\left\{\begin{matrix} 1 \\ 1 \end{matrix}\right.$	$\begin{matrix} \epsilon \\ \epsilon^* \end{matrix}$	$\left.\begin{matrix} \epsilon^* \\ \epsilon \end{matrix}\right\}$	$(T_x, T_y), (R_x, R_y)$	$(\alpha_{xx}-\alpha_{yy}, \alpha_{xy}), (\alpha_{yz}, \alpha_{xz})$

$\epsilon = e^{2\pi i/3}$

$\mathbf{C}_4$	I	C_4	C_2	C_4^3		
A	1	1	1	1	T_z, R_z	$\alpha_{xx}+\alpha_{yy}, \alpha_{zz}$
B	1	−1	1	−1		$\alpha_{xx}-\alpha_{yy}, \alpha_{xy}$
E	$\left\{\begin{matrix} 1 \\ 1 \end{matrix}\right.$	$\begin{matrix} i \\ -i \end{matrix}$	$\begin{matrix} -1 \\ -1 \end{matrix}$	$\left.\begin{matrix} -i \\ i \end{matrix}\right\}$	$(T_x, T_y), (R_x, R_y)$	$(\alpha_{yz}, \alpha_{xz})$

$\mathbf{C}_5$	I	C_5	C_5^2	C_5^3	C_5^4		
A	1	1	1	1	1	T_z, R_z	$\alpha_{xx}+\alpha_{yy}, \alpha_{zz}$
E_1	$\left\{\begin{matrix} 1 \\ 1 \end{matrix}\right.$	$\begin{matrix} \epsilon \\ \epsilon^* \end{matrix}$	$\begin{matrix} \epsilon^2 \\ \epsilon^{2*} \end{matrix}$	$\begin{matrix} \epsilon^{2*} \\ \epsilon^2 \end{matrix}$	$\left.\begin{matrix} \epsilon^* \\ \epsilon \end{matrix}\right\}$	$(T_x, T_y), (R_x, R_y)$	$(\alpha_{yz}, \alpha_{xz})$
E_2	$\left\{\begin{matrix} 1 \\ 1 \end{matrix}\right.$	$\begin{matrix} \epsilon^2 \\ \epsilon^{2*} \end{matrix}$	$\begin{matrix} \epsilon^* \\ \epsilon \end{matrix}$	$\begin{matrix} \epsilon \\ \epsilon^* \end{matrix}$	$\left.\begin{matrix} \epsilon^{2*} \\ \epsilon^2 \end{matrix}\right\}$		$(\alpha_{xx}-\alpha_{yy}, \alpha_{xy})$

$\epsilon = e^{2\pi i/5}$

$\mathbf{C}_6$	I	C_6	C_3	C_2	C_3^2	C_6^5		
A	1	1	1	1	1	1	T_z, R_z	$\alpha_{xx}+\alpha_{yy}, \alpha_{zz}$
B	1	−1	1	−1	1	−1		
E_1	$\left\{\begin{matrix} 1 \\ 1 \end{matrix}\right.$	$\begin{matrix} \epsilon \\ \epsilon^* \end{matrix}$	$\begin{matrix} -\epsilon^* \\ -\epsilon \end{matrix}$	$\begin{matrix} -1 \\ -1 \end{matrix}$	$\begin{matrix} -\epsilon \\ -\epsilon^* \end{matrix}$	$\left.\begin{matrix} \epsilon^* \\ \epsilon \end{matrix}\right\}$	$(T_x, T_y), (R_x, R_y)$	$(\alpha_{xz}, \alpha_{yz})$
E_2	$\left\{\begin{matrix} 1 \\ 1 \end{matrix}\right.$	$\begin{matrix} -\epsilon^* \\ -\epsilon \end{matrix}$	$\begin{matrix} -\epsilon \\ -\epsilon^* \end{matrix}$	$\begin{matrix} 1 \\ 1 \end{matrix}$	$\begin{matrix} -\epsilon^* \\ -\epsilon \end{matrix}$	$\left.\begin{matrix} -\epsilon \\ -\epsilon^* \end{matrix}\right\}$		$(\alpha_{xx}-\alpha_{yy}, \alpha_{xy})$

$\epsilon = e^{2\pi i/6}$

$\mathbf{S}_4$	I	S_4	C_2	S_4^3		
A	1	1	1	1	R_z	$\alpha_{xx}+\alpha_{yy}, \alpha_{zz}$
B	1	−1	1	−1	T_z	$\alpha_{xx}-\alpha_{yy}, \alpha_{xy}$
E	$\left\{\begin{matrix} 1 \\ 1 \end{matrix}\right.$	$\begin{matrix} i \\ -i \end{matrix}$	$\begin{matrix} -1 \\ -1 \end{matrix}$	$\left.\begin{matrix} -i \\ i \end{matrix}\right\}$	$(T_x, T_y), (R_x, R_y)$	$(\alpha_{xz}, \alpha_{yz})$

$S_6 \equiv C_{3i}$	I	C_3	C_3^2	i	S_6^5	S_6		
A_g	1	1	1	1	1	1	R_z	$\alpha_{xx}+\alpha_{yy}, \alpha_{zz}$
E_g	$\begin{Bmatrix} 1 \\ 1 \end{Bmatrix}$	$\begin{matrix} \epsilon \\ \epsilon^* \end{matrix}$	$\begin{matrix} \epsilon^* \\ \epsilon \end{matrix}$	$\begin{matrix} 1 \\ 1 \end{matrix}$	$\begin{matrix} \epsilon \\ \epsilon^* \end{matrix}$	$\begin{matrix} \epsilon^* \\ \epsilon \end{matrix}$	(R_x, R_y)	$(\alpha_{xx}-\alpha_{yy}, \alpha_{xy}), (\alpha_{xz}, \alpha_{yz})$
A_u	1	1	1	−1	−1	−1	T_z	
E_u	$\begin{Bmatrix} 1 \\ 1 \end{Bmatrix}$	$\begin{matrix} \epsilon \\ \epsilon^* \end{matrix}$	$\begin{matrix} \epsilon^* \\ \epsilon \end{matrix}$	$\begin{matrix} -1 \\ -1 \end{matrix}$	$\begin{matrix} -\epsilon \\ -\epsilon^* \end{matrix}$	$\begin{matrix} -\epsilon^* \\ -\epsilon \end{matrix}$	(T_x, T_y)	

$\epsilon = e^{2\pi i/3}$

D_3	I	$2C_3(z)$	$3C_2$		
A_1	+1	+1	+1		$\alpha_{xx}+\alpha_{yy}, \alpha_{zz}$
A_2	+1	+1	−1	T_z, R_z	
E	+2	−1	0	$(T_x, T_y), (R_x, R_y)$	$(\alpha_{xx}-\alpha_{yy}, \alpha_{xy}), (\alpha_{yz}, \alpha_{xz})$

D_4	I	$2C_4$	$C_2(=C_4^2)$	$2C_2'$	$2C_2''$		
A_1	1	1	1	1	1		$\alpha_{xx}+\alpha_{yy}, \alpha_{zz}$
A_2	1	1	1	−1	−1	T_z, R_z	
B_1	1	−1	1	1	−1		$\alpha_{xx}-\alpha_{yy}$
B_2	1	−1	1	−1	1		α_{xy}
E	2	0	−2	0	0	$(T_x, T_y), (R_x, R_y)$	$(\alpha_{xz}, \alpha_{yz})$

D_5	I	$2C_5$	$2C_5^2$	$5C_2$		
A_1	1	1	1	1		$\alpha_{xx}+\alpha_{yy}, \alpha_{zz}$
A_2	1	1	1	−1	T_z, R_z	
E_1	2	$2c\frac{2\pi}{5}$	$2c\frac{4\pi}{5}$	0	$(T_x, T_y), (R_x, R_y)$	$(\alpha_{xz}, \alpha_{yz})$
E_2	2	$2c\frac{4\pi}{5}$	$2c\frac{2\pi}{5}$	0		$(\alpha_{xx}-\alpha_{yy}, \alpha_{xy})$

c = cosine

D_6	I	$2C_6$	$2C_3$	C_2	$3C_2'$	$3C_2''$		
A_1	1	1	1	1	1	1		$\alpha_{xx}+\alpha_{yy}, \alpha_{zz}$
A_2	1	1	1	1	−1	−1	T_z, R_z	
B_1	1	−1	1	−1	1	−1		
B_2	1	−1	1	−1	−1	1		
E_1	2	1	−1	−2	0	0	$(T_x, T_y), (R_x, R_y)$	$(\alpha_{xz}, \alpha_{yz})$
E_2	2	−1	−1	2	0	0		$(\alpha_{xx}-\alpha_{yy}, \alpha_{xy})$

C_{3v}	I	$2C_2(z)$	$3\sigma_v$		
A_1	+1	+1	+1	T_z	$\alpha_{xx}+\alpha_{yy}, \alpha_{zz}$
A_2	+1	+1	−1	R_z	
E	+2	−1	0	$(T_x, T_y), (R_x, R_y)$	$(\alpha_{xx}-\alpha_{yy}, \alpha_{xy}), (\alpha_{yz}, \alpha_{xz})$

C_{4v}	I	$2C_4(z)$	$C_4^2 \equiv C_2''$	$2\sigma_v$	$2\sigma_d$		
A_1	+1	+1	+1	+1	+1	T_z	$\alpha_{xx}+\alpha_{yy}, \alpha_{zz}$
A_2	+1	+1	+1	−1	−1	R_z	
B_1	+1	−1	+1	+1	−1		$\alpha_{xx}-\alpha_{yy}$
B_2	+1	−1	+1	−1	+1		α_{xy}
E	+2	0	−2	0	0	$(T_x, T_y), (R_x, R_y)$	$(\alpha_{yz}, \alpha_{xz})$

C_{5v}	I	$2C_5$	$2C_5^2$	$5\sigma_v$		
A_1	1	1	1	1	T_z	$\alpha_{xx}+\alpha_{yy}, \alpha_{zz}$
A_2	1	1	1	−1	R_z	
E_1	2	$2c\frac{2\pi}{5}$	$2c\frac{4\pi}{5}$	0	$(T_x, T_y), (R_x, R_y)$	$(\alpha_{xz}, \alpha_{yz})$
E_2	2	$2c\frac{4\pi}{5}$	$2c\frac{2\pi}{5}$	0		$(\alpha_{xx}-\alpha_{yy}, \alpha_{xy})$

c = cosine

C_{6v}	I	$2C_6$	$2C_3$	C_2	$3\sigma_v$	$3\sigma_d$		
A_1	1	1	1	1	1	1	T_z	$\alpha_{xx}+\alpha_{yy}, \alpha_{zz}$
A_2	1	1	1	1	−1	−1	R_z	
B_1	1	−1	1	−1	1	−1		
B_2	1	−1	1	−1	−1	1		
E_1	2	1	−1	−2	0	0	$(T_x, T_y), (R_x, R_y)$	$(\alpha_{xz}, \alpha_{yz})$
E_2	2	−1	−1	2	0	0		$(\alpha_{xx}-\alpha_{yy}, \alpha_{xy})$

$\mathbf{C}_{\infty v}$	I	$2C_{\infty}^{\phi}$	$2C_{\infty}^{2\phi}$	$2C_{\infty}^{3\phi}$	$\cdots$	$\infty\sigma_v$		
Σ^+	+1	+1	+1	+1	$\cdots$	+1	T_z	$\alpha_{xx} + \alpha_{yy}, \alpha_{zz}$
Σ^-	+1	+1	+1	+1	$\cdots$	−1	R_z	
Π	+2	$2\cos\phi$	$2\cos 2\phi$	$2\cos 3\phi$	$\cdots$	0	$(T_x, T_y), (R_x, R_y)$	$(\alpha_{yz}, \alpha_{xz})$
Δ	+2	$2\cos 2\phi$	$2\cos 2\cdot 2\phi$	$2\cos 3\cdot 2\phi$	$\cdots$	0		$(\alpha_{xx} - \alpha_{yy}, \alpha_{xy})$
Φ	+2	$2\cos 3\phi$	$2\cos 2\cdot 3\phi$	$2\cos 3\cdot 3\phi$	$\cdots$	0		
$\cdots$	$\cdots$	$\cdots$	$\cdots$	$\cdots$	$\cdots$	$\cdots$		

ϕ = arbitrary angle

$\mathbf{C}_{3h}$	I	C_3	C_3^2	σ_h	S_3	S_3^5		
A'	1	1	1	1	1	1	R_z	$\alpha_{xx} + \alpha_{yy}, \alpha_{zz}$
A''	1	1	1	−1	−1	−1	T_z	
E'	1	ϵ	ϵ^*	1	ϵ	ϵ^*	(T_x, T_y)	$(\alpha_{xx} - \alpha_{yy}, \alpha_{xy})$
	1	ϵ^*	ϵ	1	ϵ^*	ϵ		
E''	1	ϵ	ϵ^*	−1	$-\epsilon$	$-\epsilon^*$	(R_x, R_y)	$(\alpha_{xz}, \alpha_{yz})$
	1	ϵ^*	ϵ	−1	$-\epsilon^*$	$-\epsilon$		

$\epsilon = e^{2\pi i/3}$

$\mathbf{C}_{4h}$	I	C_4	C_2	C_4^3	i	S_4^3	σ_h	S_4		
A_g	1	1	1	1	1	1	1	1	R_z	$\alpha_{xx} + \alpha_{yy}, \alpha_{zz}$
A_u	1	1	1	1	−1	−1	−1	−1	T_z	
B_g	1	−1	1	−1	1	−1	1	−1		$\alpha_{xx} - \alpha_{yy}, \alpha_{xy}$
B_u	1	−1	1	−1	−1	1	−1	1		
E_g	1	i	−1	$-i$	1	i	−1	$-i$	(R_x, R_y)	$(\alpha_{xz}, \alpha_{yz})$
	1	$-i$	−1	i	1	$-i$	−1	i		
E_u	1	i	−1	$-i$	−1	$-i$	1	i	(T_x, T_y)	
	1	$-i$	−1	i	−1	i	1	$-i$		

$\mathbf{C}_{5h}$	I	C_5	C_5^2	C_5^3	C_5^4	σ_h	S_5	S_5^7	S_5^3	S_5^9		
A'	1	1	1	1	1	1	1	1	1	1	R_z	$\alpha_{xx} + \alpha_{yy}, \alpha_{zz}$
A''	1	1	1	1	1	−1	−1	−1	−1	−1	T_z	
E_1'	1	ϵ	ϵ^2	ϵ^{2*}	ϵ^*	1	ϵ	ϵ^2	ϵ^{2*}	ϵ^*	(T_x, T_y)	
	1	ϵ^*	ϵ^{2*}	ϵ^2	ϵ	1	ϵ^*	ϵ^{2*}	ϵ^2	ϵ		
E_1''	1	ϵ	ϵ^2	ϵ^{2*}	ϵ^*	−1	$-\epsilon$	$-\epsilon^2$	$-\epsilon^{2*}$	$-\epsilon^*$	(R_x, R_y)	$(\alpha_{xz}, \alpha_{yz})$
	1	ϵ^*	ϵ^{2*}	ϵ^2	ϵ	−1	$-\epsilon^*$	$-\epsilon^{2*}$	$-\epsilon^2$	$-\epsilon$		
E_2'	1	ϵ^2	ϵ^*	ϵ	ϵ^{2*}	1	ϵ^2	ϵ^*	ϵ	ϵ^{2*}		$(\alpha_{xx} - \alpha_{yy}, \alpha_{xy})$
	1	ϵ^{2*}	ϵ	ϵ^*	ϵ^2	1	ϵ^{2*}	ϵ	ϵ^*	ϵ^2		
E_2''	1	ϵ^2	ϵ^*	ϵ	ϵ^{2*}	−1	$-\epsilon^2$	$-\epsilon^*$	$-\epsilon$	$-\epsilon^{2*}$		
	1	ϵ^{2*}	ϵ	ϵ^*	ϵ^2	−1	$-\epsilon^{2*}$	$-\epsilon$	$-\epsilon^*$	$-\epsilon^2$		

$\epsilon = e^{2\pi i/5}$

$\mathbf{C}_{6h}$	I	C_6	C_3	C_2	C_3^2	C_6^5	i	S_3^5	S_6^5	σ_h	S_6	S_3		
A_g	1	1	1	1	1	1	1	1	1	1	1	1	R_z	$\alpha_{xx} + \alpha_{yy}, \alpha_{zz}$
A_u	1	1	1	1	1	1	−1	−1	−1	−1	−1	−1	T_z	
B_g	1	−1	1	−1	1	−1	1	−1	1	−1	1	−1		
B_u	1	−1	1	−1	1	−1	−1	1	−1	1	−1	1		
E_{1g}	1	ϵ	$-\epsilon^*$	−1	$-\epsilon$	ϵ^*	1	ϵ	$-\epsilon^*$	−1	$-\epsilon$	ϵ^*	(R_x, R_y)	$(\alpha_{xz}, \alpha_{yz})$
	1	ϵ^*	$-\epsilon$	−1	$-\epsilon^*$	ϵ	1	ϵ^*	$-\epsilon$	−1	$-\epsilon^*$	ϵ		
E_{1u}	1	ϵ	$-\epsilon^*$	−1	$-\epsilon$	ϵ^*	−1	$-\epsilon$	ϵ^*	1	ϵ	$-\epsilon^*$	(T_x, T_y)	
	1	ϵ^*	$-\epsilon$	−1	$-\epsilon^*$	ϵ	−1	$-\epsilon^*$	ϵ	1	ϵ^*	$-\epsilon$		
E_{2g}	1	$-\epsilon^*$	$-\epsilon$	1	$-\epsilon^*$	$-\epsilon$	1	$-\epsilon^*$	$-\epsilon$	1	$-\epsilon^*$	$-\epsilon$		$(\alpha_{xx} - \alpha_{yy}, \alpha_{xy})$
	1	$-\epsilon$	$-\epsilon^*$	1	$-\epsilon$	$-\epsilon^*$	1	$-\epsilon$	$-\epsilon^*$	1	$-\epsilon$	$-\epsilon^*$		
E_{2u}	1	$-\epsilon^*$	$-\epsilon$	1	$-\epsilon^*$	$-\epsilon$	−1	ϵ^*	ϵ	−1	ϵ^*	ϵ		
	1	$-\epsilon$	$-\epsilon^*$	1	$-\epsilon$	$-\epsilon^*$	−1	ϵ	ϵ^*	−1	ϵ	ϵ^*		

$\epsilon = e^{2\pi i/6}$

$\mathbf{D}_{2d} \equiv \mathbf{V}_d$	I	$2S_4(z)$	$S_4^2 \equiv C_2''$	$2C_2$	$2\sigma_d$		
A_1	+1	+1	+1	+1	+1		$\alpha_{xx} + \alpha_{yy}, \alpha_{zz}$
A_2	+1	+1	+1	−1	−1	R_z	
B_1	+1	−1	+1	+1	−1		$\alpha_{xx} - \alpha_{yy}$
B_2	+1	−1	+1	−1	+1	T_z	α_{xy}
E	+2	0	−2	0	0	$(T_x, T_y), (R_x, R_y)$	$(\alpha_{yz}, \alpha_{xz})$

$\mathbf{D}_{3d}$	I	$2S_6(z)$	$2S_6^2 \equiv 2C_3$	$S_6^3 \equiv S_2 \equiv i$	$3C_2$	$3\sigma_d$		
A_{1g}	+1	+1	+1	+1	+1	+1		$\alpha_{xx} + \alpha_{yy}, \alpha_{zz}$
A_{1u}	+1	−1	+1	−1	+1	−1		
A_{2g}	+1	+1	+1	+1	−1	−1	R_z	
A_{2u}	+1	−1	+1	−1	−1	+1	T_z	
E_g	+2	−1	−1	+2	0	0	(R_x, R_y)	$(\alpha_{xx} - \alpha_{yy}, \alpha_{xy}), (\alpha_{yz}, \alpha_{xz})$
E_u	+2	+1	−1	−2	0	0	(T_x, T_y)	

$\mathbf{D}_{4d}$	I	$2S_8(z)$	$2S_8^2 \equiv 2C_4$	$2S_8^3$	$S_8^4 \equiv C_2''$	$4C_2$	$4\sigma_d$		
A_1	+1	+1	+1	+1	+1	+1	+1		$\alpha_{xx} + \alpha_{yy}, \alpha_{zz}$
A_2	+1	+1	+1	+1	+1	−1	−1	R_z	
B_1	+1	−1	+1	−1	+1	+1	−1		
B_2	+1	−1	+1	−1	+1	−1	+1	T_z	
E_1	+2	$+\sqrt{2}$	0	$-\sqrt{2}$	−2	0	0	(T_x, T_y)	
E_2	+2	0	−2	0	+2	0	0		$(\alpha_{xx} - \alpha_{yy}, \alpha_{xy})$
E_3	+2	$-\sqrt{2}$	0	$+\sqrt{2}$	−2	0	0	(R_x, R_y)	$(\alpha_{yz}, \alpha_{xz})$

$\mathbf{D}_{5d}$	I	$2C_5$	$2C_5^2$	$5C_2$	i	$2S_{10}^3$	$2S_{10}$	$5\sigma_d$		
A_{1g}	1	1	1	1	1	1	1	1		$\alpha_{xx} + \alpha_{yy}, \alpha_{zz}$
A_{1u}	1	1	1	1	−1	−1	−1	−1		
A_{2g}	1	1	1	−1	1	1	1	−1	R_z	
A_{2u}	1	1	1	−1	−1	−1	−1	1	T_z	
E_{1g}	2	$2c\frac{2\pi}{5}$	$2c\frac{4\pi}{5}$	0	2	$2c\frac{2\pi}{5}$	$2c\frac{4\pi}{5}$	0	(R_x, R_y)	$(\alpha_{xz}, \alpha_{yz})$
E_{1u}	2	$2c\frac{2\pi}{5}$	$2c\frac{4\pi}{5}$	0	−2	$-2c\frac{2\pi}{5}$	$-2c\frac{4\pi}{5}$	0	(T_x, T_y)	
E_{2g}	2	$2c\frac{4\pi}{5}$	$2c\frac{2\pi}{5}$	0	2	$2c\frac{4\pi}{5}$	$2c\frac{2\pi}{5}$	0		$(\alpha_{xx} - \alpha_{yy}, \alpha_{xy})$
E_{2u}	2	$2c\frac{4\pi}{5}$	$2c\frac{2\pi}{5}$	0	−2	$-2c\frac{4\pi}{5}$	$-2c\frac{2\pi}{5}$	0		

c = cosine

$\mathbf{D}_{3h}$	I	$2C_3(z)$	$3C_2$	σ_h	$2S_3$	$3\sigma_v$		
A_1'	+1	+1	+1	+1	+1	+1		$\alpha_{xx} + \alpha_{yy}, \alpha_{zz}$
A_1''	+1	+1	+1	−1	−1	−1		
A_2'	+1	+1	−1	+1	+1	−1	R_z	
A_2''	+1	+1	−1	−1	−1	+1	T_z	
E'	+2	−1	0	+2	−1	0	(T_x, T_y)	$(\alpha_{xx} - \alpha_{yy}, \alpha_{xy})$
E''	+2	−1	0	−2	+1	0	(R_x, R_y)	$(\alpha_{yz}, \alpha_{xz})$

$\mathbf{D}_{4h}$	I	$2C_4(z)$	$C_4^2 \equiv C_2$	$2C'_2$	$2C''_2$	σ_h	$2\sigma_v$	$2\sigma_d$	$2S_4$	$S_2 \equiv i$		
A_{1g}	+1	+1	+1	+1	+1	+1	+1	+1	+1	+1		$\alpha_{xx}, \alpha_{yy}, \alpha_{zz}$
A_{1u}	+1	+1	+1	+1	+1	−1	−1	−1	−1	−1		
A_{2g}	+1	+1	+1	−1	−1	+1	−1	−1	+1	+1	R_z	
A_{2u}	+1	+1	+1	−1	−1	−1	+1	+1	−1	−1	T_z	
B_{1g}	+1	−1	+1	+1	−1	+1	+1	−1	−1	+1		$\alpha_{xx} - \alpha_{yy}$
B_{1u}	+1	−1	+1	+1	−1	−1	−1	+1	+1	−1		
B_{2g}	+1	−1	+1	−1	+1	+1	−1	+1	−1	+1		α_{xy}
B_{2u}	+1	−1	+1	−1	+1	−1	+1	−1	+1	−1		
E_g	+2	0	−2	0	0	−2	0	0	0	+2	(R_x, R_y)	$(\alpha_{yz}, \alpha_{xz})$
E_u	+2	0	−2	0	0	+2	0	0	0	−2	(T_x, T_y)	

$\mathbf{D}_{5h}$	I	$2C_5$	$2C_5^2$	$5C_2$	σ_h	$2S_5$	$2S_5^3$	$5\sigma_v$		
A'_1	1	1	1	1	1	1	1	1		$\alpha_{xx} + \alpha_{yy}, \alpha_{zz}$
A''_1	1	1	1	1	−1	−1	−1	−1		
A'_2	1	1	1	−1	1	1	1	−1	R_z	
A''_2	1	1	1	−1	−1	−1	−1	1	T_z	
E'_1	2	$2c\frac{2\pi}{5}$	$2c\frac{4\pi}{5}$	0	2	$2c\frac{2\pi}{5}$	$2c\frac{4\pi}{5}$	0	(T_x, T_y)	
E''_1	2	$2c\frac{2\pi}{5}$	$2c\frac{4\pi}{5}$	0	−2	$-2c\frac{2\pi}{5}$	$-2c\frac{4\pi}{5}$	0	(R_x, R_y)	$(\alpha_{xz}, \alpha_{yz})$
E'_2	2	$2c\frac{4\pi}{5}$	$2c\frac{2\pi}{5}$	0	2	$2c\frac{4\pi}{5}$	$2c\frac{2\pi}{5}$	0		$(\alpha_{xx} - \alpha_{yy}, \alpha_{xy})$
E''_2	2	$2c\frac{4\pi}{5}$	$2c\frac{2\pi}{5}$	0	−2	$-2c\frac{4\pi}{5}$	$-2c\frac{2\pi}{5}$	0		

c = cosine

$\mathbf{D}_{6h}$	I	$2C_6(z)$	$2C_6^2 \equiv 2C_3$	$C_6^3 \equiv C_2$	$3C'_2$	$3C''_2$	σ_h	$3\sigma_v$	$3\sigma_d$	$2S_6$	$2S_3$	$S_6^3 \equiv S_2 \equiv i$		
A_{1g}	+1	+1	+1	+1	+1	+1	+1	+1	+1	+1	+1	+1		$\alpha_{xx} + \alpha_{yy}, \alpha_{zz}$
A_{1u}	+1	+1	+1	+1	+1	+1	−1	−1	−1	−1	−1	−1		
A_{2g}	+1	+1	+1	+1	−1	−1	+1	−1	−1	+1	+1	+1	R_z	
A_{2u}	+1	+1	+1	+1	−1	−1	−1	+1	+1	−1	−1	−1	T_z	
B_{1g}	+1	−1	+1	−1	+1	−1	−1	−1	+1	+1	−1	+1		
B_{1u}	+1	−1	+1	−1	+1	−1	+1	+1	−1	−1	+1	−1		
B_{2g}	+1	−1	+1	−1	−1	+1	−1	+1	−1	+1	−1	+1		
B_{2u}	+1	−1	+1	−1	−1	+1	+1	−1	+1	−1	+1	−1		
E_{1g}	+1	+1	−1	−2	0	0	−2	0	0	−1	+1	+2	(R_x, R_y)	$(\alpha_{yz}, \alpha_{xz})$
E_{1u}	+1	+1	−1	−2	0	0	+2	0	0	+1	−1	−2	(T_x, T_y)	
E_{2g}	+1	−1	−1	+2	0	0	+2	0	0	−1	−1	+2		$(\alpha_{xx} - \alpha_{yy}, \alpha_{xy})$
E_{2u}	+1	−1	−1	+2	0	0	−2	0	0	+1	+1	−2		

$\mathbf{D}_{\infty h}$	I	$2C_\infty^\phi$	$2C_\infty^{2\phi}$	$2C_\infty^{3\phi}$	⋯	σ_h	∞C_2	$\infty\sigma_v$	$2S_\infty^\phi$	$2S_\infty^{2\phi}$	⋯	$S_2 \equiv i$		
Σ_g^+	+1	+1	+1	+1	⋯	+1	+1	+1	+1	+1	⋯	+1		$\alpha_{xx} + \alpha_{yy}, \alpha_{zz}$
Σ_u^+	+1	+1	+1	+1	⋯	−1	−1	+1	−1	−1	⋯	−1	T_z	
Σ_g^-	+1	+1	+1	+1	⋯	+1	−1	−1	+1	+1	⋯	+1	R_z	
Σ_u^-	+1	+1	+1	+1	⋯	−1	+1	−1	−1	−1	⋯	−1		
Π_g	+2	$2\cos\phi$	$2\cos 2\phi$	$2\cos 3\phi$	⋯	−2	0	0	$-2\cos\phi$	$-2\cos 2\phi$	⋯	+2	(R_x, R_y)	$(\alpha_{yz}, \alpha_{xz})$
Π_u	+2	$2\cos\phi$	$2\cos 2\phi$	$2\cos 3\phi$	⋯	+2	0	0	$+2\cos\phi$	$+2\cos 2\phi$	⋯	−2	(T_x, T_y)	
Δ_g	+2	$2\cos 2\phi$	$2\cos 4\phi$	$2\cos 6\phi$	⋯	+2	0	0	$+2\cos 2\phi$	$+2\cos 4\phi$	⋯	+2		$(\alpha_{xx} - \alpha_{yy}, \alpha_{xy})$
Δ_u	+2	$2\cos 2\phi$	$2\cos 4\phi$	$2\cos 6\phi$	⋯	−2	0	0	$-2\cos 2\phi$	$-2\cos 4\phi$	⋯	−2		
Φ_g	+2	$2\cos 3\phi$	$2\cos 6\phi$	$2\cos 9\phi$	⋯	−2	0	0	$-2\cos 3\phi$	$-2\cos 4\phi$	⋯	+2		
Φ_u	+2	$2\cos 3\phi$	$2\cos 6\phi$	$2\cos 9\phi$	⋯	+2	0	0	$+2\cos 3\phi$	$+2\cos 4\phi$	⋯	−2		
⋯	⋯	⋯	⋯	⋯	⋯	⋯	0	0	⋯	⋯	⋯	⋯		

ϕ = arbitrary angle

$\mathbf{T}$	I	$4C_3$	$4C_3^2$	$3C_2$		
A	1	1	1	1		$\alpha_{xx} + \alpha_{yy} + \alpha_{zz}$
E	$\left\{\begin{matrix}1\\1\end{matrix}\right.$	$\begin{matrix}\epsilon\\\epsilon^*\end{matrix}$	$\begin{matrix}\epsilon^*\\\epsilon\end{matrix}$	$\left.\begin{matrix}1\\1\end{matrix}\right\}$		$(2\alpha_{zz} - \alpha_{xx} - \alpha_{yy}, \alpha_{xx} - \alpha_{yy})$
F	3	0	0	−1	$(R_x, R_y, R_z), (T_x, T_y, T_z)$	$(\alpha_{xy}, \alpha_{xz}, \alpha_{yz})$

$\epsilon = e^{2\pi i/3}$

$\mathbf{T}_h$	I	$4C_3$	$4C_3^2$	$3C_2$	i	$4S_6$	$4S_6^5$	$3\sigma_v$		
A_g	1	1	1	1	1	1	1	1		$\alpha_{xx} + \alpha_{yy} + \alpha_{zz}$
A_u	1	1	1	1	−1	−1	−1	−1		
E_g	$\left\{\begin{matrix}1\\1\end{matrix}\right.$	$\begin{matrix}\epsilon\\\epsilon^*\end{matrix}$	$\begin{matrix}\epsilon^*\\\epsilon\end{matrix}$	$\begin{matrix}1\\1\end{matrix}$	$\begin{matrix}1\\1\end{matrix}$	$\begin{matrix}\epsilon\\\epsilon^*\end{matrix}$	$\begin{matrix}\epsilon^*\\\epsilon\end{matrix}$	$\left.\begin{matrix}1\\1\end{matrix}\right\}$		$(2\alpha_{zz} - \alpha_{xx} - \alpha_{yy}, \alpha_{xx} - \alpha_{yy})$
E_u	$\left\{\begin{matrix}1\\1\end{matrix}\right.$	$\begin{matrix}\epsilon\\\epsilon^*\end{matrix}$	$\begin{matrix}\epsilon^*\\\epsilon\end{matrix}$	$\begin{matrix}1\\1\end{matrix}$	$\begin{matrix}-1\\-1\end{matrix}$	$\begin{matrix}-\epsilon\\-\epsilon^*\end{matrix}$	$\begin{matrix}-\epsilon^*\\-\epsilon\end{matrix}$	$\left.\begin{matrix}-1\\-1\end{matrix}\right\}$		
F_g	3	0	0	−1	3	0	0	−1	(R_x, R_y, R_z)	$(\alpha_{xy}, \alpha_{yz}, \alpha_{xz})$
F_u	3	0	0	−1	−3	0	0	1	(T_x, T_y, T_z)	

$\epsilon = e^{2\pi i/3}$

$\mathbf{T}_d$	I	$8C_3$	$6\sigma_d$	$6S_4$	$3S_4^2 \equiv 3C_2$		
A_1	+1	+1	+1	+1	+1		$\alpha_{xx} + \alpha_{yy} + \alpha_{zz}$
A_2	+1	+1	−1	−1	+1		
E	+2	−1	0	0	+2		$(\alpha_{xx} + \alpha_{yy} - 2\alpha_{zz}, \alpha_{xx} - \alpha_{yy})$
F_1	+3	0	−1	+1	−1	(R_x, R_y, R_z)	
F_2	+3	0	+1	−1	−1	(T_x, T_y, T_z)	$(\alpha_{xy}, \alpha_{yz}, \alpha_{xz})$

$\mathbf{O}$	I	$6C_4$	$3C_2(= C_4^2)$	$8C_3$	$6C_2$		
A_1	1	1	1	1	1		$\alpha_{xx} + \alpha_{yy} + \alpha_{zz}$
A_2	1	−1	1	1	−1		
E	2	0	2	−1	0		$(2\alpha_{zz} - \alpha_{xx} - \alpha_{yy}, \alpha_{xx} - \alpha_{yy})$
F_1	3	1	−1	0	−1	$(R_x, R_y, R_z), (T_x, T_y, T_z)$	
F_2	3	−1	−1	0	1		$(\alpha_{xy}, \alpha_{yz}, \alpha_{xz})$

$\mathbf{O}_h$	I	$8C_3$	$6C_2$	$6C_4$	$3C_4^2 \equiv 3C_2''$	$S_2 \equiv i$	$6S_4$	$8S_6$	$3\sigma_h$	$6\sigma_d$		
A_{1g}	+1	+1	+1	+1	+1	+1	+1	+1	+1	+1		$\alpha_{xx} + \alpha_{yy} + \alpha_{zz}$
A_{1u}	+1	+1	+1	+1	+1	−1	−1	−1	−1	−1		
A_{2g}	+1	+1	−1	−1	+1	+1	−1	+1	+1	−1		
A_{2u}	+1	+1	−1	−1	+1	−1	+1	−1	−1	+1		
E_g	+2	−1	0	0	+2	+2	0	−1	+2	0		$(\alpha_{xx} + \alpha_{yy} - 2\alpha_{zz}, \alpha_{xx} - \alpha_{yy})$
E_u	+2	−1	0	0	+2	−2	0	+1	−2	0		
F_{1g}	+3	0	−1	+1	−1	+3	+1	0	−1	−1	(R_x, R_y, R_z)	
F_{1u}	+3	0	−1	+1	−1	−3	−1	0	+1	+1	(T_x, T_y, T_z)	
F_{2g}	+3	0	+1	−1	−1	+3	−1	0	−1	+1		$(\alpha_{xy}, \alpha_{yz}, \alpha_{xz})$
F_{2u}	+3	0	+1	−1	−1	−3	+1	0	+1	−1		

$\mathbf{I}_h$	I	$12C_5$	$12C_5^2$	$20C_3$	$15C_2$	i	$12S_{10}$	$12S_{10}^3$	$20S_6$	$15\sigma_v$		
A_g	1	1	1	1	1	1	1	1	1	1		$\alpha_{xx} + \alpha_{yy} + \alpha_{zz}$
A_u	1	1	1	1	1	−1	−1	−1	−1	−1		
F_{1g}	3	$2c\frac{\pi}{5}$	$2c\frac{3\pi}{5}$	0	−1	3	$2c\frac{3\pi}{5}$	$2c\frac{\pi}{5}$	0	−1	(R_z, R_y, R_z)	
F_{1u}	3	$2c\frac{\pi}{5}$	$2c\frac{3\pi}{5}$	0	−1	−3	$-2c\frac{3\pi}{5}$	$-2c\frac{\pi}{5}$	0	1	(T_x, T_y, T_z)	
F_{2g}	3	$2c\frac{3\pi}{5}$	$2c\frac{\pi}{5}$	0	−1	3	$2c\frac{\pi}{5}$	$2c\frac{3\pi}{5}$	0	−1		
F_{2u}	3	$2c\frac{3\pi}{5}$	$2c\frac{\pi}{5}$	0	−1	−3	$-2c\frac{\pi}{5}$	$-2c\frac{3\pi}{5}$	0	1		
G_g	4	−1	−1	1	0	4	−1	−1	1	0		
G_u	4	−1	−1	1	0	−4	1	1	−1	0		
H_g	5	0	0	−1	1	5	0	0	−1	1		$(2\alpha_{zz} - \alpha_{xx} - \alpha_{yy}, \alpha_{xx} - \alpha_{yy}, \alpha_{xy}, \alpha_{yz}, \alpha_{xy})$
H_u	5	0	0	−1	1	−5	0	0	1	−1		

c = cosine. G and H denote four- and five-fold degenerate species, respectively.

APPENDIX II. MATRIX ALGEBRA

(1) Definition of a matrix

A matrix is an array of numbers or symbols for numbers. In general, it is written as:

$$\mathbf{A} = \begin{bmatrix} a_{11} & a_{12} & a_{13} & \cdots & a_{1n} \\ a_{21} & a_{22} & a_{23} & \cdots & a_{2n} \\ a_{31} & a_{32} & a_{33} & \cdots & a_{3n} \\ \vdots & \vdots & \vdots & & \vdots \\ a_{m1} & a_{m2} & a_{m3} & \cdots & a_{mn} \end{bmatrix} = [a_{ij}]$$

Here, the square brackets indicate that these elements constitute a matrix. **A** and $[a_{ij}]$ indicate the same matrix in abbreviated forms. In the latter, a_{ij} denotes the element in the ith row and jth column. If $m = n$, it is called a *square matrix*. In a square matrix, the set of elements a_{ij} with $i = j$ are called the *diagonal elements*. If all the diagonal elements are one and all the *off-diagonal elements* are zero, such a matrix is called a *unit (or identity) matrix*, and expressed as **E** (or **I**). A *diagonal matrix*, **D**, is similar to the unit matrix except that the diagonal elements are not necessarily equal. Thus,

$$\mathbf{E} = \begin{bmatrix} 1 & 0 & 0 \\ 0 & 1 & 0 \\ 0 & 0 & 1 \end{bmatrix} \qquad \mathbf{D} = \begin{bmatrix} d_{11} & 0 & 0 \\ 0 & d_{22} & 0 \\ 0 & 0 & d_{33} \end{bmatrix}$$

A matrix is called *symmetric* if the relationship $a_{ij} = a_{ji}$ holds for all off-diagonal elements. If $a_{ij} = -a_{ji}$, it is called *antisymmetric*.

If $m \neq n$, the matrix is called a *rectangular matrix*. Among them, the one-column matrix and one-row matrix shown below are important.

$$\mathbf{X} = \begin{bmatrix} x_1 \\ x_2 \\ x_3 \end{bmatrix} \qquad \tilde{\mathbf{X}} = [x_1 \quad x_2 \quad x_3]$$

Sometimes, the former is called a *vector*. The tilde sign over **X** in the latter indicates a *transpose* of a matrix in which the elements are interchanged across the diagonal. In the case of a 2×2 matrix, we have

$$\mathbf{A} = \begin{bmatrix} a_{11} & a_{12} \\ a_{21} & a_{22} \end{bmatrix} \qquad \tilde{\mathbf{A}} = \begin{bmatrix} a_{11} & a_{21} \\ a_{12} & a_{22} \end{bmatrix}$$

(2) Addition and subtraction

When two matrices are of same dimensions, they can be added or subtracted by the rule that $u_{ij} = a_{ij} \pm b_{ij}$. If B is

$$\mathbf{B} = \begin{bmatrix} b_{11} & b_{12} \\ b_{21} & b_{22} \end{bmatrix}$$

then

$$\mathbf{U} = \mathbf{A} \pm \mathbf{B} = \begin{bmatrix} a_{11} \pm b_{11} & a_{12} \pm b_{12} \\ a_{21} \pm b_{21} & a_{22} \pm b_{22} \end{bmatrix}$$

(3) Multiplication

Two matrices can be multiplied only if the number of columns of the first matrix **B** is equal to the number of rows of the second matrix **A**. Each element of the resulting matrix, $\mathbf{U} = \mathbf{BA}$, is given by

$$u_{ij} = \sum_{k=1}^{n} b_{ik} a_{kj}$$

For example,

$$\mathbf{BA} = \begin{bmatrix} b_{11}a_{11} + b_{12}a_{21} & b_{11}a_{12} + b_{12}a_{22} \\ b_{21}a_{11} + b_{22}a_{21} & b_{21}a_{12} + b_{22}a_{22} \end{bmatrix}$$

If **X** is a column matrix and $\tilde{\mathbf{X}}$ is its transpose,

$$\mathbf{AX} = \begin{bmatrix} a_{11} & a_{12} \\ a_{21} & a_{22} \end{bmatrix} \begin{bmatrix} x_1 \\ x_2 \end{bmatrix} = \begin{bmatrix} a_{11}x_1 + a_{12}x_2 \\ a_{21}x_1 + a_{22}x_2 \end{bmatrix}$$

$$\tilde{\mathbf{X}}\mathbf{AX} = [x_1 \quad x_2] \begin{bmatrix} a_{11}x_1 + a_{12}x_2 \\ a_{21}x_1 + a_{22}x_2 \end{bmatrix} = a_{11}x_1^2 + a_{12}x_1x_2 + a_{21}x_1x_2 + a_{22}x_2^2$$

It should be noted that matrix multiplication is generally *not commutative.*

Namely, **BA** is not necessarily equal to **AB**. For example,

$$\begin{bmatrix} 1 & 2 \\ 3 & 4 \end{bmatrix} \begin{bmatrix} 1 & 0 \\ 2 & 1 \end{bmatrix} = \begin{bmatrix} 5 & 2 \\ 11 & 4 \end{bmatrix}$$

$$\begin{bmatrix} 1 & 0 \\ 2 & 1 \end{bmatrix} \begin{bmatrix} 1 & 2 \\ 3 & 4 \end{bmatrix} = \begin{bmatrix} 1 & 2 \\ 5 & 8 \end{bmatrix}$$

However, the *associative law* holds for matrix multiplication. Thus,

$$\mathbf{A}(\mathbf{B} + \mathbf{C}) = \mathbf{AB} + \mathbf{AC} \quad \text{and} \quad (\mathbf{AB})\mathbf{C} = \mathbf{A}(\mathbf{BC})$$

If a constant λ is multiplied to **A**, each element of **A** is multiplied by λ. Namely,

$$\lambda\mathbf{A} = \begin{bmatrix} \lambda a_{11} & \lambda a_{12} \\ \lambda a_{21} & \lambda a_{22} \end{bmatrix}$$

(4) Division

Division by a matrix is accomplished as multiplication of its reciprocal (or inverse) matrix. For example, the division of **B** by **A** is written as

$$\mathbf{U} = \mathbf{B}/\mathbf{A} = \mathbf{BA}^{-1}$$

Here, $\mathbf{A}^{-1}$ is the reciprocal matrix of **A** which is defined as

$$\mathbf{AA}^{-1} = \mathbf{A}^{-1}\mathbf{A} = \mathbf{E}$$

Only the square matrix can have its reciprocal matrix. The reciprocal of the product matrix is given by

$$(\mathbf{BA})^{-1} = \mathbf{A}^{-1}\mathbf{B}^{-1}$$

If $\mathbf{A} = \tilde{\mathbf{A}}^{-1}$ (i.e., $\tilde{\mathbf{A}}\mathbf{A} = \mathbf{E}$), such a square matrix is called an *orthogonal matrix*.

For example,

$$\begin{bmatrix} 1/\sqrt{3} & 1/\sqrt{3} & 1/\sqrt{3} \\ 2/\sqrt{6} & -1/\sqrt{6} & -1/\sqrt{6} \\ 0 & 1/\sqrt{2} & -1/\sqrt{2} \end{bmatrix}, \qquad \begin{bmatrix} \cos\phi & \sin\phi & 0 \\ -\sin\phi & \cos\phi & 0 \\ 0 & 0 & 1 \end{bmatrix}$$

(5) Determinant

The determinant $|\mathbf{A}|$ of a square matrix $\mathbf{A}$ is defined as

$$|\mathbf{A}| = \begin{vmatrix} a_{11} & a_{12} & a_{13} & \cdots & a_{1n} \\ a_{21} & a_{22} & a_{23} & \cdots & a_{2n} \\ a_{31} & a_{32} & a_{33} & \cdots & a_{3n} \\ \vdots & \vdots & \vdots & & \vdots \\ a_{n1} & a_{n2} & a_{n3} & \cdots & a_{nn} \end{vmatrix} = \sum (-1)^h a_{1a} a_{2b} \cdots a_{nk}$$

Here, h is the number of exchanges necessary to bring the sequence $a, b, \ldots, k$ back to the natural order, $1, 2, \ldots, n$, and the summation is taken over all permutations of $a, b, \ldots, k$. For example,

$$\begin{vmatrix} a_{11} & a_{12} & a_{13} \\ a_{21} & a_{22} & a_{23} \\ a_{31} & a_{32} & a_{33} \end{vmatrix} = a_{11}a_{22}a_{33} + a_{12}a_{23}a_{31} + a_{13}a_{21}a_{32} - a_{13}a_{22}a_{31} - a_{11}a_{23}a_{32} - a_{12}a_{21}a_{33}$$

Here, the dotted lines indicate how to obtain positive terms. **Three different lines** were used to show how the products on the right side were obtained. Likewise, the negative terms can be obtained by changing the direction of the dotted lines by 90°. As shown above, vertical lines are used to express the determinant. It can be shown that

$$|\mathbf{AB}| = |\mathbf{A}||\mathbf{B}|$$

(6) Eigenvalues (Characteristic values)

If **A** is a square matrix of dimension n and **E** is the unit matrix of the same dimension,

$$|\mathbf{A} - \lambda \mathbf{E}| = 0$$

is called the *characteristic equation* of the matrix **A**. For example, the characteristic equation for **A** given below is written as

$$\mathbf{A} = \begin{bmatrix} a & b \\ b & a \end{bmatrix} \qquad \begin{vmatrix} a-\lambda & b \\ b & a-\lambda \end{vmatrix} = 0$$

Expansion of the characteristic equation gives

$$\lambda^2 - 2a\lambda + a^2 - b^2 = 0$$

The two eigenvalues of this equation are

$$\lambda_1 = a + b, \qquad \lambda_2 = a - b$$

More generally, the characteristic equation of a matrix A having the dimension $n \times n$ is written as

$$\lambda^n + c_1\lambda^{n-1} + c_2\lambda^{n-2} + \cdots c_{n-1}\lambda + c_n = 0$$

There are two simple relationships between the coefficients $c_1, c_2, \ldots, c_n$ and eigenvalues,

$$a_{11} + a_{22} + \cdots + a_{nn} = -c_1 = \lambda_1 + \lambda_2 + \cdots + \lambda_n$$
$$|\mathbf{A}| = \pm c_n = \lambda_1\lambda_2 \cdots \lambda_n (+ \text{ for even } n; - \text{ for odd } n)$$

These relationships are readily confirmed by the example above, given for a 2 × 2 matrix.

(7) Eigenvectors

If λ_a is an eigenvalue of **A**, a vector, $\boldsymbol{l}_a$, which satisfies the relation

$$\mathbf{A}\boldsymbol{l}_a = \boldsymbol{l}_a\lambda_a$$

is called an eigenvector of **A**. As an example, consider the 2 × 2 matrix mentioned above. The $\boldsymbol{l}_1$ for λ_1 is

$$\begin{bmatrix} a & b \\ b & a \end{bmatrix} \begin{bmatrix} l_{11} \\ l_{21} \end{bmatrix} = \begin{bmatrix} l_{11} \\ l_{21} \end{bmatrix} (a+b)$$

Then, we obtain $l_{11} = l_{21}$. Their absolute values can only be determined by using the normalization condition:

$$l_{11}^2 + l_{21}^2 = 1$$

Then,

$$l_{11} = l_{21} = 1/\sqrt{2}$$

Using the same procedure, we obtain

$$l_{12} = 1/\sqrt{2} \quad \text{and} \quad l_{22} = -1/\sqrt{2}$$

for λ_2. If we assemble these two results by columns, we have

$$\begin{bmatrix} a & b \\ b & a \end{bmatrix} \begin{bmatrix} 1/\sqrt{2} & 1/\sqrt{2} \\ 1/\sqrt{2} & -1/\sqrt{2} \end{bmatrix} = \begin{bmatrix} 1/\sqrt{2} & 1/\sqrt{2} \\ 1/\sqrt{2} & -1/\sqrt{2} \end{bmatrix} \begin{bmatrix} a+b & 0 \\ 0 & a-b \end{bmatrix}$$

More generally, this is written as

$$\mathbf{AL} = \mathbf{L\Lambda}$$

By multiplying $\mathbf{L}^{-1}$ on both sides, we obtain

$$\mathbf{L}^{-1}\mathbf{AL} = \mathbf{L}^{-1}\mathbf{L\Lambda} = \mathbf{\Lambda}$$

It is seen that the **L** matrix can transform the **A** matrix into a diagonal matrix with its eigenvalues as the diagonal elements. As shown above, the **L** matrix can be calculated once the λ values of the **A** matrix are obtained.

APPENDIX III. GENERAL FORMULAS FOR CALCULATING THE NUMBER OF NORMAL VIBRATIONS IN EACH SPECIES

Most of these tables were quoted from G. Herzberg, *Molecular Spectra and Molecular Structure*, Vol. II (Ref. 1 of Section I).

TABLE A. Point Groups Including Only Nondegenerate Vibrations

Point Group	Total Number of Atoms	Species	Number of Vibrations[a]
$\mathbf{C}_2$	$2m + m_0$	A	$3m + m_0 - 2$
		B	$3m + 2m_0 - 4$
$\mathbf{C}_s$	$2m + m_0$	A'	$3m + 2m_0 - 3$
		A''	$3m + m_0 - 3$
$\mathbf{C}_i \equiv \mathbf{S}_2$	$2m + m_0$	A_g	$3m - 3$
		A_u	$3m + 3m_0 - 3$
$\mathbf{C}_{2v}$	$4m + 2m_{xz} + 2m_{yz} + m_0$	A_1	$3m + 2m_{xz} + 2m_{yz} + m_0 - 1$
		A_2	$3m + m_{xz} + m_{yz} - 1$
		B_1	$3m + 2m_{xz} + m_{yz} + m_0 - 2$
		B_2	$3m + m_{xz} + 2m_{yz} + m_0 - 2$
$\mathbf{C}_{2h}$	$4m + 2m_h + 2m_2 + m_0$	A_g	$3m + 2m_h + m_2 - 1$
		A_u	$3m + m_h + m_2 + m_0 - 1$
		B_g	$3m + m_h + 2m_2 - 2$
		B_u	$3m + 2m_h + 2m_2 + 2m_0 - 2$
$\mathbf{D}_2 \equiv \mathbf{V}$	$4m + 2m_{2x} + 2m_{2y} + 2m_{2z} + m_0$	A	$3m + m_{2x} + m_{2y} + m_{2z}$
		B_1	$3m + 2m_{2x} + 2m_{2y} + m_{2z} + m_0 - 2$
		B_2	$3m + 2m_{2x} + m_{2y} + 2m_{2z} + m_0 - 2$
		B_3	$3m + m_{2x} + 2m_{2y} + 2m_{2z} + m_0 - 2$
$\mathbf{D}_{2h} \equiv \mathbf{V}_h$	$8m + 4m_{xy} + 4m_{xz} + 4m_{yz} + 2m_{2x} + 2m_{2y} + 2m_{2z} + m_0$	A_g	$3m + 2m_{xy} + 2m_{xz} + 2m_{yz} + m_{2x} + m_{2y} + m_{2z}$
		A_u	$3m + m_{xy} + m_{xz} + m_{yz}$
		B_{1g}	$3m + 2m_{xy} + m_{xz} + m_{yz} + m_{2x} + m_{2y} - 1$
		B_{1u}	$3m + m_{xy} + 2m_{xz} + 2m_{yz} + m_{2x} + m_{2y} + m_{2z} + m_0 - 1$
		B_{2g}	$3m + m_{xy} + 2m_{xz} + m_{yz} + m_{2x} + m_{2z} - 1$
		B_{2u}	$3m + 2m_{xy} + m_{xz} + 2m_{yz} + m_{2x} + m_{2y} + m_{2z} + m_0 - 1$
		B_{3g}	$3m + m_{xy} + m_{xz} + 2m_{yz} + m_{2y} + m_{2x} - 1$
		B_{3u}	$3m + 2m_{xy} + 2m_{xz} + m_{yz} + m_{2x} + m_{2y} + m_{2z} + m_0 - 1$

[a]Note that m is always the number of sets of equivalent nuclei not on any element of symmetry; m_0 is the number of nuclei lying on all symmetry elements present; m_{xy}, m_{xz}, m_{yz} are the numbers of sets of nuclei lying on the xy, xz, yz plane, respectively, but not on any axes going through these planes; m_2 is the number of sets of nuclei on a twofold axis but not at the point of intersection with another element of symmetry; m_{2x}, m_{2y}, m_{2z} are the numbers of sets of nuclei lying on the x, y, or z axis if they are twofold axes, but not on all of them; m_h is the number of sets of nuclei on a plane σ_h but not on the axis perpendicular to this plane.

TABLE B. Point Groups Including Degenerate Vibrations

Point Group	Total Number of Atoms	Species	Number of Vibrations[a]
$\mathbf{D}_3$	$6m + 3m_2 + 2m_3 + m_0$	A_1	$3m + m_2 + m_3$
		A_2	$3m + 2m_2 + m_3 + m_0 - 2$
		E	$6m + 3m_2 + 2m_3 + m_0 - 2$
$\mathbf{C}_{3v}$	$6m + 3m_v + m_0$	A_1	$3m + 2m_v + m_0 - 1$
		A_2	$3m + m_v - 1$
		E	$6m + 3m_v + m_0 - 2$
$\mathbf{C}_{4v}$	$8m + 4m_v + 4m_d + m_0$	A_1	$3m + 2m_v + 2m_d + m_0 - 1$
		A_2	$3m + m_v + m_d - 1$
		B_1	$3m + 2m_v + m_d$
		B_2	$3m + m_v + 2m_d$
		E	$6m + 3m_v + 3m_d + m_0 - 2$
$\mathbf{C}_{\infty v}$	m_0	Σ^+	$m_0 - 1$
		Σ^-	0
		Π	$m_0 - 2$
		$\Delta, \Phi, \ldots$	0
$\mathbf{D}_{2d} \equiv \mathbf{V}_d$	$8m + 4m_d + 4m_2 + 2m_4 + m_0$	A_1	$3m + 2m_d + m_2 + m_4$
		A_2	$3m + m_d + 2m_2 - 1$
		B_1	$3m + m_d + m_2$
		B_2	$3m + 2m_d + 2m_2 + m_4 + m_0 - 1$
		E	$6m + 3m_d + 3m_2 + 2m_4 + m_0 - 2$
$\mathbf{D}_{3d}$	$12m + 6m_d + 6m_2 + 2m_6 + m_0$	A_{1g}	$3m + 2m_d + m_2 + m_6$
		A_{1u}	$3m + m_d + m_2$
		A_{2g}	$3m + m_d + 2m_2 - 1$
		A_{2u}	$3m + 2m_d + 2m_2 + m_6 + m_0 - 1$
		E_g	$6m + 3m_d + 3m_2 + m_6 - 1$
		E_u	$6m + 3m_d + 3m_2 + m_6 + m_0 - 1$
$\mathbf{D}_{4d}$	$16m + 8m_d + 8m_2 + 2m_8 + m_0$	A_1	$3m + 2m_d + m_2 + m_8$
		A_2	$3m + m_d + 2m_2 - 1$
		B_1	$3m + m_d + m_2$
		B_2	$3m + 2m_d + 2m_2 + m_8 + m_0 - 1$
		E_1	$6m + 3m_d + 3m_2 + m_8 + m_0 - 1$
		E_2	$6m + 3m_d + 3m_2$
		E_3	$6m + 3m_d + 3m_2 + m_8 - 1$
$\mathbf{D}_{3h}$	$12m + 6m_v + 6m_h + 3m_2 + 2m_3 + m_0$	A_1'	$3m + 2m_v + 2m_h + m_2 + m_3$
		A_1''	$3m + m_v + m_h$
		A_2'	$3m + m_v + 2m_h + m_2 - 1$
		A_2''	$3m + 2m_v + m_h + m_2 + m_3 + m_0 - 1$
		E'	$6m + 3m_v + 4m_h + 2m_2 + m_3 + m_0 - 1$
		E''	$6m + 3m_v + 2m_h + m_2 + m_3 - 1$

TABLE B. ***(Continued)***

Point Group	Total Number of Atoms	Species	Number of Vibrations[a]
$\mathbf{D}_{4h}$	$16m + 8m_v + 8m_d$	A_{1g}	$3m + 2m_v + 2m_d + 2m_h + m_2 + m_2' + m_4$
	$+ 8m_h + 4m_2 + 4m_2'$	A_{1u}	$3m + m_v + m_d + m_h$
	$+ 2m_4 + m_0$	A_{2g}	$3m + m_v + m_d + 2m_h + m_2 + m_2' - 1$
		A_{2u}	$3m + 2m_v + 2m_d + m_h + m_2 + m_2' + m_4 + m_0 - 1$
		B_{1g}	$3m + 2m_v + m_d + 2m_h + m_2 + m_2'$
		B_{1u}	$3m + m_v + 2m_d + m_h + m_2'$
		B_{2g}	$3m + m_v + 2m_d + 2m_h + m_2 + m_2'$
		B_{2u}	$3m + 2m_v + m_d + m_h + m_2$
		E_g	$6m + 3m_v + 3m_d + 2m_h + m_2 + m_2' + m_4 - 1$
		E_u	$6m + 3m_v + 3m_d + 4m_h + 2m_2 + 2m_2'$
			$+ m_4 + m_0 - 1$
$\mathbf{D}_{5h}$	$20m + 10m_v + 10m_h$	A_1'	$3m + 2m_v + 2m_h + m_2 + m_5$
	$+ 5m_2 + 2m_5 + m_0$	A_1''	$3m + m_v + m_h$
		A_2'	$3m + m_v + 2m_h + m_2 - 1$
		A_2''	$3m + 2m_v + m_h + m_2 + m_5 + m_0 - 1$
		E_1'	$6m + 3m_v + 4m_h + 2m_2 + m_5 + m_0 - 1$
		E_1''	$6m + 3m_v + 2m_h + m_2 + m_5 - 1$
		E_2'	$6m + 3m_v + 4m_h + 2m_2$
		E_2''	$6m + 3m_v + 2m_h + m_2$
$\mathbf{D}_{6h}$	$24m + 12m_v + 12m_d$	A_{1g}	$3m + 2m_v + 2m_d + 2m_h + m_2 + m_2' + m_6$
	$+ 12m_h + 6m_2 + 6m_2'$	A_{1u}	$3m + m_v + m_d + m_h$
	$+ 2m_6 + m_0$	A_{2g}	$3m + m_v + m_d + 2m_h + m_2 + m_2' - 1$
		A_{2u}	$3m + 2m_v + 2m_d + m_h + m_2 + m_2' + m_6 + m_0 - 1$
		B_{1g}	$3m + m_v + 2m_d + m_h + m_2'$
		B_{1u}	$3m + 2m_v + m_d + 2m_h + m_2 + m_2'$
		B_{2g}	$3m + 2m_v + m_d + m_h + m_2$
		B_{2u}	$3m + m_v + 2m_d + 2m_h + m_2 + m_2'$
		E_{1g}	$6m + 3m_v + 3m_d + 2m_h + m_2 + m_2' + m_6 - 1$
		E_{1u}	$6m + 3m_v + 3m_d + 4m_h + 2m_2 + 2m_2'$
			$+ m_6 + m_0 - 1$
		E_{2g}	$6m + 3m_v + 3m_d + 4m_h + 2m_2 + 2m_2'$
		E_{2u}	$6m + 3m_v + 3m_d + 2m_h + m_2 + m_2'$
$\mathbf{D}_{\infty h}$	$2m_\infty + m_0$	Σ_g^+	m_∞
		Σ_u^+	$m_\infty + m_0 - 1$
		Σ_g^-, Σ_u^-	0
		Π_g	$m_\infty - 1$
		Π_u	$m_\infty + m_0 - 1$
		$\Delta_g, \Delta_u,$	0
		$\Phi_g, \Phi_u, \ldots$	0

TABLE B. (*Continued*)

Point Group	Total Number of Atoms	Species	Number of Vibrations[a]
$\mathbf{T}_d$	$24m + 12m_d + 6m_2 + 4m_3 + m_0$	A_1	$3m + 2m_d + m_2 + m_3$
		A_2	$3m + m_d$
		E	$6m + 3m_d + m_2 + m_3$
		F_1	$9m + 4m_d + 2m_2 + m_3 - 1$
		F_2	$9m + 5m_d + 3m_2 + 2m_3 + m_0 - 1$
$\mathbf{O}_h$	$48m + 24m_h + 24m_d + 12m_2 + 8m_3 + 6m_4 + m_0$	A_{1g}	$3m + 2m_h + 2m_d + m_2 + m_3 + m_4$
		A_{1u}	$3m + m_h + m_d$
		A_{2g}	$3m + 2m_h + m_d + m_2$
		A_{2u}	$3m + m_h + 2m_d + m_2 + m_3$
		E_g	$6m + 4m_h + 3m_d + 2m_2 + m_3 + m_4$
		E_u	$6m + 2m_h + 3m_d + m_2 + m_3$
		F_{1g}	$9m + 4m_h + 4m_d + 2m_2 + m_3 + m_4 - 1$
		F_{1u}	$9m + 5m_h + 5m_d + 3m_2 + 2m_3 + 2m_4 + m_0 - 1$
		F_{2g}	$9m + 4m_h + 5m_d + 2m_2 + 2m_3 + m_4$
		F_{2u}	$9m + 5m_h + 4m_d + 2m_2 + m_3 + m_4$
$\mathbf{T}_h$	$24m + 12m_v + 8m_3 + 6m_2 + m_0$	A_g	$3m + 2m_v + m_3 + m_2$
		A_u	$3m + m_v + m_3$
		E_g	$3m + 2m_v + m_3 + m_2$
		E_u	$3m + m_v + m_3$
		F_g	$9m + 4m_v + 3m_3 + 2m_2 - 1$
		F_u	$9m + 5m_v + 3m_3 + 3m_2 + m_0 - 1$
$\mathbf{I}_h$	$120m + 60m_v + 30m_2 + 20m_3 + 12m_5 + m_0$	A_g	$3m + 2m_v + m_2 + m_3 + m_5$
		A_u	$3m + m_v$
		F_{1g}	$9m + 4m_v + 2m_2 + m_3 + m_5 - 1$
		F_{1u}	$9m + 5m_v + 3m_2 + 2m_3 + 2m_5 + m_0 - 1$
		F_{2g}	$9m + 4m_v + 2m_2 + m_3$
		F_{2u}	$9m + 5m_v + 3m_2 + 2m_3 + m_5$
		G_g	$12m + 6m_v + 3m_2 + 2m_3 + m_5$
		G_u	$12m + 6m_v + 3m_2 + 2m_3 + m_5$
		H_g	$15m + 8m_v + 4m_2 + 3m_3 + 2m_5$
		H_u	$15m + 7m_v + 3m_2 + 2m_3 + m_5$

[a]Note that m is the number of sets of nuclei not any element of symmetry; m_0 is the number of nuclei on all elements of symmetry; $m_2, m_3, m_4, \ldots$ are the numbers of sets of nuclei on a twofold, threefold, fourfold, and so on, axis but not on any other element of symmetry that does not wholly coincide with that axis; m_2' is the number of sets of nuclei on a twofold axis called C_2' in the previous character tables; m_v, m_d, m_h are the numbers of sets of nuclei on planes $\sigma_v, \sigma_d, \sigma_h$, respectively, but not on any other element of symmetry.

APPENDIX IV. DIRECT PRODUCTS OF IRREDUCIBLE REPRESENTATIONS

As shown in Sec. I-10, the characters for direct products can be obtained by multiplying the corresponding characters of two representations and resolving the result into those of the irreducible representations by using Eq. 7.6. This procedure, however, can be greatly simplified if we use the following rules (Ref. 3 of Section I).

(1) General rules

$$A \times A = A, \quad B \times B = A, \quad A \times B = B, \quad A \times E = E$$
$$B \times E = E, \quad A \times F = F, \quad B \times F = F$$
$$g \times g = g, \quad u \times u = g, \quad u \times g = u$$
$$' \times ' = ', \quad '' \times '' = ', \quad ' \times '' = ''$$
$$A \times E_1 = E_1, \quad A \times E_2 = E_2, \quad B \times E_1 = E_2, \quad B \times E_2 = E_1$$

(2) Subscripts on A or B

$$1 \times 1 = 1, \quad 2 \times 2 = 1, \quad 1 \times 2 = 2 \quad \text{except for } \mathbf{D}_2 \text{ and } \mathbf{D}_{2h}$$

where $1 \times 2 = 3, \quad 2 \times 3 = 1, \quad 1 \times 3 = 2$

(3) Doubly degenerate representations

For $\mathbf{C}_3$, $\mathbf{C}_{3h}$, $\mathbf{C}_{3v}$, $\mathbf{D}_3$, $\mathbf{D}_{3h}$, $\mathbf{D}_{3d}$, $\mathbf{C}_6$, $\mathbf{C}_{6h}$, $\mathbf{C}_{6v}$, $\mathbf{D}_6$, $\mathbf{D}_{6h}$, $\mathbf{S}_6$, $\mathbf{O}$, $\mathbf{O}_h$, $\mathbf{T}$, $\mathbf{T}_d$, $\mathbf{T}_h$,

$$E_1 \times E_1 = E_2 \times E_2 = A_1 + A_2 + E_2$$
$$E_1 \times E_2 = B_1 + B_2 + E_1$$

For $\mathbf{C}_4$, $\mathbf{C}_{4v}$, $\mathbf{C}_{4h}$, $\mathbf{D}_{2d}$, $\mathbf{D}_4$, $\mathbf{D}_{4h}$, and $\mathbf{S}_4$,

$$E \times E = A_1 + A_2 + B_1 + B_2$$

For groups in the lists above which have symbols A, B, or E without subscripts, read $A_1 = A_2 = A$, and so on.

(4) Triply degenerate representations

For $\mathbf{T}_d$, $\mathbf{O}$, $\mathbf{O}_h$,

$$E \times F_1 = E \times F_2 = F_1 + F_2$$
$$F_1 \times F_1 = F_2 \times F_2 = A_1 + E + F_1 + F_2$$
$$F_1 \times F_2 = A_2 + E + F_1 + F_2$$

For T and $\mathbf{T}_h$, drop subscripts 1 and 2 from A and F.

(5) Linear molecules ($\mathbf{C}_{\infty v}$ and $\mathbf{D}_{\infty h}$)

$$\Sigma^+ \times \Sigma^+ = \Sigma^- \times \Sigma^- = \Sigma^+, \quad \Sigma^+ \times \Sigma^- = \Sigma^-$$
$$\Sigma^+ \times \Pi = \Sigma^- \times \Pi = \Pi \quad \Sigma^+ \times \Delta = \Sigma^- \times \Delta = \Delta$$
$$\Pi \times \Pi = \Sigma^+ + \Sigma^- + \Delta$$
$$\Delta \times \Delta = \Sigma^+ + \Sigma^- + \Gamma$$
$$\Pi \times \Delta = \Pi + \Phi$$

Using rule (3), we find that, in $\mathbf{C}_{3v}$, $E \times E = A_1 + A_2 + E$ (Sec. I-10). Similarly, using rules (1) and (3), we find that, in $\mathbf{D}_{4h}$, $E_u \times E_u = A_{1g} + A_{2g} + B_{1g} + B_{2g}$ (Sec. I-23).

APPENDIX V. NUMBER OF INFRARED- AND RAMAN-ACTIVE STRETCHING VIBRATIONS FOR MX_nY_m-TYPE MOLECULES

Compound	Structure	Point Group	IR or Raman	M–X Stretching	M–Y Stretching
MX_6	Octahedral	$\mathbf{O}_h$	IR	F_{1u}	
			R	A_{1g}, E_g	
MX_5Y	Octahedral	$\mathbf{C}_{4v}$	IR	$2A_1, E$	A_1
			R	$2A_1, B_1, E$	A_1
trans-MX_4Y_2	Octahedral	$\mathbf{D}_{4h}$	IR	E_u	A_{2u}
			R	A_{1g}, B_{1g}	A_{1g}
cis-MX_4Y_2	Octahedral	$\mathbf{C}_{2v}$	IR	$2A_1, B_1, B_2$	A_1, B_1
			R	$2A_1, B_1, B_2$	A_1, B_1
mer-MX_3Y_3	Octahedral	$\mathbf{C}_{2v}$	IR	$2A_1, B_2$	$2A_1, B_1$
			R	$2A_1, B_2$	$2A_1, B_1$
fac-MX_3Y_3	Octahedral	$\mathbf{C}_{3v}$	IR	A_1, E	A_1, E
			R	A_1, E	A_1, E
MX_5	Trigonal-bipyramidal	$\mathbf{D}_{3h}$	IR	A_2'', E'	
			R	$2A_1', E'$	
MX_5	Tetragonal-pyramidal	$\mathbf{C}_{4v}$	IR	$2A_1, E$	
			R	$2A_1, B_1, E$	
MX_4	Tetrahedral	$\mathbf{T}_d$	IR	F_2	
			R	A_1, F_2	
MX_3Y	Tetrahedral	$\mathbf{C}_{3v}$	IR	A_1, E	A_1
			R	A_1, E	A_1
MX_2Y_2	Tetrahedral	$\mathbf{C}_{2v}$	IR	A_1, B_1	A_1, B_2
			R	A_1, B_1	A_1, B_2
Polymeric MX_2Y_2[a]	Octahedral	$\mathbf{C}_i$	IR	$2A_u$	A_u
			R	$2A_g$	A_g
MX_4	Square-planar	$\mathbf{D}_{4h}$	IR	E_u	
			R	A_{1g}, B_{2g}	
MX_3Y	Planar	$\mathbf{C}_{2v}$	IR	$2A_1, B_1$	A_1
			R	$2A_1, B_1$	A_1
trans-MX_2Y_2	Planar	$\mathbf{D}_{2h}$	IR	B_{3u}	B_{2u}
			R	A_g	A_g
cis-MX_2Y_2	Planar	$\mathbf{C}_{2v}$	IR	A_1, B_2	A_1, B_2
			R	A_1, B_2	A_1, B_2
MX_3	Pyramidal	$\mathbf{C}_{3v}$	IR	A_1, E	
			R	A_1, E	
MX_3	Planar	$\mathbf{D}_{3h}$	IR	E'	
			R	A_1', E'	

[a]Bridging through X atoms.

APPENDIX VI. DERIVATION OF EQ. 12.3 OF SECTION I.

Using the rectangular coordinates, we write the kinetic energy as

$$2T = \tilde{\dot{\mathbf{X}}}\mathbf{M}\dot{\mathbf{X}} \tag{1}$$

where

$$\mathbf{X} = \begin{bmatrix} x_1 \\ y_1 \\ z_1 \\ x_2 \\ \vdots \\ z_N \end{bmatrix} \quad \text{and} \quad \mathbf{M} = \begin{bmatrix} m_1 & & & & & \\ & m_1 & & & & \\ & & m_1 & & & \\ & & & m_2 & & \\ & & & & \ddots & \\ & & & & & m_N \end{bmatrix}$$

By definition, the momentum p_{x_1} conjugated with x_1 is given by

$$p_{x_1} = \frac{\partial T}{\partial \dot{x}_1} = m_1 \dot{x}_1$$

$p_{y_1} \cdots p_{z_N}$ take similar forms. Using the conjugate momenta, we write T as

$$\begin{aligned} 2T &= \frac{1}{m_1} p_{x_1}^2 + \frac{1}{m_1} p_{y_1}^2 + \cdots + \frac{1}{m_N} p_{z_N}^2 \\ &= \tilde{\mathbf{P}}_x \mathbf{M}^{-1} \mathbf{P}_x \end{aligned} \tag{2}$$

where

$$\mathbf{P}_x = \begin{bmatrix} p_{x_1} \\ p_{y_1} \\ \vdots \\ p_{z_N} \end{bmatrix} \quad \text{and} \quad \mathbf{M}^{-1} = \begin{bmatrix} \mu_1 & & & \\ & \mu_1 & & \\ & & \ddots & \\ & & & \mu_N \end{bmatrix}$$

The column matrix $\mathbf{P}_x$ can be expressed as

$$\mathbf{P}_x = \mathbf{M}\dot{\mathbf{X}} \tag{3}$$

Define a set of conjugate momenta $\mathbf{P}$ associated with internal coordinates,

R. As is shown at the end of this Appendix, we have

$$\mathbf{P}_x = \tilde{\mathbf{B}}\mathbf{P} \tag{4}$$

Equations 3 and 4 give

$$\mathbf{M}\dot{\mathbf{X}} = \tilde{\mathbf{B}}\mathbf{P} \tag{5}$$

Equation 11.8 in the text gives

$$\mathbf{R} = \mathbf{B}\mathbf{X} \quad \text{and} \quad \dot{\mathbf{R}} = \mathbf{B}\dot{\mathbf{X}} \tag{6}$$

By inserting Eq. 5 into Eq. 6, we obtain

$$\dot{\mathbf{R}} = \mathbf{B}\mathbf{M}^{-1}\tilde{\mathbf{B}}\mathbf{P} \tag{7}$$

Using Eq. 4, we write Eq. 2 as

$$2T = \tilde{\mathbf{P}}\mathbf{B}\mathbf{M}^{-1}\tilde{\mathbf{B}}\mathbf{P} \tag{8}$$

If we define

$$\mathbf{G} = \mathbf{B}\mathbf{M}^{-1}\tilde{\mathbf{B}} \quad \text{(12.7 in Sec. I)}$$

Eq. 8 is written as

$$2T = \tilde{\mathbf{P}}\mathbf{G}\mathbf{P} \tag{9}$$

If Eq. 11.7 is combined with Eq. 7, we obtain

$$\dot{\mathbf{R}} = \mathbf{G}\mathbf{P}$$

or

$$\mathbf{G}^{-1}\dot{\mathbf{R}} = \mathbf{G}^{-1}\mathbf{G}\mathbf{P} = \mathbf{P} \tag{10}$$

Using Eq. 10, Eq. 9 can be written

$$\begin{aligned} 2T &= \tilde{\dot{\mathbf{R}}}\tilde{\mathbf{G}}^{-1}\mathbf{G}\mathbf{G}^{-1}\dot{\mathbf{R}} \\ &= \tilde{\dot{\mathbf{R}}}\mathbf{G}^{-1}\dot{\mathbf{R}} \quad \text{(12.3 in Sec. I)} \end{aligned}$$

Derivation of Eq. 4

The momentum p_{R_k} conjugated with the internal coordinate R_k is given by

$$p_{R_k} = \frac{\partial T}{\partial \dot{R}_k}, \qquad k = 1, 2, \ldots, s$$

If we denote the coordinates corresponding to the translational and rotational motions of the molecule by R_j^0 and its conjugate momentum by $p_{R_j}^0$

$$p_{R_j}^0 = \frac{\partial T}{\partial \dot{R}_j^0}, \qquad j = 1, 2, \ldots, 6$$

Then the momentum p_{x_1} in terms of rectangular coordinates is written as

$$\begin{aligned} p_{x_1} = \frac{\partial T}{\partial \dot{x}_1} &= \sum_k^s \frac{\partial T}{\partial \dot{R}_k} \frac{\partial R_k}{\partial x_1} + \sum_j^6 \frac{\partial T}{\partial \dot{R}_j^0} \frac{\partial R_j^0}{\partial x_1} \\ &= \sum_k^s p_{R_k} B_{k,x_1} + \sum_j^6 p_{R_j}^0 \frac{\partial R_j^0}{\partial x_1} \end{aligned}$$

The second term becomes zero since the momenta corresponding to the translational and rotational motions are zero. Thus, we have

$$\begin{aligned} p_{x_1} &= \sum_k^s p_{R_k} B_{k,x_1} \\ p_{y_1} &= \sum p_{R_k} B_{k,y_1} \\ &\;\vdots \qquad\quad \vdots \\ p_{z_N} &= \sum p_{R_k} B_{k,z_N} \end{aligned}$$

In a matrix form, this is written as

$$\mathbf{P}_x = \tilde{\mathbf{B}}\mathbf{P} \tag{4}$$

APPENDIX VII. THE G AND F MATRIX ELEMENTS OF TYPICAL MOLECULES

In the following tables, F represents F matrix elements in the GVF field, whereas F^* denotes those in the UBF field. In the latter, $F' = -\frac{1}{10}F$ was assumed for all cases, and the *molecular tension* (Refs. I-89–91 of Section I) was ignored.

(1) Bent XY_2 Molecules (C_{2v})

A_1 species—infrared and Raman active:

$$G_{11} = \mu_y + \mu_x(1 + \cos\alpha)$$

$$G_{12} = -\frac{\sqrt{2}}{r}\mu_x \sin\alpha$$

$$G_{22} = \frac{2}{r^2}[\mu_y + \mu_x(1 - \cos\alpha)]$$

$$F_{11} = f_r + f_{rr}$$

$$F_{12} = (\sqrt{2})rf_{r\alpha}$$

$$F_{22} = r^2 f_\alpha$$

$$F^*_{11} = K + 2F\sin^2\frac{\alpha}{2}$$

$$F^*_{12} = (0.9)(\sqrt{2})rF\sin\frac{\alpha}{2}\cos\frac{\alpha}{2}$$

$$F^*_{22} = r^2\left[H + F\left\{\cos^2\frac{\alpha}{2} + (0.1)\sin^2\frac{\alpha}{2}\right\}\right]$$

B_2 species—infrared and Raman active:

$$G = \mu_y + \mu_x(1 - \cos\alpha)$$

$$F = f_r - f_{rr}$$

$$F^* = K - (0.2)F\cos^2\frac{\alpha}{2}$$

(2) Pyramidal XY_3 Molecules (C_{3v})

A_1 species—infrared and Raman active:

$$G_{11} = \mu_y + \mu_x(1 + 2\cos\alpha)$$

$$G_{12} = -\frac{2}{r}\,\frac{(1 + 2\cos\alpha)(1 - \cos\alpha)}{\sin\alpha}\,\mu_x$$

$$G_{22} = \frac{2}{r^2}\left(\frac{1 + 2\cos\alpha}{1 + \cos\alpha}\right)[\mu_y + 2\mu_x(1 - \cos\alpha)]$$

$$F_{11} = f_r + 2f_{rr}$$

$$F_{12} = r(2f_{r\alpha} + f'_{r\alpha})$$

$$F_{22} = r^2(f_\alpha + 2f_{\alpha\alpha})$$

$$F^*_{11} = K + 4F\sin^2\frac{\alpha}{2}$$

$$F^*_{12} = (1.8)rF\sin\frac{\alpha}{2}\cos\frac{\alpha}{2}$$

$$F^*_{22} = r^2\left[H + F\left(\cos^2\frac{\alpha}{2} + (0.1)\sin^2\frac{\alpha}{2}\right)\right]$$

E species—infrared and Raman active:

$$G_{11} = \mu_y + \mu_x(1 - \cos\alpha)$$

$$G_{12} = \frac{1}{r}\,\frac{(1 - \cos\alpha)^2}{\sin\alpha}\,\mu_x$$

$$G_{22} = \frac{1}{r^2(1 + \cos\alpha)}[(2 + \cos\alpha)\mu_y + (1 - \cos\alpha)^2\mu_x]$$

$$F_{11} = f_r - f_{rr}$$

$$F_{12} = r(-f_{r\alpha} + f'_{r\alpha})$$

$$F_{22} = r^2(f_\alpha - f_{\alpha\alpha})$$

$$F^*_{11} = K + \left(\sin^2\frac{\alpha}{2} - (0.3)\cos^2\frac{\alpha}{2}\right)F$$

$$F^*_{12} = -(0.9)rF\sin\frac{\alpha}{2}\cos\frac{\alpha}{2}$$

$$F^*_{22} = r^2\left[H + F\left(\cos^2\frac{\alpha}{2} + (0.1)\sin^2\frac{\alpha}{2}\right)\right]$$

Here $f_{r\alpha}$ denotes interaction between Δr and $\Delta\alpha$ having a common bond (e.g., Δr_1 and $\Delta\alpha_{12}$ or $\Delta\alpha_{31}$); $f'_{r\alpha}$ denotes interaction between Δr and $\Delta\alpha$ having no common bonds (e.g., Δr_1 and $\Delta\alpha_{23}$); see Fig. I-20*c*.

(3) Planar XY_3 Molecules (D_{3h})

A_1' species—Raman active:

$$G = \mu_y$$
$$F = f_r + 2f_{rr}$$
$$F^* = K + 3F$$

A_2'' species—infrared active:

$$G = \frac{9}{4r^2}(\mu_y + 3\mu_x)$$
$$F = F^* = r^2 f_\theta$$

E' species—infrared and Raman active:

$$G_{11} = \mu_y + \frac{3}{2}\mu_x$$
$$G_{12} = \frac{3\sqrt{3}}{2r}\mu_x$$
$$G_{22} = \frac{3}{2r^2}(2\mu_y + 3\mu_x)$$
$$F_{11} = f_r - f_{rr}$$
$$F_{12} = r(f'_{r\alpha} - f_{r\alpha})$$
$$F_{22} = r^2(f_\alpha - f_{\alpha\alpha})$$
$$F^*_{11} = K + 0.675F$$
$$F^*_{12} = -(0.9)\frac{\sqrt{3}}{4}rF$$
$$F^*_{22} = r^2(H + 0.325F)$$

The symbols $f_{r\alpha}$ and $f'_{r\alpha}$ are defined in subsection (2); f_θ denotes the force constant for the out-of-plane mode (see Fig. I-20*f*).

(4) Tetrahedral XY_4 Molecules (T_d)

A_1 species—Raman active:

$$G = \mu_y$$
$$F = f_r + 3f_{rr}$$
$$F^* = K + 4F$$

E species—Raman active:

$$G = \frac{3\mu_y}{r^2}$$
$$F = r^2(f_\alpha - 2f_{\alpha\alpha} + f'_{\alpha\alpha})$$
$$F^* = r^2(H + 0.4F)$$

F_2 species—infrared and Raman active:

$$G_{11} = \mu_y + \frac{4}{3}\mu_x$$
$$G_{12} = -\frac{8}{3r}\mu_x$$
$$G_{22} = \frac{1}{r^2}\left(\frac{16}{3}\mu_x + 2\mu_y\right)$$
$$F_{11} = f_r - f_{rr}$$
$$F_{12} = (\sqrt{2})r(f_{r\alpha} - f'_{r\alpha})$$
$$F_{22} = r^2(f_\alpha - f'_{\alpha\alpha})$$
$$F^*_{11} = K + \frac{6}{5}F$$
$$F^*_{12} = \frac{3}{5}rF$$
$$F^*_{22} = r^2(H + 0.4F)$$

where $f_{\alpha\alpha}$ denotes interaction between two $\Delta\alpha$ having a common bond; $f'_{\alpha\alpha}$ denotes interaction between two $\Delta\alpha$ having no common bond.

(5) Square-Planar XY_4 Molecules (D_{4h})

A_{1g} species—Raman active:

$$G = \mu_y$$
$$F = f_r + 2f_{rr} + f'_{rr}$$
$$F^* = K + 2F$$

B_{1g} species—Raman active:

$$G = \mu_y$$
$$F = f_r - 2f_{rr} + f'_{rr}$$
$$F^* = K - 0.2F$$

B_{2g} species—Raman active:

$$G = \frac{4\mu_y}{r^2}$$
$$F = r^2(f_\alpha - 2f_{\alpha\alpha} + f'_{\alpha\alpha})$$
$$F^* = r^2(H + 0.55F)$$

E_u species—infrared active:

$$G_{11} = 2\mu_x + \mu_y$$
$$G_{12} = -\frac{2\sqrt{2}}{r}\mu_x$$
$$G_{22} = \frac{2}{r^2}(\mu_y + 2\mu_x)$$
$$F_{11} = f_r - f'_{rr}$$
$$F_{12} = (\sqrt{2})r(f_{r\alpha} - f'_{r\alpha})$$
$$F_{22} = r^2(f_\alpha - f'_{\alpha\alpha})$$
$$F^*_{11} = K + 0.9F$$
$$F^*_{12} = (\sqrt{2})r(0.45)F$$
$$F^*_{22} = r^2(H + 0.55F)$$

The symbol f_{rr} denotes interaction between two Δr perpendicular to each other; f'_{rr} denotes interaction between two Δr on the same straight line. In addition, a square-planar XY_4 molecule has two out-of-plane vibrations in the A_{2u} and B_{2u} species.

(6) Octahedral XY_6 Molecules (O_h)

A_{1g} species—Raman active:

$$G = \mu_y$$
$$F = f_r + 4f_{rr} + f'_{rr}$$
$$F^* = K + 4F$$

E_g species—Raman active:

$$G = \mu_y$$
$$F = f_r - 2f_{rr} + f'_{rr}$$
$$F^* = K + 0.7F$$

F_{1u} species—infrared active:

$$G_{11} = \mu_y + 2\mu_x$$
$$G_{12} = -\frac{4}{r}\,\mu_x$$
$$G_{22} = \frac{2}{r^2}\,(\mu_y + 4\mu_x)$$
$$F_{11} = f_r - f'_{rr}$$
$$F_{12} = 2rf_{r\alpha}$$
$$F_{22} = r^2(f_\alpha + 2f_{\alpha\alpha})$$
$$F^*_{11} = K + 1.8F$$
$$F^*_{12} = 0.9rF$$
$$F^*_{22} = r^2(H + 0.55F)$$

F_{2g} species—Raman active:

$$G = \frac{4\mu_y}{r^2}$$
$$F = r^2(f_\alpha - 2f'_{\alpha\alpha})$$
$$F^* = r^2(H + 0.55F)$$

F_{2u} species—inactive:

$$G = \frac{2\mu_y}{r^2}$$
$$F = r^2(f_\alpha - 2f_{\alpha\alpha})$$
$$F^* = r^2(H + 0.55F)$$

The symbol f_{rr} denotes interaction between two Δr perpendicular to each other, whereas f'_{rr} denotes those between two Δr on the same straight line; $f_{\alpha\alpha}$ denotes interaction between two $\Delta\alpha$ perpendicular to each other, whereas $f'_{\alpha\alpha}$ denotes those between two $\Delta\alpha$ on the same plane. Only the interaction between two $\Delta\alpha$ having a common bond was considered.

APPENDIX VIII. GROUP FREQUENCY CHARTS

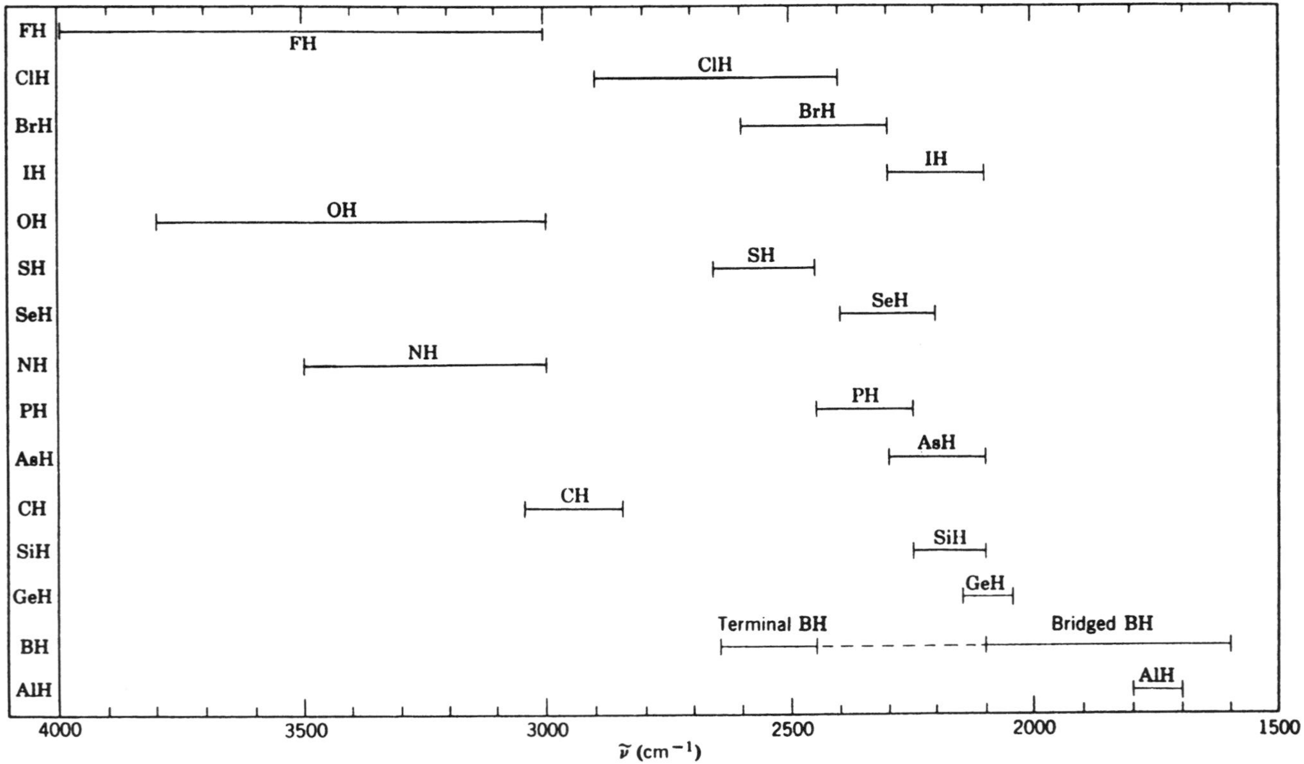

(*a*) Hydrogen stretching frequencies

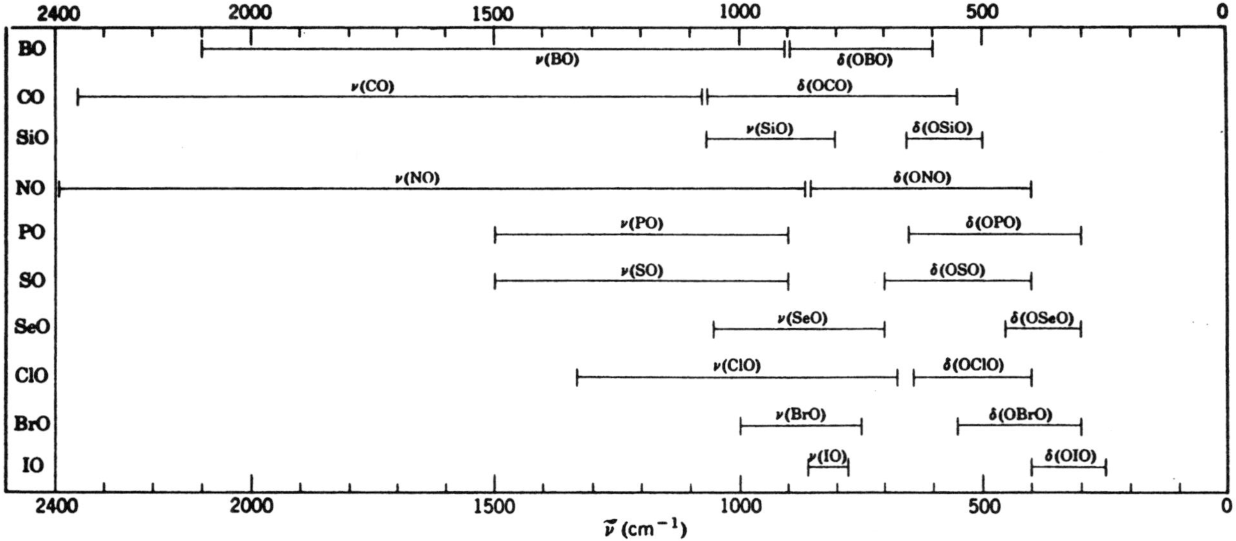

(*b*) Oxygen stretching and bending frequencies

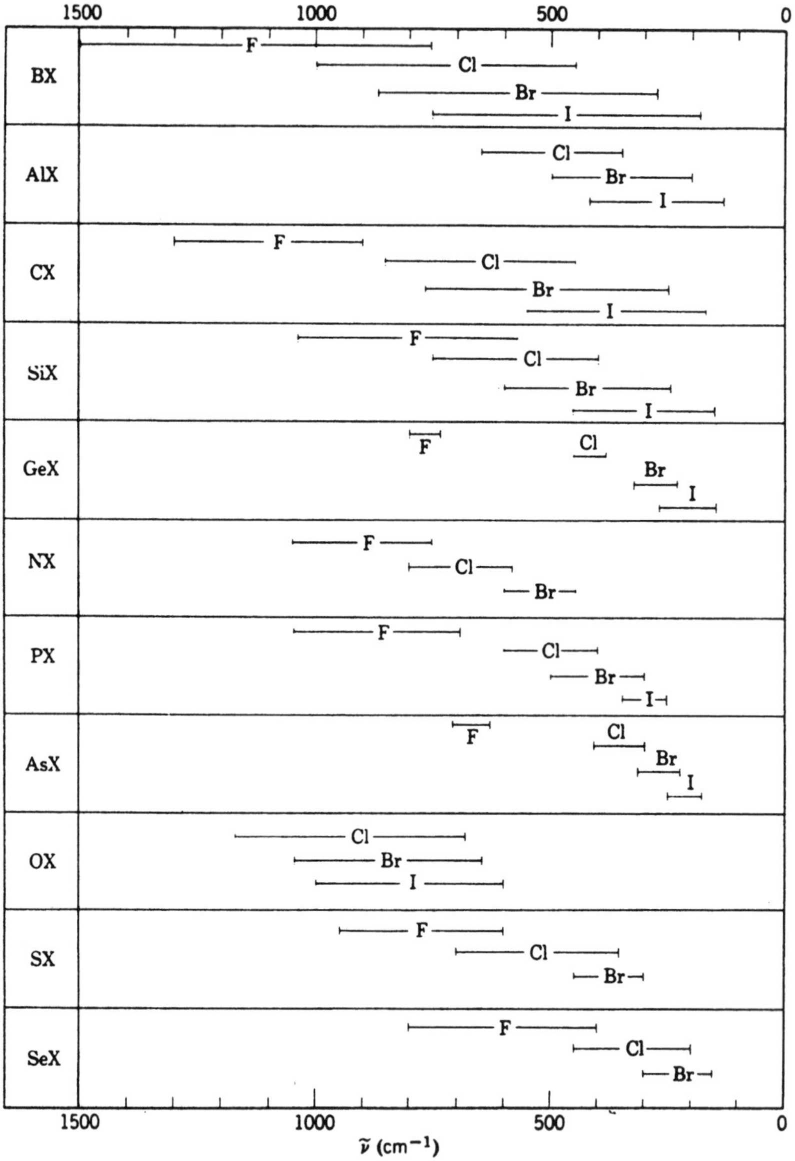

(*c*) Halogen (X) stretching frequencies

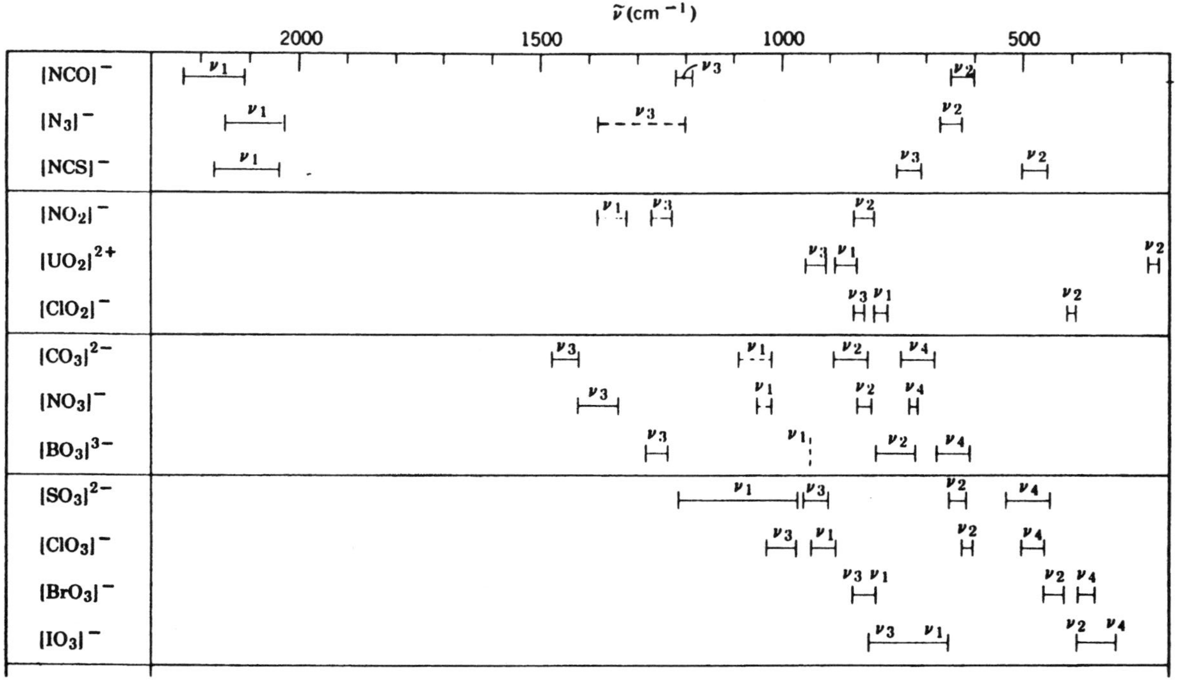

$\tilde{\nu}$ (cm^{-1})
2000
1500
1000
500
$[NCO]^-$
$[N_3]^-$
$[NCS]^-$
$[NO_2]^-$
$[UO_2]^{2+}$
$[ClO_2]^-$
$[CO_3]^{2-}$
$[NO_3]^-$
$[BO_3]^{3-}$
$[SO_3]^{2-}$
$[ClO_3]^-$
$[BrO_3]^-$
$[IO_3]^-$
ν_1
ν_2
ν_3
ν_4

(Continued)

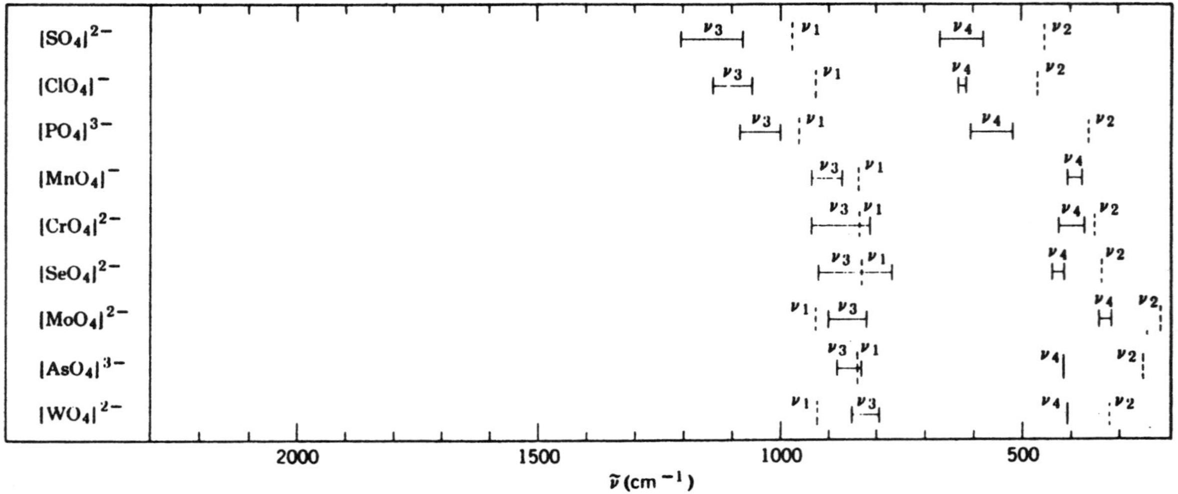

(*d*) Characteristic frequencies of some inorganic ions. (Broken lines indicate Raman active vibrations)

APPENDIX IX. CORRELATION TABLES

The following tables were reproduced with permission from the book of W. G. Fateley, F. R. Dollish, N. T. McDevitt, and F. F. Bentley (Ref. 19 of Section I). In some cases, more than one correlation is available between a pair of point groups. Then, it is necessary to specify the choice of symmetry operation from a larger group. For example, in the table of $\mathbf{C}_{4v} \rightarrow \mathbf{C}_s$ correlation, σ_v and σ_d are written above $\mathbf{C}_s$ to show two different possibilities: the first (σ_v) and second (σ_d) columns are used when the σ_v and σ_d planes, respectively, of the parent molecule become the (sole) σ plane of the $\mathbf{C}_s$ molecule.

In the $\mathbf{D}_{6h} \rightarrow \mathbf{C}_{2v}$ correlation, two different correlations exist depending upon whether the C_2' or C_2'' axis of the $\mathbf{D}_{6h}$ molecule becomes the (sole) C_2 axis of the $\mathbf{C}_{2v}$ molecule.

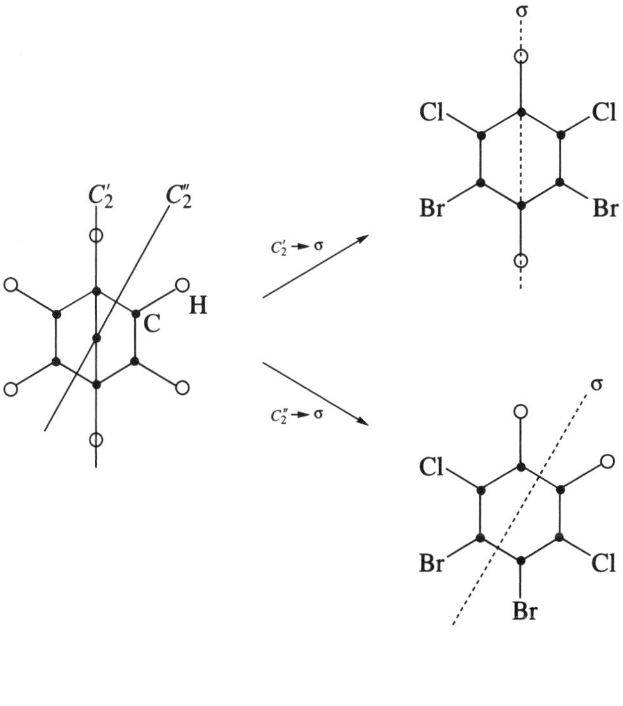

In using the correlation tables for the purpose of the correlation method (Sec. I-27), the rules given below must be followed. Those species of the point groups $\mathbf{C}_{3h}$, $\mathbf{C}_{4h}$, $\mathbf{C}_{5h}$, $\mathbf{C}_{6h}$, $\mathbf{C}_6$, $\mathbf{S}_4$, $\mathbf{S}_6$, $\mathbf{S}_8$, $\mathbf{T}$, and $\mathbf{T}_h$ marked with an asterisk (*) will not use the coefficient 2 of the E_i species in this correlation procedure *only*. Also, for those species of the point group $\mathbf{T}$ and $\mathbf{T}_h$ marked with a double dagger (‡) a coefficient 2 will be added to the E_i term related to the F_i species of the point group.

$\mathbf{C}_{2h}$	$\mathbf{C}_2$	$\mathbf{C}_s$	$\mathbf{C}_i$
A_g	A	A'	A_g
B_g	B	A''	A_g
A_u	A	A''	A_u
B_u	B	A'	A_u

$\mathbf{C}_{2v}$	$\mathbf{C}_2$	$\sigma(zx)$ $\mathbf{C}_s$	$\sigma(yz)$ $\mathbf{C}_s$
A_1	A	A'	A'
A_2	A	A''	A''
B_1	B	A'	A''
B_2	B	A''	A'

$\mathbf{D}_2$	$\mathbf{C}_2^z$	$\mathbf{C}_2^y$	$\mathbf{C}_2^x$
A	A	A	A
B_1	A	B	B
B_2	B	A	B
B_3	B	B	A

$\mathbf{D}_{2d}$	$\mathbf{S}_4$	$C_2 \rightarrow C_2(z)$ $\mathbf{D}_2$	$\mathbf{C}_{2v}$	C_2 $\mathbf{C}_2$	C_2' $\mathbf{C}_2$	$\mathbf{C}_s$
A_1	A	A	A_1	A	A	A'
A_2	A	B_1	A_2	A	B	A''
B_1	B	A	A_2	A	A	A''
B_2	B	B_1	A_1	A	B	A'
E	E	$B_2 + B_3$	$B_1 + B_2$	$2B$	$A + B$	$A' + A''$

$\mathbf{D}_{2h}$	$\mathbf{D}_2$	$C_2(z)$ $\mathbf{C}_{2v}$	$C_2(y)$ $\mathbf{C}_{2v}$	$C_2(x)$ $\mathbf{C}_{2v}$	$C_2(z)$ $\mathbf{C}_{2h}$	$C_2(y)$ $\mathbf{C}_{2h}$	$C_2(x)$ $\mathbf{C}_{2h}$	$C_2(z)$ $\mathbf{C}_2$	$C_2(y)$ $\mathbf{C}_2$	$C_2(x)$ $\mathbf{C}_2$	$\sigma(xy)$ $\mathbf{C}_s$	$\sigma(zx)$ $\mathbf{C}_s$	$\sigma(yz)$ $\mathbf{C}_s$	$\mathbf{C}_i$
A_g	A	A_1	A_1	A_1	A_g	A_g	A_g	A	A	A	A'	A'	A'	A_g
B_{1g}	B_1	A_2	B_2	B_1	A_g	B_g	B_g	A	B	B	A'	A''	A''	A_g
B_{2g}	B_2	B_1	A_2	B_2	B_g	A_g	B_g	B	A	B	A''	A'	A''	A_g
B_{3g}	B_3	B_2	B_1	A_2	B_g	B_g	A_g	B	B	A	A''	A''	A'	A_g
A_u	A	A_2	A_2	A_2	A_u	A_u	A_u	A	A	A	A''	A''	A''	A_u
B_{1u}	B_1	A_1	B_1	B_2	A_u	B_u	B_u	A	B	B	A''	A'	A'	A_u
B_{2u}	B_2	B_2	A_1	B_1	B_u	A_u	B_u	B	A	B	A'	A''	A'	A_u
B_{3u}	B_3	B_1	B_2	A_1	B_u	B_u	A_u	B	B	A	A'	A'	A''	A_u

$\mathbf{C}_{3h}$	$\mathbf{C}_3$	$\mathbf{C}_s$	$\mathbf{C}_1$
A'	A	A'	A
E'	E	$2A'*$	$2A*$
A''	A	A''	A
E''	E	$2A''*$	$2A*$

$\mathbf{C}_{3v}$	$\mathbf{C}_3$	$\mathbf{C}_s$
A_1	A	A'
A_2	A	A''
E	E	$A' + A''$

$\mathbf{D}_3$	$\mathbf{C}_3$	$\mathbf{C}_2$
A_1	A	A
A_2	A	B
E	E	$A + B$

$\mathbf{D}_{3d}$	$\mathbf{D}_3$	$\mathbf{C}_{3v}$	$\mathbf{S}_6 \equiv \mathbf{C}_{3i}$	$\mathbf{C}_3$	$\mathbf{C}_{2h}$	$\mathbf{C}_2$	$\mathbf{C}_s$	$\mathbf{C}_i$
A_{1g}	A_1	A_1	A_g	A	A_g	A	A'	A_g
A_{2g}	A_2	A_2	A_g	A	B_g	B	A''	A_g
E_g	E	E	E_g	E	$A_g + B_g$	$A + B$	$A' + A''$	$2A_g$
A_{1u}	A_1	A_2	A_u	A	A_u	A	A''	A_u
A_{2u}	A_2	A_1	A_u	A	B_u	B	A'	A_u
E_u	E	E	E_u	E	$A_u + B_u$	$A + B$	$A' + A''$	$2A_u$

$\mathbf{D}_{3h}$	$\mathbf{C}_{3h}$	$\mathbf{D}_3$	$\mathbf{C}_{3v}$	$\sigma_h \rightarrow \sigma_v(zy)$ $\mathbf{C}_{2v}$	$\mathbf{C}_3$	$\mathbf{C}_2$	σ_h $\mathbf{C}_s$	σ_v $\mathbf{C}_s$
A_1^1	A'	A_1	A_1	A_1	A	A	A_1	A'
A_2^1	A'	A_2	A_2	B_2	A	B	A'	A''
E'	E'	E	E	$A_1 + B_2$	E	$A + B$	$2A'$	$A' + A''$
A_1''	A''	A_1	A_2	A_2	A	A	A''	A''
A_2''	A''	A_2	A_1	B_1	A	B	A''	A'
E''	E''	E	E	$A_2 + B_1$	E	$A + B$	$2A''$	$A' + A''$

$\mathbf{C}_4$	$\mathbf{C}_2$
A	A
B	A
E	$2B$

$\mathbf{C}_{4h}$	$\mathbf{C}_4$	$\mathbf{S}_4$	$\mathbf{C}_{2h}$	$\mathbf{C}_2$	$\mathbf{C}_s$	$\mathbf{C}_i$	$\mathbf{C}_1$
A_g	A	A	A_g	A	A'	A_g	A
B_g	B	B	A_g	A	A'	A_g	A
E_g	E	E	$2B_u^*$	$2B*$	$2A''*$	$2A_g^*$	$2A*$
A_u	A	B	A_u	A	A''	A_u	A
B_u	B	A	A_u	A	A''	A_u	A
E_u	E	E	$2B_u^*$	$2B*$	$2A'*$	$2A_u^*$	$2A*$

$\mathbf{C}_{4v}$	$\mathbf{C}_4$	σ_v $\mathbf{C}_{2v}$	σ_d $\mathbf{C}_{2v}$	$\mathbf{C}_2$	σ_v $\mathbf{C}_s$	σ_d $\mathbf{C}_s$
A_1	A	A_1	A_1	A	A'	A'
A_2	A	A_2	A_2	A	A''	A''
B_1	B	A_1	A_2	A	A'	A''
B_2	B	A_2	A_1	A	A''	A'
E	E	$B_1 + B_2$	$B_1 + B_2$	$2B$	$A' + A''$	$A' + A''$

$\mathbf{D}_4$	C_2' $\mathbf{D}_2$	C_2'' $\mathbf{D}_2$	$\mathbf{C}_4$	$\mathbf{C}_2$	C_2' $\mathbf{C}_2$	C_2'' $\mathbf{C}_2$
A_1	A	A	A	A	A	A
A_2	B_1	B_1	A	A	B	B
B_1	A	B_1	B	A	A	B
B_2	B_1	A	B	A	B	A
E	$B_2 + B_3$	$B_2 + B_3$	E	$2B$	$A + B$	$A + B$

$\mathbf{D}_{4d}$	$\mathbf{D}_4$	$\mathbf{C}_{4v}$	$\mathbf{S}_8$	$\mathbf{C}_4$	$\mathbf{C}_{2v}$	C_2 $\mathbf{C}_2$	C_2' $\mathbf{C}_2$	$\mathbf{C}_s$
A_1	A_1	A_1	A	A	A_1	A	A	A'
A_2	A_2	A_2	A	A	A_2	A	B	A''
B_1	A_1	A_2	B	A	A_2	A	A	A''
B_2	A_2	A_1	B	A	A_1	A	B	A'
E_1	E	E	E_1	E	$B_1 + B_2$	$2B$	$A + B$	$A' + A''$
E_2	$B_1 + B_2$	$B_1 + B_2$	E_2	$2B$	$A_1 + A_2$	$2A$	$A + B$	$A' + A''$
E_3	E	E	E_3	E	$B_1 + B_2$	$2B$	$A + B$	$A' + A''$

$\mathbf{D}_{4h}$	$\mathbf{D}_4$	C_2' $\mathbf{D}_{2d}$	C_2'' $\mathbf{D}_{2d}$	$\mathbf{C}_{4v}$	$\mathbf{C}_{4h}$	C_2' $\mathbf{D}_{2h}$	C_2'' $\mathbf{D}_{2h}$	$\mathbf{C}_4$	$\mathbf{S}_4$	C_2' $\mathbf{D}_2$	C_2'' $\mathbf{D}_2$	$\underbrace{C_2, \sigma_v}$ $\mathbf{C}_{2v}$	$\underbrace{C_2, \sigma_d}$ $\mathbf{C}_{2v}$
A_{1g}	A_1	A_1	A_1	A_1	A_g	A_g	A_g	A	A	A	A	A_1	A_1
A_{2g}	A_2	A_2	A_2	A_2	A_g	B_{1g}	B_{1g}	A	A	B_1	B_1	A_2	A_2
B_{1g}	B_1	B_1	B_2	B_1	B_g	A_g	B_{1g}	B	B	A	B_1	A_1	A_2
B_{2g}	B_2	B_2	B_1	B_2	B_g	B_{1g}	A_g	B	B	B_1	A	A_2	A_1
E_g	E	E	E	E	E_g	$B_{2g} + B_{3g}$	$B_{2g} + B_{3g}$	E	E	$B_2 + B_3$	$B_2 + B_3$	$B_1 + B_2$	$B_1 + B_2$
A_{1u}	A_1	B_1	B_1	A_2	A_u	A_u	A_u	A	B	A	A	A_2	A_2
A_{2u}	A_2	B_2	B_2	A_1	A_u	B_{1u}	B_{1u}	A	B	B_1	B_1	A_1	A_1
B_{1u}	B_1	A_1	A_2	B_2	B_u	A_u	B_{1u}	B	A	A	B_1	A_2	A_1
B_{2u}	B_2	A_2	A_1	B_1	B_u	B_{1u}	A_u	B	A	B_1	A	A_1	A_2
E_u	E	E	E	E	E_u	$B_{2u} + B_{3u}$	$B_{2u} + B_{3u}$	E	E	$B_2 + B_3$	$B_2 + B_3$	$B_1 + B_2$	$B_1 + B_2$

$\mathbf{D}_{4h}$	C_2' $\mathbf{C}_{2v}$	C_2'' $\mathbf{C}_{2v}$	C_2 $\mathbf{C}_{2h}$	C_2' $\mathbf{C}_{2h}$	C_2'' $\mathbf{C}_{2h}$	C_2 $\mathbf{C}_2$	C_2' $\mathbf{C}_2$	C_2'' $\mathbf{C}_2$	σ_h $\mathbf{C}_s$	σ_v $\mathbf{C}_s$	σ_d $\mathbf{C}_s$	$\mathbf{C}_i$
A_{1g}	A_1	A_1	A_g	A_g	A_g	A	A	A	A'	A'	A'	A_g
A_{2g}	B_1	B_1	A_g	B_g	B_g	A	B	B	A'	A''	A''	A_g
B_{1g}	A_1	B_1	A_g	A_g	B_g	A	A	B	A'	A'	A''	A_g
B_{2g}	B_1	A_1	A_g	B_g	A_g	A	B	A	A'	A''	A'	A_g
E_g	$A_2 + B_2$	$A_2 + B_2$	$2B_g$	$A_g + B_g$	$A_g + B_g$	$2B$	$A + B$	$A + B$	$2A''$	$A' + A''$	$A' + A''$	$2A_g$
A_{1u}	A_2	A_2	A_u	A_u	A_u	A	A	A	A''	A''	A''	A_u
A_{2u}	B_2	B_2	A_u	B_u	B_u	A	B	B	A''	A'	A'	A_u
B_{1u}	A_2	B_2	A_u	A_u	B_u	A	A	B	A''	A''	A'	A_u
B_{2u}	B_2	A_2	A_u	B_u	A_u	A	B	A	A''	A'	A''	A_u
E_u	$A_1 + B_1$	$A_1 + B_1$	$2B_u$	$A_u + B_u$	$A_u + B_u$	$2B$	$A + B$	$A + B$	$2A'$	$A' + A''$	$A' + A''$	$2A_u$

$\mathbf{S}_4$	$\mathbf{C}_2$	$\mathbf{C}_1$
A	A	A
B	A	A
E	$2B*$	$2A*$

$\mathbf{C}_{5h}$	$\mathbf{C}_5$	$\mathbf{C}_s$	$\mathbf{C}_1$
A'	A	A'	A
E_1'	E_1	$2A'*$	$2A*$
E_2'	E_2	$2A'*$	$2A*$
A''	A	A''	A
E_1''	E_1	$2A''*$	$2A*$
E_2''	E_2	$2A''*$	$2A*$

$\mathbf{C}_{5v}$	$\mathbf{C}_5$	$\mathbf{C}_s$
A_1	A	A'
A_2	A	A''
E_1	E_1	$A' + A''$
E_2	E_2	$A' + A''$

$\mathbf{D}_5$	$\mathbf{C}_5$	$\mathbf{C}_2$
A_1	A	A
A_2	A	B
E_1	E_1	$A + B$
E_2	E_2	$A + B$

$\mathbf{D}_{5d}$	$\mathbf{D}_5$	$\mathbf{C}_{5v}$	$\mathbf{C}_5$	$\mathbf{C}_2$	$\mathbf{C}_s$	$\mathbf{C}_i$
A_{1g}	A_1	A_1	A	A	A'	A_g
A_{2g}	A_2	A_2	A	B	A''	A_g
E_{1g}	E_1	E_1	E_1	$A + B$	$A' + A''$	$2A_g$
E_{2g}	E_2	E_2	E_2	$A + B$	$A' + A''$	$2A_g$
A_{1u}	A_1	A_2	A	A	A''	A_u
A_{2u}	A_2	A_1	A	B	A'	A_u
E_{1u}	E_1	E_1	E_1	$A + B$	$A' + A''$	$2A_u$
E_{2u}	E_2	E_2	E_2	$A + B$	$A' + A''$	$2A_u$

					$\sigma_h \to \sigma(zx)$		σ_h	σ_v
$\mathbf{D}_{5h}$	$\mathbf{D}_5$	$\mathbf{C}_{5v}$	$\mathbf{C}_{5h}$	$\mathbf{C}_5$	$\mathbf{C}_{2v}$	$\mathbf{C}_2$	$\mathbf{C}_s$	$\mathbf{C}_s$
A_1'	A_1	A_1	A'	A	A_1	A	A'	A'
A_2'	A_2	A_2	A'	A	B_1	B	A'	A''
E_1'	E_1	E_1	E_1'	E_1	$A_1 + B_2$	$A + B$	$2A'$	$A' + A''$
E_2'	E_2	E_2	E_2'	E_2	$A_1 + B_2$	$A + B$	$2A'$	$A' + A''$
A_1''	A_1	A_2	A''	A	A_2	A	A''	A''
A_2''	A_2	A_1	A''	A	B_2	B	A''	A'
E_1''	E_1	E_1	E_1''	E_1	$A_2 + B_2$	$A + B$	$2A''$	$A' + A''$
E_2''	E_2	E_2	E_2''	E_2	$A_2 + B_2$	$A + B$	$2A''$	$A' + A''$

C_6	C_3	C_2	C_1
A	A	A	A
B	A	B	A
E_1	E	$2B*$	$2A*$
E_2	E	$2A*$	$2A*$

C_{6h}	C_6	C_{3h}	$S_6 \equiv C_{3i}$	C_{2h}	C_3	C_2	C_s	C_i	C_1
A_g	A	A'	A_g	A_g	A	A	A'	A_g	A
B_g	B	A''	A_g	B_g	A	B	A''	A_g	A
E_{1g}	E_1	E''	E_g	$2B_g^*$	E	$2B*$	$2A''*$	$2A_g^*$	$2A^*$
E_{2g}	E_2	E'	E_g	$2A_g^*$	E	$2A*$	$2A'*$	$2A_g^*$	$2A^*$
A_u	A	A''	A_u	A_u	A	A	A''	A_u	A
B_u	B	A'	A_u	B_u	A	B	A'	A_u	A
E_{1u}	E_1	E'	E_u	$2B_u^*$	E	$2B*$	$2A'*$	$2A_u^*$	$2A^*$
E_{2u}	E_2	E''	E_u	$2A_u^*$	E	$2A*$	$2A''*$	$2A_u^*$	$2A^*$

		σ_v	σ_d	$\sigma_v \rightarrow \sigma(zx)$			σ_v	σ_d
C_{6v}	C_6	C_{3v}	C_{3v}	C_{2v}	C_3	C_2	C_s	C_s
A_1	A	A_1	A_1	A_1	A	A	A'	A'
A_2	A	A_2	A_2	A_2	A	A	A''	A''
B_1	B	A_1	A_2	B_1	A	B	A'	A''
B_2	B	A_2	A_1	B_2	A	B	A''	A'
E_1	E_1	E	E	$B_1 + B_2$	E	$2B$	$A' + A''$	$A' + A''$
E_2	E_2	E	E	$A_1 + A_2$	E	$2B$	$A' + A''$	$A' + A''$

		C_2'	C_2''			C_2	C_2'	C_2''
D_6	C_6	D_3	D_3	D_2	C_3	C_2	C_2	C_2
A_1	A	A_1	A_1	A	A	A	A	A
A_2	A	A_2	A_2	B_1	A	A	B	B
B_1	B	A_1	A_2	B_2	A	B	A	B
B_2	B	A_2	A_1	B_3	A	B	B	A
E_1	E_1	E	E	$B_2 + B_3$	E	$2B$	$A + B$	$A + B$
E_2	E_2	E	E	$A + B_1$	E	$2A$	$A + B$	$A + B$

$\mathbf{D}_{6d}$	$\mathbf{D}_6$	$\mathbf{C}_{6v}$	$\mathbf{C}_6$	$\mathbf{D}_{2d}$	$\mathbf{D}_3$	$\mathbf{C}_{3v}$	$\mathbf{D}_2$	$\mathbf{C}_{2v}$	$\mathbf{S}_4$	$\mathbf{C}_3$	C_2' $\mathbf{C}_2$	C_2'' $\mathbf{C}_2$	$\mathbf{C}_s$
A_1	A_1	A_1	A	A_1	A_1	A_1	A	A_1	A	A	A	A	A'
A_2	A_2	A_2	A	A_2	A_2	A_2	B_1	A_2	A	A	A	B	A''
B_1	A_1	A_2	A	B_1	A_1	A_2	A	A_2	B	A	A	A	A''
B_2	A_2	A_1	A	B_2	A_2	A_1	B_1	A_1	B	A	A	B	A'
E_1	E_1	E_1	E_1	E	E	E	$B_2 + B_3$	$B_1 + B_2$	E	E	$2B$	$A + B$	$A' + A''$
E_2	E_2	E_2	E_2	$B_1 + B_2$	E	E	$A + B_1$	$A_1 + A_2$	$2B$	E	$2A$	$A + B$	$A' + A''$
E_3	$B_1 + B_2$	$B_1 + B_2$	$2B$	E	$A_1 + A_2$	$A_1 + A_2$	$B_2 + B_3$	$B_1 + B_2$	E	$2A$	$2B$	$A + B$	$A' + A''$
E_4	E_2	E_2	E_2	$A_1 + A_2$	E	E	$A + B_1$	$A_1 + A_2$	$2A$	E	$2A$	$A + B$	$A' + A''$
E_5	E_1	E_1	E_1	E	E	E	$B_2 + B_3$	$B_1 + B_2$	E	E	$2B$	$A + B$	$A' + A''$

$\mathbf{D}_{6h}$	$\mathbf{D}_6$	C_2' $\mathbf{D}_{3h}$	C_2'' $\mathbf{D}_{3h}$	$\mathbf{C}_{6v}$	$\mathbf{C}_{6h}$	C_2' $\mathbf{D}_{3d}$	C_2'' $\mathbf{D}_{3d}$	$\sigma_h \to \sigma(xy)$, $\sigma_v \to \sigma(yz)$ $\mathbf{D}_{2h}$	$\mathbf{C}_6$	$\mathbf{C}_{3h}$	C_2' $\mathbf{D}_3$	C_2'' $\mathbf{D}_3$	σ_v $\mathbf{C}_{3v}$	σ_d $\mathbf{C}_{3v}$	$\mathbf{S}_6$	$\mathbf{D}_2$
A_{1g}	A_1	A_1'	A_1'	A_1	A_g	A_{1g}	A_{1g}	A_g	A	A'	A_1	A_1	A_1	A_1	A_g	A
A_{2g}	A_2	A_2'	A_2'	A_2	A_g	A_{2g}	A_{2g}	B_{1g}	A	A'	A_2	A_2	A_2	A_2	A_g	B_1
B_{1g}	B_1	A_1''	A_2''	B_2	B_g	A_{1g}	A_{2g}	B_{2g}	B	A''	A_1	A_2	A_2	A_1	A_g	B_2
B_{2g}	B_2	A_2''	A_1''	B_1	B_g	A_{2g}	A_{1g}	B_{3g}	B	A''	A_2	A_1	A_1	A_2	A_g	B_3
E_{1g}	E_1	E''	E''	E_1	E_{1g}	E_g	E_g	$B_{2g} + B_{3g}$	E_1	E''	E	E	E	E	E_g	$B_2 + B_3$
E_{2g}	E_2	E'	E'	E_2	E_{2g}	E_g	E_g	$A_g + B_{1g}$	E_2	E'	E	E	E	E	E_g	$A + B_1$
A_{1u}	A_1	A_1''	A_1''	A_2	A_u	A_{1u}	A_{1u}	A_u	A	A''	A_1	A_1	A_2	A_2	A_u	A
A_{2u}	A_2	A_2''	A_2''	A_1	A_u	A_{2u}	A_{2u}	B_{1u}	A	A''	A_2	A_2	A_1	A_1	A_u	B_1
B_{1u}	B_1	A_1'	A_2'	B_1	B_u	A_{1u}	A_{2u}	B_{2u}	B	A'	A_1	A_2	A_1	A_2	A_u	B_2
B_{2u}	B_2	A_2'	A_1'	B_2	B_u	A_{2u}	A_{1u}	B_{3u}	B	A'	A_2	A_1	A_2	A_1	A_u	B_3
E_{1u}	E_1	E'	E'	E_1	E_{1u}	E_u	E_u	$B_{2u} + B_{3u}$	E_1	E'	E	E	E	E	E_u	$B_2 + B_3$
E_{2u}	E_2	E''	E''	E_2	E_{2u}	E_u	E_u	$A_u + B_{1u}$	E_2	E''	E	E	E	E	E_u	$A + B_1$

$\mathbf{D}_{6h}$	C_2 $\mathbf{C}_{2v}$	C'_2 $\mathbf{C}_{2v}$	C''_2 $\mathbf{C}_{2v}$	C_2 $\mathbf{C}_{2h}$	C'_2 $\mathbf{C}_{2h}$	C''_2 $\mathbf{C}_{2h}$	$\mathbf{C}_3$	C_2 $\mathbf{C}_2$	C'_2 $\mathbf{C}_2$	C''_2 $\mathbf{C}_2$	σ_h $\mathbf{C}_s$	σ_d $\mathbf{C}_s$	σ_v $\mathbf{C}_s$	$\mathbf{C}_i$
A_{1g}	A_1	A_1	A_1	A_g	A_g	A_g	A	A	A	A	A'	A'	A'	A_g
A_{2g}	A_2	B_2	B_1	A_g	B_g	B_g	A	A	B	B	A'	A''	A''	A_g
B_{1g}	B_1	A_2	B_2	B_g	A_g	B_g	A	B	A	B	A''	A'	A''	A_g
B_{2g}	B_2	B_1	A_2	B_g	B_g	A_g	A	B	B	A	A''	A''	A'	A_g
E_{1g}	$B_1 + B_2$	$A_2 + B_1$	$B_2 + A_2$	$2B_g$	$A_g + B_g$	$A_g + B_g$	E	$2B$	$A + B$	$A + B$	$2A''$	$A' + A''$	$A' + A''$	$2A_g$
E_{2g}	$A_1 + A_2$	$A_1 + B_2$	$A_1 + B_1$	$2A_g$	$A_g + B_g$	$A_g + B_g$	E	$2A$	$A + B$	$A + B$	$2A'$	$A' + A''$	$A' + A''$	$2A_g$
A_{1u}	A_2	A_2	A_2	A_u	A_u	A_u	A	A	A	A	A''	A''	A''	A_u
A_{2u}	A_1	B_1	B_2	A_u	B_u	B_u	A	A	B	B	A''	A'	A'	A_u
B_{1u}	B_2	A_1	B_1	B_u	A_u	B_u	A	B	A	B	A'	A''	A'	A_u
B_{2u}	B_1	B_2	A_1	B_u	B_u	A_u	A	B	B	A	A'	A'	A''	A_u
E_{1u}	$B_2 + B_1$	$A_1 + B_2$	$B_1 + A_1$	$2B_u$	$A_u + B_u$	$A_u + B_u$	E	$2B$	$A + B$	$A + B$	$2A'$	$A' + A''$	$A' + A''$	$2A_u$
E_{2u}	$A_2 + A_1$	$A_2 + B_1$	$A_2 + B_2$	$2A_u$	$A_u + B_u$	$A_u + B_u$	E	$2A$	$A + B$	$A + B$	$2A''$	$A' + A''$	$A' + A''$	$2a_u$

$C_{3i} \equiv S_6$	C_3	C_i	C_1
A_g	A	A_g	A
E_g	E	$2A_g^*$	$2A*$
A_u	A	A_u	A
E_u	E	$2A_u^*$	$2A*$

S_8	C_4	C_2	C_1
A	A	A	A
B	A	A	A
E_1	E	$2B*$	$2A*$
E_2	$2B*$	$2A*$	$2A*$
E_3	E	$2B*$	$2A*$

$C_{\infty v}$ †	C_{6v}	C_{4v}	C_{3v}	C_{2v}
$A_1 \equiv \Sigma^+$	A_1	A_1	A_1	A_1
$A_2 \equiv \Sigma^-$	A_2	A_2	A_2	A_2
$E_1 \equiv \Pi$	E_1	E	E	$B_1 + B_2$
$E_2 \equiv \Delta$	E_2	$B_1 + B_2$	E	$A_1 + A_2$
$E_3 \equiv \Phi$	$B_2 + B_1$	E	$A_1 + A_2$	$B_1 + B_2$
$E_4 \equiv \Gamma$	E_2	$A_1 + A_2$	E	$A_1 + A_2$
...				

$D_{\infty h}$ †	D_{6h}	C_{6v}	C_{3v}	D_{4h}	C_{4v}	C_{2v}	$C_{\infty v}$
Σ_g^+	A_{1g}	A_1	A_1	A_{1g}	A_1	A_1	$\Sigma^+ \equiv A_1$
Σ_g^-	A_{2g}	A_2	A_2	A_{2g}	A_2	A_2	$\Sigma^- \equiv A_2$
Π_g	E_{1g}	E_1	E	E_g	E	$B_1 + B_2$	$\Pi \equiv E_1$
Δ_g	E_{2g}	E_2	E	$B_{1g} + B_{2g}$	$B_1 + B_2$	$A_1 + B_2$	$\Delta \equiv E_2$
...							
Σ_u^+	A_{2u}	A_1	A_1	A_{2u}	A_1	A_1	$\Sigma^+ \equiv A_1$
Σ_u^-	A_{1u}	A_2	A_2	A_{1u}	A_2	A_2	$\Sigma^- \equiv A_2$
Π_u	E_{1u}	E_1	E	E_u	E	$B_1 + B_2$	$\Pi \equiv E_1$
Δ_u	E_{2u}	E_2	E	$B_{1u} + B_{2u}$	$B_1 + B_2$	$A_1 + A_2$	$\Delta \equiv E_2$
...							

† The z axis of $C_{\infty v}$ and $D_{\infty h}$ groups must coincide with z axis of point group.

T	D_2	C_3	C_2	C_1
A	A	A	A	A
E	$2A^*$	E	$2A^*$	$2A^*$
F	$B_1 + B_2 + B_3$	$A + E$‡	$A + 2B$	$3A$

$\mathbf{T}_d$	$\mathbf{T}$	$\mathbf{D}_{2d}$	$\mathbf{C}_{3v}$	$\mathbf{S}_4$	$\mathbf{D}_2$	$\mathbf{C}_{2v}$	$\mathbf{C}_3$	$\mathbf{C}_2$	$\mathbf{C}_s$
A_1	A	A_1	A_1	A	A	A_1	A	A	A'
A_2	A	B_1	A_2	B	A	A_2	A	A	A''
E	E	$A_1 + B_1$	E	$A + B$	$2A$	$A_1 + A_2$	E	$2A$	$A' + A''$
F_1	F	$A_2 + E$	$A_2 + E$	$A + E$	$B_1 + B_2 + B_3$	$A_2 + B_1 + B_2$	$A + E$	$A + 2B$	$A' + 2A''$
F_2	F	$B_2 + E$	$A_1 + E$	$B + E$	$B_1 + B_2 + B_3$	$A_1 + B_1 + B_2$	$A + E$	$A + 2B$	$2A' + A''$

$\mathbf{T}_h$	$\mathbf{T}$	$\mathbf{D}_{2h}$	$\mathbf{S}_6 \equiv \mathbf{C}_{3i}$	$\mathbf{D}_2$	$\mathbf{C}_{2v}$	$\mathbf{C}_{2h}$	$\mathbf{C}_3$	$\mathbf{C}_2$	$\mathbf{C}_s$	$\mathbf{C}_i$	$\mathbf{C}_1$
A_g	A	A_g	A_g	A	A_1	A_g	A	A	A'	A_g	A
E_g	E	$2A_g^*$	E_g	$2A^*$	$2A_1^*$	$2A_g^*$	E	$2A^*$	$2A'^*$	$2A_g^*$	$2A^*$
F_g	F	$B_{1g} + B_{2g} + B_{3g}$	$A_g + E_g$‡	$B_1 + B_2 + B_3$	$A_2 + B_1 + B_2$	$A + 2B_g$	$A + E$‡	$A + 2B$	$A' + 2A''$	$3A_g$	$3A$
A_u	A	A_u	A_u	A	A_2	A_u	A	A	A''	A_u	A
E_u	E	$2A_u^*$	E_u	$2A^*$	$2A_2^*$	$2A_u^*$	E	$2A^*$	$2A''^*$	$2A_u^*$	$2A^*$
F_u	F	$B_{1u} + B_{2u} + B_{3u}$	$A_u + E_u$‡	$B_1 + B_2 + B_3$	$A_1 + B_1 + B_2$	$A_u + 2B_u$	$A + E$‡	$A + 2B$	$2A' + A''$	$3A_u$	$3A$

$\mathbf{O}$	$\mathbf{T}$	$\mathbf{D}_4$	$\mathbf{D}_3$	$\mathbf{C}_4$	$3C_2$ $\mathbf{D}_2$	$C_2, 2C_2'$ $\mathbf{D}_2$	$\mathbf{C}_3$	C_2 $\mathbf{C}_2$	C_2' $\mathbf{C}_2$
A_1	A	A_1	A_1	A	A	A	A	A	A
A_2	A	B_1	A_2	B	A	B_1	A	A	B
E	E	$A_1 + B_1$	E	$A + B$	$2A$	$A + B_1$	E	$2A$	$A + B$
F_1	F	$A_2 + E$	$A_2 + E$	$A + E$	$B_1 + B_2 + B_3$	$B_1 + B_2 + B_3$	$A + E$	$A + 2B$	$A + 2B$
F_2	F	$B_2 + E$	$A_1 + E$	$B + E$	$B_1 + B_2 + B_3$	$A + B_2 + B_3$	$A + E$	$A + 2B$	$2A + B$

$\mathbf{O}_h$	$\mathbf{O}$	$\mathbf{T}_d$	$\mathbf{T}_h$	$\mathbf{T}$	$\mathbf{D}_{3d}$	$\mathbf{D}_{4h}$	$\mathbf{C}_{3v}$	$\mathbf{D}_3$	$\mathbf{C}_{3i} \equiv \mathbf{S}_6$
A_{1g}	A_1	A_1	A_g	A	A_{1g}	A_{1g}	A_1	A_1	A_g
A_{2g}	A_2	A_2	A_g	A	A_{2g}	B_{1g}	A_2	A_2	A_g
E_g	E	E	E_g	E	E_g	$A_{1g} + B_{1g}$	E	E	E_g
F_{1g}	F_1	F_1	F_g	F	$A_{2g} + E_g$	$A_{2g} + E_g$	$A_2 + E$	$A_2 + E$	$A_g + E_g$
F_{2g}	F_2	F_2	F_g	F	$A_{1g} + E_g$	$B_{2g} + E_g$	$A_1 + E$	$A_1 + E$	$A_g + E_g$
A_{1u}	A_1	A_2	A_u	A	A_{1u}	A_{1u}	A_2	A_1	A_u
A_{2u}	A_2	A_1	A_u	A	A_{2u}	B_{1u}	A_1	A_2	A_u
E_u	E	E	E_u	E	E_u	$A_{1u} + B_{1u}$	E	E	E_u
F_{1u}	F_1	F_2	F_u	F	$A_{2u} + E_u$	$A_{2u} + E_u$	$A_1 + E$	$A_2 + E$	$A_u + E_u$
F_{2u}	F_2	F_1	F_u	F	$A_{1u} + E_u$	$B_{2u} + E_u$	$A_2 + E$	$A_1 + E$	$A_u + E_u$

		C_2, σ_d	C'_2, σ_h					
$\mathbf{O}_h$	$\mathbf{C}_3$	$\mathbf{D}_{2d}$	$\mathbf{D}_{2d}$	$\mathbf{C}_{4v}$	$\mathbf{D}_4$	$\mathbf{C}_{4h}$	$\mathbf{S}_4$	$\mathbf{C}_4$
A_{1g}	A	A_1	A_1	A_1	A_1	A_g	A	A
A_{2g}	A	B_1	B_2	B_1	B_1	B_g	B	B
E_g	E	$A_1 + B_1$	$A_1 + B_2$	$A_1 + B_1$	$A_1 + B_1$	$A_g + B_g$	$A + B$	$A + B$
F_{1g}	$A + E$	$A_2 + E$	$A_2 + E$	$A_2 + E$	$A_2 + E$	$A_g + E_g$	$A + E$	$A + E$
F_{2g}	$A + E$	$B_2 + E$	$B_1 + E$	$B_2 + E$	$B_2 + E$	$B_g + E_g$	$B + E$	$B + E$
A_{1u}	A	B_1	B_1	A_2	A_1	A_u	B	A
A_{2u}	A	A_1	A_2	B_2	B_1	B_u	A	B
E_u	E	$A_1 + B_1$	$A_2 + B_1$	$A_2 + B_2$	$A_1 + B_1$	$A_u + B_u$	$A + B$	$A + B$
F_{1u}	$A + E$	$B_2 + E$	$B_2 + E$	$A_1 + E$	$A_2 + E$	$A_u + E_u$	$B + E$	$B + E$
F_{2u}	$A + E$	$A_2 + E$	$A_1 + E$	$B_1 + E$	$B_2 + E$	$B_u + E_u$	$A + E$	$B + E$

$\mathbf{O}_h$	$3C_2$ $\mathbf{D}_{2h}$	$C_2, 2C'_2$ $\mathbf{D}_{2h}$	C_2, σ_h $\mathbf{C}_{2v}$	C_2, σ_d $\mathbf{C}_{2v}$	C'_2, σ_h $\mathbf{C}_{2v}$	$3C_2$ $\mathbf{D}_2$	$C_2, 2C'_2$ $\mathbf{D}_2$
A_{1g}	A_g	A_g	A_1	A_1	A_1	A	A
A_{2g}	A_g	B_{1g}	A_1	A_2	B_1	A	B_1
E_g	$2A_g$	$A_g + B_{1g}$	$2A_1$	$A_1 + A_2$	$A_1 + B_1$	$2A$	$A + B_1$
F_{1g}	$B_{1g} + B_{2g} + B_{3g}$	$B_{1g} + B_{2g} + B_{3g}$	$A_2 + B_1 + B_2$	$A_2 + B_1 + B_2$	$A_2 + B_1 + B_2$	$B_1 + B_2 + B_3$	$B_1 + B_2 + B_3$
F_{2g}	$B_{1g} + B_{2g} + B_{3g}$	$A_{1g} + B_{2g} + B_{3g}$	$A_2 + B_1 + B_2$	$A_1 + B_1 + B_2$	$A_1 + A_2 + B_2$	$B_1 + B_2 + B_3$	$A + B_2 + B_3$
A_{1u}	A_u	A_u	A_2	A_2	A_2	A	A
A_{2u}	A_u	B_{2u}	A_2	A_1	B_2	A	B_1
E_u	$2A_u$	$A_u + B_{1u}$	$2A_2$	$A_1 + A_2$	$A_2 + B_2$	$2A$	$A + B_1$
F_{1u}	$B_{1u} + B_{2u} + B_{3u}$	$B_{1u} + B_{2u} + B_{3u}$	$A_1 + B_1 + B_2$	$A_1 + B_1 + B_2$	$A_1 + B_1 + B_2$	$B_1 + B_2 + B_3$	$B_1 + B_2 + B_3$
F_{2u}	$B_{1u} + B_{2u} + B_{3u}$	$A_u + B_{2u} + B_{3u}$	$A_1 + B_1 + B_2$	$A_2 + B_1 + B_2$	$A_1 + A_2 + B_1$	$B_1 + B_2 + B_3$	$A + B_2 + B_3$

$\mathbf{O}_h$	C_2, σ_h $\mathbf{C}_{2h}$	C'_2, σ_h $\mathbf{C}_{2h}$	σ_h $\mathbf{C}_s$	σ_d $\mathbf{C}_s$	C_2 $\mathbf{C}_2$	C'_2 $\mathbf{C}_2$	$\mathbf{C}_i$	$\mathbf{C}_1$
A_{1g}	A_g	A_g	A'	A'	A	A	A_g	A
A_{2g}	A_g	B_g	A'	A''	A	B	A_g	A
E_g	$2A_g$	$A_g + B_g$	$2A'$	$A' + A''$	$2A$	$A + B$	$2A_g$	$2A$
F_{1g}	$A_g + 2B_g$	$A_g + 2B_g$	$A' + A''$	$A' + 2A''$	$A + 2B$	$A + 2B$	$3A_g$	$3A$
F_{2g}	$A_g + 2B_g$	$2A_g + B_g$	$A' + 2A''$	$2A' + A''$	$A + 2B$	$2A + B$	$3A_g$	$3A$
A_{1u}	A_u	A_u	A''	A''	A	A	A_u	A
A_{2u}	A_u	B_u	A''	A'	A	B	A_u	A
E_u	$2A_u$	$A_u + B_u$	$2A''$	$A' + A''$	$2A$	$A + B$	$2A_u$	$2A$
F_{1u}	$A_u + 2B_u$	$A_u + 2B_u$	$2A' + A''$	$2A' + A''$	$A + 2B$	$A + 2B$	$3A_u$	$3A$
F_{2u}	$A_u + 2B_u$	$2A_u + B_u$	$2A' + A''$	$A' + 2A''$	$A + 2B$	$2A + B$	$3A_u$	$3A$

$\mathbf{I}_h$	$\mathbf{I}$	$\mathbf{C}_5$	$\mathbf{C}_3$	$\mathbf{C}_2$	$\mathbf{C}_1$
A_g	A	A	A	A	A
A_u	A	A	A	A	A
F_{1g}	F_1	$A + E_1$	$A + E$	$A + 2B$	$3A$
F_{1u}	F_1	$A + E_1$	$A + E$	$A + 2B$	$3A$
F_{2g}	F_2	$A + E_2$	$A + E$	$A + 2B$	$3A$
F_{2u}	F_2	$A + E_2$	$A + E$	$A + 2B$	$3A$
G_{1g}	G_1	$E_1 + E_2$	$2A + E$	$2A + 2B$	$4A$
G_{1u}	G_1	$E_1 + E_2$	$2A + E$	$2A + 2B$	$4A$
H_g	H	$A + E_1 + E_2$	$A + 2E$	$3A + 2B$	$5A$
H_u	H	$A + E_1 + E_2$	$A + 2E$	$3A + 2B$	$5A$

APPENDIX X. SITE SYMMETRY FOR THE 230 SPACE GROUPS

The tables of site symmetry for the 230 space groups shown below were reproduced with permission from the book of J. R. Ferraro and J. S. Ziomek (Ref. 17 of Section I). The number in front of the point group notation represents the number of distinct sets of sites, and those in parentheses indicates the number of equivalent sites for each distinct set. Since the number of sites for $\mathbf{C}_p$, $\mathbf{C}_{pv}$ ($p = 1, 2, 3, \cdots$) and $\mathbf{C}_s$ is infinite, no coefficients are given in front of these point group notations.

Space group		Site symmetries
1 $P1$	$\mathbf{C}_1^1$	$\mathbf{C}_1(1)$
2 $P\bar{1}$	$\mathbf{C}_i^1$	$8\mathbf{C}_i(1); \mathbf{C}_1(2)$
3 $P2$	$\mathbf{C}_2^1$	$4\mathbf{C}_2(1); \mathbf{C}_1(2)$
4 $P2_1$	$\mathbf{C}_2^2$	$\mathbf{C}_1(2)$
5 $B2$ or $C2$	$\mathbf{C}_2^3$	$2\mathbf{C}_2(1); \mathbf{C}_1(2)$
6 Pm	$\mathbf{C}^1$	$2\mathbf{C}_s(1); \mathbf{C}_1(2)$
7 Pb or Pc	$\mathbf{C}_s^2$	$\mathbf{C}_1(2)$
8 Bm or Cm	$\mathbf{C}_s^3$	$\mathbf{C}_s(1); \mathbf{C}_1(2)$
9 Bb or Cc	$\mathbf{C}^4$	$\mathbf{C}_1(2)$
10 $P2/m$	$\mathbf{C}_{2h}^1$	$8\mathbf{C}_{2h}(1); 4\mathbf{C}_2(2); 2\mathbf{C}_s(2); \mathbf{C}_1(4)$
11 $P2_1/m$	$\mathbf{C}_{2h}^2$	$4\mathbf{C}_i(2); \mathbf{C}_s(2); \mathbf{C}_1(4)$
12 $B2/m$ or $C2/m$	$\mathbf{C}_{2h}^3$	$4\mathbf{C}_{2h}(1); 2\mathbf{C}_i(2); 2\mathbf{C}_2(2); \mathbf{C}_s(2); \mathbf{C}_1(4)$
13 $P2/b$ or $P2/c$	$\mathbf{C}_{2h}^4$	$4\mathbf{C}_i(2); 2\mathbf{C}_2(2); \mathbf{C}_1(4)$
14 $P2_1/b$ or $P2_1/c$	$\mathbf{C}_{2h}^5$	$4\mathbf{C}_i(2); \mathbf{C}_1(4)$
15 $B2/b$ or $C2/c$	$\mathbf{C}_{2h}^6$	$4\mathbf{C}_i(2); \mathbf{C}_2(2); \mathbf{C}_1(4)$
16 $P222$	$\mathbf{D}_2^1$	$8\mathbf{D}_2(1); 12\mathbf{C}_2(2); \mathbf{C}_1(4)$

Space group		Site symmetries
17 $P222_1$	$\mathbf{D}_2^2$	$4\mathbf{C}_2(2); \mathbf{C}_1(4)$
18 $P2_12_12$	$\mathbf{D}_2^3$	$2\mathbf{C}_2(2); \mathbf{C}_1(4)$
19 $P2_12_12_1$	$\mathbf{D}_2^4$	$\mathbf{C}_1(4)$
20 $C222_1$	$\mathbf{D}_2^5$	$2\mathbf{C}_2(2); \mathbf{C}_1(4)$
21 $C222$	$\mathbf{D}_2^6$	$4\mathbf{D}_2(1); 7\mathbf{C}_2(2); \mathbf{C}_1(4)$
22 $F222$	$\mathbf{D}_2^7$	$4\mathbf{D}_2(1); 6\mathbf{C}_2(2); \mathbf{C}_1(4)$
23 $I222$	$\mathbf{D}_2^8$	$4\mathbf{D}_2(1); 6\mathbf{C}_2(2); \mathbf{C}_1(4)$
24 $I2_12_12_1$	$\mathbf{D}_2^9$	$3\mathbf{C}_2(2); \mathbf{C}_1(4)$
25 $Pmm2$	$\mathbf{C}_{2v}^1$	$4\mathbf{C}_{2v}(1); 4\mathbf{C}_s(2); \mathbf{C}_1(4)$
26 $Pmc2_1$	$\mathbf{C}_{2v}^2$	$2\mathbf{C}_s(2); \mathbf{C}_1(4)$
27 $Pcc2$	$\mathbf{C}_{2v}^3$	$4\mathbf{C}_2(2); \mathbf{C}_1(4)$
28 $Pma2$	$\mathbf{C}_{2v}^4$	$2\mathbf{C}_2(2); \mathbf{C}_s(2); \mathbf{C}_1(4)$
29 $Pca2_1$	$\mathbf{C}_{2v}^5$	$\mathbf{C}_1(4)$
30 $Pnc2$	$\mathbf{C}_{2v}^6$	$2\mathbf{C}_2(2); \mathbf{C}_1(4)$
31 $Pmn2_1$	$\mathbf{C}_{2v}^7$	$\mathbf{C}_s(2); \mathbf{C}_1(4)$
32 $Pba2$	$\mathbf{C}_{2v}^8$	$2\mathbf{C}_2(2); \mathbf{C}_1(4)$
33 $Pna2_1$	$\mathbf{C}_{2v}^9$	$\mathbf{C}_1(4)$
34 $Pnn2$	$\mathbf{C}_{2v}^{10}$	$2\mathbf{C}_2(2); \mathbf{C}_1(4)$
35 $Cmm2$	$\mathbf{C}_{2v}^{11}$	$2\mathbf{C}_{2v}(1); \mathbf{C}_2(2); 2\mathbf{C}_s(2); \mathbf{C}_1(4)$
36 $Cmc2_1$	$\mathbf{C}_{2v}^{12}$	$\mathbf{C}_s(2); \mathbf{C}_1(4)$
37 $Ccc2$	$\mathbf{C}_{2v}^{13}$	$3\mathbf{C}_2(2); \mathbf{C}_1(4)$
38 $Amm2$	$\mathbf{C}_{2v}^{14}$	$2\mathbf{C}_{2v}(1); 3\mathbf{C}_s(2); \mathbf{C}_1(4)$
39 $Abm2$	$\mathbf{C}_{2v}^{15}$	$2\mathbf{C}_2(2); \mathbf{C}_s(2); \mathbf{C}_1(4)$
40 $Ama2$	$\mathbf{C}_{2v}^{16}$	$\mathbf{C}_2(2); \mathbf{C}_s(2); \mathbf{C}_1(4)$
41 $Aba2$	$\mathbf{C}_{2v}^{17}$	$\mathbf{C}_2(2); \mathbf{C}_1(4)$
42 $Fmm2$	$\mathbf{C}_{2v}^{18}$	$\mathbf{C}_{2v}(1); \mathbf{C}_2(2); 2\mathbf{C}_s(2); \mathbf{C}_1(4)$
43 $Fdd2$	$\mathbf{C}_{2v}^{19}$	$\mathbf{C}_2(2); \mathbf{C}_1(4)$
44 $Imm2$	$\mathbf{C}_{2v}^{20}$	$2\mathbf{C}_{2v}(1); 2\mathbf{C}_s(2); \mathbf{C}_1(4)$
45 $Iba2$	$\mathbf{C}_{2v}^{21}$	$2\mathbf{C}_2(2); \mathbf{C}_1(4)$
46 $Ima2$	$\mathbf{C}_{2v}^{22}$	$\mathbf{C}_2(2); \mathbf{C}_s(2); \mathbf{C}_1(4)$
47 $Pmmm$	$\mathbf{D}_{2h}^1$	$8\mathbf{D}_{2h}(1); 12\mathbf{C}_{2v}(2); 6\mathbf{C}_s(4); \mathbf{C}_1(8)$
48 $Pnnn$	$\mathbf{D}_{2h}^2$	$4\mathbf{D}_2(2); 2\mathbf{C}_i(4); 6\mathbf{C}_2(4); \mathbf{C}_1(8)$
49 $Pccm$	$\mathbf{D}_{2h}^3$	$4\mathbf{C}_{2h}(2); 4\mathbf{D}_2(2); 8\mathbf{C}_2(4); \mathbf{C}_s(4); \mathbf{C}_1(8)$
50 $Pban$	$\mathbf{D}_{2h}^4$	$4\mathbf{D}_2(2); 2\mathbf{C}_i(4); 6\mathbf{C}_2(4); \mathbf{C}_1(8)$
51 $Pmma$	$\mathbf{D}_{2h}^5$	$4\mathbf{C}_{2h}(2); 2\mathbf{C}_{2v}(2); 2\mathbf{C}_2(4); 3\mathbf{C}_s(4); \mathbf{C}_1(8)$
52 $Pnna$	$\mathbf{D}_{2h}^6$	$2\mathbf{C}_i(4); 2\mathbf{C}_2(4); \mathbf{C}_1(8)$
53 $Pmna$	$\mathbf{D}_{2h}^7$	$4\mathbf{C}_{2h}(2); 3\mathbf{C}_2(4); \mathbf{C}_s(4); \mathbf{C}_1(8)$
54 $Pcca$	$\mathbf{D}_{2h}^8$	$2\mathbf{C}_i(4); 3\mathbf{C}_2(4); \mathbf{C}_1(8)$

Space group		Site symmetries
55 *Pbam*	$\mathbf{D}_{2h}^{9}$	$4\mathbf{C}_{2h}(2); 2\mathbf{C}_2(4); 2\mathbf{C}_2(4); \mathbf{C}_1(8)$
56 *Pccn*	$\mathbf{D}_{2h}^{10}$	$2\mathbf{C}_i(4); 2\mathbf{C}_2(4); \mathbf{C}_1(8)$
57 *Pbcm*	$\mathbf{D}_{2h}^{11}$	$2\mathbf{C}_i(4); \mathbf{C}_2(4); \mathbf{C}_s(4); \mathbf{C}_1(8)$
58 *Pnnm*	$\mathbf{D}_{2h}^{12}$	$4\mathbf{C}_{2h}(2); 2\mathbf{C}_2(4); \mathbf{C}_s(4); \mathbf{C}_1(8)$
59 *Pmmn*	$\mathbf{D}_{2h}^{13}$	$2\mathbf{C}_{2v}(2); 2\mathbf{C}_i(4); 2\mathbf{C}_s(4); \mathbf{C}_1(8)$
60 *Pbcn*	$\mathbf{D}_{2h}^{14}$	$2\mathbf{C}_i(4); \mathbf{C}_2(4); \mathbf{C}_1(8)$
61 *Pbca*	$\mathbf{D}_{2h}^{15}$	$2\mathbf{C}_i(4); \mathbf{C}_1(8)$
62 *Pnma*	$\mathbf{D}_{2h}^{16}$	$2\mathbf{C}_i(4); \mathbf{C}_s(4); \mathbf{C}_1(8)$
63 *Cmcm*	$\mathbf{D}_{2h}^{17}$	$2\mathbf{C}_{2h}(2); \mathbf{C}_{2v}(2); \mathbf{C}_i(4); \mathbf{C}_2(4); 2\mathbf{C}_s(4); \mathbf{C}_1(8)$
64 *Cmca*	$\mathbf{D}_{2h}^{18}$	$2\mathbf{C}_{2h}(2); \mathbf{C}_i(4); 2\mathbf{C}_2(4); \mathbf{C}_s(4); \mathbf{C}_1(8)$
65 *Cmmm*	$\mathbf{D}_{2h}^{19}$	$4\mathbf{D}_{2h}(1); 2\mathbf{C}_{2h}(2); 6\mathbf{C}_{2v}(2); \mathbf{C}_2(4); 4\mathbf{C}_s(4); \mathbf{C}_1(8)$
66 *Cccm*	$\mathbf{D}_{2h}^{20}$	$2\mathbf{D}_2(2); 4\mathbf{C}_{2h}(2); 5\mathbf{C}_2(4); \mathbf{C}_s(4); \mathbf{C}_1(8)$
67 *Cmma*	$\mathbf{D}_{2h}^{21}$	$2\mathbf{D}_2(2); 4\mathbf{C}_{2h}(2); \mathbf{C}_{2v}(2); 5\mathbf{C}_2(4); 2\mathbf{C}_s(4); \mathbf{C}_1(8)$
68 *Ccca*	$\mathbf{D}_{2h}^{22}$	$2\mathbf{D}_2(2); 2\mathbf{C}_i(4); 4\mathbf{C}_2(4); \mathbf{C}_1(8)$
69 *Fmmm*	$\mathbf{D}_{2h}^{23}$	$2\mathbf{D}_{2h}(1); 3\mathbf{C}_{2h}(2); \mathbf{D}_2(2); 3\mathbf{C}_{2v}(2); 3\mathbf{C}_2(4); 3\mathbf{C}_s(4); \mathbf{C}_1(8)$
70 *Fddd*	$\mathbf{D}_{2h}^{24}$	$2\mathbf{D}_2(2); 2\mathbf{C}_i(4); 3\mathbf{C}_2(4); \mathbf{C}_1(8)$
71 *Immm*	$\mathbf{D}_{2h}^{25}$	$4\mathbf{D}_{2h}(1); 6\mathbf{C}_{2v}(2); \mathbf{C}_i(4); 3\mathbf{C}_s(4); \mathbf{C}_1(8)$
72 *Ibam*	$\mathbf{D}_{2h}^{26}$	$2\mathbf{D}_2(2); 2\mathbf{C}_{2h}(2); \mathbf{C}_i(4); 4\mathbf{C}_2(4); \mathbf{C}_s(4); \mathbf{C}_1(8)$
73 *Ibca*	$\mathbf{D}_{2h}^{27}$	$2\mathbf{C}_i(4); 3\mathbf{C}_2(4); \mathbf{C}_1(8)$
74 *Imma*	$\mathbf{D}_{2h}^{28}$	$4\mathbf{C}_{2h}(2); \mathbf{C}_{2v}(2); 2\mathbf{C}_2(4); 2\mathbf{C}_s(4); \mathbf{C}_1(8)$
75 *P*4	$\mathbf{C}_4^1$	$2\mathbf{C}_4(1); \mathbf{C}_2(2); \mathbf{C}_1(4)$
76 $P4_1$	$\mathbf{C}_4^2$	$\mathbf{C}_1(4)$
77 $P4_2$	$\mathbf{C}_4^3$	$3\mathbf{C}_2(2); \mathbf{C}_1(4)$
78 $P4_3$	$\mathbf{C}_4^4$	$\mathbf{C}_1(4)$
79 *I*4	$\mathbf{C}_4^5$	$\mathbf{C}_4(1); \mathbf{C}_2(2); \mathbf{C}_1(4)$
80 $I4_1$	$\mathbf{C}_4^6$	$\mathbf{C}_2(2); \mathbf{C}_1(4)$
81 $P\bar{4}$	$\mathbf{S}_4^1$	$4\mathbf{S}_4(1); 3\mathbf{C}_2(2); \mathbf{C}_1(4)$
82 $I\bar{4}$	$\mathbf{S}_4^2$	$4\mathbf{S}_4(1); 2\mathbf{C}_2(2); \mathbf{C}_1(4)$
83 $P4/m$	$\mathbf{C}_{4h}^1$	$4\mathbf{C}_{4h}(1); 2\mathbf{C}_{2h}(2); 2\mathbf{C}_4(2); \mathbf{C}_2(4); 2\mathbf{C}_s(4); \mathbf{C}_1(8)$
84 $P4_2/m$	$\mathbf{C}_{4h}^2$	$4\mathbf{C}_{2h}(2); 2\mathbf{S}_4(2); 3\mathbf{C}_2(4); \mathbf{C}_s(4); \mathbf{C}_1(8)$
85 $P4/n$	$\mathbf{C}_{4h}^3$	$2\mathbf{S}_4(2); \mathbf{C}_4(2); 2\mathbf{C}_i(4); \mathbf{C}_2(4); \mathbf{C}_1(8)$
86 $P4_2/n$	$\mathbf{C}_{4h}^4$	$2\mathbf{S}_4(2); 2\mathbf{C}_i(4); 2\mathbf{C}_2(4); \mathbf{C}_1(8)$
87 $I4/m$	$\mathbf{C}_{4h}^5$	$2\mathbf{C}_{4h}(1); \mathbf{C}_{2h}(2); \mathbf{S}_4(2); \mathbf{C}_4(2); \mathbf{C}_i(4); \mathbf{C}_2(4); \mathbf{C}_s(4); \mathbf{C}_1(8)$
88 $I4_1/a$	$\mathbf{C}_{4h}^6$	$2\mathbf{S}_4(2); 2\mathbf{C}_i(4); \mathbf{C}_2(4); \mathbf{C}_1(8)$
89 *P*422	$\mathbf{D}_4^1$	$4\mathbf{D}_4(1); 2\mathbf{D}_2(2); 2\mathbf{C}_4(2); 7\mathbf{C}_2(4); \mathbf{C}_1(8)$
90 $P42_12$	$\mathbf{D}_4^2$	$2\mathbf{D}_2(2); \mathbf{C}_4(2); 3\mathbf{C}_2(4); \mathbf{C}_1(8)$
91 $P4_122$	$\mathbf{D}_4^3$	$3\mathbf{C}_2(4); \mathbf{C}_1(8)$
92 $P4_12_12$	$\mathbf{D}_4^4$	$\mathbf{C}_2(4); \mathbf{C}_1(8)$

Space group		Site symmetries
93 $P4_222$	$\mathbf{D}_4^5$	$6\mathbf{D}_2(2); 9\mathbf{C}_2(4); \mathbf{C}_1(8)$
94 $P4_22_12$	$\mathbf{D}_4^6$	$2\mathbf{D}_2(2); 4\mathbf{C}_2(4); \mathbf{C}_1(8)$
95 $P4_322$	$\mathbf{D}_4^7$	$3\mathbf{C}_2(4); \mathbf{C}_1(8)$
96 $P4_32_12$	$\mathbf{D}_4^8$	$\mathbf{C}_2(4); \mathbf{C}_1(8)$
97 $I422$	$\mathbf{D}_4^9$	$2\mathbf{D}_4(1); 2\mathbf{D}_2(2); \mathbf{C}_4(2); 5\mathbf{C}_2(4); \mathbf{C}_1(8)$
98 $I4_122$	$\mathbf{D}_4^{10}$	$2\mathbf{D}_2(2); 4\mathbf{C}_2(4); \mathbf{C}_1(8)$
99 $P4mm$	$\mathbf{C}_{4v}^1$	$2\mathbf{C}_{4v}(1); \mathbf{C}_{2v}(2); 3\mathbf{C}_s(4); \mathbf{C}_1(8)$
100 $P4bm$	$\mathbf{C}_{4v}^2$	$\mathbf{C}_4(2); \mathbf{C}_{2v}(2); \mathbf{C}_s(4); \mathbf{C}_1(8)$
101 $P4_2cm$	$\mathbf{C}_{4v}^3$	$2\mathbf{C}_{2v}(2); \mathbf{C}_2(4); \mathbf{C}_s(4); \mathbf{C}_1(8)$
102 $P4_2nm$	$\mathbf{C}_{4v}^4$	$\mathbf{C}_{2v}(2); \mathbf{C}_2(4); \mathbf{C}_s(4); \mathbf{C}_1(8)$
103 $P4cc$	$\mathbf{C}_{4v}^5$	$2\mathbf{C}_4(2); \mathbf{C}_2(4); \mathbf{C}_1(8)$
104 $P4nc$	$\mathbf{C}_{4v}^6$	$\mathbf{C}_4(2); \mathbf{C}_2(4); \mathbf{C}_1(8)$
105 $P4_2mc$	$\mathbf{C}_{4v}^7$	$3\mathbf{C}_{2v}(2); 2\mathbf{C}_s(4); \mathbf{C}_1(8)$
106 $P4_2bc$	$\mathbf{C}_{4v}^8$	$2\mathbf{C}_2(4); \mathbf{C}_1(8)$
107 $I4mm$	$\mathbf{C}_{4v}^9$	$\mathbf{C}_{4v}(1); \mathbf{C}_{2v}(2); 2\mathbf{C}_s(4); \mathbf{C}_1(8)$
108 $I4cm$	$\mathbf{C}_{4v}^{10}$	$\mathbf{C}_4(2); \mathbf{C}_{2v}(2); \mathbf{C}_s(4); \mathbf{C}_1(8)$
109 $I4_1md$	$\mathbf{C}_{4v}^{11}$	$\mathbf{C}_{2v}(2); \mathbf{C}_s(4); \mathbf{C}_1(8)$
110 $I4_1cd$	$\mathbf{C}_{4v}^{12}$	$\mathbf{C}_2(4); \mathbf{C}_1(8)$
111 $P\bar{4}2m$	$\mathbf{D}_{2d}^1$	$4\mathbf{D}_{2d}(1); 2\mathbf{D}_2(2); 2\mathbf{C}_{2v}(2); 5\mathbf{C}_2(4); \mathbf{C}_s(4); \mathbf{C}_1(8)$
112 $P\bar{4}2c$	$\mathbf{D}_{2d}^2$	$4\mathbf{D}_2(2); 2\mathbf{S}_4(2); 7\mathbf{C}_2(4); \mathbf{C}_1(8)$
113 $P\bar{4}2_1m$	$\mathbf{D}_{2d}^3$	$2\mathbf{S}_4(2); \mathbf{C}_{2v}(2); \mathbf{C}_2(4); \mathbf{C}_s(4); \mathbf{C}_1(8)$
114 $P\bar{4}2_1c$	$\mathbf{D}_{2d}^4$	$2\mathbf{S}_4(2); 2\mathbf{C}_2(4); \mathbf{C}_1(8)$
115 $P\bar{4}m2$	$\mathbf{D}_{2d}^5$	$4\mathbf{D}_{2d}(1); 3\mathbf{C}_{2v}(2); 2\mathbf{C}_2(4); 2\mathbf{C}_s(4); \mathbf{C}_1(8)$
116 $P\bar{4}c2$	$\mathbf{D}_{2d}^6$	$2\mathbf{D}_2(2); 2\mathbf{S}_4(2); 5\mathbf{C}_2(4); \mathbf{C}_1(8)$
117 $P\bar{4}b2$	$\mathbf{D}_{2d}^7$	$2\mathbf{S}_4(2); 2\mathbf{D}_2(2); 4\mathbf{C}_2(4); \mathbf{C}_1(8)$
118 $P\bar{4}n2$	$\mathbf{D}_{2d}^8$	$2\mathbf{S}_4(2); 2\mathbf{D}_2(2); 4\mathbf{C}_2(4); \mathbf{C}_1(8)$
119 $I\bar{4}m2$	$\mathbf{D}_{2d}^9$	$4\mathbf{D}_{2d}(1); 2\mathbf{C}_{2v}(2); 2\mathbf{C}_2(4); \mathbf{C}_s(4); \mathbf{C}_1(8)$
120 $I\bar{4}c2$	$\mathbf{D}_{2d}^{10}$	$\mathbf{D}_2(2); 2\mathbf{S}_4(2); \mathbf{D}_2(2); 4\mathbf{C}_2(4); \mathbf{C}_1(8)$
121 $I\bar{4}2m$	$\mathbf{D}_{2d}^{11}$	$2\mathbf{D}_{2d}(1); \mathbf{D}_2(2); \mathbf{S}_4(2); \mathbf{C}_{2v}(2); 3\mathbf{C}_2(4); \mathbf{C}_s(4); \mathbf{C}_1(8)$
122 $I\bar{4}2d$	$\mathbf{D}_{2d}^{12}$	$2\mathbf{S}_4(2); 2\mathbf{C}_2(4); \mathbf{C}_1(8)$
123 $P4/mmm$	$\mathbf{D}_{4h}^1$	$4\mathbf{D}_{4h}(1); 2\mathbf{D}_{2h}(2); 2\mathbf{C}_{4v}(2); 7\mathbf{C}_{2v}(4); 5\mathbf{C}_s(8); \mathbf{C}_1(16)$
124 $P4/mcc$	$\mathbf{D}_{4h}^2$	$\mathbf{D}_4(2); \mathbf{C}_{4h}(2); \mathbf{D}_4(2); \mathbf{C}_{4h}(2); \mathbf{C}_{2h}(4); \mathbf{D}_2(4);$ $2\mathbf{C}_4(4); 4\mathbf{C}_2(8); \mathbf{C}_s(8); \mathbf{C}_1(16)$
125 $P4/nbm$	$\mathbf{D}_{4h}^3$	$2\mathbf{D}_4(2); 2\mathbf{D}_{2d}(2); 2\mathbf{C}_{2h}(4); \mathbf{C}_4(4); \mathbf{C}_{2v}(4);$ $4\mathbf{C}_2(8); \mathbf{C}_s(8); \mathbf{C}_1(16)$
126 $P4/nnc$	$\mathbf{D}_{4h}^4$	$2\mathbf{D}_4(2); \mathbf{D}_2(4); \mathbf{S}_4(4); \mathbf{C}_4(4); \mathbf{C}_i(8); 4\mathbf{C}_2(8); \mathbf{C}_1(8)$
127 $P4/mbm$	$\mathbf{D}_{4h}^5$	$2\mathbf{C}_{4h}(2); 2\mathbf{D}_{2h}(2); \mathbf{C}_4(4); 3\mathbf{C}_{2v}(4); 3\mathbf{C}_s(8); \mathbf{C}_1(16)$
128 $P4/mnc$	$\mathbf{D}_{4h}^6$	$2\mathbf{C}_{4h}(2); \mathbf{C}_{2h}(4); \mathbf{D}_2(4); \mathbf{C}_4(4); 2\mathbf{C}_2(8); \mathbf{C}_s(8); \mathbf{C}_1(16)$

Space group		Site symmetries
129 $P4/nmm$	$\mathbf{D}_{4h}^{7}$	$2\mathbf{D}_{2d}(2); \mathbf{C}_{4v}(2); 2\mathbf{C}_{2h}(4); \mathbf{C}_{2v}(4); 2\mathbf{C}_{2}(8); 2\mathbf{C}_{s}(8); \mathbf{C}_{1}(16)$
130 $P4/ncc$	$\mathbf{D}_{4h}^{8}$	$\mathbf{D}_{2}(4); \mathbf{S}_{4}(4); \mathbf{C}_{4}(4); \mathbf{C}_{i}(8); 2\mathbf{C}_{2}(8); \mathbf{C}_{1}(16)$
131 $P4_2/mmc$	$\mathbf{D}_{4h}^{9}$	$4\mathbf{D}_{2h}(2); 2\mathbf{D}_{2d}(2); 7\mathbf{C}_{2v}(4); \mathbf{C}_{2}(8); 3\mathbf{C}_{s}(8); \mathbf{C}_{1}(16)$
132 $P4_2/mcm$	$\mathbf{D}_{4h}^{10}$	$\mathbf{D}_{2h}(2); \mathbf{D}_{2d}(2); \mathbf{D}_{2h}(2); \mathbf{D}_{2d}(2); \mathbf{D}_{2}(4); \mathbf{C}_{2h}(4);$ $4\mathbf{C}_{2v}(4); 3\mathbf{C}_{2}(8); 2\mathbf{C}_{s}(8); \mathbf{C}_{1}(16)$
133 $P4_2/nbc$	$\mathbf{D}_{4h}^{11}$	$3\mathbf{D}_{2}(4); \mathbf{S}_{4}(4); \mathbf{C}_{i}(8); 5\mathbf{C}_{2}(8); \mathbf{C}_{1}(16)$
134 $P4_2/nnm$	$\mathbf{D}_{4h}^{12}$	$2\mathbf{D}_{2d}(2); 2\mathbf{D}_{2}(4); 2\mathbf{C}_{2h}(4); \mathbf{C}_{2v}(4); 5\mathbf{C}_{2}(8); \mathbf{C}_{s}(8); \mathbf{C}_{1}(16)$
135 $P4_2/mbc$	$\mathbf{D}_{4h}^{13}$	$\mathbf{C}_{2h}(4); \mathbf{S}_{4}(4); \mathbf{C}_{2h}(4); \mathbf{D}_{2}(4); 3\mathbf{C}_{(}8); \mathbf{C}_{s}(8); \mathbf{C}_{1}(16)$
136 $P4_2/mnm$	$\mathbf{D}_{4h}^{14}$	$2\mathbf{D}_{2h}(2); \mathbf{C}_{2h}(4); \mathbf{S}_{4}(4); 3\mathbf{C}_{2v}(4); \mathbf{C}_{2}(8); 2\mathbf{C}_{s}(8); \mathbf{C}_{1}(16)$
137 $P4_2/nmc$	$\mathbf{D}_{4h}^{15}$	$2\mathbf{D}_{2d}(2); 2\mathbf{C}_{2v}(4); \mathbf{C}_{i}(8); \mathbf{C}_{2}(8); \mathbf{C}_{s}(8); \mathbf{C}_{1}(16)$
138 $P4_2/ncm$	$\mathbf{D}_{4h}^{16}$	$\mathbf{D}_{2}(4); \mathbf{S}_{4}(4); 2\mathbf{C}_{2h}(4); \mathbf{C}_{2v}(4); 3\mathbf{C}_{2}(8); \mathbf{C}_{s}(8); \mathbf{C}_{1}(16)$
139 $I4/mmm$	$\mathbf{D}_{4h}^{17}$	$2\mathbf{D}_{4h}(1); \mathbf{D}_{2h}(2); \mathbf{D}_{2d}(2); \mathbf{C}_{4v}(2); \mathbf{C}_{2h}(4);$ $4\mathbf{C}_{2v}(4); \mathbf{C}_{2}(8); 3\mathbf{C}_{s}(8); \mathbf{C}_{1}(16)$
140 $I4/mcm$	$\mathbf{D}_{4h}^{18}$	$\mathbf{D}_{4}(2); \mathbf{D}_{2d}(2); \mathbf{C}_{4h}(2); \mathbf{D}_{2h}(2); \mathbf{C}_{2h}(4); \mathbf{C}_{4}(4);$ $2\mathbf{C}_{2v}(4); 2\mathbf{C}_{2}(8); 2\mathbf{C}_{s}(8); \mathbf{C}_{1}(16)$
141 $I4_1/amd$	$\mathbf{D}_{4h}^{19}$	$2\mathbf{D}_{2d}(2); 2\mathbf{C}_{2h}(4); \mathbf{C}_{2v}(4); 2\mathbf{C}_{2}(8); \mathbf{C}_{s}(8); \mathbf{C}_{1}(16)$
142 $I4_1/acd$	$\mathbf{D}_{2h}^{20}$	$\mathbf{S}_{4}(4); \mathbf{D}_{2}(4); \mathbf{C}_{i}(8); 3\mathbf{C}_{2}(8); \mathbf{C}_{1}(16)$
143 $P3$	$\mathbf{C}_{3}^{1}$	$3\mathbf{C}_{3}(1); \mathbf{C}_{1}(3)$
144 $P3_1$	$\mathbf{C}_{3}^{2}$	$\mathbf{C}_{1}(3)$
145 $P3_2$	$\mathbf{C}_{3}^{3}$	$\mathbf{C}_{1}(3)$
146 $R3$	$\mathbf{C}_{3}^{4}$	$\mathbf{C}_{3}(1); \mathbf{C}_{1}(3)$
147 $P\bar{3}$	$\mathbf{C}_{3i}^{1}$	$2\mathbf{C}_{3i}(1); 2\mathbf{C}_{3}(2); 2\mathbf{C}_{i}(3); \mathbf{C}_{1}(6)$
148 $R\bar{3}$	$\mathbf{C}_{3i}^{2}$	$2\mathbf{C}_{3i}(1); \mathbf{C}_{3}(2); 2\mathbf{C}_{i}(3); \mathbf{C}_{1}(6)$
149 $P312$	$\mathbf{D}_{3}^{1}$	$6\mathbf{D}_{3}(1); 3\mathbf{C}_{3}(2); 2\mathbf{C}_{2}(3); \mathbf{C}_{1}(6)$
150 $P321$	$\mathbf{D}_{3}^{2}$	$2\mathbf{D}_{3}(1); 2\mathbf{C}_{3}(2); 2\mathbf{C}_{2}(3); \mathbf{C}_{1}(6)$
151 $P3_112$	$\mathbf{D}_{3}^{3}$	$2\mathbf{C}_{2}(3); \mathbf{C}_{1}(6)$
152 $P3_121$	$\mathbf{D}_{3}^{4}$	$2\mathbf{C}_{2}(3); \mathbf{C}_{1}(6)$
153 $P3_212$	$\mathbf{D}_{3}^{5}$	$2\mathbf{C}_{2}(3); \mathbf{C}_{1}(6)$
154 $P3_221$	$\mathbf{D}_{3}^{6}$	$2\mathbf{C}_{2}(3); \mathbf{C}_{1}(6)$
155 $R32$	$\mathbf{D}_{3}^{7}$	$2\mathbf{D}_{3}(1); \mathbf{C}_{3}(2); 2\mathbf{C}_{2}(3); \mathbf{C}_{1}(6)$
156 $P3ml$	$\mathbf{C}_{3v}^{1}$	$3\mathbf{C}_{3v}(1); \mathbf{C}_{s}(3); \mathbf{C}_{1}(6)$
157 $P31m$	$\mathbf{C}_{3v}^{2}$	$\mathbf{C}_{3v}(1); \mathbf{C}_{3}(2); \mathbf{C}_{s}(3); \mathbf{C}_{1}(6)$
158 $P3cl$	$\mathbf{C}_{3v}^{3}$	$3\mathbf{C}_{3}(2); \mathbf{C}_{1}(6)$
159 $P31c$	$\mathbf{C}_{3v}^{4}$	$2\mathbf{C}_{3}(2); \mathbf{C}_{1}(6)$
160 $R3m$	$\mathbf{C}_{3v}^{5}$	$\mathbf{C}_{3v}(1); \mathbf{C}_{s}(3); \mathbf{C}_{1}(6)$
161 $R3c$	$\mathbf{C}_{3v}^{6}$	$\mathbf{C}_{3}(2); \mathbf{C}_{1}(6)$
162 $P\bar{3}1m$	$\mathbf{D}_{3d}^{1}$	$2\mathbf{D}_{3d}(1); 2\mathbf{D}_{3}(2); \mathbf{C}_{3v}(2); 2\mathbf{C}_{2h}(3); \mathbf{C}_{3}(4);$ $2\mathbf{C}_{2}(6); \mathbf{C}_{s}(6); \mathbf{C}_{1}(12)$

Space group		Site symmetries
163 $P\bar{3}1c$	$\mathbf{D}_{3d}^2$	$\mathbf{D}_3(2)$; $\mathbf{C}_{3i}(2)$; $2\mathbf{D}_3(2)$; $2\mathbf{C}_3(4)$; $\mathbf{C}_i(6)$; $\mathbf{C}_2(6)$; $\mathbf{C}_1(12)$
164 $P\bar{3}m1$	$\mathbf{D}_{3d}^3$	$2\mathbf{D}_{3d}(1)$; $2\mathbf{C}_{3v}(2)$; $2\mathbf{C}_{2h}(3)$; $2\mathbf{C}_2(6)$; $\mathbf{C}_s(6)$; $\mathbf{C}_1(12)$
165 $P\bar{3}c1$	$\mathbf{D}_{3d}^4$	$\mathbf{D}_3(2)$; $\mathbf{C}_{3i}(2)$; $2\mathbf{C}_3(4)$; $\mathbf{C}_i(6)$; $\mathbf{C}_2(6)$; $\mathbf{C}_1(12)$
166 $R\bar{3}m$	$\mathbf{D}_{3d}^5$	$2\mathbf{D}_{3d}(1)$; $\mathbf{C}_{3v}(2)$; $2\mathbf{C}_{2h}(3)$; $2\mathbf{C}_2(6)$; $\mathbf{C}_s(6)$; $\mathbf{C}_1(12)$
167 $R\bar{3}c$	$\mathbf{D}_{3d}^6$	$\mathbf{D}_3(2)$; $\mathbf{C}_{3i}(2)$; $\mathbf{C}_3(4)$; $\mathbf{C}_i(6)$; $\mathbf{C}_2(6)$; $\mathbf{C}_1(12)$
168 $P6$	$\mathbf{C}_6^1$	$\mathbf{C}_6(1)$; $\mathbf{C}_3(2)$; $\mathbf{C}_2(3)$; $\mathbf{C}_1(6)$
169 $P6_1$	$\mathbf{C}_6^2$	$\mathbf{C}_1(6)$
170 $P6_5$	$\mathbf{C}_6^3$	$\mathbf{C}_1(6)$
171 $P6_2$	$\mathbf{C}_6^4$	$2\mathbf{C}_2(3)$; $\mathbf{C}_1(6)$
172 $P6_4$	$\mathbf{C}_6^5$	$2\mathbf{C}_2(3)$; $\mathbf{C}_1(6)$
173 $P6_3$	$\mathbf{C}_6^6$	$2\mathbf{C}_3(2)$; $\mathbf{C}_1(6)$
174 $P\bar{6}$	$\mathbf{C}_{3h}^1$	$6\mathbf{C}_{3h}(1)$; $3\mathbf{C}_3(2)$; $2\mathbf{C}_s(3)$; $\mathbf{C}_1(6)$
175 $P6/m$	$\mathbf{C}_{6h}^1$	$2\mathbf{C}_{6h}(1)$; $2\mathbf{C}_{3h}(2)$; $\mathbf{C}_6(2)$; $2\mathbf{C}_{2h}(3)$; $\mathbf{C}_3(4)$; $\mathbf{C}_2(6)$; $2\mathbf{C}_s(6)$; $\mathbf{C}_1(12)$
176 $P6_3/m$	$\mathbf{C}_{6h}^2$	$\mathbf{C}_{3h}(2)$; $\mathbf{C}_{3i}(2)$; $2\mathbf{C}_{3h}(2)$; $2\mathbf{C}_3(4)$; $\mathbf{C}_i(6)$; $\mathbf{C}_s(6)$; $\mathbf{C}_1(12)$
177 $P622$	$\mathbf{D}_6^1$	$2\mathbf{D}_6(1)$; $2\mathbf{D}_3(2)$; $\mathbf{C}_6(2)$; $2\mathbf{D}_2(3)$; $\mathbf{C}_3(4)$; $5\mathbf{C}_2(6)$; $\mathbf{C}_1(12)$
178 $P6_122$	$\mathbf{D}_6^2$	$2\mathbf{C}_2(6)$; $\mathbf{C}_1(12)$
179 $P6_522$	$\mathbf{D}_6^3$	$2\mathbf{C}_2(6)$; $\mathbf{C}_1(12)$
180 $P6_222$	$\mathbf{D}_6^4$	$4\mathbf{D}_2(3)$; $6\mathbf{C}_2(6)$; $\mathbf{C}_1(12)$
181 $P6_422$	$\mathbf{D}_6^5$	$4\mathbf{D}_2(3)$; $6\mathbf{C}_2(6)$; $\mathbf{C}_1(12)$
182 $P6_322$	$\mathbf{D}_6^6$	$4\mathbf{D}_3(2)$; $2\mathbf{C}_3(4)$; $2\mathbf{C}_2(6)$; $\mathbf{C}_1(12)$
183 $P6mm$	$\mathbf{C}_{6v}^1$	$\mathbf{C}_{6v}(1)$; $\mathbf{C}_{3v}(2)$; $\mathbf{C}_{2v}(3)$; $2\mathbf{C}_s(6)$; $\mathbf{C}_1(12)$
184 $P6cc$	$\mathbf{C}_{6v}^2$	$\mathbf{C}_6(2)$; $\mathbf{C}_3(4)$; $\mathbf{C}_2(6)$; $\mathbf{C}_1(12)$
185 $P6_3cm$	$\mathbf{C}_{6v}^3$	$\mathbf{C}_{3v}(2)$; $\mathbf{C}_3(4)$; $\mathbf{C}_s(6)$; $\mathbf{C}_1(12)$
186 $P6_3mc$	$\mathbf{C}_{6v}^4$	$2\mathbf{C}_{3v}(2)$; $\mathbf{C}_s(6)$; $\mathbf{C}_1(12)$
187 $P\bar{6}m2$	$\mathbf{D}_{3h}^1$	$6\mathbf{D}_{3h}(1)$; $3\mathbf{C}_{3v}(2)$; $2\mathbf{C}_{2v}(3)$; $3\mathbf{C}_s(6)$; $\mathbf{C}_1(12)$
188 $P\bar{6}c2$	$\mathbf{D}_{3h}^2$	$\mathbf{D}_3(2)$; $\mathbf{C}_{3h}(2)$; $\mathbf{D}_3(2)$; $\mathbf{C}_{3h}(2)$; $\mathbf{D}_3(2)$; $\mathbf{C}_{3h}(2)$; $3\mathbf{C}_3(4)$; $\mathbf{C}_2(6)$; $\mathbf{C}_s(6)$; $\mathbf{C}_1(12)$
189 $P\bar{6}2m$	$\mathbf{D}_{3h}^3$	$2\mathbf{D}_{3h}(1)$; $2\mathbf{C}_{3h}(2)$; $\mathbf{C}_{3v}(2)$; $2\mathbf{C}_{2v}(3)$; $\mathbf{C}_3(4)$; $3\mathbf{C}_s(6)$; $\mathbf{C}_1(12)$
190 $P\bar{6}2c$	$\mathbf{D}_{3h}^4$	$\mathbf{D}_3(2)$; $3\mathbf{C}_{3h}(2)$; $2\mathbf{C}_3(4)$; $\mathbf{C}_2(6)$; $\mathbf{C}_s(6)$; $\mathbf{C}_1(12)$
191 $P6/mmm$	$\mathbf{D}_{6h}^1$	$2\mathbf{D}_{6h}(1)$; $2\mathbf{D}_{3h}(2)$; $\mathbf{C}_{6v}(2)$; $2\mathbf{D}_{2h}(3)$; $\mathbf{C}_{3v}(4)$; $5\mathbf{C}_{2v}(6)$; $4\mathbf{C}_s(12)$; $\mathbf{C}_1(24)$
192 $P6/mcc$	$\mathbf{D}_{6h}^2$	$\mathbf{D}_6(2)$; $\mathbf{C}_{6h}(2)$; $\mathbf{D}_3(4)$; $\mathbf{C}_{3h}(4)$; $\mathbf{C}_6(4)$; $\mathbf{D}_2(6)$; $\mathbf{C}_{2h}(6)$; $\mathbf{C}_3(8)$; $3\mathbf{C}_2(12)$; $\mathbf{C}_s(12)$; $\mathbf{C}_1(24)$
193 $P6_3/mcm$	$\mathbf{D}_{6h}^3$	$\mathbf{D}_{3h}(2)$; $\mathbf{D}_{3d}(2)$; $\mathbf{C}_{3h}(4)$; $\mathbf{D}_3(4)$; $\mathbf{C}_6(4)$; $\mathbf{C}_{2h}(6)$; $\mathbf{C}_{2v}(6)$; $\mathbf{C}_3(8)$; $\mathbf{C}_2(12)$; $2\mathbf{C}_s(12)$; $\mathbf{C}_1(24)$
194 $P6_3/mmc$	$\mathbf{D}_{6h}^4$	$\mathbf{D}_{3d}(2)$; $3\mathbf{D}_{3h}(2)$; $2\mathbf{C}_{3v}(4)$; $\mathbf{C}_{2h}(6)$; $\mathbf{C}_{2v}(6)$; $\mathbf{C}_2(12)$; $2\mathbf{C}_s(12)$; $\mathbf{C}_1(24)$

Space group		Site symmetries
195 $P23$	$\mathbf{T}^1$	$2\mathbf{T}(1); 2\mathbf{D}_2(3); \mathbf{C}_3(4); 4\mathbf{C}_2(6); \mathbf{C}_1(12)$
196 $F23$	$\mathbf{T}^2$	$4\mathbf{T}(1); \mathbf{C}_3(4); 2\mathbf{C}_2(6); \mathbf{C}_1(12)$
197 $I23$	$\mathbf{T}^3$	$\mathbf{T}(1); \mathbf{D}_2(3); \mathbf{C}_3(4); 2\mathbf{C}_2(6); \mathbf{C}_1(12)$
198 $P2_13$	$\mathbf{T}^4$	$\mathbf{C}_3(4); \mathbf{C}_1(12)$
199 $I2_13$	$\mathbf{T}^5$	$\mathbf{C}_3(4); \mathbf{C}_2(6); \mathbf{C}_1(12)$
200 $Pm3$	$\mathbf{T}_h^1$	$2\mathbf{T}_h(1); 2\mathbf{D}_{2h}(3); 4\mathbf{C}_{2v}(6); \mathbf{C}_3(8); 2\mathbf{C}_s(12); \mathbf{C}_1(24)$
201 $Pn3$	$\mathbf{T}_h^2$	$\mathbf{T}(2); 2\mathbf{C}_{3i}(4); \mathbf{D}_2(6); \mathbf{C}_3(8); 2\mathbf{C}_2(12); \mathbf{C}_1(24)$
202 $Fm3$	$\mathbf{T}_h^3$	$2\mathbf{T}_h(1); \mathbf{T}(2); \mathbf{C}_{2h}(6); \mathbf{C}_{2v}(6); \mathbf{C}_3(8); \mathbf{C}_2(12); \mathbf{C}_s(12); \mathbf{C}_1(24)$
203 $Fd3$	$\mathbf{T}_h^4$	$2\mathbf{T}(2); 2\mathbf{C}_{3i}(4); \mathbf{C}_3(8); \mathbf{C}_2(12); \mathbf{C}_1(24)$
204 $Im3$	$\mathbf{T}_h^5$	$\mathbf{T}_h(1); \mathbf{D}_{2h}(3); \mathbf{C}_{3i}(4); 2\mathbf{C}_{2v}(6); \mathbf{C}_3(8); \mathbf{C}_s(12); \mathbf{C}_1(24)$
205 $Pa3$	$\mathbf{T}_h^6$	$2\mathbf{C}_{3i}(4); \mathbf{C}_3(8); \mathbf{C}_1(24)$
206 $Ia3$	$\mathbf{T}_h^7$	$2\mathbf{C}_{3i}(4); \mathbf{C}_3(8); \mathbf{C}_2(12); \mathbf{C}_1(24)$
207 $P432$	$\mathbf{O}^1$	$2\mathbf{O}(1); 2\mathbf{D}_4(3); 2\mathbf{C}_4(6); \mathbf{C}_3(8); 3\mathbf{C}_2(12); \mathbf{C}_1(24)$
208 $P4_232$	$\mathbf{O}^2$	$\mathbf{T}(2); 2\mathbf{D}_3(4); 3\mathbf{D}_2(6); \mathbf{C}_3(8); 5\mathbf{C}_2(12); \mathbf{C}_1(24)$
209 $F432$	$\mathbf{O}^3$	$2\mathbf{O}(1); \mathbf{T}(2); \mathbf{D}_2(6); \mathbf{C}_4(6); \mathbf{C}_3(8); 3\mathbf{C}_2(12); \mathbf{C}_1(24)$
210 $F4_132$	$\mathbf{O}^4$	$2\mathbf{T}(2); 2\mathbf{D}_3(4); \mathbf{C}_3(8); 2\mathbf{C}_2(12); \mathbf{C}_1(24)$
211 $I432$	$\mathbf{O}^5$	$\mathbf{O}(1); \mathbf{D}_4(3); \mathbf{D}_3(4); \mathbf{D}_2(6); \mathbf{C}_4(6); \mathbf{C}_3(8); 3\mathbf{C}_2(12); \mathbf{C}_1(24)$
212 $P4_332$	$\mathbf{O}^6$	$2\mathbf{D}_3(4); \mathbf{C}_3(8); \mathbf{C}_2(12); \mathbf{C}_1(24)$
213 $P4_132$	$\mathbf{O}^7$	$2\mathbf{D}_3(4); \mathbf{C}_3(8); \mathbf{C}_2(12); \mathbf{C}_1(24)$
214 $I4_132$	$\mathbf{O}^8$	$2\mathbf{D}_3(4); 2\mathbf{D}_2(6); \mathbf{C}_3(8); 3\mathbf{C}_2(12); \mathbf{C}_1(24)$
215 $P\bar{4}3m$	$\mathbf{T}_d^1$	$2\mathbf{T}_d(1); 2\mathbf{D}_{2d}(3); \mathbf{C}_{3v}(4); 2\mathbf{C}_{2v}(6); \mathbf{C}_2(12); \mathbf{C}_s(12); \mathbf{C}_1(24)$
216 $F\bar{4}3m$	$\mathbf{T}_d^2$	$4\mathbf{T}_d(1); \mathbf{C}_{3v}(4); 2\mathbf{C}_{2v}(6); \mathbf{C}_s(12); \mathbf{C}_1(24)$
217 $I\bar{4}3m$	$\mathbf{T}_d^3$	$\mathbf{T}_d(1); \mathbf{D}_{2d}(3); \mathbf{C}_{3v}(4); \mathbf{S}_4(6); \mathbf{C}_{2v}(6); \mathbf{C}_2(12); \mathbf{C}_s(12); \mathbf{C}_1(24)$
218 $P\bar{4}3n$	$\mathbf{T}_d^4$	$\mathbf{T}(2); \mathbf{D}_2(6); 2\mathbf{S}_4(6); \mathbf{C}_3(8); 3\mathbf{C}_2(12); \mathbf{C}_1(24)$
219 $F\bar{4}3c$	$\mathbf{T}_d^5$	$2\mathbf{T}(2); 2\mathbf{S}_4(6); \mathbf{C}_3(8); 2\mathbf{C}_2(12); \mathbf{C}_1(24)$
220 $I\bar{4}3d$	$\mathbf{T}_d^6$	$2\mathbf{S}_4(6); \mathbf{C}_3(8); \mathbf{C}_2(12); \mathbf{C}_1(24)$
221 $Pm3m$	$\mathbf{O}_h^1$	$2\mathbf{O}_h(1); 2\mathbf{D}_{4h}(3); 2\mathbf{C}_{4v}(6); \mathbf{C}_{3v}(8); 3\mathbf{C}_{3v}(12); 3\mathbf{C}_s(24); \mathbf{C}_1(48)$
222 $Pn3n$	$\mathbf{O}_h^2$	$\mathbf{O}(2); \mathbf{D}_4(6); \mathbf{C}_{3i}(8); \mathbf{S}_4(12); \mathbf{C}_4(12); \mathbf{C}_3(16);$ $2\mathbf{C}_2(24); \mathbf{C}_1(48)$
223 $Pm3n$	$\mathbf{O}_h^3$	$\mathbf{T}_h(2); \mathbf{D}_{2h}(6); 2\mathbf{D}_{2d}(6); \mathbf{D}_3(8); 3\mathbf{C}_{2v}(12);$ $\mathbf{C}_3(16); \mathbf{C}_2(24); \mathbf{C}_s(24); \mathbf{C}_1(48)$
224 $Pn3m$	$\mathbf{O}_h^4$	$\mathbf{T}_d(2); 2\mathbf{D}_{3d}(4); \mathbf{D}_{2d}(6); \mathbf{C}_{3v}(8); \mathbf{D}_2(12);$ $\mathbf{C}_{2v}(12); 3\mathbf{C}_2(24); \mathbf{C}_s(24); \mathbf{C}_1(48)$
225 $Fm3m$	$\mathbf{O}_h^5$	$2\mathbf{O}_h(1); \mathbf{T}_d(2); \mathbf{D}_{2h}(6); \mathbf{C}_{4v}(6); \mathbf{C}_{3v}(8);$ $3\mathbf{C}_{2v}(12); 2\mathbf{C}_s(24); \mathbf{C}_1(48)$
226 $Fm3c$	$\mathbf{O}_h^6$	$\mathbf{O}(2); \mathbf{T}_h(2); \mathbf{D}_{2d}(6); \mathbf{C}_{4h}(6); \mathbf{C}_{2v}(12); \mathbf{C}_4(12);$ $\mathbf{C}_3(16); \mathbf{C}_2(24); \mathbf{C}_s(24); \mathbf{C}_1(48)$

Space group		Site symmetries
227 $Fd3m$	O_h^7	$2T_d(2)$; $2D_{3d}(4)$; $C_{3v}(8)$; $C_{2v}(12)$; $C_s(24)$; $C_2(24)$; $C_1(48)$
228 $Fd3c$	O_h^8	$T(4)$; $D_3(8)$; $C_{3i}(8)$; $S_4(12)$; $C_3(16)$; $2C_2(24)$; $C_1(48)$
229 $Im3m$	O_h^9	$O_h(1)$; $D_{4h}(3)$; $D_{3d}(4)$; $D_{2d}(6)$; $C_{4v}(6)$; $C_{3v}(8)$; $2C_{2v}(12)$; $C_2(24)$; $2C_s(24)$; $C_1(48)$
230 $Ia3d$	O_h^{10}	$C_{3i}(8)$; $D_3(8)$; $D_2(12)$; $S_4(12)$; $C_3(16)$; $2C_2(24)$; $C_1(48)$

Note the following equivalent nomenclatures:

$C_i \equiv S_2$
$C_2 \equiv C_{1h}$
$D_2 \equiv V$
$D_{2h} \equiv V_h$
$D_{2d} \equiv V_d$
$C_{3i} \equiv S_6$

Index

Since the number of compounds included in this book is numerous, entries for most of individual compounds are collected under general entries such as diatomic molecules and sulfur compounds. Infrared and (resonance) Raman spectra, normal modes of vibration, and vibrational frequencies of individual molecules are found under respective general entries. Boldface numbers refer to diagrams, figures, and tables.

Accidental degeneracy, 62
Acoustical branch, 133
Alkali halide vapor, 160
Aluminosilicates, 258
Ammonia(NH_3), **5**
Ammonium chloride(NH_4Cl), **192**
Anharmonicity, 12
Anharmonicity corrections, 110, 153, **154–156**
Anisotropy of polarizability, 99
Anomalous polarization (ap), 101, **113, 221**
Antimony compounds, 271
Anti-Stokes line, 8
Antisymmetric vibration, 31
Aragonite, 125
Arsenic compounds, 271
A-term resonance, 106, 109
Axis of symmetry, p-fold(C_p), 24

Badger's rule, 15
Band assignments, 88
Bismuth compounds, 271
B matrix, 65
Bohr's frequency condition, 1
Borazine($B_3N_3H_6$), **247**
Boron compounds, 243, **244**
Boron hydrides(boranes), **246**
Bravais lattices, **119**
Brillouin zone, 134
B-term resonance, 106, 111
Buckminsterfullerene(C_{60}), **250, 253**

Calcite, 125, 126, **128, 130–132, 137–140**
Carbonate radical(CO_3), 181

Carbon clusters, 248, 250
Carbon compounds, 248, 256
Carbon dioxide(CO_2), **20, 31**
Carbon oxide anions, **257**
Carbon tetrachloride(CCl_4), **10, 100**
Center of symmetry, *i*, 24
Ceramic superconductors, 141
Character, χ, 40
Character tables, 40, **321–334**
Chemical synthesis, 116
Chlorate, potassium($KClO_3$), **178**
Chlorine isotope pattern, 196
Chromium hexacarbonyl($Cr(CO)_6$), **115**
Class, 38
Combination bands, 5
Correlation method, 127
Correlation tables, 124, **361–374**
Crystallographic point groups, 117, **118**
Crystals, vibrational analysis, 124
Cubic potential, **11,** 12
C_{60}, **250, 253**
C_{70}, **253**
$C_{60}Br_{24}$, 251
$C_{60}K_3$, 252

Decius formulas, 69
Degenerate vibration, 28
Depolarization ratio, 97, 98
Diamond, 254, **255**
Diatomic molecules, 153, **158, 159**
Diazene(N_2H_2), 188
Dichroism, infrared, 137
Dipole moment, changes in, **33**
Dispersion curves, **134, 137**
Doubly degenerate vibration, 28

Electromagnetic spectrum, **3**
E matrix, 65
Energy level diagram, **4**
Excitation profile, 112

Factor group analysis, 126
Fermi resonance, 62
F matrix, 64, **350–355**
Force constant, 10, 13, **14, 15**
Force field, 77
Franck–Condon factor, 107
Fundamentals, 5

Gaseous phase, infrared spectra, 113
Generalized coordinate, *q*, 16
Generalized valence force(GVF) field, 77
Germanium compounds, 256, 261
Germanium tetrachloride($GeCl_4$), 196, **197**
GF matrix method, 63
Glide plane, 120
G matrix, 64, **350–355**
Gordy's rule, 15
Graphite, 254, **255**
Group, 38
Group frequencies, 88
Group frequency charts, **356–360**
Group theory, 37
g-Vibration, 37

Halogeno compounds, **263**
Halogens, 161
Harmonic oscillator, 5, 10
Hermann–Mauguin(H–M) notation, 117
Hermite polynomial, 12
Herschbach–Laurie equation, 16
High-temperature species, 115
Hot bands, 5
Hydrogen halides, 161
Hydronium ion(OH_3^+), 174
Hydroxyl ion(OH^-), 161

Ice, 170
Identity, *I*, 24
Infrared spectra:
  $B_3N_3H_6$(borazine), **247**
  $[B_6X_6]Cs_2$ (X = Cl, Br, and I), **245**
  $CaCO_3$(calcite), **137**
  $[ClO_3]K$, **178**
  $Cr(CO)_6$, **115**
  C_{60}, **251**
  C_{70}, **251**
  FClO, **173**
  $GeCl_4$, **197**
  H_2NCN(cyanamide), **264**
  $[IO_3]K$, **178**
  MF_6 (M = Mo, Tc, Ru and Rh), **222**
  NH_3, **5**
  NH_4Cl, **192**
  $Ni(CO)_n$ ($n = 1 \sim 4$), **116**
  NiF_2, **165**

$[OsF_nCl_{6-n}]^{2-}$ ($n = 2 \sim 4$), **227**
$P_4N_4Cl_8$, **270**
Si_6H_{12}, **259**
SiO_2 (α-quartz), **260**
XeF_4, **208**
Improper rotation, 46
Inert gas matrices, 113
Infrared dichroism, 137
Intensity:
Infrared absorption, 95
Raman scattering, 102
Interhalogeno compounds, **169**
Internal coordinate, 49, **51**
Internal tension, 79
Inverse polarization (ip), 101
Inversion doubling, 173, **175**
Iodate, potassium(KIO_3), **178**
Iodine(I_2), **109**
Iridium hexachloride anion($[IrCl_6]^{2-}$), **221**
Irreducible representation, 40
Direct products of, **344–345**
Isotope splitting, 115

Kronecker's delta (δ_{ij}), 41
Lagrange's equation, 17, 18
Laser lines, **7**
L matrix, 91
Lattice modes, 124
Lattice point (LP), 118
Lattice vibrations, 132

Matrix algebra, 335–340
Matrix co-condensation technique, 116
Matrix effect, 117
Matrix isolation technique, 114
Maxwell–Boltzmann distribution law, 5, 9, 104
Mean value of polarizability, 99
Metal cluster compounds, 240, **241**
Metal isotope spectroscopy, 85, **87**
Molecular spectra, origin of, 1
Multiplication table, 38
Mutual exclusion rule, 37

Nickel carbonyls($Ni(CO)n$, $n = 1 \sim 4$), **116**
Nickel difluoride(NiF_2), **165**
Nickel octaethylporphyrin(Ni(OEP)), **112**
Nitrogen compounds, 261–**263**
Nonsymmetric vibration, 32
Normal coordinate, *Q*, 18
Normal coordinate analysis, 63, 139
Normal modes of vibration:
CH_3X molecules, **89**
CH_2X_2 molecules, **90**
WXYZ(bent) molecules, **189**
XY_2(bent) molecules, **31**
XY_2(linear) molecules, **31**
XY_3(planar) molecules, **180**
XY_3(pyramidal) molecules, **173**
XY_4(square-planar) molecules, **207**
XY_4(tetrahedral) molecules, **191**
XY_5(trigonal-bipyramidal) molecules, **210**
XY_6(octahedral) molecules, **215**
X_2Y_2(bent) molecules, **186**
X_2Y_6(bridged) molecules, **230**
X_2Y_6(ethane-type) molecules, **234**
XYZ(linear) molecules, **162**
ZXY_4(tetragonal-pyramidal) molecules, **212**
Normal Raman scattering, 9
Normal vibration, 16, 19
number for each species, 42, **341–344**

Octahedral molecules, 214
Optical branch, 133
Orbital valence force (OVF) field, 80
Order of group, *h*, 40
Overtone bands, 5, 12

Parabolic potential, 10
Pentagonal-bipyramidal molecules, 226
Pentagonal-planar molecules, 214
Phosphorus compounds, 265, **267–269**
group frequency chart, **266**
Planar four-atom molecules, 180
Planck's constant, *h*, 1
Plane of symmetry, σ, 24
Point groups, 25, **27–30**
Polarizability, 8, 33
principal axes of, 35
Polarizability ellipsoid, 34, **35, 36**
Polarizability tensor, 34
Polarization of Raman lines, 97
Polarized infrared spectra, **137**
Polarized Raman spectra, **139**
Potassium nitrate(KNO_3), **183**

Potassium sulfate(K_2SO_4), **200**
Potential energy curve, **11**
Potential energy distribution (PED), 93
Potential field, 77
Pressure broadening, 96
Product rule, 83
Proper rotation, 45
Pyramidal four-atom molecules, 173

Quartz, **260**

Radicals, 114
Raman scattering, 6
Raman spectra:
  $[B_6X_6]Cs_2$ (X = Cl, Br and I), **245**
  $CaCO_3$ (calcite), **139**
  CCl_4, **10, 100**
  C_{60}, **252**
  C_{70}, **254**
  $[IrCl_6]^{2-}$, **221**
  $[NO_3]K$, **183**
  N_4X_4 (X = S and Se), **265**
  $[OsF_nCl_{6-n}]^{2-}$ ($n = 2 \sim 4$), **227**
  $Pb_6O(OH)_6(ClO_4)_4H_2O$, **243**
  $P_4N_4Cl_8$, **270**
  Si_n (n = 4, 5 and 6), **258**
  Si_6H_{12}, **259**
  SiO_2 (α-quartz), **260**
  $(SND)_4$, **275**
  $(SNH)_4$, **275**
  $[SnX_3]^-$ (X = Cl, Br and I), **177**
  $[SO_4]K_2$, **200**
  XeF_4, **208**
  $YBa_2Cu_3O_{7-\delta}$, **143**
Rayleigh scattering, 6
Redlich–Teller product rule, 84
Reduced mass, μ, 10
Reducible representation, 40
Redundant condition, 50
Redundant coordinate, 77
Representation of the group, Γ, 39, 40
Resonance fluoresence, 8
Resonance Raman scattering, 8, 9, 106
Resonance Raman (RR) spectra:
  I_2, **109**
  Ni(OEP), **113**
  $[Re_2F_8]^{2-}$, **238**
  TiI_4, **111**
R matrix, 63
Rotational spectrum, 5
Rotation-reflection axis, S_p, 24

Salt-molecule reaction, 168
Schönflies(S) notation, 117
Schrödinger wave equation, 11, 23
Screw axis, 120
Secular equation, 80
Selection rules, 27, 53, 54, 58
Selenium compounds, 263, 271, 275
Silanone(H_2SiO), 184
Silicates, 257, **259**
Silicon clusters, 256, **258**
Silicon compounds, 256
Similarity transformation, 39
Simultaneous transition, 188
Site group analysis, 125
Site symmetry, 141, **142, 374–381**
S matrix, 67
Space group, 120, **121**
Square-planar five-atom molecules, 204
Stokes lines, 8
Structure determination, 60–62
Subgroup, 124
Sulfate radical(SO_4), 201
Sulfur compounds, 263, 271, **272–274**
Sum rule, 84
S vector, 67
Symmetric vibration, 31
Symmetry coordinate, 72
Symmetry elements, 24, **26**
Symmetry in crystals, 117
Symmetry of normal vibration, 27
Symmetry operation, 24
Symmetry selection rules, 53

Taylor's series, 17
Tetragonal-pyramidal molecules, 211
Tetrahedral five-atom molecules, 189
Time-resolved resonance Raman(TR^3) spectroscopy, 238
Titanium tetrachloride(CCl_4), **111**
Totally symmetric vibration, 32
Triatomic molecules, 162, **163, 164, 166**
Trigonal-bipyramidal molecules, 209
Triply degenerate vibration, 29
T-shaped molecules, 182

U matrix, 72
Urey–Bradley force(UBF) field, 78
u-Vibration, 37

Vibrational coupling, 91
Vibrational frequencies:
  triatomic halogeno compounds, **169**
  WXYZ(bent) molecules, **190**
  X_2 molecules, **154–156, 158**
  X_3 (bent) molecules, **166**
  X_3 (linear) molecules, **171**
  X_4 (tetrahedral) molecules, **271**
  X_4 (square-planar) molecules, **273**
  XH_2 (bent) molecules, **170**
  XH_3 (pyramidal) molecules, **174**
  XXY (bent, linear) molecules, **166**
  XY molecules, **154–156, 159, 160**
  XY_2 (bent) molecules, **163, 164, 166, 167**
  XY_2 (linear) molecules, **163, 164, 166, 167**
  XY_3 (planar) molecules, **181, 182**
  XY_3 (pyramidal) molecules, **176, 177**
  XY_4 (distorted tetrahedral) molecules, **198**
  XY_4 (square-planar) molecules, **207**
  XY_4 (tetrahedral) molecules, **191, 193, 194, 199, 200**
  XY_5 (tetragonal pyramidal) molecules, **213**
  XY_5 (trigonal bipyramidal) molecules, **210**
  XY_6 (octahedral) molecules, **216–218, 223**
  XY_7 (pentagonal bipyramidal) molecules, **228**
  X_2Y_2(bent) molecules, **187**
  X_2Y_6 (ethane-type) molecules, **235**
  X_2Y_6 (non-planar, bridged) molecules, **232**
  X_2Y_6 (planar, bridged) molecules, **233**
  X_2Y_7 molecules, **236**
  X_2Y_9 molecules, **239**
  XY_4WZ (octahedral) molecules, **224, 225**
  XYZ (bent) molecules, **172**
  XYZ(linear) molecules, **171**
  XY_5Z (octahedral) molecules, **224, 225**
  XY_5Z_2 (pentagonal bipyramidal) molecules, **228**
  $ZWXY_2$ (tetrahedral) molecules, **206**
  ZXY_2 (planar) molecules, **185**
  ZXY_2(pyramidal) molecules, **179**
  ZXY_3(tetrahedral) molecules, **202**
  ZXY_4(tetragonal pyramidal) molecules, **213**
  Z_2XY_2 (tetrahedral) molecules, **205**
  ZXYW(planar) molecules, **185**
Vibrational quantum number, *v*, 11

Water, liquid, 168

Xenon hexafluoride(XeF_6), 220, **222**
Xenon pentafluoride anion($[XeF_5]^-$), 62
Xenon tetrafluoride(XeF_4), 61, **208**
X matrix, 65

Zero-point energy, 3